UNIVERSITY of MICHIGAN
GENERAL LIBRARY
OCTAVIA WILLIAMS BATES
BEQUEST

I0836692

COAL-BEDS ON RIO CANDIOTA, RIO GRANDE DO SUL.

SCIENTIFIC RESULTS

OF

A JOURNEY IN BRAZIL.

By LOUIS AGASSIZ,

AND HIS TRAVELLING COMPANIONS.

GEOLOGY AND PHYSICAL GEOGRAPHY OF BRAZIL.

By CH. FRED. HARTT,

PROFESSOR OF GEOLOGY IN CORNELL UNIVERSITY.

WITH ILLUSTRATIONS AND MAPS.

BOSTON:
FIELDS, OSGOOD, & CO.

LONDON:
TRÜBNER & CO., 8 & 60, PATERNOSTER ROW.

1870.

To PROFESSOR LOUIS AGASSIZ,
DIRECTOR OF THE THAYER EXPEDITION.

MY DEAR SIR:—

I have the honor to offer you this volume on the Physical Geography and Geology of Brazil as a summary of the scientific results of my explorations as an *attaché* of the Thayer Expedition, together with those of a second private expedition,—the natural outgrowth of the former,—made to continue investigations which I had been obliged to leave unfinished.

I take this opportunity of acknowledging my deep indebtedness to you for the interest you have taken in my scientific studies, for your constant wise counsel and advice, and for a thousand kindnesses received at your hands.

With the highest consideration and respect, I have the honor to be,

My dear Sir,

Your former pupil,

CH. FRED. HARTT.

CORNELL UNIVERSITY, ITHACA, N. Y.,
May 30, 1870.

PREFATORY NOTE.

THIS volume is the result of two journeys made by myself in Brazil. The first was undertaken under the direction of Professor Agassiz in connection with the Thayer Expedition in the years 1865 and 1866. On this journey I studied very carefully the Geology and Physical Geography of the coast between Rio and Bahia, going over a very large part of the ground on horseback or in canoe. As a sketch of the journey is to be found in the "Journey in Brazil," it is not necessary to repeat it here. My companion was Mr. Edward Copeland, of Boston, one of the volunteer aids of the Expedition, and I take pleasure in acknowledging my indebtedness to him for the very valuable assistance he rendered me. On this journey Mr. Copeland and I made very large collections of marine invertebrates and fishes, though we did not neglect to secure other objects of natural history. I hope that these collections will throw much light on the fresh-water and marine animals of the coast, and, as they were made at frequent intervals between Bahia and Rio de Janeiro, that they will prove valuable in the study of the distribution of these animals. My studies of the stone and coral reefs and of the geology of Brazil proved so very attractive to me that on the year following I returned to Brazil, and spent my vacation, several months, in examining the coast between Pernambuco and Rio, exploring more particularly the vicinity of Bahia and the islands and coral reefs of the Abrolhos. I was aided pecuniarily on this second expedition by Mr. John Lockwood, of the Adelphi Academy, Brooklyn, the New York Association for the Advancement of Science and Art, and the Cooper Institute; but my thanks are especially due to Miss Chadeayne, the principal of an excellent ladies' school in Jersey City, to Mr. Van Nostrand, of Newark, to my friend Major O. C. James, of Bergen, without whose generous aid I should not have accomplished my expedition and brought home my voluminous collection, and to Mr. J. E. Mills, and Mr. R. L. Dugdale, of New York.

I wish that I could adequately acknowledge the innumerable kindnesses and attentions I have received from Captains Tinklepaugh and Slocomb, and the officers on the splendid steamer of the generous Messrs. Garrison and Allen, and from a hundred kind friends in Brazil. The hospitality with which I was everywhere received in Brazil, and the assistance offered me wherever I went in the prosecution of my studies of the country, have made me love the land of the *Sabiá*, and it is my sincerest wish in acknowledgment of so much kindness to be to some humble degree instrumental in removing false impressions so current about Brazil, and to make the resources of the Empire better known in America. In the course of the following pages I shall have frequently opportunity to acknowledge the aid not only of Brazilians, but of foreigners resident in the country.

This volume was intended at first as a report to Professor Agassiz as the Director of the Thayer Expedition, embracing simply the results of my explorations as a geologist of that expedition, together with those of my second independent journey, both reports to be published among the contributions of the Thayer Expedition. During the preparation of these reports, and in consequence of the delay in publication, I have had the opportunity of examining more or less critically the works of the majority of the writers on Brazil, and the volume, from a simple report of my own investigations, has grown to a general work, in which I have incorporated the best results of others who have written on the Geology and Physical Geography of Brazil. I have to acknowledge valuable contributions to this volume from Messrs. J. A. Allen, Orestes H. St. John, and Thomas Ward, all of whom were employed on the Thayer Expedition. I am indebted to Professor Jeffries Wyman for an interesting communication with reference to a Botocudo skull, and to Professor Alpheus Hyatt for his valuable paper on the Cretaceous Fossils of Maroïm; Professor O. C. Marsh has kindly examined and described in the Journal of Science a few reptilian remains I collected at Bahia, Professor Verrill described the radiates of my second journey, and Mr. S. J. Smith has published a valuable paper on the Crustacea of the same journey. Professor Marsh's paper I have given almost entire. From Professor Verrill's I have drawn largely, and I have given the general results of that of Mr. Smith. A part of the chapter on the Coral Reefs appeared in the American Naturalist, together with several of the wood-cuts, which have been kindly lent me for this volume. I must express my thanks to Messrs. Putnam, Packard, Morse, and Hyatt, of the Peabody Academy, for valuable assistance rendered.

Professor Agassiz has generously allowed me the use of books, maps, and photographs, and has assisted me in various ways. I regret exceedingly that his illness has prevented him from preparing for the work the paper on the fishes which he contemplated.

In the recent excellent work "On the Highlands of the Brazil," by Captain Burton, the celebrated African traveller, now English Consul at Santos in the Province of São Paulo, the author uses throughout the definite article prefixed to the name of the Empire. Burton says, "I do not call the country 'Brazil,' which she does not; nor does any other nation but our own."* Captain Burton's reasoning does not seem to me quite conclusive. It is true that the Portuguese say "*o Brazil, the Brazil*," but the article is not prefixed to give any particular definiteness to the name. They say also *a Inglaterra*, *a França*, *o Paraguay*, &c., the definite article being applied to almost all names of places. So we have in French *le Brésil*, *la France*, *l'Angleterre*. In Italian the definite article is used before the name of a country when the whole of it is meant, but ordinarily it is not employed. In Spanish it is used much as in French.

Since the Brazilians use the article before the names of other countries as well as their own, in accordance with a custom followed by other Romance languages, there would seem to be no better reason for saying *The Brazil*, contrary to English usage, than "*The France*," because the French do. It is true that many English writers on Brazil have used the article, but the majority have not, and no one but Burton uses it to-day. "*The Brazils*," as that author has remarked, is an anachronism occasionally seen, but only proper between the years 1572 and 1576, when the country was divided into two governments. It is more high-sounding than the ordinary form, and so probably remains in use. In the North European languages, German, Hollandish,

* Vol. I. p. 3, note.

Danish, or Swedish, the name is derived from the Latin form *Brasilia;* * in Danish and German, *Brasilien.* In like manner we have the German *Italien, Sicilien,* &c. In none of these languages is the article used. I have followed in this work the common usage, and have omitted the article.

The use of the article before the names of places in Brazil is very puzzling, and foreigners writing on the country are sure to commit blunders. The names of the provinces are especially difficult. Several of them are derived from rivers, as Amazonas, Pará, Parahyba, &c., and these take the article in Portuguese, as do Ceará, Alagôas (plural), Bahia, and Espirito Santo; but those bearing the names of saints, together with Pernambuco, Sergipe, Minas Geraes, Matto Grosso, and Goyaz, do not. The same difficulty is met with in the names of rivers, serras, &c., and it is impossible to give a general rule to guide one in writing them. Most writers seem to be impressed with the idea that Portuguese is only bad Spanish, and that it will do just as well to write Brazilian names in the Spanish form; so we find some of our best authors on Brazil using San Francisco for São Francisco. The only safe way seems to be to give geographical names exactly as used by the Brazilians themselves.

Brazil, in its climate, people, and productions of all kinds, was at the time of its discovery totally different from Europe. The European colonists in Brazil therefore had no names to give to the things they saw about them. Such was not the case in North America, where the early explorers found animals and plants resembling those of Europe, and they recognized the bear, the wolf, the codfish, the herring, the oak, &c., &c. But in Brazil all was new, and the euphoni-

* Hakluyt used the form Bresilia, and it is to be found in the works of other old English writers.

ous indigenous names were adopted and incorporated, and to-day the Portuguese of Brazil is full of them, and places, too, bear to a very large extent their Indian names. Far and wide, up and down the coast, and through the country, was found distributed the great nation of the Tupís, speaking everywhere the same general language, and it is from this language, now spoken over a large part of Brazil, that these names were taken. These names have been written with the Portuguese pronunciation, and have often been much corrupted, so that their orthography varies very much. Foreigners murder them fearfully. Among the writers on Brazil Bates and Burton are perhaps the most accurate in their use of Portuguese and Tupí names. In this work I have followed the best authorities, and I have taken particular pains to insure correctness in the geographical names; but since there is no standard of orthography for these names even in Brazil, and since one is frequently obliged to depend on an ignorant guide for the name of a place or an object, absolute uniformity and accuracy are out of the question.

In looking up the derivation of Brazilian geographical names, I was led into a study of the word "Brazil," but I soon found that Humboldt had preceded me, and I am able to add little to what he has said on the subject. Since there are afloat so many incorrect ideas concerning the derivation of the name, it seems not unadvisable to give here in brief the result of Humboldt's researches.*

According to Humboldt the name, under the various forms of *Bracie*, *Brazil*, *Berzil*,† appears on Italian maps from

* *Géographie du nouveau Continent,* Tom. II. p. 214, to which the reader is referred for the details of the discussion.

† Among the many curious old forms of the name Brazil may be mentioned

1351 to 1459, applied to one or more of the islands of the Açores, and more particularly to a point of the island of Terceira, which still bears the name.

For three centuries before the discovery of the route to the Indies around the Cape of Good Hope there was known in Europe a dye-wood called *bresill*, *brasilly*, *bresilji*, *braxilis* or *brasile*, which appears to have been derived from one or more East Indian species of *Cæsalpina* and *Pterocarpus*.

As to how this dye-wood came to bear the name of Brazil I know not, and I fear that any attempts to derive it from Sanskrit or other roots will lead to nothing satisfactory.* Anghiera speaks of the occurrence of Brazil wood in Haiti, which was known long before the discovery of Brazil. Grinæus speaks in 1499 of the Brazil-wood seen at Paria (Payra).

Humboldt says: "In proportion as discoveries extended themselves to the south of Cape Santa Augustinho, especially after Pedro Alvarez Cabral, in May, 1500, had taken possession of the Terra de Santa Cruz, the commerce in the red-wood of Continental America became more active. On the fourth expedition of Vespucius, in which one of the ships was lost on the shoals around Fernando Noronha, a cargo of Brazil-wood was taken in near the Bahia. All the world

that of *Presill*, found in an ancient publication, described by Humboldt in his *Géographie du nouveau Continent*, Tome V. p. 239, entitled *Copia der Newen Zeytung auss Presillig Landt*.

* See Humboldt, *op. cit.* Tome II. p. 222. It is always unsafe to proceed in the investigation of the etymological derivation of proper names, unless one has historical evidence of some sort to guide him.

† Decade I. Liv. 4. p, 11, quoted by Humboldt. Anghiera says that there were found at Haiti "sylvas immensas, quæ arbores nullas nutriebant alias prætaquam coccineas quarum lignum mercatores Itali *verzinum*, Hispani *brasilum* appellant."

knows that little by little, in the first half of the sixteenth century, this same abundance of dye-wood has caused the name of *Terra de Santa Cruz*, given by Cabral, to be changed into the *Terra de Brasil*, 'changement inspiré par le démon, dit l'historiographe Barros, car le vil bois qui teint le drap en rouge ne vant pas le sang versé pour notre salut.' So from the Asiatic Archipelago the name *Brasil* has passed to a cape of the island of Terceira, and then to the southern shores of the New Continent."

CONTENTS.

CHAPTER I.

THE PROVINCE OF RIO DE JANEIRO.

PAGE

The Serra do Mar. — The Serra da Mantiqueira, and the Pico do Itatiaiossú. — The Rio Parahyba do Sul and its Tributaries. — Description of the Bay of Rio, its Islands, Tides, &c. — The Sugar-Loaf. — The Corcovado. — The Gneiss of Rio. — The Organ Mountains. — Geological Observations along the Cantagallo Railroad. — Drift Phenomena at Rio, Tijuca, and on the Dom Pedro II. Railroad. — Decomposition of Gneiss *in sitû* and its Effect on the Forms of the Hills. — Recent Rise of the Coast. — The Coast between Rio and Cape Frio, its Lakes, Salines, &c. — Cape Frio. — Os Buzios. — Islands of Santa Anna. — Frade de Macahé. — Campos dos Goitacazes, their Lagoons, Canal, &c. — Rio Parahyba. — São João da Barra. — Sugar Plantations. — City of Campos. — The Rio Muriahé. — Tertiary Beds. — Sugar Fazendas. — São Fidelis. — The Gold-Mines of Cantagallo. — Geological Notes on the Country between São Fidelis and Bom Jesus. — The Rio Itabapuana. — The Serra d'Itabapuana. — The Garrafão. — Barra do Itabapuana . . . 1

CHAPTER II.

PROVINCE OF ESPIRITO SANTO.

Barreiras do Sirí. — Itapémerim. — Coast between Itapémerim and Benevente. — Benevente. — Guarapary ; Consolidated Beach, Corals, &c. — Rio Jecú. — Bay of Espirito Santo. — Nossa Senhora da Penha. — Victoria. — Decomposition of Gneiss and Formation of Boulders of Decomposition. — Recent Rise of the Coast. — Corals, &c. of the Bay of Victoria. — Rio Santa Maria. — German Colonies. — Fisheries. — Sand Plains. — Tertiary Plain at Carapina. — Mestre Alvaro. — Serra. — Nova Almeida. — Rio Reis Magos. — Santa Cruz. — Basin of the Rio Doce. — Description of the River. — Guandú ; its Colony and Agricultural Resources. — Porto de Souza. — Geology of Vicinity. — Luxuriance of Vegetation on the Doce. — Woods. — Game. — Francylvania. — Climate of the Doce. — Linhares. — Lagôa Juparanãa. — The Future of the Doce. — American Colonists. — Salt Trade. — Barra Secca. — Sea-Turtles. — Consolidated Beaches and the Mode of their Formation. — Character of Coast between the Rivers Doce and São Matheos. — Rio São Matheos described. — Geological Features. — Fertility of its Lands. — Cocoa-Palms and their Distribution. — City of São Matheos. — Rio Itahúnas. — Cliffs of Os Lençôes. — Coast between Itahúnas and Rio Mucury . . . 56

CHAPTER III.

PROVINCE OF MINAS GERAES. — THE MUCURY AND JEQUITINHONHA BASINS.

The Basin of the Mucury. — Porto Alegre. — Description of the River below Santa Clara. — Luxuriance of Forest Vegetation. — Santa Clara. — Minas Geraes a Land-locked Province. — Want of Roads. — The Philadelphia Road and the Mucury Colonies. — Difference in Topography and Soils between the Tertiary and Gneiss Lands west of Santa Clara. — Urucú, its Dutch Colony, Soils, Climate, &c. — Philadelphia and its German Colonies. — Great Fertility of the Mucury Basin. — Character of Country between Philadelphia and the Head-waters of the Mucury. — The Basin of the Jequitinhonha. — The Rio Pardo. — General Geological Structure of the Jequitinhonha-Pardo Basin. — The Head-waters of the Setubal, their Geological Features and Catinga Forests. — Geological Excursion from the Fazenda de Santa Barbara to Alto dos Bois. — Difficulty of geologizing in Brazil. — The Brazilian Campos. — The Chapadas between Itinga and Calháo. — The great Calháo-Arassuahy Valley. — Magnificent View over the Valley from the Chapada at Agua da Nova. — Calháo and the Geology of its Vicinity. — Description of the Country between Calháo and Sucuriú. — The Chapadas. — Minas Novas, its Geology, Gold-Mines, &c. — Occurrence of Gold in Drift. — Gold-Mines of the Arraial da Chapada ; their former Richness : not yet worked out. — Decomposition of Clay Slates in the Minas Novas Region. — The Rio Arassuahy. — The Rio Jequitinhonha from its Confluence with the Arassuahy to the Sea described : its Geology, Vegetation, Commerce. — The Salto Grande 125

CHAPTER IV.

THE ISLANDS AND CORAL REEFS OF THE ABROLHOS.

The Geology of the Abrolhos. — Trap-bed, Fossil Plants, &c. — Land Fauna and Flora; Spiders, Lizards, and Sea-Birds. — The Cemetery of the Frigate-Birds. — The Whale and Garoupa Fisheries. — Importance of these Fisheries. — The mythical Brazilian Reef. — The Coral Reefs and Consolidated Beaches confounded by Travellers and Writers. — The Author's Discovery of the Porto Seguran Coral Reef. — Coral-building Corals found almost wholly to the north of Cape Frio. — The Fringing Reef of Santa Barbara ; its Structure and Life. — Corals found on the Reef. — Star-fishes, Ophiurans, &c. — Resemblance between the Echinoderms of the Abrolhos and West Indies. — The Chapeirões. — The Parcel dos Abrolhos ; its Appearance ; forms a serious Obstacle to Navigation. — Safe Canal west of the Islands. — The Parcel dos Parédes. — The Recife do Lixo. — Its great Extent. — The Submerged Border and its Coral Growth. — The Coral Fauna of Brazil. — The Millepores and their Stinging Properties. — The Reefs of Timbebas, Itacolumí, Porto Seguro, Santa Cruz, Camamú, Bahia, Maceió, and Pernambuco. — The Roccas 174

CHAPTER V.

PROVINCE OF BAHIA. — COAST SOUTH OF SÃO SALVADOR.

The Tertiary Lands between the Rivers Mucury and Peruhype ; their Vegetation, &c. — Colonia Leopoldina and its Coffee Plantations. — Villa Viçosa. — The Canal joining the Rivers Peruhype and Caravellas. — Formation of Beaches and Beach Ridges. — Coast between Caravellas and Pòrto Seguro. — Monte Pascoal. — Porto Seguro and its *Recife*, or Consolidated Beach. — Santa Cruz and its Reef. — Coast northward to the Jequitinhonha ; the Lagôa do Braço, Campos, &c. — The Canal Po-assú and the Rio da Salsa. — Mangrove Swamps between the Jequitinhonha and Pardo. — Cannavieiras — The Salt Trade of the Jequitinhonha. — Description of the lower Part of the Rio Pardo ; Cacáo Plantations, &c. — Coast northward to Ilhéos. — Prince Neuwied's Description of the Country between Ilhéos and Conquista, Possões, and Cachoeira ; the Forests, Campos, Social Plants, &c. — Ilhéos. — Rio and Lagôa Itahype. — Dead Coral Banks. — Rio das Contas. — Bay of Camamú. — Turva Deposits. — Villa de Camamú. — Coast northward to the Bahia de Todos or Santos. — The Bay of All Saints described. — Ilha Itaparica. — Rio Jaguaripe and Nazareth. — Rio Paraguassú. — Description of River below Cachoeira. — The Tram-road. — Sant' Amaro and the Agricultural Institute 215

CHAPTER VI.

THE SÃO FRANCISCO BASIN.

The Explorations of Halfeld, Liais, St. John, Allen, Ward, Burton, &c. — General Shape of the Basin. — Its uniform Width. — The São Francisco Valley hollowed out of a Series of horizontal Beds of Limestone and Sandstone. — The Chapadas. — The so-called "Serra" separating the São Francisco from the Tocantins Basin an irregular Strip or Table-land of Sandstone. — The Serras of Araripe and Dous Irmãos. — Table-topped Hills in the Valley of the São Francisco: Outliers of the Chapadas. — Doubts about the Age of the Sandstones and Limestones. — Limestones of the Rio das Velhas. — Remains of Extinct Quadrupeds in Brazil, spoken of by Cazal, Spix and Martius, &c. — Claussen's Discoveries in the Caves at Curvelo. — Dr. Lund's exhaustive Researches at Lagôa Santa. — Caves described; their Number, Extent, Stalactites and Deposits of Bones in Saltpetre Earth. — Immense Quantities of small Bones brought in by Owls, &c. — Large Number of Fossil Animals discovered by Lund. — Former Existence of Megatheria, Mylodons, Mastodons, immense Armadillos and Cats, Horses, &c., in Brazil. — Remains of a Race of Man of high Antiquity. — Reinhardt's Generalizations. — The Rio de São Francisco of the Sixteenth Class among the Rivers of the World, but third in Rank in Brazil. — General Description of the Stream. — Its Affluents, the Rios Pará, Paraopeba, and Das Velhas. — The Rio das Velhas alone capable of being made navigable for Steamers. — The São Francisco navigable with but few Interruptions for two hundred and sixty-four Leagues below the Rio das Velhas. — Cost of removing Obstructions. — Proposed Railway from Joazeiro to Piranhas. — Fertility of Low Lands of São Francisco Valley. — Liais's Picture of the Campos 274

CHAPTER VII.

THE PROVINCE OF BAHIA, — INTERIOR.

Journeys of Spix and Martius, Nicolay and Lacerda, Allen, and other Explorers. — Geological and Physical Features of Country between Malhada and Cachoeira, described by Von Martius. — Sandstones. — Remains of Mastodons found near Villa do Rio de Contas. — Immense Copper Boulder from Cachoeira. — Rev. Mr. Nicolay's Report of Journey from Cachoeira to the Chapada Diamantina. — Occurrence of Diamonds in Sandstones. — Limestones. — Sterile Plains. — Diamantiferous Sands of the Chapada. — The Diamond Mines of Sincorá and Lencões. — Annual Yield of the Provinces in Diamonds. — Mr. Allen's Report of a Journey from Chique-Chique, viâ Jacobina to Cachoeira. — Country between Chique-Chique and Jacobina an immense Limestone Plain. — The Chapada at Jacobina a detached flat-topped Mass of Sandstone. — Gneiss Hills. — "Lake Plain," east of Jacobina. — Knobs. — Potholes, probably of Glacial Origin. — Eastern Sandstone Plain. — Climate, Vegetation, &c. of Route. — Difference in Topography between Gneiss Regions of Bahia and the Mucury described and accounted for. — Former greater Extension of Forests. — Von Martius's Description of the Country between Cachoeira and Joazeiro. — Country near Feira da Conceição. — Serra do Rio Peixe. — Rio Itapicurú. — Want of Rain at Queimados. — Serra de Tiuba. — Tanques and Fossil Bones near Coche d'Agua, Barriga Molle, and Neighborhood. — Monte Santo. — The Great Meteorolite of Bemdêgo. — Rock Inscriptions. — Villa Nova da Rainha. — Joazeiro to be the Terminus of Bahia and São Francisco Railroad. — Rio de Salitre. — Salt Licks. — Mr. Allen's Note on the Salt of the São Francisco Valley. — Saltpetre. — Geology of Country between Carunhanha and Urubú. — Change in Geological Structure, Climate, Vegetation, &c., below Urubú . 294

CHAPTER VIII.

PROVINCE OF BAHIA, GEOLOGY OF THE VICINITY OF SÃO SALVADOR AND THE BAHIA AND SÃO FRANCISCO RAILROAD.

Topography of the Vicinity of São Salvador da Bahia. — The Upper and Lower Cities. — The Population, &c. — The Harbor. — The Commerce of the City and Province. — The Climate, &c. — The Bahia Steam Navigation Company. — The Bahia and São Francisco Railroad. — The Paraguassú Steam Tram-road. — The Gneiss of Bahia. — Decomposition. — Drift Deposits. — Consolidation of Beaches. — Stone Reef at Rio Vermelho. — Blown Sands covering the Drift of the Hills. — Mr. Allport's Description of the Cretaceous Beds of Monserrate and Plataforma. — Fossil Fishes, Crocodiles, &c. — Description of several Species of Fossil Mollusks. — Cretaceous Beds of Plataforma and Vicinity. — Prof. Marsh's Notice of the Reptilian Remains. — Fossil Fishes at Agua Comprida. — Gneiss at the Rio Johannes. — Taboleiros and Sand Plains of Camassarí. — Peculiarities of the Topography of the Tertiary Hills. — Tabatinga Clay. — Sand Plains and Taboleiros of the Imbuçahy. — Peat-Bog. — Drift. — Diamond-washings at Pitanga. — Cretaceous Strata at Pojuca. — Piassabas. — Campos of Alagoinhas. — Tertiary Hills. — Character of Vegetation 333

CHAPTER IX.

THE PROVINCES OF SERGIPE AND ALAGÔAS, AND THE RIVER SÃO FRANCISCO BELOW THE FALLS.

The Province of Sergipe. — Its Division into *Mattas* and *Agrestes*. — The Rio Real. — Estancia; New Red Sandstone, Sugar Plantations, &c. — Sand Dunes. — The Rio Vasabarris. — The Rio Cotinguiba. — Aracajú. — Cretaceous Beds with Inocerami at Sapucahy. — Maroïm. — Cretaceous Limestone with Ammonites. — "Fossil Turtles." — Sugar Plantations. — Messrs. Schramm and Company. — The Bar of the São Francisco. — Sand Dunes of the Pontal. — Character of the River below Penêdo. — Aracaré. — Villa Nova and its Cretaceous Sandstones. — The City of Penêdo and its Geology. — Its Commerce and Fair. — Notes on the Piranha and its Habits. — Propriá. — Morro do Chaves and Cretaceous Fossils. — Traipú. — Iron Ore. — Campos, Vegetation, Cactuses, &c. — Páo de Assucar. — Cattle Fazendas. — Piranhas. — County flat and covered by Boulders. — The River Valley a Narrow Gorge in a Gneiss Plain. — The Falls of Paulo Affonso. — Halfeld's Description. — Liais's Description. — Comparison between Paulo Affonso and Niagara. — Mastodon Remains from near the Falls. — Climate of the São Francisco below the Falls. — Steam Navigation. — Character of the Coast of the Province of Alagôas, South of Maceió. — The Lagôas. — The City of Maceió and the Geology of its Vicinity. — Tertiary Beds. — Harbor and Reefs 378

CHAPTER X.

PROVINCE OF PERNAMBUCO.

The Limits, Area, &c. of the Province. — Its Topography, Geology, Climate, Soils, &c. — The Rivers. — Productions of the Province. — The City of Pernambuco or Recife. — Derivation of these Names. — Situation of the City. — The Stone Reef. — The Port formed by it. — Shallowness of Water along this part of the Coast. — Pernambuco a Calling Station for Foreign Shipping. — The Pernambuco and São Francisco Railroad. — Table of Heights along the Line. — Island of Itamaracá. — Fossiliferous Limestones. — Fertility of the Island, Cocoa-Palm Groves, &c. — Fernando de Noronha. — Darwin's Description of the Geology of the Island. — Its Dryness and Sterility . . 427

CHAPTER XI.

THE PROVINCE OF PARAHYBA DO NORTE.

Limits of the Province. — The Serra or Plateau of the Cairirís Velhos. — The Climate, Productions, &c. of the Province. — Fertile Lands found only along the Coast. — The Rio Parahyba do Norte, its Navigability. — The City of Parahyba. — The Consolidated Beach at the Mouth of the River. — The River and Town of Mamanguape. — The Geology of the Vicinity of Parahyba. — Cretaceous Limestone with Fossils. — Observations of Professor Agassiz and Mr. Williams. — Mr. Williams's Observations on the Geology of the Country between Parahyba and the Gold-Mines of Pianco. — Mode of Occurrence of the Gold. — "The Tasso Brazilian Gold-Mining Company (Limited)" 440

CHAPTER XII.

THE PROVINCE OF RIO GRANDE DO NORTE.

Limits of the Province, its Position, Mountain, and River Systems, &c. — The Rio Piranhas. — Vegetation. — Productions. — The Carnahuba Palm and its Uses. — Cochineal. — Cattle. — Climate. — Natal. — Geology of the Province 451

CHAPTER XIII.

THE PROVINCE OF CEARÁ.

Geographical Position and Limits of the Province. — The Serra da Ybiapaba. — Its Topography and Geological Structure. — Serra de Araripe. — River Basins. — General Sketch of the Geology of the Province. — Climate. — Character of the Soil. — Productions. — City of Fortaleza — Population of the Province. — Gardner's Sketch of the Geology of Ceará. — Character of Country in the Vicinity of Aracaty. — Description of Country between Aracaty and Icó. — Serra de Pereira. — Villa do Icó and Vicinity. — Country between Icó and Crato. — Gold Washings. — Crato. — Serra de Araripe. — Villa da Barra do Jardim. — Description of Fossil Fish Locality. — The Fishes noticed by Spix and Martius and Others, and described by Professor Agassiz. — Glacial Phenomena of Vicinity of Fortaleza spoken of by Professor Agassiz. — Mammalian Remains. — Minerals. — Meteorolites 456

CHAPTER XIV.

PROVINCE OF PIAUHY.

Geographical Position, Limits, &c., of the Province. — The Rio Parnahyba and its Tributaries. — Description of its Basin. — General Geological Structure and Topography of the Province. — Table-topped Hills of Sandstone. —

The Serra dos Dous Iramãos and its Structure. — Discussion of Gardner's Observations on the Geology of Piauhy and Ceará. — Gardner mistaken in referring the great Sandstone Sheet to the Cretaceous. — Sandstones probably Tertiary. — Their great Extension over Brazil. — Distribution of the Cretaceous Beds in Brazil. — Climate of Piauhy. — The Campos Mimosos and the Campos Agrestes. — Peculiarities of their Vegetation. — Productions of the Province, Population, &c. 473

CHAPTER XV.

THE PROVINCES OF MARANHÃO, PARÁ, AND AMAZONAS.

Sandstones of the Coast of Maranhão. — The Interior composed of Metamorphic Rocks. — Gold-Mines of Turí and Maracassumé. — Climate of the Province. — Rains. — Cities of Maranhão, Caxias, &c. — Pororoca or Bore at the Mouth of the River Mearim. — Professor Agassiz's Sketch of the Geology of the Amazonian Valley. — His Theory of the Mode of Deposition of the Amazonian Beds. — Discussion of this Question. — Cretaceous Rocks in the Amazonian Valley 484

CHAPTER XVI.

PROVINCES OF GOYAZ AND MATTO GROSSO.

The Geographical Position of the Province of Goyaz. — The Chapada da Mangabeira. — Geology of the Vicinity of Natividade. — Gold-Washings of the Serra da Natividade, the Arraial da Chapada, and the Arraial da Conceição. — Structure of the Serra at the Town of Arrayas. — The Serra Geral. — Subterranean Streams. — Western and Southern Goyaz composed of Metamorphic Rocks. — Distribution of Gneissose and Granite Rocks in Western Brazil. — The Montes Pyreneos and their Height. — The Rio Araguaya and its Navigation. — Dr. Couto de Magalhães. — Ilha do Bananal. — Note on Piranhas. — Gold, Diamonds, Iron, and Chrome Ores. — Climate, Forests, Population, &c. — The Western Part of the Plateau of Brazil composed of undisturbed Beds of Sandstone, &c. — The Amazonas-Paraguay Water-shed a Plain without Serras 496

CHAPTER XVII.

PROVINCES OF SÃO PAULO, PARANÁ, SANTA CATHARINA, AND RIO GRANDE DO SUL.

The Serra do Mar of São Paulo a Plateau. — Its Character. — Drainage in the Provinces of São Paulo and Paraná to the Westward. — Eastward-flowing Streams of no Importance — The São Paulo Railroad. — Description of the

Country, along the Railway between Santos and São Paulo by Major O. C. James. — Geology of Vicinity of São Paulo. — Mawe's Description of the Gold-Mines of Jaraguá, and the Method of extracting the Gold. — Country westward to Campinas. — Iron-Mines at Ypanema. — Serra Arassoiava or Guaraçoiava. — Climate, Products, &c. of the Province of São Paulo. — General Topographical Features of the Province of Paraná, its Climate, Productions, &c. — Matte or Paraguayan Tea. — Tea-Planting in Brazil. — Rivers. — Colonies. — Paranaguá. — Coal Basin on the Rio Tuberão in the Province of Santa Catharina. — General Description of the Physical Features of the Province and that of Rio Grande do Sul. — History of the Coal-Mines of Brazil. — Observations of Perigot, Bouleich, Avé-Lallemant, Plant, &c. — Description of the Coal-Fields of the River Jaguarão. — Engineer McGinty's Report on the Candiota Coal. — Coal Basin on Rio São Sepé. — Basin near São Jeronymo 505

CHAPTER XVIII.

THE GOLD-MINES OF BRAZIL.

Geological Distribution of Gold in Brazil. — Gold in Gneiss, at Jaraguá, Cantagallo, Pianco, and elsewhere. — The richest Deposits found in Veins traversing Clay Slates. — Character of Auriferous Quartz. — Granular Quartz, or Cacó. — Gold when associated with Sulphides rarely visible. — The auriferous Iron Ore, Jacutinga. — Gold-Mines of São João d'El-Rei. — The Morro Velho Mine, Mode of Occurrence of the Gold, Method of Extraction, Yield, &c. — The Gongo Soco Mine. — The Rossa Grande Gold-Mining Company. — Mines at Morro de Santa Anna, Congonhas do Campo, São Vicente, Cata Branca. — The Gold-Mines of Brazil not yet fairly developed . 532

CHAPTER XIX.

RÉSUMÉ OF THE GEOLOGY OF BRAZIL.

Eozoic Rocks and their Distribution in Brazil. — Absence of Limestones. — The Silurian Age in Brazil. — The auriferous Clay-Slates of Minas probably Lower Silurian. — Note on the Silurian of the Andes. — The Distribution of Marine Animals in the Palæozoic. — The Devonian Age in Brazil and South America. — The Carboniferous of Brazil and Bolivia. — The New Red Sandstone. — The Cretaceous, its Distribution in Brazil and South America. — Several distinct Periods represented. — Tertiary Rocks. — Drift. — The Glacial Phenomena of Patagonia. — Tapanhoacanga. — The Drift of Rio and of the Region of Decomposition. — The Drift of the Dry Region of Bahia, Sergipe, and Alagôas. — Examination into the Merits of the various Theories proposed to account for the Formation of the Brazilian Drift. — The Theory of Subaerial Decomposition. — Wave Action during a Subsidence. — Wave Action during Elevation. — All these Theories unsatisfactory. — The Glacial Hypothesis the most reasonable 547

APPENDIX.

ON THE BOTOCUDOS.

Origin of the Name Botocudo. — Stature. — Physical Form and Characteristics. — Manner of Wearing Hair. — Lip and Ear Ornaments. — Professor Wyman's Description of Skull of Botocudo from São Matheos. — Comparison with other described Botocudo Skulls. — Color of Botocudo. — Manner of Painting the Body. — Dislike to being clothed. — Bows and Arrows described. — Gerber's Enumeration of the Tribes. — Von Tschudi's Description of the Tribes. — Ranchos and Huts. — Food. — Mode of procuring Fire. — Manufactures. — Marriage Customs. — The Botocudos cruel Husbands. — Facility with which Wounds heal. — Treatment of Children. — Religious Ideas. — Belief in the Bad Spirit, Janchon. — No Belief in a Supreme God. — Burial Customs. — War Customs. — Cannibalism. — Dance. — The Botocudos fast disappearing. — Botocudo Character. — Geographical Distribution of the Botocudos. — Peculiarities of their Language, Pronunciation, Grammatical Structure, &c. — Botocudo Vocabularies 577

LIST OF ILLUSTRATIONS.

Coal-Beds on Rio Candiota, Rio Grande do Sul . . . Frontispiece.
Map of Part of Province of Rio de Janeiro . . . . . . Page 5
The Sugar-Loaf, Corcovado, and Gavia, from São Domingo . . 10
Section at Praia Grande 13
The Organs 16
Section along the Line of the Cantagallo Railroad 22
The Cascatinha at Tijuca 27
Section of Drift at Bennett's, Tijuca 28
Boulders at Tijuca 30
Ideal Section of Drift-covered Gneiss Hill 31
Diagram to show Contrast in Moulding between Drift-covered and Bare Surfaces 34
Coast just East of Rio 37
Cape Frio 39
One of the Islands of Santa Anna 42
Frade de Macahé 43
Looking up the Rio Parahyba from above Campos 49
The Pedra Lisa 54
Serra of Itapémerim, seen from the Sea 58
Coast South of Guarapary 61
Bay of Esprito Santo, looking up to the Pão de Assucar, and City of Victoria 66
Monte Jutuquara and Gneiss Hills near Victoria 67
Fortaleza de Perитininga 68
Boulder of Decomposition, Victoria 69
Decomposing Surface, Ilho do Boi 70
Nossa Senhora da Penha, Victoria 70
Ancient Sea-Level on Pão d'Assucar, Victoria 71
Arrangement of Littoral Fauna at Villa Velha 73
Morro de Mestre Alvaro 80
Coast between Victoria and Rio Doce 85
Rio Doce at Porto de Souza 92
View on Rio Doce 94
Looking up the Doce from near Linhares 97

Lagôa Juparanãa, looking towards the Outlet 102
Lagôa do Aviso 103
Fazenda of Campo Redondo, São Matheos 120
Santa Rita River Valley 143
Corrugated Quartz Veins at Minas Novas 158
Island of Santa Barbara dos Abrolhos 175
The Islands of the Abrolhos from the South 176
Recife do Lixo 202
Section across Border of Lixo Reef 210
Profile Sketch of River Valley near Porto Seguro 225
Monte Pascoal from the Sea 226
Section across Stone Reef at Porto Seguro 229
Bays of Santa Cruz and Cabral 233
View of the Coast South of the Rio de Contas 260
Map of Camamú Bay 262
Map of Bahia and Vicinity 267
Ground View of the Lapa Vermelha 282
Section of Rio São Francisco Valley 310
View of Valley Walls near Jacobina 313
Knobs near the Serra da Terra Dura 315
Sections of several Cuttings on the Bahia and São Francisco Railroad 355-367
Tertiary Hills near Pojuca Tunnel 371
The Taboleiros near Alagoinhas 376
Sand-dunes at the Mouth of the Rio Real 380
Bar of Rio Cotinguiba 382
Maroïm 384
Villa Nova and Penêdo 397
Traipú from near Marcação 407
Looking down the River from Pão d'Assucar 410
Cattle Fazenda and Garden near Pão d'Assucar 411
View near Allegria 412
View looking up towards Piranhas 413
Cachoeira de Paulo Affonso 415
Pernambuco 434
Island of Fernando de Noronha 438
Mouth of Rio Parahyba do Norte 443
Sketch-Map of the Rio Parahyba do Norte 444
Mouth of Rio Grande do Norte 455
Cutting on the São Paulo Railway, showing Drift lying on Decomposed Rock 508
Botocudo Man and Woman 581
Skull of Botocudo 585

GEOLOGY AND PHYSICAL GEOGRAPHY OF BRAZIL.

CHAPTER I.

THE PROVINCE OF RIO DE JANEIRO.

The Serra do Mar. — The Serra da Mantiqueira, and the Pico do Itatiaiossú. — The Rio Parahyba do Sul and its Tributaries. — Description of the Bay of Rio, its Islands, Tides, &c. — The Sugar-Loaf. — The Corcovado. — The Gneiss of Rio. — The Organ Mountains. — Geological Observations along the Cantagallo Railroad. — Drift Phenomena at Rio, Tijuca, and on the Dom Pedro II. Railroad. — Decomposition of Gneiss *in sitû* and its Effect on the Forms of the Hills. — Recent Rise of the Coast. — The Coast between Rio and Cape Frio, its Lakes, Salines, &c. — Cape Frio. — Os Buzios. — Islands of Santa Anna. — Frade de Macahé. — Campos dos Goitacazes, their Lagoons, Canal, &c. — Rio Parahyba. — São João da Barra. — Sugar Plantations. — City of Campos. — The Rio Muriahé. — Tertiary Beds. — Sugar Fazendas. — São Fidelis. — The Gold Mines of Cantagallo. — Geological Notes on the Country between São Fidelis and Bom Jesus. — The Rio Itabapuana. — The Serra d'Itabapuana. — The Garrafão. — Barra do Itabapuana.

THE province of Rio de Janeiro is almost entirely composed of gneiss, and this gneiss region is mountainous and high. The Serra do Mar, skirting the coast of the province of São Paulo, enters the southwest corner of the province of Rio, and, composed of a great number of parallel ridges, often much broken, traverses it from one end to the other. These mountains form the edge of the great Brazilian plateau, which consists along its eastern border of a broad band of gneiss. The course of the Serra do Mar is approximately east-northeast, so that, as the coast of the province of Rio de Janeiro runs eastward from São Paulo to Cape Frio, the Serra do Mar, on entering the province of Rio, trends gradually away from the coast, passing by the head of the Bay of Rio. The gneiss plateau is bordered in the eastern part of the province by low plains, tertiary and recent. The minor ridges of the grand Serra do Mar break down abruptly on the edge of these low grounds. In the western part of the province, where the Serra skirts the coast, that coast is often high, bold, very irregular in outline, and bordered by numerous rocky islands. South of the Serra do Mar, at Rio, lie several isolated mountains and ranges of hills, really belonging to the same great mountain system, but separated from the plateau by low plains similar to those north of Cape Frio.

To the northwest of the Serra do Mar, and separated from it in part by the valley of the river Parahyba do Sul, is another great mountain range called the Serra da Mantiqueira, which is also composed of gneiss, and belongs to the same system of upheaval as the Serra do Mar. This range separates itself from the coast range near the city of São Paulo, and, lying inside the Serra do Mar, skirts the coast to a much greater distance to the north than the latter does. The ridge properly called the Serra da Mantiqueira accompanies the northern part of the province of Rio for a few miles, when the province boundary line leaves it and runs off to the eastward.

At a distance of four or five miles from the northwest corner of the province, in the Serra da Mantiqueira, is the Pico do Itatiaiossú, which appears to be the highest point in Brazil, and, according to the *Revista Trimensal do Instituto Historico e Geographico Brasileiro*, has an altitude of about 10,300 feet.* It is said to be volcanic in structure, and two craters are reported to exist on it, together with sulphur springs and sulphur deposits.† I have never seen the Itatiaiossú, but I have the strongest doubt as to its being a volcano. Snow occasionally falls on this mountain during the winter, and is said to remain sometimes for several days. There can be no reasonable doubt about this being the highest peak in Brazil. Itacolumí, so long famed, is, according to Burton, only about 6,400 feet in height. According to Eschwege, this last peak is 5,720 feet; Gerber says 1,112 metres (3,650 feet); while the highest point cited by Gerber is the Alto da Serra da Piedade, in the Municipio de Sabará, which Liais makes only 1,783 metres (5,853 feet). Gardner estimates the height of the Organ Mountains at from 7,500 to 7,800 feet. It is interesting to observe, as Burton has remarked, that the summit line is not in the interior in Brazil, but close to the coast. Almost precisely south of the Itatiaiossú, in the province of São Paulo, is a high point in the Serra do Mar, which gives rise to two rivers, — one, a small stream, the Rio Pirahy, flowing east and then northeast between two ridges of the Serra do Mar; the other, the Rio Parahyba do Sul, under the name of Pirahytinga, which flows southwestward and westward for about eighty miles, when it escapes northward around the end of the ridge which has so far formed its barrier on the north, and, doubling upon itself, flows thence, with a general east-northeast direction, behind the Serra do Mar, traversing the whole

* Dr. Candido Mendes de Almeida, in his *Atlas do Imperio do Brazil*, makes the height only 2,994 metres, or 9,829 feet.

† Burton, Explorations, &c., Vol. I. p. 61. He does not here speak from personal observation.

province of Rio de Janeiro, until, some forty-five miles from the sea, the mountains break down on the edge of the plateau, when it reaches the Campos dos Goitacazes, and empties into the sea after a course of 102 Brazilian leagues (Gerber), or 408 miles.

According to Gerber, the altitude of the river above the sea, at the mouth of the river Parahybuna, is 272 metres. The stream is so much broken up by rapids above São Fidelis that steam navigation ends there, and in the rest of its course it gives passage only to canoes and large montarias propelled by the setting-pole or towing-rope.

The principal affluents of the Parahyba are the Rio Pirahy, already mentioned, which affords steam navigation for a short distance above its mouth; the Rio Preto, which rises in the Itatiaiossú, and is a much larger stream, falling into the Parahyba north of Rio; the Pirapitinga, which rises in the Serra of the Pardo; the Piabanha, which descends from the Serra do Mar from near Petropolis, and enters the Rio Parahyba at Entre Rios, a little above the Parahybuna; (this latter descends from the Mantiqueira, on the opposite side of the basin, in the same meridian with the Piabanha); the Pomba, which rises in the Serra da Mantiqueira, a few miles east of Barbacena; and the Muriahé, which takes its source in the same range, a few leagues to the east. The gneiss region of Rio, where uncultivated, bears a most vigorous virgin forest growth, and its soils are particularly favorable for the cultivation of coffee; the great valley of the Parahyba, above São Fidelis, and the valleys of its affluents, are largely occupied by coffee plantations. The same is true of the gneiss regions of the north, whose topography and soils are, over large districts, favorable for coffee planting.

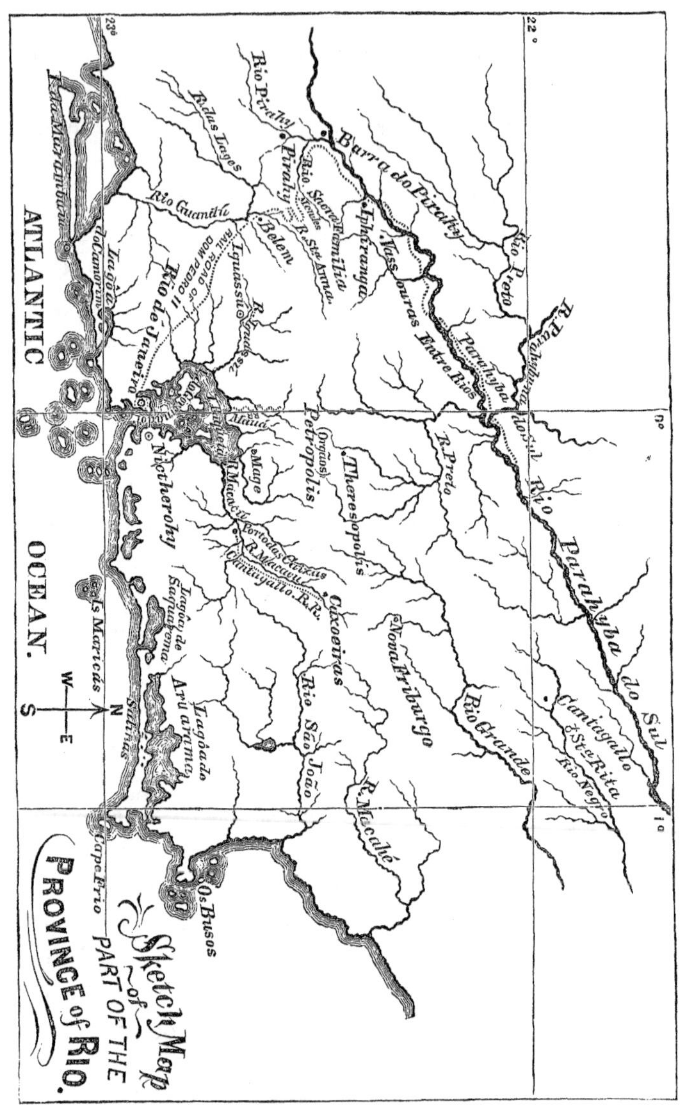
Sketch Map
of
PART OF THE
PROVINCE of RIO.
ATLANTIC
OCEAN.
Barra do Pirahy
Vassouras
Entre Rios
Rio Parahyba do Sul
Petropolis
Theresopolis
Nova Friburgo
Cantagallo
Rio de Janeiro
Nictherohy
Maricás
Cape Frio
Os Busos
Rio Grande
R. Macahé
Rio São João
Belem
Pirahy
Rio Pirahy
Mauá
Magé
Cachoeiras
Iguassú
Rio Preto
R. Preto
N
S
E
W
22°
23°

With these words of introduction, let us now look at the general points of interest to the geologist and physical geographer in the vicinity of Rio, and then examine the province to the east.

The Bay of Rio is a basin penetrating inland about twenty miles. It is only about a mile wide at the mouth, or Barra, but widens immediately. The shores on each side run in and out with deep, sweeping curves, making beautiful bays; and, three miles inside, passing the city of Nictherohy on the east and Rio de Janeiro on the west, the bay widens rapidly, and then, with the same irregular shore line, attains its greatest width about six miles from the mouth, when it contracts and runs off inland toward the northeast. It contains several islands; those near the city, as the Ilha das Cobras and Ilha Enxada, are of gneiss, like that of the adjacent hills.

The Ilha do Governador, a large island which I have examined on the eastern side, is composed partially of gneiss; but I strongly suspect that it is largely made up of the same tertiary clays which I have found on the Cantagallo Railroad, near Porto das Caixas. It seems too level to consist of gneiss alone. This island possesses fertile soils, but the ants are a terrible plague. Burton says that there are kjoekkenmoeddings upon it. Ancient shell-heaps have been reported from Brazil by Dr. Henry Naegeli of Rio, and they occur at various places along the coast. At Santos there are some very extensive ones, and St. Hilaire speaks of numerous shell-heaps on the coast of the province of Espirito Santo near Santa Cruz.

The smaller islands are gneiss, though, as in Paquetá, we find isolated gneiss masses united by stretches of sands containing recent shells. Spix and Martius have called attention to the great number of palms growing on the islands in the Bay of Rio, owing to the dampness and heat of the climate. The bay is nowhere very deep; along the shores it is very shallow; and Mouchez's map shows the soundings progressing quite uniformly from all sides, from one to fourteen or eighteen metres. Near the middle of the bay, off Rio, we find the greatest depth, — thirty-one metres.

Near the mouth of the bay the shores are rocky and sandy; the water is clear, and the bottom is composed of sand and shells; but around the whole head of the bay, where the shores are low, and a host of little rivers bring down great quantities of silt, the basin is bordered by extensive mangrove swamps, and the bottom is shallow and muddy.

In the turbid and somewhat brackish waters of the head of the bay oysters of an immense size flourish, often growing attached to the roots of the mangroves. Fish are very abundant, and the *curraes*, or weirs, for taking them are conspicuous objects in the shore scenery. The waters of the interior of the bay are exceedingly clear and bright, and off the islands of Paquetá and Governador a deposit of shells, with a calcareous mud, is in process of accumulation. Almost all the shells are small, and consist chiefly of species of *Arca*, *Venus*, *Murex*, *Cardium*, *Dentalium*, &c. I have looked diligently for corals, but the only Madreporians I have seen in the Bay of Rio consist of a couple of species of *Astrangia*,* found growing on the shells, and contributing really nothing to the deposits there accumulating. The tide in the Bay of Rio, as near as I can learn, rises from three to five feet,† but is very uncertain, as has been remarked by

* Professor Agassiz tells me that a species of Porites has been collected by Dr. Naegeli.

† On the map of Rio, by M. Barral, it is set down as three feet. Spix and Martius are in error in making it fourteen or fifteen feet.

Spix and Martius.* I have seen the tide stand at the same level for a whole day at the docks in Rio. On one occasion I visited Paquetá, and during a day and a half the tide was high; but, shortly after my return to Rio, it fell to an extraordinarily low level, and remained so for many hours. This appears to be owing to the fact that the bay opens to the southward, while the mouth is very contracted, so that the waters may be piled up on the coast by a southerly wind in such a way as to prevent the tide from flowing out, while other winds may depress the level of the waters for as long a period.

The entrance to the harbor is very bold on both sides, and is sentinelled by steep gneiss hills, — rounded or conical, wooded or bare, — presenting lichen-blacked precipices, whose faces are smoothed and rounded in a most remarkable manner. On the eastern side some of these hills are more than a thousand feet in height. They are clustered closely together, and stretch off in a sea of hills along the bay for a few miles on the eastern side, while along the coast, eastward toward Cape Frio, they form a line of irregular mountains of much grandeur and beauty. On the west side of the entrance to the bay is the conical peak of the Pão de Assu-

* Spix and Martius, *Reise*, Vol. I. p. 95. "The interior basin of Rio de Janeiro has its tide as well as the ocean. At new and full moon, high water, which rises fourteen or fifteen feet, sets in at thirty minutes past four; the ebb sometimes continues a whole day without intermission, at which time the current is the strongest on the west side of the bay. On the other hand, when the flood begins, a whirling current is remarked on the east side. The flood continues a shorter time than the ebb, and usually runs at the rate of three or four sea miles per hour. This strong flood has more than once led the captains of the ships into error, and caused them to cast anchor too close to the shore."

The same authors remark that the saltness of the water of the bay is rather less than that of the ocean. This is to be expected, from the large drain of fresh water into it.

car (*Pot de Beurre* of the early French colonists), with its smooth, precipitous sides. This celebrated rock belongs to a short range of hills that runs westward to the Lagôa das Freitas, and which is separated from the Corcovado, and the hills of which the latter forms a part, by the valley of Botafogo. The Sugar-loaf is said to be over a thousand feet in height.* The other hills are much lower, and are rounded, with precipitous sides facing the valley. They are composed of heavy beds of gneiss, which is very homogeneous, coarse-grained, and often very porphyritic in structure, with large crystals of flesh-colored orthoclase feldspar. These beds have very nearly the same strike as the trend of the hills they compose, and dip southward at a moderate angle, as may be observed in the cliffs along the pass leading from Botafogo to the Praia de Copocabana. The Corcovado is a sharp, angular peak, which separates itself from the great mass of hills lying back of Rio, rising by a long, narrow incline on the northwestern side to a point only a few rods in area on top. On three sides it drops off in a splendid sheer precipice, which on the south is several hundred feet high. Below this precipice the mountain presents toward the south a very steep wooded slope. A similar slope, not a *talus*, runs below the cliffs on the opposite side of the valley, and may be observed in very many hills of the gneiss region. In the Corcovado we find rocks similar to those of the Sugar-loaf, but varying considerably in character, being generally, near the top, well laminated, sometimes with large crystals of feldspar, and not infrequently full of garnets. The dip varies somewhat, and I find in my field-book two notes of a slight southerly

* Burmeister says 1,212 feet. (*Reise nach Brasilien*, p. 57.) Dr. Almeida makes it 373 metres, or about 1,426 feet.

THE SUGAR-LOAF, CORCOVADO AND GAVIA, FROM SÃO DOMINGO.

dip near the top, while an observation taken some distance below the reservoir on the aqueduct gave a marked northward dip. The general dip of the gneiss in the peak, as well as in the hills east, is certainly northward, and one sees the same dip in certain ledges of rock in the harbor, which lie in the line of strike of the beds of the Corcovado. The valley which separates the hills of the Corcovado from those of the Sugar-loaf is therefore an anteclinal valley. Professor Agassiz has independently made the same observations, and has come to the same conclusions.*

Three miles southwest of the Corcovado are the Tres Irmãos, a group of conical peaks very interesting to the student of topography, and a couple of miles west of these is the Gavia, an isolated, tower-like, flat-topped mountain, said to be 3,000 feet high. Within the irregular ring formed by these moun-

* Pissis has called attention to the resemblance borne between the fine-grained gneiss, with garnets overlying the porphyritic variety forming the base of the Corcovado, and that of the rocks at Copocabana. (*Mém. de l'Instit. de France*, Tom. X. pp. 362, 363.)

tains is the beautiful Lagôa de Freitas, a sheet of water held in a basin among the hills, and shut out from the sea — like the lagôas so common along the coast east and west of Rio — by the throwing up of a sand-beach across its mouth. Travellers all speak of the romantic beauty of this spot, and it is worthy of their praise; for though clothed in the warm verdure of the tropics, it is really Swiss-like in the character of its scenery. If the geologist has any soul, any love for the beautiful, there is no scene which, with all his cold analysis of topographical and geological elements, is more likely to impress him as an *artist's* work. I know of no view which has affected me so much — not only as a scientific observer, but as a man — as that of the vicinity of Rio from the top of the Corcovado. There are a thousand subjects for observation and study; and, with all, there comes over one a feeling akin to, but infinitely more deep and impressive than, that which one experiences when in some old cathedral he sits down to study the sublime creation of one of the old masters. He who can lean over the parapet that crowns the Corcovado, and look down more than 2,000 feet † on the temple of palms of the Botanical Garden, and on the silent Lagôa de Freitas, — "another sky," in whose blue depths sail soft fleecy clouds, — who can gaze on the proud encircling peaks, green with an everlasting spring, and shivering with silvery reflections from the Cecropias, — who can look out over the island and sail dotted sea, and the surges creeping up on the long, curving sea-beaches, and then over the bay, with the city fringing widely its sweeping curves, the sea of hills beyond, the majestic Serra dos Orgãos heaving its great back, in the exquisite blue distance, far above the level line of the clouds, its great minarets

* 2,179 feet is the exact height. Burmeister says 2,164.

sharply defined against the purple ether, — and can intelligently take into consideration all the geological, climatic, and other natural laws which have determined the elements of beauty and usefulness in the scene, and not have his whole soul moved within him in homage to the Artist whose hand has moulded continents, carved out their lineaments, spread over them their mantle of vegetation, and peopled them with living forms, has not gone beyond the alphabet and grammar of his science, and has no idea of the literature of Nature.

The Corcovado is only one peak of a mass of hills which occupies a large area west of Rio. It is united by a pass 1,000 feet in height to another mass lying northward of the last, and which culminates in a sharp conical peak called Tijuca, the latter being about 3,447 feet in height. The Tijuca range is connected with a group of hills which extends several leagues to the west. Westward of the Gavia is a stretch of plain only a few feet in elevation above the sea, and broken up by very numerous lakes, some of which are of considerable size. Between the hills which I have just described and the Serra do Mar there is a wide extent of low country, in some parts perfectly flat and very low, in others somewhat diversified with hills. I have seen this country only from the railway-train, but I would suggest the more than probability that a considerable part of it may be made up of beds of stratified clays, like those of the Cantagallo Railroad, presently to be described. The Serra do Mar, as well as the whole range, including the Organ and Cantagallo Mountains, as has been already stated, is composed of gneiss.

At Rio the gneiss varies very much in texture. As a general thing it occurs in very thick homogeneous beds, varying from an exceedingly coarse porphyritic kind containing large crystals of black mica, and crystals of pink feldspar several inches long, to a fine, even-grained, compact, light-gray variety. Sometimes it is very distinctly laminated, fissile, flaggy, or schistose. Garnets are very common in it. It is largely quarried for building purposes in all directions, and the finer-grained kinds are much used for paving, not only in the streets of Rio, but in other towns on the coast. The numerous quarries afford very excellent sections. The unbroken compactness and the little jointing of rock, even on the surface, strike the observer as remarkable. Many of the hills are monoliths. The dip and strike are, owing to the homogeneity of the rock, generally very indistinct. Immense granite veins, with the minerals composing them often very coarsely crystallized, traverse the rock, together with veins of milky quartz and occasional dykes of diorite or greenstone. Faults are numerous. I copy from my note-book the following section made from the barracks north of Praia Grande on the eastern side of the harbor and extending southward to the city: —

SECTION AT PRAIA GRANDE.

At the barracks, the gneiss, *a*, is dark gray with a fine lamination. The crystals of feldspar are very small, and there are a great many red garnets. Strike N. 40° E. Dip 30° S.

About 800 feet southward of the barracks is a large quarry, in a bed of very compact gneiss, *d*, showing scarcely a trace of stratification. It is composed of feldspar in large crystals, mica, finely crystallized, and very little quartz. Garnets are abundant, and the rock is traversed by veins, a number of which I have noted as dipping steeply southward. In one of these veins, which was four inches in width, the sides consisted almost wholly of large, coarse crystals of a light, flesh-colored feldspar with a little green mica, while the middle portion was made up of clear, glassy quartz of a light reddish color, with an occasional large crystal of black mica.

Above these beds are others, in which I have noted that the crystals of feldspar lie with their longer sides parallel to the plane of stratification. In a large vein here we find the same arrangement of the materials as in that above described.

f. Gneiss with very large crystals of feldspar.

g. Thin beds of compact mica-slate, or extremely fine-grained and distinctly bedded gneiss. Strike N. 35° E. Dip 35° S.

Here the section fails on reaching the city. In a hill immediately east of the church of São João the gneiss is principally composed of large feldspar crystals. Strike N. 80° E. Dip 35° S. In the point between Praia Grande and São Domingo the rock is like that last seen, the strike being N. 80° E., and the dip southward. Just south of Fort São Domingo there are seen veins of iron ore in the gneiss; the dip is about 45° S., and the strike about N. 80° E.

I have carefully examined the cuttings on the Dom Pedro II. Railroad, from Belem, at the southern base of the Serra do Mar, to Ypiranga, in the valley of the Parahyba do Sul, and I have studied them in detail, on foot, from the mouth of the great tunnel, which pierces the crest of the Serra, to Ypiranga. The whole ridge is composed of gneiss, which varies very much in character, as a general rule being dark gray in color, well laminated, often fine-grained and schistose. The strike varies, according to my observations, from N. 45° E. to N. 80° E., and the mean of thirty-four observations of strike taken along the road from the southern entrance of the great tunnel to the Barra do Pirahy, would give N. 62° E.* The dip is almost invariably northward from Belem to the Parahyba, so that the Serra is here a monoclinal ridge, but it is very probable that the same strata are repeated. There seems to have been much dislocation of the strata, and faults are common. In some cases the beds are much plicated, though, as above remarked, the general dip is remarkably uniform. Quartz and granite veins are very numerous, and trap dykes are not uncommon. At Ypiranga is a thin bed of crystalline limestone, which is exposed near the railway, and is more or less quarried for burning into lime.

The Serra dos Orgãos and the Serra da Estrella, so frequently described, is a heavy ridge lying at the head of the Bay of Rio, and belonging to the Serra do Mar. It is a magnificent mountain pile, which, to the east, runs up into a series of picturesque *aiguilles*. I have never myself visited the Organs, and I quote Professor Agassiz's remarks on the structure of this ridge. He says: † "The chain is formed by the sharp folding up of the strata, sometimes quite vertically, in other instances with a slope more or less steep, but always rather sudden. To one standing on the hill to the east of Theresopolis, the whole range presents itself in perfect profile; the axis, on either side of which dip the almost vertical beds

* Of eight reliable observations of strike, taken in the vicinity of Rio, the mean is N. 55° 30′ E.; but these observations vary from N. 10° E. to N. 80° E.

† Journey in Brazil, p. 491.

THE ORGANS.

of metamorphic rocks composing the chain, occupies about the centre of the range.

"To the north, though very steeply inclined, the beds are not so vertical as in the southern prolongation of the range. The consequence of this difference is the formation of more massive and less disconnected summits on the north side, while on the south side, where the strata are nearly or quite vertical, the harder set of beds alone have remained standing, the softer intervening beds having been gradually disintegrated. By this process have been formed those strange peaks which appear from a distance like a row of organ-pipes, and have suggested the name by which the chain is known. They consist of vertical beds, isolated from the general mass in consequence of the disappearance of contiguous strata. The aspect of these mountains from Rio is much the same as from Theresopolis, only that from the two points of view — one being to the northeast, the other to the southwest of the range — their summits present themselves in reverse order. When seen in complete profile, their slender appearance is most striking. Viewed from the side, the broad surface of the strata, though equally steep, exhibits a triangular form rather than that of vertical columns. It is strange that the height of the Organ Mountain peaks, so conspicuous a feature in the landscape of Rio, should not have been accurately measured. The only precise indication I have been able to find is recorded by Liais, who gives 7,000 feet as the maximum height observed by him." *

"These abrupt peaks frequently surround closed basins, very symmetrical in shape, but without any outlet. On account of this singular formation, the glacial phenomena which abound in the Organ Mountains are of a peculiar character." †

Pissis ‡ has described the Organs in very much the same

* Gardner makes them 7,500 to 7,800 feet.

† Similar aiguilles are common elsewhere on the coast in the gneiss belt, and present the same features as are described by Professor Agassiz. I may cite here, as examples, the Frade de Macahé, the Pedra Lisa and the Garrafão, near Itabapuana, the peaks of Itapémerim, &c., all of which I shall describe hereafter.

‡ "Le leptinite s'y trouve fortement redressé et présente des lames colossales qui, vues de Rio de Janeiro dans le sens de leur épaisseur, apparaissent comme de hautes et minces aiguilles placées les unes à côté des autres et ressemblant assez à des tuyaux d'orgues." — *Mém. de l'Inst. de France,* Tom. X. p. 360.

terms as Agassiz. The same author calls attention to the great steepness of the southern slope of the Organ Mountains, and speaks of the little streams which, precipitating themselves over it, are visible, as I can testify, from Rio, — a distance of more than thirty miles, — appearing like silvery lines drawn down the blue flank of the Serra. The needles of the Organs are usually represented as seen from their thinnest side, which would give one the idea that they were tower or chimney-like masses. Professor Agassiz has figured one of the most noted of the peaks as seen in this way. Through his kindness I am enabled to present a side view of the same peak, from a photograph in his possession. The views of the Organs in Gardner's Travels, and in the majority of works on Brazil, are very poor.

The gneiss of the vicinity of Rio and of the Serra do Mar is remarkably unproductive in useful minerals. Indeed, I do not know of any mineral deposits of economic value in the region, except gold, which occurs in these gneisses in São Paulo, at Cantagallo, and elsewhere, but not very abundantly. The almost entire absence of limestone is remarkable. I have nowhere seen any trace of graphite.

The Rio Macacú is one of the little streams entering the Bay of Rio near its northeastern extremity. The whole country bordering the bay, as above stated, is one great mangrove swamp, and from the mouth of the river to the Porto da Villa Nova this is the character of its banks. At this place the soil consists, like that of the rest of the swamp, of a soft, dark-blue mud, or clay, containing a little decomposed mica in silvery flakes. This mud at Villa Nova is twenty-four feet deep, and is underlaid by sand and gravel containing recent shells; but farther back from the river, as might be expected, it is not so thick, being only about six feet.

The general surface of the country here is perfectly flat, and only about one foot above water level at spring-tides. The river * is one hundred feet wide at the Porto, but it is very deep, and its banks are bordered with rush-like plants, — the *quina*, a plant with a triangular stem, and the *periperí*, or Brazilian papyrus, so much used for mats, while the swamps are covered by mangroves, *tapibuia* trees, and a dense growth of plants, such as love the salt marshes.

Leaving the river at Porto da Villa Nova, and following along the extension of the line of the Cantagallo Railway, one soon leaves the swamps, and, rising a few feet, finds himself on a plain of coarse white sand, in which are seen exposed in excavations an abundance of recent shells, like those which lie on the beaches along the bay, *Venus flexuosa* being especially common. These plains are sparsely wooded, and support a vegetation quite different from what we have observed in the swamps.† Conspicuous among the trees, or rather large shrubs, is the *pitangueira* (*Eugenia*), noted for its refreshingly acid, red fruit. Bromeliaceous plants are common, together with cactuses, &c. Where, however, a soil has accumu-

* Burmeister remarks upon the coffee-brown color of the water of this river. The vicinity is very unhealthy, intermittent fevers of a typhoid character prevailing.

† Spix and Martius describe a similar sand-plain on the road from Rio to the Imperial Fazenda of Santa Cruz: "On the way hither we remarked a stretch of ground composed of coarse dry granite sand. The low but very pleasing wood covering it resembled, in its shining green, stiff foliage, our laurel woods, but, as a token of the tropical climate, it was characterized by the multiplicity of the flower-forms." The same authors mention the following plants as occurring on this ground: "*Schinus Aroeira, terebinthifolia* Raddi; *Pohlana* (*Langsdorffia* Leandr.) *instrumentaria* Mart.; *Spixia heteranthera* Leandr.; *Byrsonima nitidissima* Humb.; *Sapium ilicifolium* W.; *Alsodea Physiphora* Mart.; *Petroea racemosa* Nees; *Solena grandifolia; Serianæ, Paulinicæ, sp.*, &c." (Spix and Martius, *Reise*, Vol. I. p. 181.)

lated, the vegetation assumes the dignity of a forest growth. Presently we come upon some low hillocks, very rounded in their outline, which stand like islands in the sand-plain. Some of these are cut through by the railway, and we see that they are isolated and denuded outliers of a formation underlying the sands. Rising above the general level of the sands, they are bathed by them round about, like islands. Some of these hillocks are composed of a bed of white or reddish arenaceous clay, obscurely stratified, like a kaolin mixed with sand, with an occasional quartz pebble, and irregularly tinged by yellow or red ferric oxide. Over the evenly rounded surface of this clay is spread a thin sheet of quartz pebbles, generally well rounded, following all the sinuosities of the surface on which it lies, though the bed varies much in thickness. Over this pebble sheet is a concentric coat several feet thick, of a perfectly structureless, arenaceous clay, consisting of decomposed feldspar and fragments of quartz, deeply colored by ferric oxide, and resembling the unwashed and unassorted product which would result from the mechanical trituration of decomposed gneiss, with a mixture of the clays just described. All the hillocks were covered by the same material. At Porto das Caixas we rise by a steep incline some thirty feet, more or less (I have no note of the exact height), to a level plain of large extent. Cuttings at the railroad dépôt show that it is composed of a horizontal deposit of the same tinted sandy clays we found occupying the centre of the hillocks just described, and that it is overlaid by a thick bed of structureless red clay, separated from the underlying deposit by a layer of quartz pebbles. Taking the train, we go westward over this plain some ten miles.*

* Being unfurnished with a reliable map, I can only give distances approximately.

Part of it is dry and sparsely wooded, but there are large areas occupied by swamps, in which grows a more or less luxuriant swamp vegetation, — a dismal scene, presenting sickly-looking trees loaded with orchids, ferns, and other parasites, and draped heavily with a species of *Tillandsia*. By and by we reach the end of a spur from the Serra do Morro Queimado, and, skirting the western base of the hills, gradually ascend, passing from the plain to the gneiss valley of the Rio Macacú, and, at the present terminus of the railroad at Cachoeiras, find ourselves in a narrow valley, among gneiss hills, at the foot of the Serra do Morro Queimado, an eastward continuation of the Serra dos Orgãos. The gneiss here is of the same general character as that at Rio. An observation by the riverside, near the residence of Mr. Williams, the superintendent of the railroad, gave the strike N. 70° E. The dip was vertical. Garnets were very abundant.

On examining the soil, whether high up on the slopes, in the hills, or elsewhere in the numerous cuttings of the roads, &c., the rock is seen to be covered by the same red clay that we have observed forming the surface-coating of the plain at Porto das Caixas; and only a superficial examination of the intervening country is necessary to show that the same red clay covers, not only the clay hillocks of the railway extension and the plains described, but the hills also, descending everywhere to the level of the sand-plain. In addition, it is to be seen frequently underlaid by an irregular sheet of quartz pebbles, and it very often contains angular fragments of quartz and sometimes masses of gneiss, though these last are rare. It is absolutely structureless, and, much as it may resemble the stratified clays above described, or soft, decomposed gneiss, it needs only a very short experience to enable one to distinguish from a hand specimen whether it be one or the other. Generally this structureless clay is deep red; but on the surface it is often more or less yellowish. The following ideal section illustrates the geology of the country between the Bay of Rio and the Serra do Morro Queimado, along the route over which we have travelled.

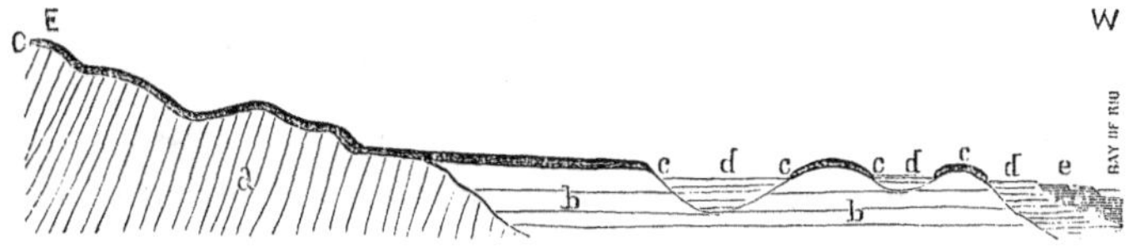

SECTION ALONG THE LINE OF THE CANTAGALLO RAILROAD.

a. Gneiss, Eozoic.
b. Stratified clays, Tertiary. (Drift, Agassiz)
c. Drift.
d. Raised beaches with recent shells.
e. Modern mud deposit.
f. Bay of Rio.

This section tells a very clear story. Late in the tertiary period, long after the hills of Rio were upheaved, and when the country stood at a slightly lower level, the stratified clays which I have described were deposited over the whole area of the basin of the Bay of Rio, and the adjoining flat country. These were afterward upheaved, most probably, as I shall attempt to show hereafter, to a much greater height than at present, and subjected to denudation by water and glacial action. Owing to the shape of the basin and the great number of streams flowing into it, as well as to the fact that it must have been the focal point toward which the glaciers of the encircling hills converged, it is not to be wondered at that, along this part of the coast, the clays have been so completely denuded and swept away, leaving only small patches fringing the shores in certain somewhat sheltered positions. It seems an argument in favor of the prevalence of glacial action in the region, that the denudation is so complete, else we should have expected to find the sheet largely cut through by valleys, leaving more numerous outlying masses. As to the age of these clays, more hereafter. I have referred them to the Tertiary, though Professor Agassiz is inclined to regard them as drift. The superficial red clay deposit I believe, with Professor Agassiz, to be drift. The stratified sands were deposited in shallow water when the sea stood only a few feet higher than at present, and they have been elevated by a recent rise of the coast, — a rise which I believe to be still going on. The recent muds, now accumulating from sediment brought down by the streams, form a marshy fringe outside the raised beaches.

Having introduced the reader to the general geological and topographical features of the vicinity of Rio de Janeiro, let us now examine somewhat closely the drift phenomena observable there. In the following remarks I have purposely confined myself to the results of my own personal observations made during extended excursions over the country near Rio, and a detailed examination of every cutting on the Dom Pedro II. Railroad, from the Great Tunnel to Ypiranga, on the results of which survey I have made a long report to Professor Agassiz. In connection with this subject we will take into consideration some of the topographical elements so remarkable in the hills of Rio and the Serra do Mar. If we examine the gneiss hills at Rio de Janeiro and the vicinity, we find that they are invariably covered, where the slopes are not too steep, by the same coat of red soil which we have observed on the Cantagallo Railroad. This may vary more or less in the coarseness or fineness of its ingredients, but it invariably presents everywhere the same general character of a sheet of structureless, unarranged material, composed of ground-up gneiss, perfectly devoid of stratification, and always of a deep red color passing into yellow near the surface, especially where the material is sandy and light. There is rarely any humus, because the decay of vegetable matter is too rapid to allow of its accumulating as a soil, as in northern countries. This clay sheet varies in thickness from a few feet to one hundred; sometimes it is stiff and bakes very hard, in other cases it is more sandy and light. Usually it is quite free from admixture with boulders, but sometimes angular fragments of quartz of considerable size occur in it, together with rounded or angular masses of gneiss or diorite. The latter rocks are almost always in a decomposed state, and, except to the experienced eye, are recognizable only in fresh cuttings. Under this clay one sometimes finds, as in the cuttings at Tijuca, a thin layer of quartz pebbles like that seen on the Cantagallo Railroad, but this is not always present. The surface of the gneiss on which the drift rests is always *moutonnée* and remarkably evenly rounded down, and the sheet of quartz pebbles lies immediately upon it, following all its curves; but the pebble sheet may be wanting over large areas, or vary very suddenly and irregularly in thickness. The gneiss *in sitû* is almost invariably decomposed beneath the drift to a depth varying from a few inches to one hundred feet. The feldspar has been converted into clay, the mica has parted with its iron, &c., but the altered crystals of the gneiss still occupy their relative position with reference to one another. The planes of stratification are well marked, and the veins of quartz, though cracked up, remain in place.

This extraordinary decomposition of the Brazilian gneiss and other rocks has long attracted attention, and Darwin has described it very accurately in his Geological Observations. He was of the opinion that it had taken place under the sea before the present valleys had been excavated.* Pissis also described it; but it appears to me that he has greatly over-estimated the depth to which the softening of the rocks has extended, when he says that, in the gneiss region between São Fidelis and the Serra dos Orgãos, the gneiss has been decomposed to a depth, in some places, of 300 metres!! †

This decomposition results, in my opinion, from the action of the warm rain-water soaking through the rock, and carrying with it carbonic acid, derived not only from the air, but from the vegetation decaying upon the soil, together with organic acids, nitrate of ammonia, &c. I believe that the remarkable decomposition of the rocks in Brazil has taken place only in regions anciently, or at present, covered by forest. Heusser and Claraz have suggested that it is aided by nitric acid. They say: "It is, without doubt, determined by the violence and frequency of the tropical rains, and by the dissolving action of water, which increases with the temperature. It is necessary to observe, moreover, that this water contains some nitric acid, on account of the thunder-storms which follow each other with great regularity during many months of the year." ‡

Professor Agassiz has discussed this subject at some

* Darwin, Geological Observations, p. 144.

† Pissis, *Mém. de l'Inst. de France*, Tom. X. p. 358. It is well to remember that, before the glacial origin of the clays overlying the decomposed rock was pointed out by Professor Agassiz, the thickness of these clays was included in the estimate of the depth to which the decomposition had taken place.

‡ MM. Ch. Heusser and G. Claraz, Ann. des Mines, 5me Série, Mém., Tom. XVII. p. 291.

length in the Journey in Brazil, his opinion being that the softening of the rock is due to the action of warm rain-water. It has been objected to this theory that the stone employed in Brazil for building purposes endures remarkably well, showing, after the lapse of centuries, very little change. This argument, in my opinion, is of very little weight, for a smooth, naked surface from which the water runs off rapidly, and which is, the greater part of the time, dry, is placed under conditions very different from those of the gneiss overspread by a thick coating of wet drift paste, and constantly soaked with water.

Brazil is not the only country in which the rocks have softened to a great depth. The same phenomenon has been observed in the Southern States of the Union and in India.* I have seen gneiss decomposed to a depth of several feet in the vicinity of New York.

When the gneiss is fine-grained, homogeneous, and not distinctly stratified, it is often difficult to distinguish the rock decomposed *in situ* from the drift; but the line of demarcation between the two, even when not marked by the pebble sheet, is usually easily distinguishable even in old cuttings. The slope of a railway cutting is apt to gully out along the line of junction of the drift and the underlying decomposed rock. This line is invariably gently undulating, and one never sees jagged edges of strata or angular masses projecting upward into the drift.

Professor Agassiz has spoken of the valley of Tijuca below Bennett's as a locality where the drift is very beautifully exhibited. The mountain mass of Tijuca is separated from the mass of the Corcovado group by the pass of Boa Vista,

* Dr. Benza says that in the Neelgherries granite is sometimes decomposed to a depth of forty feet. (*Madras Journal of Literature*, &c., Oct. 1836, p. 246.)

which is about 1,000 feet above the sea. Eastward runs the valley of Andarahy downward toward the city, while to the westward one descends a most romantic valley to a great alluvial plain, in a sort of bay or amphitheatre among the hills. This valley, occupied by a tumbling mountain stream, descends very rapidly, ending abruptly below the Cascate Grande at some height above the plain.

THE CASCATINHA AT TIJUCA.

Minor valleys from Tijuca and the mountains to the south descend and join this valley. At Bennett's the drift clay is full of boulders of quartz, gneiss, and greenstone. If we ascend the brook which flows through Mr. Bennett's fazenda for a few rods, we shall find that it has cut its bed through the general clay sheet which everywhere covers the hills, and it is perfectly easy to see that this sheet is in no place a deposit thrown down by the brook. This loose material consists of a brownish or reddish earth without the slightest signs of stratification, in which are buried boulders of gneiss, usually rounded and of many qualities, together with rounded masses of quartz.

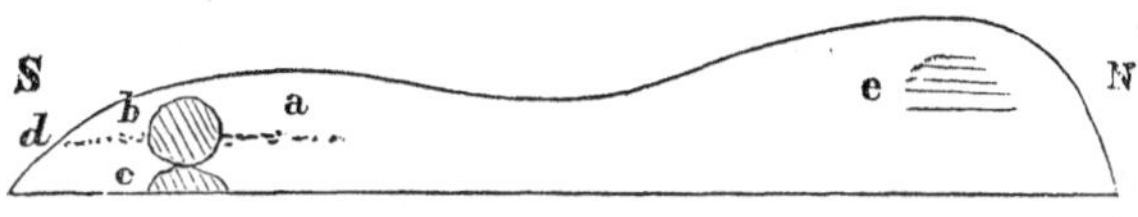

SECTION OF DRIFT AT BENNETT'S, TIJUCA.

At *c* is a very large boulder of homogeneous, unlaminated, fine-grained gneiss, from around which the drift has been washed away, and this rests on a mass, *in sitû*, of gneiss coarsely crystallized (porphyritic) and well laminated. At *e* we see the gneiss *in sitû*, much decomposed, and laid bare in a cutting.

The solid gneiss beds are well seen in the bed of the brook, and the rock is generally quite coarse and porphyritic in structure. I find a note of a strike N. 55° W., which, from the generally uniform northeastward strike seen elsewhere, might be suspected to be an incorrect observation ; but at the bridge, a little farther down the brook, I found the strike to be N. 10° W. If we follow down the valley we shall find the soil full of boulders, and some of these are many feet in diameter.

As we descend the valley still further these boulders are seen lying bare, not only in the brook where the water has washed away the loose material, but on the hillsides. I think that no geologist familiar with drift phenomena, who should suddenly find himself in this valley, would have even the slightest suspicion of their being anything else than the most ample testimony of the former prevalence of glacial action over the region; yet, in the beginning of his drift studies in Brazil, he is almost sure to commit some gross blunders, for it is not even a general rule that the loose boulders found on the surface are erratics. On my first visit to Tijuca, very soon after my arrival in Brazil, and after Professor Agassiz had announced the discovery of drift at Rio, I was struck with the appearance of some trap masses on a hillside near Bennett's, which looked remarkably like erratics; but a close study of them satisfied me that they resulted from the surface decomposition of a great trap dyke. Not descending far enough into the valley, and satisfying myself that a great proportion of the gneiss masses that I examined at the time were not erratics, I came most decidedly to the conclusion that the surface deposits of Rio were not drift, but were in some way due to the decomposition of the rock, as had been heretofore supposed. I desire to record here the fact, that I began my studies of the Brazilian drift with a conviction that Professor Agassiz was wrong, and I feel much gratified that my independent observations have so fully confirmed the results of his own. If one descends the valley towards the Cascate Grande, he will see that the valley is heaped with a confusion of immense boulders tumbled one upon another; masses of greenstone, weighing hundreds of tons, piled up with those of gneiss of all qualities. Where these are bare they are always rounded, as is seen in the engraving, but I believe that this is referable, to a very large extent, to a concentric and even decomposition of the surface; but

BOULDERS AT TIJUCA.

there is no resisting the conclusion that we have here a morainic deposit from a glacier which anciently occupied the valley. The above woodcut is from a stereograph published by Leuzinger at Rio, and represents the boulder masses above the Cascate Grande. Descending to the plain below the Cascate, which is also seen in the woodcut, one traces the drift clays and boulders quite down to the plain, when they end abruptly, and the flat lands are seen to be of alluvial origin resting on sea sands, of the same age as the sands of Paquetá and the Cantagallo Railroad extension. One cannot find in the plain, nor did I ever see anywhere on the beach sands, either drift clays or boulders.

Returning to the Dom Pedro II. Railroad, we may trace the drift sheet everywhere from Belem to the Barra do Pirahy, over the whole Serra do Mar; and one may here study its structure in the most detailed way. From Belem to the Parahyba River the same red clay entirely covers the surface, lying even on very high slopes. Nowhere is there the slightest sign of stratification, and it is sharply defined from the alluvial deposits of the river. The same pebble sheet is seen almost everywhere, though in cutting after cutting it may sometimes be wanting. Boulders are rare, and are almost invariably so decomposed as to be seen only in fresh cuttings. The rock on which the drift rests is always smoothly and evenly rounded down. The following diagram will illustrate the structure of one of the gneiss hillocks of the Serra do Mar.

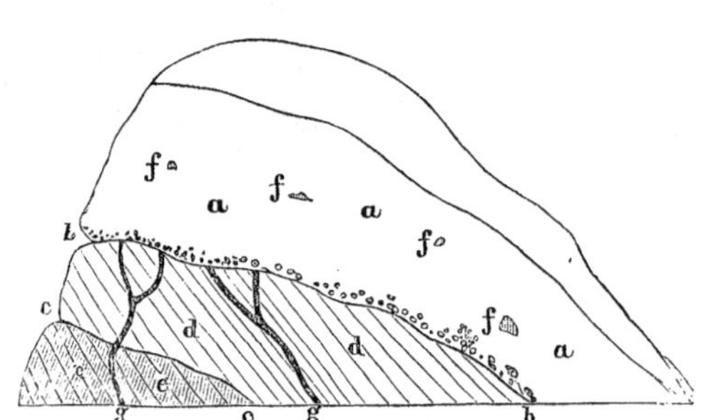

IDEAL SECTION OF DRIFT-COVERED GNEISS HILL.

a a. Drift clay.
f f. Angular fragments of quartz.
b b. Pebble sheet lying on rounded surface of gneiss.
d d. Gneiss *in sitû*, but decomposed.
e e. Gneiss undecomposed.
g g. Quartz and granite veins traversing both solid and decomposed gneiss.

Among the first elements in the Brazilian mountain scenery which attract the attention of the observer on approaching the coast near Rio are the bare rock slopes which, instead of forming ragged precipices, as in northern latitudes, are most remarkably smooth, and devoid of irregularities; and these surfaces, where the rock is homogeneous in texture, usually have a high slope, and frequently descend and dip beneath the sea level, the sea washing over their even faces.* Sometimes the top of a hill is bare, evenly moulded, and round, or a rounded boss projects from the side of a hill. These smooth cliffs and rock-slopes are heavily striped above the sea by broad lines drawn down their faces, blackened by a sort of cryptogamic growth; so that these slopes, of a rich, rather purplish, black tint, show very strangely, especially when set into the mantle of verdure which covers the hills. Of this character are the bare mass of the Sugar-loaf, the precipices of the Corcovado, and the steep slopes of hundreds of hills in the vicinity. Approaching nearer to one of these slopes, we find that its surface is often scored by a system of little rain-courses, and is covered by cactuses and bromeliaceous plants.

The Pedra Bonita is a bare mass of rock opposite the Gavia, but not so high. It is partly surrounded by almost vertical slopes, and on top, over a large area, is bare and rounded off. Here, though one may observe the same thing elsewhere, the way in which this rounding and smoothing down of the rock is produced may be studied, and one is soon forced to believe, that, whatever the glaciers may have contributed to the shaping of the topo-

* These evenly rounded, wave-washed rock slopes are very interesting, and have already been called attention to by Darwin (Geological Observations, p. 144). Such slopes may be seen not only on the shores of the quiet bays, but exposed to the full wash of the Atlantic waves, as at the mouth of the Bay of Rio and elsewhere. In describing, farther on, the Bay of Espirito Santo, I shall show how these wave-washed slopes originate.

graphical features of the country, the general moulding of the hills has been due primarily to subaerial denudation. The gneiss of the Pedra Bonita is decomposing. Where the rock is level the decomposed feldspar, &c. has been washed away by the rain, and we have the rock covered by a thin coating of loose quartz grains, which I hardly need say are angular. Where there is a decided slope, the loose sand is washed or blown away. The rock itself is much softened on the surface, and, to a considerable depth, the feldspar has been more or less altered. This semi-decomposed layer forms a concentric coat over the whole rock. Sometimes this is continuous and unbroken, but if the area exposed be large, we usually find that there is a tendency for it to separate itself from the undecomposed rock below, and to crack up. Sometimes this layer is only a few inches in thickness, in others it may be several feet. When the surface is horizontal, or nearly so, the tendency is for it to break up finally into small angular pieces, which, wasting away and rounding down by decomposition, cover the rock with loose, boulder-like masses, or are entirely removed, leaving a smooth, unencumbered surface. If the slope be very steep or vertical, the decomposed mass may fall or slide off. On very steep slopes, as those of the Sugar-loaf, Corcovado, Morro de Santa Theresa, and elsewhere, this sheet scales off and falls, breaking up below. The Brazilian gneiss cliffs rarely have a *talus* of broken fragments at their base. The decomposition going on evenly all over their face gives only sand and clay, washed down by the rains and distributed over the earthy slope below, and the half-decomposed fragments soften and finally decompose entirely. One must be very careful in examining a cutting that runs under a high cliff, for the earth resulting from the decomposition of the face of the cliff resembles more or less the drift-earth spread over the surface of the ground. Oftentimes a mass of half-decomposed rock separates itself from the face of the cliff in a great lenticular sheet. This may crack across horizontally, particularly if the plane of stratification cuts the surface of the cliff in this way, and the lower half may drop off, leaving an overhanging portion attached to the cliff. One may observe hanging masses of this kind attached to the precipice of the Sugar-loaf and Corcovado, and in innumerable other gneiss localities. If we examine one of the rounded gneiss hills, — as, for instance, one of those just behind the Dom Pedro II. Hospital at Botafogo, where part of the hill is bare and steep, — looking at the hill in a cross-section, we may observe that the general rounded curves of the drift-covered portion are quite out of harmony with those of the bare portions, which are usually flatter, and we may further notice that the very steep slopes or precipices are usually on the side of the hill away from which the strata dip. This is the case on both sides of the Botafogo valley. In the following diagram I have tried to represent the difference in moulding between a glaciated surface and that of one of these bare cliffs.

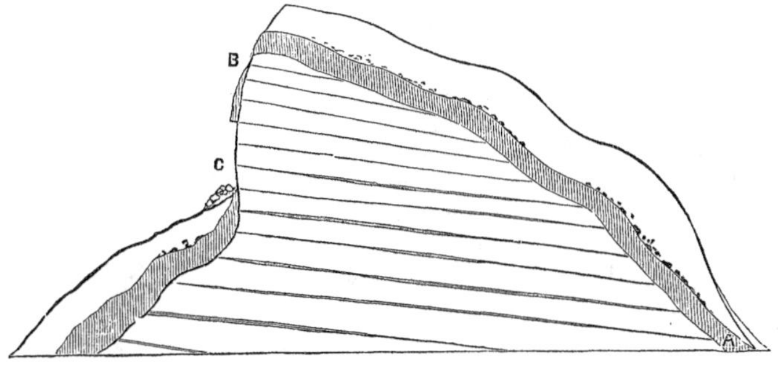

DIAGRAM TO SHOW CONTRAST IN MOULDING BETWEEN DRIFT-COVERED AND BARE SURFACES.

The outline of the drift-covered portion of the hill, *a b*, has been unchanged since the drift; for though the rock has been protected by the drift, the gneiss has decomposed, but has remained *in sitû*. Over the face, *b c*, the decomposition has gone on also; but since the slope has not been protected, it has worn constantly backwards, growing gradually more regular in outline as each new decomposed coat was thrown off. If one will take the pains to examine the curves resulting from the decomposition of bare surfaces, and those which result from decomposition where, as under a drift-sheet, the decomposed rock remains *in sitû*, he will be convinced that the moulding of the surface of the gneiss on which the drift rests is of an altogether different character from that resulting from simple subaerial denudation.

The sands containing recent shells along the Bay of Rio, and elsewhere, and which rise to the height of a few feet above high-water mark, bespeak a recent elevation of the coast, as has been observed independently by his Majesty the Emperor, Dr. Capanema,* and myself; but there are other proofs of the same upheaval in the holes excavated by sea-urchins, which are found in the vicinity, in many places many feet above high-water mark.

The islands of Maricás lie a few miles off the coast

* Dr. G. S. de Capanema has written more or less on the geology of Brazil. One of his papers mentioned by Burton bears the title, "Decomposição dos Penedos no Brasil," and was published at Rio in 1866. I am very sorry that I have never seen any of the papers of this geologist, who has travelled very extensively over the Empire. I know of his work only through quotations or references. From a MS. of Major Coutinho, placed in my hands by Professor Agassiz, I should infer that Dr. Capanema is a disbeliever in the glacial origin of the surface deposits claimed by Professor Agassiz and myself to be drift, and that he rather considers them to be the result of decomposition alone.

between Rio and Cape Frio. They are composed of gneiss in beds of unequal hardness, and a most excellent opportunity is afforded for the comparison of glaciated rock surfaces and surfaces denuded by subaerial decomposition, or wave action. The general surface of the islands is rounded, and is covered by drift clay. Some bare rocks are evenly rounded, but along the shore, where the waves beat, the softer beds are worn away more deeply than the hard ones, and the surface is very irregular. On the rocks a species of *Echinometra* (*E. Michelini* Desor), a sea-urchin with moderately long, dark-purple spines, is very common, living in a zone extending downwards from some distance below high-water mark. Here they are found, each in a cup-shaped depression worn in the rock, and in some places the rock is fairly honeycombed on the surface by their "nests." Above the zone of living sea-urchins the nests are found, but they are empty, and in protected localities, as, for instance, in narrow openings between rocks, they may be observed to extend to a height of several feet above high-water level, showing that the elevation of the coast has been very recent. I have observed that the nests appear less and less well preserved the higher we examine them, which has convinced me that the upheaval has been a gradual one, and I believe that it is still in progress.

The coast between Rio and Cape Frio is, for a large part, high and exceedingly picturesque. Many of the hills are bare and strangely shaped. The following sketch represents the coast as seen from near the island of Maricás. To the left are the hills of Rio.

Long sea-beaches stretching across bays formed by the hills have formed lagoons along this coast, and the low grounds between the mouth of Rio and Ponta Negra

COAST JUST EAST OF RIO.

are largely occupied by swamps and lagoons. Of the latter the most notable is the Lagôa de Maricá, which is some six miles long, salt, and separated from the sea, like the Lagôa de Freitas, by a sand-beach, through which the inhabitants are obliged occasionally to cut, in order to give passage to the waters of the lake during times of freshets. This lake, as well as the others along this coast, is extraordinarily rich in fish.

A sandy shore extends from west to east from Ponta Negra to Cape Frio. "It may be eight or ten leagues wide from the sea to the cordilheira, being roughened in this interval by several serras, and occupied in parts by various lagôas. All the flat part of this plain is useless for agricultural purposes on account of the depth of the sand, and its being overflowed a part of the year."* This low country appears to extend along the Rios São João and Una to the sea north of Cape Frio.

Lagôa Saquarema is a large lake lying east of Ponta Negra, and separated from the sea, like the Lagôa Maricá, by a sand strip. "It is three (Portuguese) miles long, and three

* *Diccionario Geografico,* art. *Maricá.* This work is a mere compilation, and is full of inaccuracies. In quoting from it I have done so with much care.

quarters of a league in its greatest width. It is salt, full of fish, and separated from the sea by a narrow tract of sandy ground. When the adjoining plains begin to be covered by the floods of the streams which empty into it, the inhabitants of the vicinity open an outlet to the ocean at the eastern extremity, which remains an unfordable river during the winter, at the end of which the surf closes it up." *

Lagôa Araruama is a narrow strip of salt water about twenty miles long, and with a varying width, in some places several miles, lying on the coast west of Cape Frio, and separated from the sea by a narrow strip of sand thrown up by the waves. Cazal says † that "it empties itself from its eastern extremity by a canal twenty-six braças large at its mouth, which is situated at a distance of a league and a half north of Cape Frio. Notwithstanding that a number of streams empty into it, its waters are salted by the communication which they have with the ocean. The tide makes itself felt as far as Ponta Grossa, which is situated at about its middle. Thence westward the waters go with the winds. It abounds in fish of various kinds. In some parts it is many braças in depth,‡ in others one may wade from one shore to the other." Milliet says that "between the sea, the city, and the Lake Ararauma are natural salines, which were prohibited by royal letters on the 26th of February, 1690, and 18th of January, 1691, the Portuguese government intending thereby to favor the

* *Corografia Brazilica,* Tom. II. p. 38. This old work by Cazal, published in 1818, is very much more reliable than the Diccionario Geografico.

† *Corografia Brazilica,* Tom. II. p. 38.

‡ Prinz Max. zu Neu-Wied, *Reise nach Brasilien,* 1er Band, 85te Seite. "Wir fanden das Wasser der Lagôa von geringer Tiefe und so klar, dass wir den weissen Sandboden des Grundes mit seinen Korallengewächsen deutlich wahrnehmen konnten; bei der geringen Tiefe sassen wir oft fest."

commerce of salt of its own European possessions. Notwithstanding the royal decrees, Domingos da Silva Ribeiro, judge-in-ordinary of this city, ordered, in 1768, that the communication between Lake Araruama and the salines of Maçabamba be closed, and in the following year the above salines furnished in six months 50,000 alqueires of salt."

Cape Frio is the name given to the southernmost point of a high, precipitous gneiss island, situated at the angle where the coast line, coming eastward from Rio, bends northward toward Cape São Thomé. It is only about three miles long, very irregular in outline, and is almost divided into two parts. The northern is, according to Mouchez, 394 metres in altitude. A lighthouse was erected on this point, but it

CAPE FRIO.

proved to be at too great a height, being above the level of the clouds, so that it had to be abandoned. The present lighthouse is situated at a lower level on the southern point. The island is separated from the main-land by a narrow but rather deep channel. The land opposite consists of a group of gneiss hills, formerly islands, which have been united together by sand beaches and sand plains, which extend northwestward, joining the beach of Maçambamba, or Massambamba, and running northward to the Rio Itajurú.

Between the city of Cape Frio and Os Buzios the coast is very irregular in outline, rather low, with small gneiss hills along the shore, forming occasional rocky promontories which are united by curving sea-beaches. I find in my notes the query whether the low flat lands seen from the sea may not be Tertiary. Towards Armação the coast is bordered by low gneiss hills, which, owing to the northwest (landward) dip of the rocks, and the correspondence of their strike with the general trend of the coast, as well as from the way in which they have been worn by the waves, present to the sea perpendicular, rugged cliffs of no great altitude. Along this coast are quite a number of little rocky islets of gneiss.

At Os Buzios* the coast line suddenly bends off to the northwestward, and runs towards São João, when it curves round and sweeps off with a northeast trend to Cape São Thomé. The gneiss shows itself on the shore of the point just east of the town. It is well laminated, much plicated, and has a general low dip to the northwestward. The gneiss of the point on the western side of the town has very much the same character. The gneiss behind the town seems to be overlaid by tertiary clays, which show themselves in the point west of the town; but I had no opportunity of examining them with much care. They are so much denuded that they are not easily recognizable by topograph-

* *Os Buzios* takes its name from its richness in shells. Among the species I collected there were *Cassis Madagascariensis* and *Cypræa exanthema*. Woodward in his Manual says that no Cypræas occur on the Brazilian coast. *Cypræa exanthema* is not at all rare, and occurs also at Bahia, where I have found another little species in great abundance. Several species of corals occur at Os Buzios; *Millepora alcicornis* is especially abundant, and the rocks are covered with patches of the common *Palythoa*, together with a *Zoanthus* with an emerald disk, common elsewhere on the coast.

ical features alone. There is a small hill just east of the village which bears a little church. It is cut away on the shore by the waves, and forms a little bluff composed of rounded quartz pebbles derived from the rocks of the vicinity, cemented together loosely by a soft, greenish clay. I have seen nothing like it elsewhere. Westward of Os Buzios is quite a range of gneiss hills, lying between the head-waters of the Rios Garcia and Trapiche. The shore beyond the rocky point west of Os Buzios appears, for some distance, to be tertiary, but there are some patches of recent sands. These deposits appear to terminate on the shore with a bright red cliff, very conspicuous from a distance, when the tertiary bluffs recede from the coast, and a sandy flat, backed by low plains, extends on to the gneiss hills of the Serra de São João. Thence to Macahé much of the shore is low, with stretches of tertiary clays, more or less denuded, and gneiss hills. North of Macahé the tertiary plains soon recede from the coast running off toward Campos on the Rio Parahyba do Sul, the land bordering the coast being flat, more or less swampy, and diversified by numberless shallow lagoons, some of which are of large extent. Off Macahé, and distant a few miles from the shore, lie the little gneiss islands of Santa Anna, known to coasters as their only retreat, north of Os Buzios, when storms or northeast winds prevent them from passing Cape São Thomé. There is usually a strong current off the Cape, but it is very variable. During the prevalence of a long northeast blow it runs southward with such rapidity that it is impossible for coasting vessels to tack against it. They may succeed in beating up from the islands of Santa Anna, close to the shore, until they reach the Cape, but they are then swept back. I was once nearly a fortnight beating off this coast during the prevalence of a northeast breeze, and I visited the islands many times. They are of no very especial interest, but one may observe here the sea-urchins' nests, raised above high-water mark, and may study some of the topographical features developed by decomposition. I add a little sketch of the westernmost island, as seen from the northeast, to show the smoothly rounded character of its steep sides.

ONE OF THE ISLANDS OF SANTA ANNA.

The Serras, always clothed with the virgin forest, stretch along, at varying distances from the coast, in a magnificent range of hills, with steep slopes toward the sea, forming one of the grandest panoramas of mountain scenery on the coast of Brazil. Of the altitude of these hills I have no precise information, but I should estimate some of them as at least 6,500 feet. There is one very conspicuous, obelisk-like peak, lying behind Macahé, but standing somewhat in advance of the general range of the Serra, and called the Frade de Macahé. Lieutenant Mouchez* gives its altitude on one of his charts as 1,750 metres (5,745 feet), which

* Lieutenant Mouchez has on his charts given the heights of various hills along the coast, but I find no note as to whether they are the results of actual measurements.

FRADE DE MACAHÉ.

would make it almost as high as the Serra da Piedade in Minas. The Serras break down on reaching the valley of the Parahyba River, just below São Fidelis, when, in detached Serras, they recede somewhat from the coast, cross the Muriahé and Itabapuana rivers, tying in with the Serras of Itabapuana and Itapémerim, and forming some very picturesque mountain scenery.

All the flat, sandy, and swampy land, interspersed with lagoons, which borders the Rio Parahyba almost to Campos, a city some miles above its mouth, as well as that which stretches southward to Macahé, or thereabouts, is of very recent formation, and is composed principally of sands and silt brought down by the river. Off the coast of São Thomé, as well as for some distance northward, the water is very shallow, and much discolored. These lands are bordered by a long, dreary sand-beach.* The country behind is, to a considerable extent, covered by shrubbery and trees, but there are extensive open plains where herds of cattle graze. By means of a ditch uniting the lagoons, and dignified by the name of a canal, water communication with Macahé has been opened,† and a considerable trade is carried on between the two places, or the settlements on the route, by means of canoes. Much of this swampy ground is excellent for rice, — an important product of this part of the country.

The lakes of this region are very numerous. They are all shallow, but some are several leagues in diameter. The

* Owing to the northeast trend of the beach south of Cape Frio, and the prevalence of northeast winds, the waves strike the beach obliquely, and there is a tendency for the sands to move southward. There are no dunes here.

† I have understood that this canal was finished, but Pompéo says that in 1864 it was only two thirds completed.

largest is Lagôa Feia, an irregular lake some twenty miles long, lying about ten miles south of Campos. Owing to its great area and its very small depth, its waters are kept constantly turbid through their agitation by the winds. It receives from the west quite a little stream, the Rio Macabú, which rises near the Serra do Frade. This river is navigable for canoes for some twenty miles above the lake. Another mountain stream of considerable importance rises among the Serras just north of the Macabú, and uniting in its course several large lakes, also enters Lagôa Feia. This lake is united on all sides with a multitude of lagoons of greater or less size by a perfect network of little channels, so that its waters flow partly to the north into the Parahyba, while in part they escape into a system of long narrow lagoons that stretch along just behind the beach ridges of the shore at Cape São Thomé, and communicate by channels across the beach with the sea. One of these lagoons, which passes by the name of the Rio Iguassú, is some fifteen miles long. It has evidently been formed by the throwing up in very recent times, probably during the prevalence of some very heavy storm, of a line of sand-beach just outside of the shore. Similar lagoons are found elsewhere along the coast, as for instance just south of Belmonte, on the Jequitinhonha, and I believe that the great line of sand-beaches stretching along the coast was thrown up, to begin with, by an extraordinarily heavy storm which prevailed along the whole coast, and which, in many instances, where the water was very shallow, disturbed the bottom at some little distance outside of the shore line, throwing up a sand barrier which, through the drifting of sands by the winds, as well as by the action of the waves, has since reached its present dimensions. During the overflow of the Parahyba its waters back up over much of the plain on both sides of the river, and the country becomes largely submerged. At the time of the enchente, or annual freshet, the inhabitants cut outlets across the sea-beaches for some of the lakes in the southern part of the campos. North of the Parahyba, and near the city of Campos, there is another large lake called Lagôa do Campello, and the country thence northward to the Guaxindiba is full of lagoons and cut up by little channels.

The Parahyba empties into the sea by two mouths, distant some two miles from one another, and between which is the Ilha do Lima. The delta of the Parahyba projects two or three miles beyond the general line of the coast. The mouths of the river are obstructed by bars, over which the waves at times break fearfully, and an entrance can ordinarily be effected only at high tide ; yet small coasting steamers and vessels do enter, and small river steamers, and sometimes even schooners, ascend as far as Campos. At the mouth on the south side is the miserable little town of São João da Barra, built on a sand-bank which admits of no cultivation whatever. It contains some two thousand inhabitants, who subsist principally by fishery, shipbuilding, and commerce. It owes its importance to the sole fact that vessels are often long delayed either off the bar waiting to enter, or inside waiting to go out, and this keeps up a little trade. In its lower course the river is wide and shallow. Mangrove swamps and low grounds, sometimes covered with bushes and trees, often waving with the tasselly spikes of the *Ubá* (*Gynerium parvifolium* Nees), border it for a few miles, but by and by the banks, which are composed of the richest alluvial clay, grow higher, and thence to Campos the beautiful river is bordered by immense sugar plantations, and the scenery is enlivened by frequent fazendas and the fabricas with their tall smoke-stacks. The plains are covered with thick beds of silt, deposited by the river during the annual overflows.

The waters of the Parahyba, as well as of all the mountain streams of the province flowing through gneiss regions, are very turbid and usually of a milky brownish tinge, throwing down a copious sediment, even in dry times; but when the river is swollen by rains, the quantity of silt is very much increased. This material, whether derived from the gneiss rock itself or from the drift, consists chiefly of decomposed feldspar and mica, and the water of the river is glistening with the minute silvery flakes of the latter mineral. The soil deposited by the river is very productive in sugar-cane, and the region round about Campos manufactures a very large amount of sugar and rum, the former of a very good quality; and this is the principal product of the plains. At Campos the country, though flat, is somewhat higher, and one may see, from an inspection of the river-bank, that the alluvial deposits are underlaid by tertiary clays which have been more or less denuded.

Notwithstanding the turbidness of the water of the Parahyba, it is, when the sediment has been deposited, very potable, and may be preserved for a long time. The usual custom is to keep the water in great earthen jars, sometimes weeks or months before using it.

The city of Campos is a respectable town of about twenty thousand * inhabitants, built on the right bank of the river. Its trade consists principally in sugar and coffee, and it is a place of extraordinary stir. The vicinity is flat and fertile,

* At least so says Pompéo in his Geography. As there has been no regular census, it is impossible to give with accuracy the number of inhabitants of Brazilian towns.

and largely cultivated for sugar. In the vicinity of Campos the Goyabeira, or Guava-tree of the West Indies (*Psidium Guaiava Raddi*), is very largely cultivated, and the fruit is manufactured into a sweetmeat which is exported in great quantity from Campos. There are extensive low tracts which in the wet season — in part for the whole year — form shallow lagoons and marshes. In these marshes, as well as in the ditch, dignified by the name of canal, which runs Macahé-ward, Mr. Copeland and I collected a great abundance of ampullariæ, planorbes, &c. The former I found laying in June. The eggs were large and salmon-colored, and were attached in bunches to the grass. These swamps are also rich in fish, Piabas, Acaras, Trahiras, &c.

The Rio Muriahé is a little stream entering the Parahyba from the north a short distance above Campos, taking its rise in the province of Minas. At its mouth it is perhaps 400 or 500 feet wide. It is navigable for a few miles to the first falls. The following observations were made during an ichthyological excursion up the river: The lands in the vicinity of the mouth of the river along the left bank of the Parahyba are alluvial and flat, so far as I could see. Two miles up the river, near the fazenda of the Baroness of Muriahé, the ground rises somewhat above its general level, is hummocky, diversified by immense ant-hills, and covered by a red drift soil.

Ascending a little farther, higher grounds of the same character are cut through, and they are seen to be composed of tertiary clays and sandstones, as may be observed in a bluff near Jundiá. At Pestrella the land rises to a height of seventy feet, and is composed below of a thick bed of coarse dark-red sandstone, extending below water-level, over which are beds of whitish and red sandy clays.

These hills are outliers of the tertiary beds of the coast, much denuded by glacial action. A large part of the lands bordering the river are low, and above the fazenda of the Barão d'Itabapuana they consist of beds of horizontally stratified sands above, with irregularly stratified brown clayey earth below. In this last were layers of a dark material, which appeared to be made up of leaves. I may here state that it has been reported that coal, or lignite, occurs in the vicinity of Campos. I have seen no signs of palæozoic or secondary rocks anywhere in the province of Rio de Janeiro, and at Campos I could not learn of the existence of any such deposits. The report may have originated from these recent vegetable deposits on the Muriahé. Gneiss shows itself at the fazenda of Piranga. Above the fazenda of Oiteiro the plains cease, and the river winds among gneiss hills, some of which are of considerable altitude; the Morro do Sapateiro, distant a few miles northwest of the Santa Rita, being perhaps 2,000 feet in altitude, while the Serra da Onça, on the opposite side of the river, is set down by Mouchez as 1,400 metres in height.

The soils of the higher lands I examined on the Muriahé were not good, but those of the lower were. The principal product of the region is sugar, and there are some immense fabricas on the river, as, for instance, those of Taepebas * and the one belonging to Senhor João Caldas Vianna, Jr., which were the only ones we visited.

At the warehouse of Senhor Amaral, at the head of navigation, the country consists of gneiss, with rather low rounded hills bordering the river, and higher ones on the southwest. At the warehouse I found the strike N. 65° E., dip 85° W.,

* This factory is worked by steam, and the molasses is separated from the sugar by centrifugal motion.

and a short distance up the river at the fazenda de Santa Maria das Taepebas I observed a well laminated gray and fine-grained gneiss, with strike N. 60° – 62° E., dip nearly vertical. Garnets are abundant in the rock here.

In ascending the Parahyba from Campos to São Fidelis, the head of steam navigation, we have before us one of the finest pieces of river and mountain scenery on the coast.

LOOKING UP THE RIO PARAHYBA FROM ABOVE CAMPOS.

For about one third of the distance above Campos the country bordering the river is flat. It then becomes higher and more uneven, and gneiss is seen in the river-banks. What appears to be the highest point of the Sapateiro range is on the Parahyba River, on the left bank. It is a large dome-shaped hill, with more or less bare, precipitous sides. Between it and the river are several other quite prominent hills, in the first of which are quarries, from which a large part of the gneiss used for building purposes in Campos is obtained. On the western side of the river opposite the Sapateiro range stretch westward the serras of São Fidelis in a series of sharp peaks. The height of some of the peaks must be at least 3,500 feet, probably more. The sides of the mountains are all regularly rounded like those of Rio. Like the Organs, some of the hills are very sharp, but they are not so prominent as to give to this landscape so striking a character as the Organs do to the scenery of the serra of that name. The gneiss of São Fidelis is similar to that of Rio, and contains a large quantity of garnets, some of the crystals of which are an inch and a half in diameter. At a rocky place by the river-side I observed a strike of N. 64° E., the dip being vertical. At São Fidelis navigation ends. Above that point the river is full of rapids, obstructed by rocks, and is navigable only for canoes and the like. The lands of the immediate vicinity are not largely cultivated.

At Cantagallo, during the reign of the first Vice-King of Brazil, gold was discovered by certain seekers and smugglers of gold, garimpeiros, who, quietly taking possession of the place for many years, extracted gold in secret, and it was a long time before the Brazilian government discovered the region whence so much gold found its way to the capital. Mawe says that the rock of the locality is granite composed of feldspar, hornblende, quartz, mica, sometimes holding garnets, — evidently gneiss, like that of the Serra do Mar to the westward. He states that the gold comes from the lowest bed of cascalho, or gravel, occurring always in rounded grains, and that he never saw a crystallized specimen. Gold and ferric oxide were the only metallic substances found here. At the time of Mawe's visit (1808) so little gold was extracted, that the quinto, or royalty, paid to the government would scarcely suffice to pay the officers and soldiers appointed to collect it. Von Tschudi says * that the gold of Cantagallo came from the bed of a stream.

Another locality where gold was formerly washed is Santa

* *Reisen durch Süd-America,* Dritter Band, 176te Seite. This volume contains a lengthy description of Cantagallo and Nova Friburgo.

Rita, a place about five Brazilian leagues northeast of Cantagallo. Mawe * describes the gold as occurring in a bed of cascalho,† or gravel, overlaid by earth. The layer of cascalho varies in thickness from two feet to seven or eight inches, and lies under a thickness of four or five feet of earth.

The Cantagallo region was never very rich in gold. At Santa Rita and in the vicinity Mawe found heavy deposits of limestone.

From São Fidelis I made a horseback journey in company with Mr. Copeland across the country northward by way of the Vallão Grande, to Bom Jesus on the Itabapuana River, the dividing line between the provinces of Rio and Espirito Santo.

For the greater part of the distance to the Rio Muriahé the road led through the most dense and luxuriant virgin forest. Little was to be seen of the rock, or even of the soil, but I observed that the drift clay, where exposed, often contained boulders of gneiss; and masses of rock of large size were sometimes seen resting upon it. On

* "Travels in the Interior of Brazil, particularly in the Gold and Diamond Districts of that Country. By Authority of the Prince Regent of Portugal. By John Mawe." I do not know the date of the English edition. I have a copy of the American reprint, which appeared in 1816, and was published in Philadelphia. It has as a frontispiece a large steel engraving, representing a number of negroes at work under a long thatched shed washing for diamonds, — an engraving which has been copied over and over again, and is familiar to every young student of geography. A German edition in my possession, entitled *Reisen in das Innere von Brasilien,* was published in 1816 at Leipzig, while Burton mentions a French work bearing the title *Voyages dans l'Intérieur du Brésil en* 1809 *et* 1810, published in the same year, and which I suppose to be a translation of the above "Travels," though Mawe began his explorations in Brazil in September, 1807. This is a work of much interest, and contains many valuable geological facts.

† The word *cascalho,* pronounced *cascalyo,* means gravel in Portuguese. In Brazil the auriferous cascalho is almost invariably composed of quartz pebbles.

the Vallão Grande, about a league west of the river Muriahé, I saw well-laminated gneiss with a strike of N. 64° E., and a northward dip. The valley of the Muriahé I found well cultivated where I crossed it, and furnishing large crops of sugar-cane. Leaving the river, our course lay over a serra, which our guide called Matúca. It is composed of gneiss, and must be over 2,000 feet high. On our descent on the northern side, I observed thick beds of a kind of gneiss composed almost entirely of quartz, and in the drift I saw boulders of this rock mixed up with boulders of common gray gneiss.* From this serra to Bom Jesus the country is all gneiss, with low rounded hills, the whole being covered with a most vigorous forest growth. The Rio Itabapuana is a little stream comparable to the Rio Pirahy on the Dom Pedro II. Railroad. Between Bom Jesus and the Ribeirão do Jardim the land is rather low, and diversified by rounded gneiss hills of inconsiderable elevation. The river is bordered by flat, alluvial lands, often marshy, the resort of great numbers of water-birds, piaçocas (*Parra Jacana*), cranes, &c.

On the Espirito Santo side, between the Ribeirão do Jardim and the Ribeirão Formoso, begins the Serra de São Romão e Santa Paz, or the Serra de Itabapuana, which rises abruptly from the river and, more or less broken, runs off in a northeast direction to the Rio Itapémerim. The hills on the Rio Itabapuana are more than a thousand feet high, and are composed of gneiss which dips southward at a moderate angle and with its usual strike. They are very precipitous on the southern side, the rocks being covered by an abundant growth of cactuses, &c. On the same side of the river, in the angle between the Ribeirão Formoso and

* I was told that there was limestone in this serra, but I saw none.

the Itabapuana, is a solitary irregular conical peak called the Pedra Formosa, which, standing alone in front of the amphitheatre formed by the Serra of São Romão, and that of Santa Paz, forms a fine piece of mountain scenery. Thence onward to the fazenda of São Pedro the country is still gneiss, the hills are low, covered by a most fertile red drift soil, and are heavily forest-clothed. The river is rocky and swift, and just below the fazenda there are some considerable rapids which extend for nearly a mile, and would furnish an abundant water-power. The soils of this vicinity are very good, and excellent cane, coffee, and cotton are raised. The cotton which I examined on the fazenda of Senhor Martinho Fr. Medino was the finest I ever saw on the coast. The gneiss of this vicinity, and of Porto da Limeira, lies remarkably horizontally. At the rapids at the fazenda of Senhor Martinho it has a northward dip of some ten degrees only. At São Pedro and Porto da Limeira it is well laminated, but it has a very irregular dip and strike, though the dip is generally northward. The rock is full of granite and quartz veins. About three miles south of Limeira is a remarkable isolated peak called the Garrafão,* or the "*demijohn*," which forms a very conspicuous landmark, visible from a considerable distance off the coast at sea. It is precipitous on all sides, and as it is long from east to west, and very narrow, it presents very different aspects according to the position from which it is seen. From some points of the compass it appears dome-shaped, from others like an immense tower or pillar rising out of the generally plain country. Mouchez estimates its height at 910 metres; I should set it down as between 2,500 and 3,000 feet.

* It has almost precisely the same structure and nearly the same form as the Garrafão among the "Organs."

Between the Rio Itabapuana and the Parahyba, eight or ten miles south of the Garrafão, and in a line with the Serra da Onça, is a remarkable group of gneiss hills, which is visible from Cape São Thomé, a distance of at least forty miles. One of these is a very sharp, conical needle, called Pedra Lisa.* This needle is seen in the following sketch.

THE PEDRA LISA.

Descending the river from Porto da Limeira, one soon leaves the gneiss region, and comes upon a flat country, for the most part very heavily wooded, and more or less diversified by shallow lagoons, one of which, Lagôa Feia, is quite extensive. The river is very narrow and tortuous, and only navigable for very small steamers. Much of the land is very low, and must be frequently overflowed ; but there are some considerable patches of tertiary, which are however much denuded. At the fazendas of Senhores Pedro Mendes and Antonio Martim these lands rise to a height of perhaps sixty feet, and on the river they are seen to be composed of the characteristic tertiary sandstones and clays. The little village of the Barra do Itabapuana, principally inhabited by

* I have seen this remarkable peak from all sides. It always appears as a needle or sharp cone. Mouchez makes it 1,150 metres in height, and I do not think he has over-estimated it.

fishermen, is built on a strip of sand on the right bank of the river, near the mouth. It is separated from the shore by a narrow, shallow channel, or lagoon, which runs southward from the river, parallel with the shore, and just behind the beach ridge. This lagoon communicates with a marshy tract covered by mangroves,* south of which red tertiary sandstones are exposed on the edge of the marsh, and the land rises apparently some twenty feet, forming to the southward a large patch of tertiary. Opposite the town there is a large sandy island, which is separated from the beach by a lagoon which stretches northward along the shore for some distance.

A league or more to the south of the Barra do Itabapuana † are two or three rocky points of tertiary sandstone, presenting low red cliffs. The same rock is said to occur at Manguinhos. Isolated masses of this rock, covered at high tide, occur off the Barra of the Itabapuana, and also at Manguinhos.

The mouth of the Itabapuana is, like the Parahyba, obstructed by a sand-bar, and is entered with difficulty. The water is shallow off the coast, and vessels sometimes anchor outside the bar and take in cargoes of wood, &c.

* The mud of the mangrove swamps is very soft, being composed of the finest kind of silt, and it is black and stinking with decaying matter. It is full of shells, leaves, the exuviæ of crabs, &c.

† *Ita,* in Tupy, means stone, and *poam* or *puam* an island; and I suspect that the name Itabapuana may have been given from the little rocky islands mentioned below. Cazal gives the name of the river as Camapuan or Cabapuanna, and says that the savages called it Reritigbá. (*Corografia,* Vol. I. p. 61.)

CHAPTER II.

PROVINCE OF ESPIRITO SANTO.

Barreiras do Sirí. — Itapémerim. — Coast between Itapémerim and Benevente. — Benevente. — Guarapary; Consolidated Beach, Corals, &c. — Rio Jecú. — Bay of Espirito Santo. — Nossa Senhora da Penha. — Victoria. — Decomposition of Gneiss and Formation of Boulders of Decomposition. — Recent Rise of the Coast. — Corals, &c. of the Bay of Victoria. — Rio Santa Maria. — German Colonies. — Fisheries. — Sand Plains. — Tertiary Plain at Carapina. — Mestre Alvaro. — Serra. — Nova Almeida. — Rio Reis Magos. — Santa Cruz. — Basin of the Rio Doce. — Description of the River. — Guandú: its Colony and Agricultural Resources. — Porto de Souza. — Geology of Vicinity. — Luxuriance of Vegetation on the Doce. — Woods. — Game. — Francylvania. — Climate of the Doce. — Linhares. — Lagôa Juparanãa. — The Future of the Doce. — American Colonists. — Salt Trade. — Barra Secca. — Sea-Turtles. — Consolidated Beaches and the Mode of their Formation. — Character of Coast between the Rivers Doce and São Matheos. — Rio São Matheos Described. — Geological Features. — Fertility of its Lands. — Cocoa-Palms and their Distribution. — City of São Matheos. — Rio Itahúnas. — Cliffs of Os Lençôes. — Coast between Itahúnas and Rio Mucury.

A SHORT distance northward of the Rio Itabapuana, and not far from the sea-shore, is Lake Marobá, from which flows the river of the same name. Between the Itabapuana and this river the coast lands are low and marshy. Just south of the Barra do Marobá the lands rise somewhat along the shore, and tertiary red sandstone shows itself in the beach. The tertiary bluffs of the Itabapuana sweep round back of the lake, and come down to the shore north of the Barra, and are continued thence northward in a fine range of bluffs called the Barreiras do Sirí, which, from the bright red colors of the clays and sandstones composing them, present a very picturesque appearance from the sea. These cliffs are seventy to eighty feet in height, and the country lying back of them is a wooded plain.

The lowest bed seen in the cliffs of Sirí is a coarse, dark-red sandstone, with indistinct stratification, and, where exposed on the beach, full of holes, presenting an appearance more like that of the surface of a lava stream than anything else. This mass of sandstone is penetrated by deep, perpendicular, pipe-like holes, which, in many cases, communicate with one another. This sandstone rises to a height of about twelve feet above low-water mark, and is overlaid by a bed, about twenty feet thick, of a sandy clay, whitish and reddish,* which penetrates into the cavities of the sandstone. The sandstone seems to result in part from the irregular cementing of the sandy clays by oxide of iron. The clays are soft and show no distinct stratification. The proportion of sand varies very much, some of the clays being exceedingly fine in texture, like kaolin. They are not at all plastic. The color varies from a pure white to a bright red, and sometimes the clay is variegated with curved lines of red or yellow, so as to look like fancy Castile soap. Over the clay is an irregular deposit of very dark-red sandstone, which is well stratified, and sometimes forms lenticular masses ; and over this in turn lies a bed of red clay, which I could not well examine. Between the clay and the soil, which is usually light brown, there is a layer of pebbles and iron-stone nodules. A few miles below Itapémerim the tertiary lands

* Prinz Max. zu Neu Wied, *Reise nach Brasilien,* Vol. I. p. 169, speaks of these cliffs, and gives the following note, which I leave in his own words : —

"Der Untersuchung des Herrn Professor Hausmann zu Göttingen zufolge gehört dieses Fossil, welches einen Hauptbestandtheil eines grossen Theils dieser Küste von Brasilien ausmacht, zum verhärteten Steinmark, wohin man auch die sächsische Wunder-Erde zählt. Es stimmt in allen Kennzeichen mit dem Steinmarke überein."

recede from the beach, and are broadly denuded on both sides of the river. The Itapémerim is a much larger stream than the Itabapuana. It rises near the frontier of Minas Geraes, west of Victoria, behind the Serra do Pombal, and has a course of about eighty miles. It is shallow in its lower course, and of little importance. There is an extensive alluvial plain bordering the river on the south side for a few miles above the town, in part belonging to the fazenda of the Barão de Itapémerim; this tract of land is very fertile, and a considerable part is cultivated for sugar. There are a few fazendas farther up the river. That of Muquí, belonging to the Baron, is built on a gneiss hill.

The serras approach nearer the coast in going northward, and in the neighborhood of Itapémerim are very high, presenting the same topographical peculiarities as in the south.

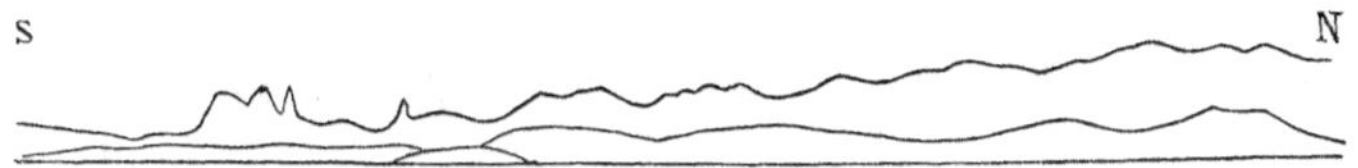

SERRA OF ITAPÉMERIM, SEEN FROM THE SEA.

About twenty miles west of the town of Itapémerim, and south of the river, is a very remarkable irregular peak called the Frade, while a few miles to the northeast is a group of needles, which presents an exceedingly strange appearance when seen from certain points of the compass, appearing sometimes like the fingers of a giant hand held up against the sky. Some of these needles are at least 3,000 feet high. The whole country lying behind them, even beyond the limits of the province, is very mountainous, and is composed of gneiss. I was informed that specular iron occurs in the serras of Itapémerim, but I had no way of verifying the report.

In 1723 the first settlers on the Itapémerim discovered gold in one of its affluents on the north, called the Rio do Castello.* "A decree of the 17th September of the following year, 1724, determined that the auriferous lands should be divided into small portions among all those who wished to employ themselves in the labor of mining, with the condition that they should subject themselves to the laws of the *sesmarias* and to the impost on the gold." † A gold-washing town was established at the confluence of the Castello with the Itapémerim, but the hopes of a rich yield proved deceptive,‡ and with the incursion of the Indians the place went down.

North of the Itapémerim the coast is bordered by extensive plains of coarse white sand, like those of the Island of Paquetá and of the extension of the Cantogallo Railroad. These plains are quite destitute of grass, and are covered sparsely by clumps of stunted trees, among which is the Pitangueira (*Eugenia pedunculata*), with an abundance of cactuses (*Cereus*), and bromeliaceous plants. About half way to the little town of Piuma, distant some eight miles north of Itapémerim, gneiss hills stretch along the coast. Among these is the Morro de Agah, one of the noted landmarks of the coast, — a sharp, saddle-shaped ridge, appearing pyramidal when seen from the north. Gneiss shows itself occasionally on the beaches, and there are a few little gneiss islands off the coast.§ The Piuma is a very small river of little or no importance.

* Dicc. Geog., Art. *Castello.*

† The *Corografia Brasilica*, published in 1817, speaks of these mines as having been abandoned on account of the incursions of the Indians.

‡ Von Tschudi, *Reisen durch Süd America*, Dritter Band, 60ste Seite.

§ I observed on the shore, just south of the point of the Agah, a rock exposed, which appeared to be a sandstone. It was much decomposed, and I was unable to examine it very carefully.

From Piuma to Benevente, which is a ride of only about two hours, and cannot be more than from four to six miles, the road leads over low gneiss hills, because the shore is rocky and the beach has to be abandoned. The Rio Iritiba, or Benevente, is a small stream, navigable for some eight leagues above its mouth, which is narrow, situated between gneiss hills, and unobstructed by a bar, — a circumstance owing to the protection of this part of the coast by a rocky point extending southward some two miles east of the mouth. The water at the entrance is deep, and vessels of considerable tonnage may enter at any time of the tide. Just inside, the river spreads out into a broad sheet of water, so that it forms one of the best and most frequented harbors on the coast of Espirito Santo. Vessels frequently find refuge in the little bay behind the point east of Benevente, where they are well sheltered from northeast storms. The town is a small one, and situated at the mouth of the river, on the northern side, at the base of a low gneiss hill.* Besides ship-building, its trade consists in wood and a little coffee. According to Von Tschudi, the lands lying back of Benevente are very fertile, and the place possesses natural advantages which might, if properly used, make it of much importance. Leaving Benevente, and going northward, the path crosses obliquely a projecting point, and passes over and among low hills of red sandstone and clays, the sandstone cropping out frequently at the base of the hills. The soil covering these hills is gray, and appears to be very rich. In some parts it is of a deep pinkish-red color. On leaving the hills a plain of white sand is reached, like

* Gneiss, gray, well laminated. Strike N. 55° E. Dip varying from vertical to 88° N. Rock intersected by numerous granite veins.

that of Itapémerim. This occupies a sort of bay in the tertiary lands, which soon reach the coast again and form a long line of red cliffs, extending for several miles along the shore, almost to the little fishing village of Miahype. This village is situated on a narrow sand-bank uniting a small mass of gneiss to the shore, off which a similar mass forms an island. Thence northward the tertiary lands extend along the shore, with narrow patches of sands in front, and with occasional interruptions, to the village of Guarapary,* when the shore becomes very much indented, and the distribution of the various formations are too complicated for description here. Many of the low hills of the vicinity have a basis of gneiss, but are capped with tertiary clays. The gneiss shows itself in a number of rocky points and ledges, and at the mouth of the river there are several islands. The gneiss is gray, but it is very micaceous, the mica being in moderately large black crystals. I find in my note-book an observation of strike N. 80° E., dip 80° S.

COAST SOUTH OF GUARAPARY.

In a little cove just south of the town is a large detached mass of sandstone, lying just in front of the beach, and at

* The country lying between the coast and the Serra do Pero Cão, distant some six miles from Guarapary, is composed of low hills, some of which are undoubtedly tertiary, interspersed with higher ones, as the Orobó, which are gneiss.

low tide washed by the waves. It consists of quartz sand cemented by carbonate of lime, and the rock is exceedingly hard. It is well stratified, and the layers are nearly horizontal. The rock is cut up by joints, which the sea has widened, so that it is much broken up, while the washing out of the calcareous cement by the sea has left the surface very ragged. This mass is only part of a solidified beach which has been laid bare by an encroachment of the sea. At low water it furnishes to the naturalist a very rich harvest of marine animals, for in the holes near its edge grow masses of *Siderastræa stellata* Verr., while *Acanthastræas* of considerable size grow attached to its sides. *Favias* and *Agaricias*, &c., occur in great abundance, and the rocks are covered with great patches of the common brown *Palythoa* of the coast, and of a spreading, green-disked *Zoanthus*, not determined. Several beautiful species of sea-anemones, deeply tinted, are very common, but the species have not yet been made out. At low tide there may be obtained from the rocks beautiful specimens of *Hymenogorgia quercifolia*, *Eunicia humilis*, and *Plexaurella dichotoma*. Sea-urchins (*Echinometra Michelini*) are exceedingly abundant here, and are used for food by the natives. They live packed securely away in deep holes, not only in the sandstone of the reef, but also in the gneiss of the adjacent points. Here also the beautiful little crimson star-fish, *Echinaster crassispina*, is very abundant, occurring among the sea-weeds and in little pools betwixt tide-marks; and, in the tide-pools, *Ophiura cinerea*, together with a number of other species of the same order, occurs. Holothurians, some of them a foot long, are very abundant, packed into crevices, in tide-pools, and under rocks; and a pretty comatula, *Antedon Dubenii* Bölsche, or *Braziliensis* Lütk., may be seen through the water covering the rocks like rosettes of brown feathers.*

This locality is rich in crustaceans, but not especially so in mollusks; indeed, the whole coast has a rather poor molluscan fauna. A large octopod is common here, and is hooked out from the crevices in the rocks by the inhabitants, who use it for food. Guarapary is an excellent collecting ground for marine invertebrates, though not so good as Victoria. Inside the mouth of the bay, in water which is brackish and impure, occurs a slender-branching, tender, nodose Halcyonoid, undetermined. The entrance to the harbor of Guarapary is good and secure, and shelter for shipping is furnished by the little islands, Escalvada and Raza, lying off the coast; but the marshy lands in the vicinity make the place exceedingly unhealthy, and notwithstanding the lands to the west of the town are good, and woods and valuable balsams and fish are in abundance in the vicinity, the unhealthiness of the climate has placed a ban upon its growth. The river is a little one, with a course of only a few leagues, and takes its rise in the Serra do Pero Cão. It is said to offer navigation as far as the coast serra, and to unite in its course a number of little lakes.

In going northward from Guarapary the path leads, first, over a gneiss district bordering the northern side of the river, and then descends to a plain of white sand, sparsely covered by trees. Crossing this, low tertiary hills are reached, with more or less gneiss, bare in places, especially along the shore, and you come upon the little brook Pero Cão, beyond which the shores are sandy and flat as

* Of these Radiates, as well as the other marine invertebrates of the reef, Mr. Copeland and I made a considerable collection, which is in the Museum of Comparative Zoölogy, in Cambridge, but has not yet been worked up.

far as a small river called by my guide Una, but which does not seem to be represented on the chart of Mouchez. This little stream escapes into the sea just south of a projecting point of gneiss. Between this point and the serra the country is low. The distance of the mountains from the sea is only about five miles. A point or two passed, and one reaches a long sand-beach, which extends some eight miles north to a low gneiss point called Ponta da Fructa. Along this beach are, in some places, sand-dunes twenty to twenty-five feet in height.

From the Ponta da Fructa, northward, to the Ponta de Jecú — a distance of seven or eight miles — stretches an almost straight sand-beach, behind which are plains, sandy and marshy, — a perfect batrachian paradise.

The Ponta de Jecú is a gneiss hill, somewhat similar to that of the Fructa, and formerly an island, but now joined to the mainland by a sand-beach. There are other smaller hills in the vicinity. According to a sketch in my own note-book, the river enters the sea to the south of the point, but Mouchez's chart shows it entering on the north side, which was probably the case at the time his chart was made, the mouth having been closed on the south by a storm. The river Jecú rises among the serras to the west, and is an insignificant stream, apparently smaller than the Muriahé. It is with difficulty navigable for canoes for only a short distance. Some five miles above its mouth a canal, cut long ago by the Jesuits, runs off northward and communicates with the port of Victoria, distant about five miles. This was done to facilitate the transport of the products of the country to Victoria, as well as to avoid the dangerous passage by sea from the mouth of the river around the reefs and rocky point on the south of the bay of Espirito Santo.

The cutting of this canal is said to have improved the health of the region of the Jecú.

On this river, some thirty miles from its mouth, and somewhat farther from the city of Victoria, was established, in the year 1847, a German colony, Santa Isabel, among the gneiss hills lying east of the serra, in a region healthy and fertile, and proper for the culture of coffee, cotton, &c.; but the colony has not been prosperous, owing to bad management and the want of roads.*

From the mouth of the Jecú to the bay of Espirito Santo the sand-beaches continue, backed by the sandy and marshy plains of the Campos de Piratinanga.†

The Bay of Espirito Santo is about two and a half miles wide, and irregular in shape. On the north is the Ponta do Tubarão, with a rocky sandstone shore, whence sweeps around westward and southward a long sand-beach, joining a rocky point, near which enters the channel of the Rio da Serra. Thence southward, for a mile or more, the land is high and irregular, and the shore consists of sea-beaches between projecting gneiss points. We then reach the entrance of a narrow, irregular channel, — the estuary or bay of the Rio Santa Maria, — that extends westward among gneiss hills. On the south side of the bay is an irregular conical gneiss hill, some 700 feet high, called Monte Morêno, forming a rocky point. West of this, and separated only by a short sand-beach and a small stream coming from the swamps to the southward, is another conical hill, some four hundred feet high, crowned by the pictu-

* Tschudi (*Reisen*, etc., Dritter Band, 8te Seite) gives as the mean annual temperature for the locality +18 Réaumur = 70.50 Fahr.

† St. Hilaire, in speaking of the sandy plains between Jecú and Victoria, says that the vegetation covering them resembles in many points that of the elevated plateaux of Minas Novas. (Tome II. 2de Partie, p. 229.)

resque pile of the convent of Nossa Senhora da Penha. West of this, between the Morro de Nossa Senhora da Penha and high gneiss hills, is the deep cove of Villa Velha, with the ancient village of the same name built on the edge of the sand-plain. Westward of this cove the shore stretches to the Pão de Assucar, along the southern side of the channel of the Santa Maria. It consists of a number of high rocky points, united by mud-flats and sand-beaches. The opposite shore of the channel is of the same general character. North of Monte Morêno and of the mouth of the channel are two more high gneiss islands, lying one north of the other. There are, besides, many smaller ones, together with a number of rocks and skerries, and the channel is obstructed near the Pão de Assucar by islands and rocks. The Pão de Assucar is a precipitous, irregularly conical gneiss hill, 400 to 500 feet high, falling off to the north, presenting to the channel a smooth, almost vertical, face. Here, by the projection of a point from the northern side, the channel is suddenly narrowed down to a width of only about 600 feet. Passing the Pão, the channel widens out into a most spacious harbor, and on the northern side, in a fine amphitheatre among the hills, is built the city of Victoria. This basin extends only about a couple of miles west of the Pão. At its head it receives the waters of the canal from the Jecú, and of the Rios Crubixa and Santa Maria. A channel extends northward, and, uniting the mouths of several rivers, passes round the hills of Victoria and enters the Bay of Espirito Santo, thus rendering them an island. This island is composed of gneiss, is very high, rugged, and clothed with forest. To the north and west the country is a plain, while the shore alone, along the south side of the bay and channel, is hilly. The hills of the island

LOOKING UP TO THE PÃO DE ASSUCAR AND CITY OF VICTORIA.

and adjoining mainland then form an isolated group, of which the main mass just behind the city must be fully 1,000 feet in height. The channel of the harbor is a narrow valley, which, owing to its rocky sides, has been easily kept open. The gneiss of the locality is very homogeneous, porphyritic, and of the same general character as that of the coast southward. As a general thing the hills are dome-shaped and regularly rounded, as is represented in the following sketch, but sometimes they are conical. In some cases one or more sides, or the top, is bare and smooth, as is the case with the Pão de Assucar. These bare surfaces are almost

MONTE JUTUQUARA AND GNEISS HILLS NEAR VICTORIA.

always rounded with remarkable regularity, and are never jagged and angular, like our northern precipices, or the cliffs on the São Francisco, below the falls of Paulo Affonso. This is owing to the uniform decomposition of a surface unbroken by joints or planes of stratification; for many of these mountains are actually formed of a single, unbroken mass of gneiss. The cliffs are rarely vertical, and not unfrequently form bare places on a mountain-side, set in a framework of verdure. Such a bare slope is represented in the following sketch of the Fortaleza de Peritininga, below Villa Velha, Bay of Espirito Santo. They are usually

FORTALEZA DE PERITININGA.

stained by perpendicular lines, or bands, of a rich, deep, inky, purplish-black color, being some minute lichen growth, and covered with scattered tufts of beautiful bromeliaceous plants, orchids, cactuses, &c., which give them a very picturesque appearance. Just below the city of Victoria proper there is one of these bare hillsides, which forms an exceedingly attractive element in the romantic scenery of the island.

Standing in an amphitheatre among the beautiful hills, and in full view from the sea, forming a most valuable landmark for the sailor, is a conical mountain, bearing on top a tower-like mass, the face of which is excavated on the eastern side by a considerable cavern, in which, it is said, in old times fugitive slaves took refuge. This mountain is called Jutuquara,* or Frade de São Leopardo, according to Mouchez. Its height must be 700 feet at least, probably more. It is represented in the sketch on the preceding page. The gneiss

* Prince Neu Wied gives a wretched drawing of this mountain in his work.

BOULDER OF DECOMPOSITION, VICTORIA.

hills, down to a certain level, are covered, as in the south, by drift clay, in which are imbedded rounded and angular fragments of quartz and gneiss. This forms a rather coarse and arenaceous soil, which is not so fertile as the drift soils of Rio. Decomposition obtains here as elsewhere. The rocky shores and islands of the bay are lined with rounded masses of gneiss, often of immense size, and which, lying loosely about, have all the appearance of erratic boulders. Similar masses we have already found on the shores of Paquetá and many of the other islands in Rio harbor. Sometimes these *boulders of decomposition* are seen perched insecurely on the tops of other rocks, as is the case near the Pão de Assucar. I have seen no locality where the formation of these boulders is better exemplified than here. On the hillsides the surface of projecting rock-masses undergoes a sort of softening, which causes it to separate from the undecomposed rock beneath, and break up into irregular fragments by the formation of a system of cracks. Through these fissures the water finds access. Each one of the fragments decomposes all round, and the loose decomposed material being washed away, these masses become rounded, separate more and more from one another, and sometimes fall over, so as to lie on the surface of the soil. One must therefore be exceedingly careful not to make blunders in examining them. The loose rocks lying on the side of the Morro de Nossa Senhora da Penha, and carefully represented in the accompanying view, are *boulders of decomposition.* Where the rock undergoing decomposition is on the sea-shore, and the action of the waves assists in the removal of the decomposed material as soon as formed, the effects produced may be still more striking. The easternmost extremity of the Ilha do Boi, just opposite the Penha, is a projecting, sloping mass of compact gneiss, as represented in the following rough sketch. The whole surface of the gneiss is soft-

DECOMPOSING SURFACE, ILHA DO BOI.

ened to a depth of several feet, and has shrunk entirely away from the undecomposed rock. This sheet has cracked through perpendicularly to the surface, and covers the rock below like a pavement. The action of the waves has, as represented, removed these loose fragments from over a considerable area, which is left very regularly rounded and uncracked. In other localities this decomposition and denudation have gone on until only a few of the heavier blocks are

NOSSA SENHORA DA PENHA, VICTORIA.

left on the surface, presenting the appearance of erratics. On the rocks known as the Pacotes, lying a little to the southward of the bay, off the coast, several large boulders of this kind are seen lying, presenting the appearance of buildings. It might be objected to our theory of the glacial origin of the Brazilian surface-clays and pebbles that they were formed by a decomposition of this kind along the shore of a slowly sinking continent. This would never produce such a coating of clay as forms the drift of Brazil, and there would certainly be associated with the deposits stratified sands and gravels and silts, which are nowhere to be found. Farther on we shall discuss in detail the whole subject of Brazilian drift.

Along the northwestern face of the Pão de Assucar runs an irregular horizontal line, as represented in the woodcut below. This line consists of a series of very shallow hollows,

ANCIENT SEA-LEVEL ON PÃO D'ASSUCAR, VICTORIA.

sometimes continuous, and evidently worn by wave-action within comparatively recent times. This old wave-line is not traceable along the whole extent of the cliff. I first observed it in the latter part of August, 1865. In September, 1867, I revisited the locality shortly after a tide of the full moon, which had left a well-marked muddy line running around the base of the Pão de Assucar, and which the succeeding tides had not reached. Measuring as nearly as I could from the middle of the wave-line, the mean of two measurements gave me as the height of the old water-level above the high-tide level of the 13th September 3.16 metres, or a little more than seven feet. From the old line to the upper edge of the zone of oysters is 3.56 metres. On the face of the cliff of the Pão, in a little cove on the western side, I cut a groove with my chisel indicating the height reached by the tide of the 13th September, 1867. This same old water-line may be seen in several places on the rocks on the opposite side of the channel below the Pão de Assucar, as well as on the face of a cliff at the western end of the beach at Villa Velha, where, as nearly as I could judge, it had the same height above the sea. This wave-line marks a period of rest when the continent, standing for some time at the same level, gave an opportunity for the little waves of the sheltered port to excavate the line. No such line marks the present sea-level, and I infer, from that and other facts, that the land is at present rising.

The water in the middle of the bay is very shallow, and the bottom appears to be a bank that comes so near to the surface that the sea sometimes breaks over it. Between the Morêno and Point Tubarão the average depth is about fifteen metres. The depth decreases on entering the channel between the Ilha do Boi and Monte Morêno, where it is from four to nine metres. As the channel contracts the depth increases, and just below the Pão de Assucar it reaches twenty-four metres, while in the narrowest point, at the Pão de Assucar, it is sixteen metres. In front of the town the depth varies from six to ten metres, and an excellent and spacious anchorage-ground is offered. The water opposite the town is turbid, and the littoral fauna is characterized by an abundance of oysters, covering the rocks and piers and mangroves, to within a few inches of high-water mark. This is a very small species, with exceedingly sharp, wavy edges, and used for food in Victoria. The coves between the rocky points along the channel above the bay of Villa Velha are muddy, and often lined by mangroves. Going down the channel the oysters grow less numerous, and give way to barnacles and mussels. On the north side of the cove at Villa Velha the arrangement of the principal elements of the littoral fauna is as is represented in the following diagram.

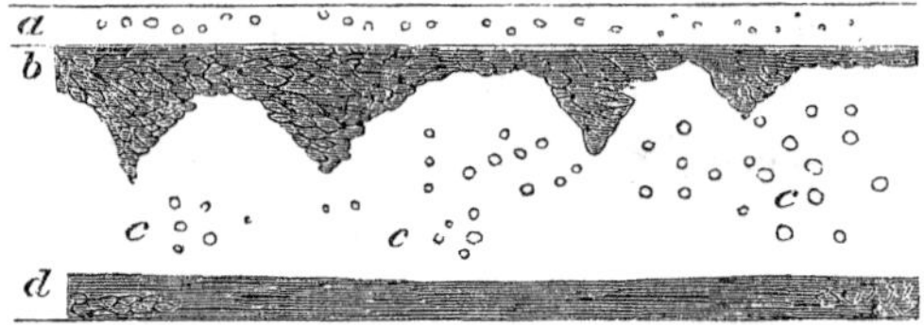

a. Zone of small barnacles, breadth three to four inches.

b. Little black mussels, all reaching the same upper level, but forming irregular patches, extending in some places as far as a metre below high-tide mark.

c. Large scattered barnacles, and green sea-weeds (*Ulvæ*).

d. Coarse brown sea-weeds, corallines, &c., and oysters.

The sea-urchins begin above *d*, and extend below low water mark, but their untenanted nests occur up to, and beyond high-water mark.

On the beach at Villa Velha, particularly near the western end, dead corals are thrown up in great numbers. Indeed, they are more abundant there than on any other South American beach I have seen. The commonest species is the *Mussa Harttii* Verrill, which is thrown up by the waves in great quantities, and evidently grows in the immediate vicinity; but owing to the muddiness of the waters of the cove I could not, in the absence of a dredge, find it alive. It is usually drifted ashore attached to sea-weeds. So abundant is it in the muddy beach at low tide that it is collected by the inhabitants of the Villa for lime-making, and is called *cachimbo*, or " pipe-stem." This species is almost invariably incrusted with beautiful Bryozoa. On the same beach occurs a species of *Siderastræa*, and a *Pectinia*, or *Symphyllia*, none of which I have observed living in the bay. I have seen in the sands underlying the muddy shores farther up the bay shells and corals which cannot now live in the turbid and brackish water. Before the rise of the land the bay extended farther on both sides. The river has since narrowed its channel, and the turbid waters have driven down the coral fauna nearer the mouth of the bay. I doubt whether the *Mussæ* are now to be found living at Villa Velha. I believe that the specimens on the shore have been long dead, and are thrown up by storms. Leaving Villa Velha, and going down the bay, the oysters disappear as the shore becomes more exposed. Outside the harbor, and at the base of Monte Morêno, is a little island called Ilha Balêeiro. The tide-pools of this island and of the adjoining shore are rich in corals, and at low tide one may collect *Hymenogorgia*, *Eunicia*, *Plexaurellæ*, and all the species common at Guarapary. I am not aware of the existence of any coral-banks in the bay and vicinity. The Bay of Victoria would be a rich ground for dredging, and so would be the banks lying off the harbor. In fishing I have brought up on the hook masses of nullipores, &c., loaded with life, and at the mouth of the bay I captured in this way a large many-rayed star-fish. Victoria appears to have a reputation as a locality for shells; but the littoral mollusk fauna of Brazil is very poor, and the shells of the beaches are badly broken and worn.*

The Rio Crubixa enters the harbor of Victoria just above the city. According to the *Diccionario Geografico*, "it descends from the Cordilheira dos Aimorés among rocks, amongst which is found a certain species of coral of a dark color and fragile, with which the Botocudo women are accustomed to decorate their heads, necks, arms, and legs."

The Rio Santa Maria is a much larger stream, which rises some fifty miles, more or less, northwest of Victoria, in the Serra dos Aimorés.

On the Rio Santa Maria is located the colony of Santa Leopoldina, and as the history of the colony and of the country where it is situated has an especial interest at this

* I find in one of the letters of the celebrated Joseph de Anchieta a statement that the manati occurred in the bay of Espirito Santo, and his description seems to me of sufficient interest to warrant my giving it in his own words: —

"De Bove Marino, — Hac quoad rationem temporis, jam ad alia transeamus. Piscis quidam est quem Bovem marinum dicimus, Indi *Iguaraguâ* nommant frequens in oppido Spirito Sancto et aliis versus Boream habitationibus, ubi aut nulla est, aut exigua admodum, et minor quam apud nos frigores injuria: hic ingentis est magnitudinis herbis pascitur, quod ipse gramina depasta scopulis, quos æstuaria alluunt, inhærentia indicant. Bovem mola corpore superat, cute obtegitur dura, elephanti colorem referenti; duo vellut brachia quibus natat, habet ad pectus sub quibus et ubera, ad quæ proprios fœtus nutrit, os bovi per omnia similis. Esui est congruentissimus, ita ut ducernere nequeas, utrumque carnis, an potius piscis loco haberi debeant; ex cujus pinguedine, quæ cuti ex maxime circa caudam inhæret, ad modo igni fit liquamen, quod jure butyro comparari et haud scio an possit antecellere; cujus ad omnia cibaria condienda olei vice usus est: Ossibus solidis, et durissimis quæ, possint eboris vices gerere, totum corpus est compactum." — *Collecção de noticias para a historia e geographia das nações ultramarinas. Lisboa*, 1812.

time, when efforts are being made to colonize the Brazilian coast, I translate the following from an account of a visit to the colony by Von Tschudi.* He says: "We crossed swiftly the Lameirão of the bay and steered into the river Santa Maria. Its current is quite slow, and offers, consequently, no particular hindrance to the ascent. Not far from its mouth it receives from its left bank the little river Carapina, and somewhat farther west lies, on the right bank, the Porto da Pedra, consisting of a couple of houses and a large *venda.* Up to this point there is sufficient water, even for steamers of moderate draft. . . . . Following the many windings of the river in ascending it to the north-northwest, we reached, after an eight hours' journey, the junction of the river Mangarahyba with the Rio Santa Maria. The locality becomes, the farther up the stream one goes, more and more hilly. The banks of the river themselves, where the character of the lands allows, are in part inhabited by Brazilians, who have here laid out little fazendas and occupy themselves principally with the raising of horses and cattle. The affluents of the Santa Maria are very unimportant. From the south there empty into it the Rio Curipé, the Rio Tauhá, the Rio Una, and several other brooks whose names I have forgotten; on the north the Rio Jacuhy, Rio Tramerim, (Jatamerim?) and a couple of entirely unimportant little rivers near the settlements Murinho, Aruaba, and Pendiuca. A rather extensive property is that of Senhor José de Queimado on the left bank, several miles up the stream from Porto da Pedra; over against it several little islands stand out upon the surface of the stream. Likewise on the northern bank lies the hamlet of Santa Maria, distant some seven to eight leagues above the mouth of the stream, from

* Von Tschudi, *Reisen durch Brasilien,* Dritter Band, Cap. I.

which it takes its name. Where the Rio Mangarahy unites, or makes *barra* with the Santa Maria, lies the extensive fazenda of José Claudio de Freitas. From this estate up the river the Rio Mangarahy is navigable for boats only a short distance. The navigability of the Rio Santa Maria ends at about one league from the mouth of the Rio Mangarahy at the Cachoeira de José, above which rocks make the channel of the river impassable. . . . .

"The Rio Santa Maria rises westward from the colony of Santa Leopoldina in a mountain range, on whose western slope are the sources of several of the tributaries of the Rio Doce. It first becomes important after receiving the river Mangarahy. This last is formed of a number of mountain brooks, of which the greater part rise in the southern part of the colony. It receives two larger tributaries on the south, one of which, the Rio do Medio, bounds the colony on the south, and the other, considerably smaller, on the east; the Brazo do Sul takes its rise beyond the colony. Its northern tributaries are very numerous, but are only worthy of the name of brooks. The most considerable are the Riberão da Sumaca, Corrego Isabel," &c., &c.

"The margins of the Rio Mangarahy, from its union with the Rio Santa Maria to the colony, are quite thickly inhabited by an agricultural population, and much more considerably than those of the main river, because the locality here repays labor better. It appears that on certain places on the head-waters of the Rio Mangarahy traces of wash-gold were found; at least the names California de Dentro (in the colony), and California de Fora on the southern bank of the Rio do Meio, would indicate it. The colony was founded in 1857 by Germans, who settled on the Santa Maria, and its branch, the Riberão das Farinhas. This choice of locality proving injudicious, another spot farther south, near the Quartel Braganza, was chosen, and settled also with Germans; but, through the worst possible management on the part of government employés, as well as the bad quality of the lands, the colony became demoralized, and has been a failure."

"The territory of Santa Leopoldina is composed of quite high, for the most part steep, mountains and narrow valleys, rarely broader than the channel of the river which flows through it. The soil consists principally of quartz sand, the surface soil, usually two to three inches thick, is held together by a network of fine roots with some humus. Only in certain places, where one of the valleys widens somewhat near a stream, layers of rich soil brought down from the mountains are found, and here, naturally, is also the greatest fertility. The national custom of preparing the soil for the first cultivation, by axe and fire, is, for situations like those of Santa Leopoldina, the most destructive. The heavy fire from the burning of the felled forests destroys partially the layer of humus and organic substances, and although ashes remain as nourishment for the future harvest, it is at the same time deprived of quite a deep layer of soil in which it can take root, and which, in addition, the moisture binds together. Through the cutting down of the trees the steep mountain slopes are exposed to the full influence of the tropical rains, and through these the best part of the cultivated fields is washed away and carried to the Rio Santa Maria, which finally deposits it in the Lameirão of the Bay of Victoria. It is a well-established fact that in Santa Leopoldina the soil, through culture, becomes more quickly unfruitful than in any other colony. The forests with which the mountains of Santa Leopoldina are covered do not present the same majestic appearance as those of the north and south of the province. They have much more the appearance of a weak second growth (Capoeiras) than of a virgin forest. All those plants which, to the practised eye of the Brazilian husbandman, bespeak a fruitful soil, — such as the Pão d'Alho (Garlic-tree), Jacarandá (Rosewood), Taquara-assú (Bamboo), &c., — either do not appear, or are represented by very feeble specimens. The cultivated plants correspond in their development with the forest vegetation. The corn remains low, and ordinarily produces small ears. The stalks often dry up before the ears make their appearance. The roots of the mandioca are smaller in the second year than they are in the other colonies, and frequently become, according to the testimony of the colonists, black and useless, which fact is to be accounted for by the want of an adequate thickness of humus. The black beans fail entirely. Equally unfavorable is the character of the soil for the growth of the coffee-tree ; in the first year, while it yet needs but little nourishment, it grows very favorably ; but in the second it sickens, and, as a rule, gives out. Colonists who had set out 1,000 to 1,200 coffee-trees, possessed at the end of the second year only a couple of hundred, and so soon as these, on the following year, had bloomed, and the fruit had set, the leaves rolled up, fell, and the little trees, without exception, gave out."

While in Victoria I met numbers of colonists from Santa Leopoldina, and all told the same story.* There can be no doubt that a lamentable mistake was committed in establishing the colony in so wretched a region. The lands of the

* There is a strong prejudice against Germans in this part of the country. They are represented as idle and given to drinking, and I am very sorry to say that it is fully confirmed by my acquaintance with those colonists I met in Victoria. Elsewhere the Germans make good colonists.

central part of the province of Espirito Santo are very poor, whereas those of the north and south are very fertile; and Victoria, though possessed of a most excellent harbor, will never in itself, in all probability, become a place of much importance because of the want of fertility of the surrounding country, and the impracticability of making it one of the ports of the province of Minas Geraes. There are some good lands, as I shall have occasion to show, lying to the northeast, and the as yet undeveloped agricultural regions of the Rio Doce are distant only some sixty miles. The cutting of a canal to unite the Doce by water with Victoria has been advocated, but in my judgment it is not practicable. In the event of the successful colonization of the Doce, a railroad to some point on that river could easily be constructed, so far as the physical difficulties to be encountered are to be considered. The mouth of the Doce is so dangerous to enter that it will never answer as a port, and the Rio São Matheos, lying to the north, though it may be entered by small vessels and steamers, is, after all, ill suited to be the port of the Doce. By making Victoria the outlet of the commerce of the Doce many advantages will be gained. São Matheos will never answer as a port for a foreign trade. The products of that region, as well as of the Doce, are more likely to go to Rio than elsewhere for final shipment to foreign ports, and this trade is now carried on in small vessels and coasting steamers. The voyage from Rio to São Matheos, though often very short, is uncertain, owing to the prevalence of northeast winds, and the passage of the point of the Doce is often difficult. By making Victoria the port, the voyage to Rio would be shortened one hundred miles at least, and made very much more easy, while Victoria, being a port admitting ships of large ton-

MORRO DE MESTRE ALVARO.

nage, might be made the centre of a direct trade with foreign ports.

At present the province of Espirito Santo, though possessing abundantly the sources of wealth, is one of the poorest and most wretched of the Empire. The water bordering the coast is very shallow, and just off Victoria are the extensive banks of Victoria, which are very rich in fish, especially *garoupas*, *pargos*, *vermelhos* (species of *Serranus*), &c., and they are much resorted to by the fishermen of the coast, especially from Guarapary; but such is the sloth of the fishing population that nothing is done to develop this mine of wealth, and the Victorienses eat *codfish* and European *sardines* when they might export fish to Europe!

I had an opportunity of fishing one day on these banks, and can testify to the abundance of the fish, but I had no opportunity of dredging; occasionally, as above remarked, the hooks brought up masses of nullipore rich in life, and from a depth of some fifty feet I collected in this way a species of *Pterogorgia*, apparently new.

On the northern side of the island of Victoria is, as already remarked, a tidal channel running westward and communicating with the Rio Santa Maria. This channel is seen in the accompanying engraving of the Morro de Mestre Alvaro. It receives and carries to the sea a part at least of the water of the Santa Maria and of the river westward. It is but an estuary, its waters ebbing and flowing with the tide. It is very turbid, and oysters grow along its banks for a long distance above the Passagem. I am not aware that it is ever used for the purposes of navigation. Northward of this channel are a few gneiss hills, and thence to the grand mountain mass of the Morro da Serra, or Mestre Alvaro, stretches a sand-plain which extends eastward to the Ponta do Tubarão and westward to the Rio Santa Maria, bounded on the north by the serra and the steep slope of the edges of the tertiary strata, which lie between the serra and the sea-coast, as seen in the engraving. This plain consists on the surface of coarse white sand without shells. The absence of sea-shells from these marine deposits appears remarkable at first sight, but it is no doubt largely owing to the fact that they have been dissolved out by rains. Where these plains have been cut through by rivers, shells are seen in the lowest beds, as, for instance, on the left bank of the Itabapuana, a few miles above the Barra. The plain north of Victoria bears the same kind of vegetation as that which characterizes the plains farther south. Just previous to the last rise of the land these plains were under water, and the hills of the Victoria group stood as islands in the mouth of a bay.

Crossing the sand-plains, a steep ascent of about fifty feet at Carapina brings one to the level of the tertiary plains, where one leaves the sands. These plains are covered by a clayey soil varying much in fertility. Near Carapina there are some lands available for cultivation, and part of the plain is covered by trees. At Carapina the soil is clayey, with very little sand, and of a slate-blue color; but going eastward toward the sea the soil grows drier and more sterile. The trees are very scattered and coarse-barked, and when in clumps are free from undergrowth. The open plain is covered by a scanty growth of tall coarse grass, and is diversified by great numbers of ant-hills, — irregularly dome-shaped structures of clay, often quite round, as hard as stone, and resembling boulders scattered over the plain. I presume that the character of the vegetation over this region is, as elsewhere, in part due to fires, which, periodically set by the inhabitants, have killed all except the most hardy trees and other plants. The Rio Carahype is a little stream which rises, according to Gerber's map, to the northwest of the serra; but according to my notes and observations it takes its origin in the hills north of the serra, and, traversing the plain, reaches the sea a few miles north of the Ponta do Tubarão. It is only a respectable brook, of no service for navigation, unless it be for canoes. It has cut down through the tertiary beds, and has a rather deep channel. Its valley is narrow, with steep sides. Near the town of Serra the valley is about one hundred feet in depth, showing that the tertiary clays, lying against a sloping bottom, probably also thickening towards the hills, form beds that slope gently from the hills to the sea. In some localities near Serra gneiss is exposed in the bottom of the river. Near the hills the surface is irregular, and the plain becomes undulating and broken, — the result, I believe, of the action of glaciers, as well as of streams, from the hills. The soils of the tertiary lands bordering the high grounds, as in the vicinity of Serra and along the Rio Reis Magos, a few miles from its mouth, are quite good, and are used for the culture of coffee, cotton, &c. The forests of these regions are more luxuriant than is elsewhere the case on the plains. Lying to the north of the river, and not far from the shore, is the large and shallow Lagôa Jacuné, which during heavy rains overflows and sends its waters into the Carahype, at which time that river widens its channel and opens its bar.

The Mestre Alvaro, or the Morro da Serra, is an isolated, irregularly pyramidal mountain-mass of gneiss standing in the gneiss plain a few miles northwest of Victoria, and presenting on all sides very similar outlines. Its height I should estimate at about 3,500 feet. It stands like a pyramid on the plain, majestic and alone. More pleasing and symmetrical in outline, and more isolated in its position, it is more beautiful than Tijuca; and, seen from the sea, it looms up grand and blue against the tropical sky and the far-off line of serras, which lie along the horizon like the front of an approaching storm. It is densely covered with forest, but on its slopes are extensive coffee-plantations, which yield well. The *Diccionario Geografico* says that formerly emeralds and magnetic iron were found there.* In very striking contrast with the quiet, *sans souci* air of Victoria, with its grass-grown streets, is the little village of Serra, which, owing to its being situated in an agricultural region, is one of the most business-like towns in the province of Espirito Santo. This little town is built near the base of the hills on the northeast side of the Mestre Alvaro. I was told that water communication existed between it and the Bay of Victoria, but my visit to the locality was rather hasty, and this may not be correct. Northward of the Mestre Alvaro are a few gneiss hills stretching in a line toward the Rio dos Reis Magos. West of the Mestre there is much low land. The appearance of the coast from Victoria to within a few miles of the Rio Doce, a distance of nearly fifty miles, as seen from a point at sea about ten miles east of Riacho, is given in the accompanying illustration. The main Serra dos Aimorés is seen in the background, stretching along like

* Saint Hilaire ascended the Mestre Alvaro. He speaks of meeting with the bamboo *Taquara-assú* in the forest, at a considerable elevation above the plain, and remarks that these plants require humidity and considerable elevation. (SAINT-HILAIRE, *Voyages sur le littoral du Brésil*, 2nd Partie, p. 275.) Von Martius says that the bamboos flourish principally at a height of 1,800 to 2,000 feet above the sea.

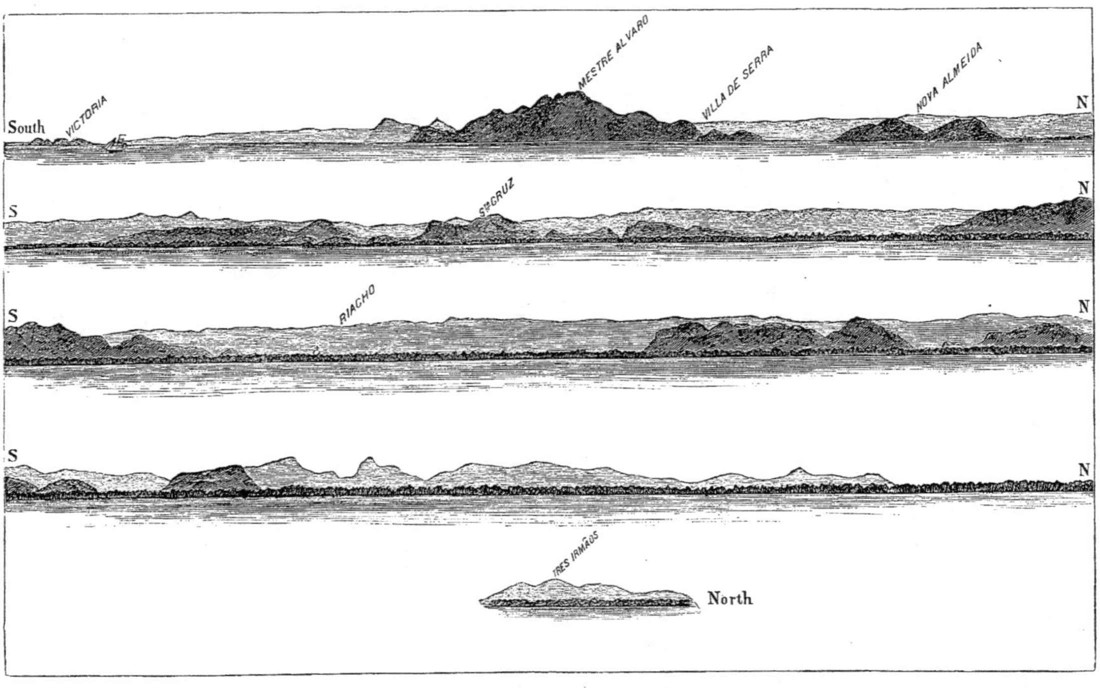

COAST BETWEEN VICTORIA AND RIO DOCE.

a wall at a distance of fifteen to twenty-five miles, while in front are the isolated groups of hills of Victoria, Mestre Alvaro, Nova Almeida, &c.

The Rio Reis Magos, or Apiapitanga, is a little stream which rises in the serra northwest of the Mestre and empties into the sea fifteen miles north of the Bay of Espirito Santo. Canoes ascend it only twenty miles. It empties into a little bay, on the south side of which, situated partly on the level of the tertiary plain and partly beneath the slope, is the ancient village of Nova Almeida.* The tertiary beds continue northward to the port of Aldêa Velha, or Santa Cruz, north of which they disappear from the coast, and give way to marshes and swamps that occupy the coast thence to the Rio Doce. Between Nova Almeida and Santa Cruz, as well as to the north, the sandstone beds form in some parts rocky shores, off which lie isolated skerries, which obstruct both bays.

The river Santa Cruz is, like the Reis Magos, a little stream affording navigation for canoes only. Its principal affluents are the Piriqui-assú † and Piriqui-mirim. It opens into a little bay like that of Nova Almeida, and which affords a harbor for small vessels. The bays of Nova Almeida and Santa Cruz are noteworthy in that they open broadly to the sea, and, unlike the mouths of the other rivers, are unobstructed by bars, — a circumstance which is, perhaps, owing to the fact that the rivers emptying into them bring down but little sediment.

* Here was established anciently a Jesuit missionary station, with a school for furnishing instruction in the Tupí language.

† Saint-Hilaire describes heaps of oyster and other shells bordering the river Piriqui-assú, near Aldêa Velha, which are without doubt *Kjœkkenmœddings.* Similar shell-heaps, or *ostreiras,* as they are called in Brazil, are found on the coast of São Paulo, and on the Ilha do Governador, in the Bay of Rio. They often contain human remains, pottery, &c.

I have sailed several times close along the shore from Santa Cruz to the Rio Doce, but I have never landed, and I am unable to describe it in detail, as well as the flat lands lying behind it, nor have I any trustworthy information concerning them. The shore is bordered by a sand-beach like that stretching south of the Parahyba do Sul, and the lands in its rear are plains diversified by swamps, shallow lagoons, and channels which, having never been explored, are laid down incorrectly on all maps. Much of this low ground is suitable for pasturage, and, so far as my observations go, the greater part is very heavily wooded.

The great Serra da Mantiqueira, separating itself from the Serra do Mar near São Paulo, runs to the north of that serra, and almost parallel with it, on the northern confines of the provinces of São Paulo and Rio de Janeiro, into Minas, passing near Barbacena, whence it continues with the same trend, under various names, beyond the Rio Doce. To the southeast it gives rise to a host of little rivers which flow into the Parahyba do Sul, while on the northeastern side the Rio Grande, one of the tributaries of the Paraná system, and the Rio Doce take their rise. From Barbacena a cordilheira runs northward, separating the waters of the Rio Doce from those of the Rio Grande and the São Francisco, while another line of serras, with a general northeast course, leaving the Serra do Espinhaço near Diamantina, forms a water-shed between the rivers Doce and Jequitinhonha. The basin of the Doce forms an unequal-sided quadrilateral figure, whose sides measure very nearly as follows: the northeastern side 120, the southeastern 230, the western 190, and the northern 90 miles, which would make the area drained by the river very much larger than that of the Parahyba do Sul. Gerber estimates the area embraced within the basin of the Doce at 2,300 square leagues. The Rio Doce, under the name of Chopotó, takes its rise only a few miles from Barbacena, in the angle formed by the union of the Serra da Mantiqueira and the Serra do Espinhaço. From this point the waters flow westward into the Rio das Mortes, a tributary of the Paraná system, eastward into the Pomba, one of the branches of the Parahyba, and northward into the Doce. Its course is at first about north-northeast up to latitude 19°, when it bends abruptly around the Serra das Ibiturunas, and then flows in a southeast direction to the confines of the province, when, after passing a series of rapids called the Escadinhas, it reaches lower grounds, and is thereafter navigable to its mouth.

This river is of so much importance that I add a translation of a description of it from the *Diccionario Geografico*, Art. *Doce*.

"The ancient geographers considered the Ribeirão do Carmo as the principal origin of the Rio Doce, while others assert that it rises near Sabará, in the Ribeiro Santa Barbara, which empties into the Piracicaba. But if one means by the source of a river that point which is at the greatest distance from its mouth, we ought to place that of the Rio Doce at twelve leagues * to the east of the Villa of Barbacena, in the spot where the Rio Chopotó begins. This river runs about twenty leagues in a northward direction, receiving on its left bank the Rio das Pirangas, besides various streams from both sides. It only offers easy navigation for canoes, when, having watered the settlement of Santa Anna do Deserto, it inclines toward the northwest, being augmented by the stream Turvo on the right and the Rio

* According to Gerber's map, the source of the Rio Doce (Chopotó) lies at about five leagues east of Barbacena.

Gualacho * on the left. The waters of this river make it change its course a little towards the east, and both united precipitate themselves over the fall called Inferno ; below this fall the Rio Doce takes the name it bears, and flows gently on, receiving on the right the little Rio da Casca, and on the left the Piracicaba, and, six leagues farther on, it passes over reefs blackened by time, — whence the name *Escuro* given to this fall. Three leagues lower down, on its left, are the mouths of the rivers Santa Antonio and Correntes, at a distance of eight leagues from one another.† Below the last of these river-mouths is the Cachoeira Bagauriz, where a pointed rock divides the waters of the Doce, which subdivide again and again before uniting anew in a species of basin, formed, as it appears, by several islets. This basin extends for a distance of two leagues, and, because of the current, much dexterity in the government of a canoe is required to reach it. At the end of this series of islets the Rio Doce takes a more quiet course, and, the mouth of the Saçuhípequeno passed, it becomes once more turbulent, and is successively impeded by the little Cachoeira d'Ilha-Braba, with that of Fiqueira, much more dangerous, in the Serra Beteruna, where it is necessary to transport the canoes by land a distance of fifteen *braças*, and with that of the Rebojo-do-Capim ; five leagues farther down stream the Rio Saçuhí-grande comes to swell it on the left with its tribute of waters, after having watered the comarca of Serro Frio. Passing this tributary, the Rio Doce receives from different quarters an innumerable number of limpid brooks, and

* According to Eschwege, quoted by Gerber, the Barra do Gualacho is 341 metres above the sea-level.

† On Gerber's map the mouth of the Correntes is only two and a half leagues below that of the Santa Antonio.

makes many turns before arriving at the fall denominated *Cachoeirinha*, over which canoes pass without being unloaded; still farther down it receives on its right the Ribeirão Laranjeira, and a little beyond, to the right, the Cuiaté, which empties into it, when the main river becomes majestic for the distance of two leagues, below which various reefs, with some falls of little importance, and whirlpools make navigation very difficult, without entirely destroying it. These different obstacles are known under the names of Rebojo de João Pinto and Rebojo da Onça, distant two leagues from one another. The second of these obstacles passed, the current flows to the right in the summer, but to the left in the rainy season. Half a league onward the bed of the river describes some diagonal lines, which imitate a capital M,—the name which is commonly given to it; and one league lower down various reefs, called the Cachoeirão, intercept navigation, so that it is necessary to transport by land the canoes and goods. Two leagues below the fall of Cachoeirão are encountered three other whirlpools, which are not dangerous, and the island of Natividade, which divides the Rio Doce into two unequal arms. In the dry season the canoes are unloaded at this island to pass the great Cachoeira called the Escadinhas, because for one league it is formed of steps of stone. The canoe-men carry the goods on their backs as far as Porto de Souza; but when the waters abound the canoes descend without difficulty to the Registo de Lorena, near the confluence of the river Mandú (Guandú), which comes from the south and serves as the limit to the provinces of Minas Geraes and Espirito Santo, and ascend it also with cargo, albeit with some difficulty. Between the mouth of the Mandú and Porto de Souza several rapids are encountered, which are descended with ease, but which cannot be ascended except by hard rowing, or by towing the canoes."

The Rio Manhuassú is quite a considerable little stream, which enters the Doce from the south, opposite the island of Natividade. Very little is known about it, as it flows through a forest region inhabited by the Botocudos. The Guandú is, as above described, only a little stream, and of very little importance. The country bordering the river near its junction with the Doce is very flat, with deep alluvial, clayey soils. It is heavily timbered, and affords most fertile lands for the agriculturist; but the Botocudos, who have been until late years hostile, have prevented the successful colonization of the country. A little colony of a few families has been established on the Guandú, but it was not, at the time of my visit, flourishing.

The rocks exposed here in the river channel are gneiss. I observed some heavy veins of milky-white quartz near that place, but I had not time to examine them. Opposite Guandú, on the north side of the river, there are several quite high gneiss hills, and opposite Porto de Souza is a bare hill not far from the river. Gerber has laid down on his map a little river entering the Doce opposite the Guandú, which appears to be a mistake. At Porto de Souza the river valley is very narrow, and the river was, at the time of my visit in December, 1865, not more than 250 feet wide opposite the port.

The port is situated at the foot of a series of rapids, at the head of navigation. Even in the dry season it would be possible at all times, I was assured, for a small steamer to reach it. Gneiss is exposed in the bed of the river and along the banks. It is gray, coarse, and homogeneous. Strike N. 60° E. Dip 45° Northeastward. In the river-banks the surface of the gneiss is rugged, and not smooth as in glaciated surfaces, showing that it has been determined by water action. Above it are beds of coarse, yellowish sands and quartz gravel, the latter often very coarse. At the Quartel, or Barracks, these beds reach a level of about twenty feet above that of the river. Above these is a layer of brown, clayey earth, with scarcely any admixture of sand, but full of little silvery mica flakes. This affords a good, deep, fertile soil, suitable for coffee, corn, beans, castor-beans, &c., &c. The forests of this region are quite luxuriant, and are rich in valuable woods and game.* All this region is finely adapted for agricultural purposes, and some day must become the home of an agricultural population; but the Indians have so far held sway over it.

At the port the left-margin of the river is occupied by some immense sand-banks, covered during the rainy season, when the river rises some twenty feet above its ordinary level. Descending the river from Porto de Souza, the river continues very narrow, deep, and rapid, with rocky banks for a mile or more, the banks becoming lower as one descends. The gneiss is exposed in the bottom of the banks on both sides of the river, and over this, as at the Porto, are alluvial deposits, making the banks eight to fifteen feet in height. At a distance from the river are gneiss hills. A quarter of a mile below the Ilha da Esperança the river widens to at least 600 feet. The Rio Mutum,† laid down

* I observed in the forest between Porto de Souza and Guandú great numbers of Barrigudo trees (Bombax), some of considerable size.

† The Rio Doce is very rich in fish, a large collection of which was made by my companion and myself for Professor Agassiz. Prinz Max. zu Neu Wied speaks of the occurrence of a species of sawfish (Pristis Serra?) in the lower waters of the Doce, and says that it enters the Lagôa Juparanãa. Mr. Copeland

RIO DOCE AT PORTO DE SOUZA.

on Gerber's map as being some three or four leagues below the Porto, is not more than three miles. It is but a brook; but it has a fall near the mouth, and, being in the midst of a very fertile country, may be of importance by and by for its water-power. On the Mutum the land is in some places flat, and thirty to forty feet high, and may be in part tertiary; but I doubt it. On the north side of the river are many forest-clothed gneiss hills, the most of which range from 600 to 700 feet in height. East of these, a few miles from the Mutum, and standing back from the Doce, is a prominent hill called the Morro do Padre, the greater part of which is bare, and which must be at least 2,000 feet in height. On the opposite side of the river is the Morro do

and I took specimens at or near the mouth of the Mutum. They certainly ascend to Porto de Souza. The canoe-men said that they ascended to the Guandú. The fishermen all said that this fish secured its prey by striking a lateral blow with its long snout or jaw. On one of the saw-teeth of a large specimen we took there was impaled the large, tough scale of a curumatã, which could only have been pierced by a smart blow, when attached to the side of the fish. The fisherman said that a true shark, *caissão de dentes*, was found in the Doce.

Lage among gneiss hills.* Leaving the gneiss hills below the little river São João, the land becomes lower, and the hills flat-topped, or with flat outlines, and a rather gentle slope toward the river. Some of the hills appear to be gneiss, others tertiary. The low lands bordering the river are alluvial. So exceedingly dense is the forest covering this whole country, that, except it be occasionally a bare gneiss rock by the river-side, a bare gneiss hill which is never to be mistaken, or the exposures of the alluvial deposits of the river, there is no guide for the working out of geological features other than the general topographical outlines of the country. The clothing of forest tends very much to exaggerate the height of the lower lands. Nowhere in Brazil, not even at Pará, have I seen a more luxuriant forest growth than that of the Doce. The trees, all bound together by llianas, and filled in with a dense undergrowth of palms and shrubs, crowd down to the water's edge, and stretch their great vine-draped branches out over the river, as if in want of light and air. The forest forms a dense wall along the river, — so dense that the eye does not penetrate into its shade, — and one must be armed with a strong wood-knife who would enter it.

In these forests flourishes the Jacarandá, or rosewood (*Bignonia Brasiliana* Lam.), which once abounded along the river, but, having been extensively cut for exportation, is now to be met with of sufficient size for cutting only at a distance from the stream. The principal article of export from the Doce has been rosewood, which has the reputation of being of a fine quality. The Cupiuba (*Copaifera officinalis*), furnishing a valuable wood and an abundant oil used in medicine

* Right bank opposite Ilha do Lage. Gneiss bands very siliceous. Strike N. 80° E. Dip northward 50°.

and the arts, abounds here, together with the Páo Brazil (*Cæsalpinia echinata* Fr. All.), noted for its once costly tint; the Sapucaia (*Lecythis*), furnishing food to the Botocudos; the Cedro (*Cedrela*), Ipé (*Tecoma*), Páo d'Arco (*Bignonia*), Peroba (*Aspidospermum*), Putumujú (*Putumujú*), Vinhatico (*Acacia*), and species of *Genipa*, *Machærum*, *Ingá*, *Bowditchia*, &c., abound. The names of the trees furnishing valuable woods for construction and cabinet-work, many of great beauty and durability, are legion, and when the country becomes inhabited, these must become a source of wealth.

Several species of palms, among them the Airí (*Astrocaryum*) and Palmetto (*Euterpe*), flourish in the forest, while the Embaüba (*Cecropia*), the food of the sloth, with the Ubás and Heliconias, form one of the marked features of the vegetation of the river-banks.*

Game is exceedingly abundant, among which may be mentioned the Anta (*Tapirus Americanus*), whose tracks, together with those of the Capabara (*Hydrochærus Capabara*), are everywhere seen by the river margins. The Paca (*Cœlogenys Paca*) and Cutia (*Dasyprocta*) are very common, and are valued for food. There are, at least, two species of sloth found here, — *Bradypus tridactylus* and *B. torquatus*, — together with species of opossum (*Didelphys*) and Cuati (*Nasua*). Armadillos (*Dasypus*) abound as elsewhere, and I have seen at least two species. Wild pigs Caititús and Queixadas (*Dicotyles*), are found in herds in

* Prince Max. zu Neu Wied gives a plate representing a view on the Rio Doce. It was evidently not drawn from nature, but it gives quite a good idea of the shore vegetation. An immense alligator is represented in the foreground. One may spend a month on the Doce and not see a single alligator, and those of the Doce are very small.

VIEW ON RIO DOCE.

the forest, and are hunted for food. At least four species of *Felis* occur here, — *Felis Onça*, *F. concolor*, *F. pardalis*, *F. macroura*, — together with the fierce black jaguar, which may be only a variety. One or two species of *Cervus* are not uncommon, together with hares and squirrels. Of monkeys there are the following: *Ateles hypoxanthus*, *Mycetes ursinus*, species of *Cebus* and *Callithrix*, and *Jacchus (Hapale) leucocephalus*. Jacupembas (*Penelope marail* Linn.), Mutuns (*Crax*), Araras (*Psittacus macoa* Linn.) and other parrots, are very abundant.

The river, after leaving the gneiss lands, widens very much, and is in some places 800 to 1,000 feet wide, occasionally diversified by wooded islands, and affording stretches of river scenery of very great beauty. Francylvania is the name of a settlement established on the northern bank of the Doce, at a short distance back from the river, opposite the mouth of the little river Santa Maria, and not far above the mouth of the Rio Panca. Here, not many years ago, was established a Brazilian colony, under the direction of one Dr. França Leite,* which proved a failure; its site being now marked only by a luxuriant second growth (*capoeira*), which on the Brazilian coast

* The history of this colony is briefly as follows: Dr. França Leite, some fifteen or more years ago, conceived the idea of establishing a colony on the Rio Doce, and at Francylvania, aided by government, he formed a settlement. Mills for sawing lumber and grinding mandioca were erected, and a considerable amount of ground was put under cultivation. Establishments were opened at the Povoação, at Monserras, and Ipyranga, where many cattle were raised. But Dr. Leite's plan proved visionary. He failed to get the aid he demanded; the colonists, dissatisfied, and harassed by the Botocudos, who were very troublesome neighbors, began to withdraw, and after an existence of some three years, the Botocudos put an end to this colony by killing Ervalina, Leite's brother-in-law, and a slave, and burning the settlement; and thus failed another attempt to settle the Doce.

rapidly springs up on abandoned lands. The soils in the vicinity of Francylvania are of the most productive kind, and the locality was wisely chosen for the establishment of a colony.

Descending from Francylvania to the hills of Santa Antonio, the lands bordering the river are still lower, and the hills have long, gentle slopes, but no rocks are seen, except gneiss, on the borders of the river. The islands all consist of sand, overlaid by a thick bed of brown soil. Alluvial lands of the same character occur on both sides of the river, forming irregular strips. These lands, which are liable to be overflowed during the *enchente*, are of the highest fertility, and are especially proper for the culture of sugar-cane.

A few leagues south of Francylvania the river passes through a region diversified by hills of gneiss several hundred feet high. On the south side of the river the westernmost of these hills form a range which, under the name of the Serra de Sant' Antonio, is seen stretching off southward, tying in with the coast mountains of Santa Cruz and Nova Almeida. In this belt of country are many lagôas, some of them of considerable size. Among these may be mentioned the Lagôa Sant' Antonio do Norte, on the north side, and the Lagôas Páo Gigante and Limão, on the south, all of which communicate with the river by small, blackwater streams. These lakes and streams have been very incorrectly laid down on the maps, owing to the fact that no trustworthy survey of the river has ever been published.* A few miles farther east, and opposite the Barra do Rio Limão, is a range of high hills, which stretches off as far as

* I am informed that the river was surveyed and mapped for the English Company of the Doce by Mr. Fred. Wilner, but I am not aware that the map was ever published.

one can see from the river towards the northeast. This hilly country ends on the river, just to the east of these hills, in a sharp ridge about a mile long, escarped on the eastern side, and known as the Morro da Terra Alta.

LOOKING UP THE DOCE FROM NEAR LINHARES.

Passing through this country the river is much diversified by islands, and is in some places more than 1,000 feet wide. The lands here vary much in character, and are noted for their richness. Over the whole country spreads the heaviest forest growth, but at the time of my visit it was unbroken by a fazenda, and in the hands of the Indians and rosewood-cutters. On leaving the hills the river soon widens to at least half a mile, and is full of beautiful wooded islands, while it is bordered by alluvial lands which, during the *enchente*, are liable to be overflowed. These lands extend to Linhares. About two leagues above this town there are, on the southern bank, a few clearings. I think that nowhere in Brazil have I ever seen so rank and luxuriant a vegetation. These lands are covered with the same brownish, clayey soil we have observed farther up the river.

This soil, which is called *maçapé*, or *massapé*, is found on all the alluvial lands bordering the rivers on the Brazilian coast, and is noted for its great productiveness.

On the Rio Doce those massapé lands which are of sufficient height to escape the effects of the *enchente* may be used for the cultivation of almost any of the products of the country, — such as sugar-cane, tobacco, coffee, cotton, mandioca, &c., — but a large part is likely to be submerged every year when the river is full. The annual overflow begins in December with the daily thunder-storms, and lasts usually until March. During its prevalence the river margins are overflowed for a greater or less length of time, the freshets of March often being as high as those of December. In the year 1833 occurred an extraordinarily heavy freshet, since when nothing like it has been known. The water of the river, even in the dry season, is very turbid with sediment, and of a light yellowish-brown color. During the *enchente* it becomes very much more turbid, and a thin deposit of mud is thrown down over the flat lands every year. On the subsidence of the waters, the vegetation left decaying over the wet country is apt to breed fevers, and the Rio Doce has had a very bad reputation for being very unhealthy. It certainly is feverish, but I could not learn that it was any worse than São Matheos, or even so bad. My companion and I suffered nothing from our visit. The river waters are bad, but if allowed to deposit their sediment, and stand some time, become very potable and safe. The climate of the Doce is warm and very damp, and it is upon the distribution of rain throughout the entire year that the luxuriance of the vegetation seems to depend. The climate is damper than at Victoria or São Matheos. *Carne secca*, or dried salted beef, which keeps well elsewhere, soon spoils here in certain seasons of the year.

The massapé grounds which are liable to be overflowed are used for the culture of almost all the products of the country, except mandioca, cotton, and coffee. The root of the mandioca requiring more than one season to mature, is likely to be injured by a freshet, so that its culture is confined to the higher grounds. These massapé lands are especially good for sugar-cane, which, as at Campos, is not injured by an overflow, and also for beans, Indian corn, rice, bananas, &c. The corn and beans are planted usually in March or April, so as to become ripe before the *enchente*. Lands of this character are very extensive, and the Doce region is adapted to sustain a very large population. There can be no doubt that with the clearing of the forests and the tilling of the ground the region would become more healthy.

Linhares is built on the left bank of the river, on the top of a bluff formed by the projection southward of a point from the great tertiary plain lying north of the Doce. The bluff is, if I remember rightly, about eighty feet high, and exposes beds of white and red clays of the ordinary type. On the Doce these clays occupy near Linhares only a small extension on the northern bank. Their boundary line trends off northward to São Matheos and northwestward to the Lagôa Juparanãa,* which is held in a basin scooped out of the beds of this formation. The lagôa lies at a distance of about two miles to the north-northwest of Linhares, and communicates with the Doce at Linhares by a very narrow and tortuous, but deep channel, called the Rio Juparanãa, which flows over the low wooded ground close to the edge of the bluffs. This channel is about fifty feet wide, and, according to Senhor

* The word "Juparanãa," according to Prinz Neu Wied, is Tupí, and means "sea."

Raffael P. do Carvalho, furnishes at all times at least four to five feet of water, which would be sufficient for a little steamer. It is exceedingly tortuous, but it would be a very easy matter to cut off some of the bends, and thus shorten very much the distance between the Doce and the lake.

At the foot of the lake the river is bordered on the west side by a narrow strip of low ground, north of which the tertiary bluffs begin; and, with the exception of this little stretch of alluvial land, and of a similar stretch at the head of the lake, the lake is bordered by the bluffs. Freireiss * gives the length of the lake as seven leagues, width half a league, and circumference sixteen to eighteen leagues. The *Diccionario Geografico* (*vide* Art. *Juparanan*), which is here manifestly inaccurate, gives the circumference as only five leagues. According to my estimate, and the information of Senhor Raffael, it must be at least twenty miles long and in some places four miles wide. It is very deep, and in some places, according to Freireiss, the depth is at least eight to twelve *klafter* (fathoms). The water of the lake and river is of a light milky color. The banks of the lake are eighty to one hundred and fifty feet high, as near as I could judge, the height being greater at the head of the lake. Along the eastern side, between the foot and the Fazenda do Guaxe, white and pink clays are exposed in the bluffs, and in many places the coarse red sandstone of the tertiary crops out at water-level or thereabouts. About a mile north of Guaxe, on the east shore of the lake, and opposite a very small island of gneiss, a point called the Ponta de Ouro, a quarter of a mile long, juts out into the lake. This point is about fifty feet in height, and forms on the southern side a line of picturesque cliffs. The strata composing this point

* Quoted by Neu Wied, *Reise nach Brasilien*, Vol. I. p. 214.

are perfectly horizontal, and consist of a white or pink feldspathic clay, in some localities with no admixture of sand, but for the most part with a large percentage of coarse, angular, or slightly rounded sand and gravel, these materials being scattered through the mass apparently with no order whatever. The lowest beds are the most sandy. The red color is due to ferric oxide, which is distributed through the mass very unequally, sometimes cementing together portions of the beds into stalagmite-like masses penetrating the clays. The tint is sometimes yellow. Under the clays is the coarse red sandstone which occurs in very solid and compact masses, in which case the rock is regularly and evenly cemented. The bluffs are steep and wooded, as also is the plain above. Small farms are located along the lake, with an occasional little fazenda. The slopes of the bluffs are found to yield very abundantly, and to produce excellent coffee. The Lagôa Juparanãa is, like the Doce, very full of fish. I found here two species of Cagados, probably the *Emys depressa* and *Emys radiolata* of Max. zu Neu Wied. A species of Unio is abundant in the lake, and is said to be used for food. I saw heaps of the shells lying in front of a fisherman's hut, but the animal may have been used for bait.

At the head of the lake there enters a little river called the São Raffael. It rises in the forest, in the country of the Botocudos, and has never been explored.* Gerber's map represents a Rio Preto as flowing into the Juparanãa, but no such river exists. The lake also, as laid down on his map, is too small, and the island is too large. It is really only a rock. The head of the lake cannot be distant from

* Senhor Raffael says that above it divides into three branches, and that it is rich in fish, but that it contains no surubims (*Platystoma*).

LAGÔA JUPARANÃA, LOOKING TOWARDS THE OUTLET.

the city of São Matheos, according to the best information I have received, more than three to four leagues. The bar of the Doce is so bad as practically to forbid the entry of vessels, although they sometimes cross it; but the river is navigable for a little steamer, during the whole year, from its mouth to Porto de Souza, a distance of ninety miles; so also is Lake Juparanãa. Until a railroad is built to Victoria it would seem best to construct a good wagon-road through the forest, from the head of lake Juparanãa, over the plain, to São Matheos, and make São Matheos the port of the Doce; but São Matheos can never offer the same advantages as a port that Victoria does.

The lands surrounding the lake are plains covered by forest, but owing to the dryness and little fertility of the soil, it is not very luxuriant. The soils of these high lands, however, vary very much in quality, in some localities being excellent for cotton, mandioca, &c., in others sandy and barren. Rosewood abounds in these forests, and is quite extensively cut.

The higher lands near Linhares appear to be fertile, and I have nowhere seen more vigorous and luxuriant crops of sugar-cane, bananas, &c., than I saw growing on the plantation of Senhor Alexandre, by the river-side, just below the town, and situated on alluvial grounds. The river here appears to be bordered by three terraces, but I had no time to examine them closely.

Lying almost parallel with the river, and to the northeast of Linhares, is a very beautiful, narrow lagoon, which looks like an old river valley, and is called the Lagôa do Aviso. It is said to empty into the Doce to the east of Linhares. There are other lakes in the vicinity.

LAGÔA DO AVISO.

From Linhares the river runs off in a southeasterly direction to the sea. The stream is wide and interspersed with many islands. The shores are low, forest-clothed, and overflowed during the annual freshet. At the mouth the river widens very much and enters the sea obliquely, from behind a long sand-spit which extends southward from the left bank. The river just inside the mouth forms a fine sheet of water, but it is very shallow, the bottom being shifting, so that the soundings vary with the quantity of water in the river. So great is the amount of water, and so shallow is the river, that the tide is not felt inside the mouth, and the water is always fresh. Great numbers of drift trees are brought down during the freshets, and the sea-beaches of the vicinity are strewn with them. The mouth is wide, shallow, and obstructed by a bar, on which the waves break fearfully. It is always difficult, and sometimes for weeks together impossible, to enter the Doce, and very many vessels have been lost in the attempt.* Just northeast of the bar are extensive banks, which extend two or three miles out to sea. There is a little hamlet on the right bank of the river, near its mouth, and a short distance above, on the left bank, is a small settlement called the Povoação. The lands here are sandy, but they are, for such soils, quite fertile, producing mandioca, cotton, mamona, sugar-cane, &c. The coast at the mouth of the Doce projects considerably to the eastward, and there is usually a strong current near the shore, depending for its direction upon the wind.

Many years ago an English company was formed to open the river Doce as a highway for navigation into Minas, but, from the above description of the river, any one can see how unfit it is for navigation above the falls. The enterprise of course proved a failure, and a curse has undeservedly fallen on the Doce. At present a very small commerce in salt is carried on between the coast and various points along the river, the salt being transported in canoes. The journey from the sea to Correntes, a town situated on the river of the same name, an affluent of the Doce, a distance

* The coasting steamers occasionally enter the Doce.

in all of about two hundred miles, consumes over forty days. From the Barra to Porto de Souza the journey is easily performed; but above that point the canoes must be towed and poled with the greatest difficulty, and at very short intervals the load has to be removed, so as to allow the empty canoe to pass a waterfall. On the coast a bag of salt of about sixty pounds costs 2$000 (two milreis, or about a dollar). In Minas it costs eight milreis or more. As nearly as I could learn, the whole enterprise of the English company was conducted with great extravagance and want of good management, so that it is no wonder it failed. It is useless to think of making the Rio Doce the highway to Minas, and of using the mouth of the river as a port. The country bordering the Rio Doce and the Lake Juparanãa, and extending westward for some distance beyond Porto de Souza, must be treated as a great agricultural region by itself.

Since my visit to the Doce quite a number of American families have settled on the river, forming a little colony, which, so far as I can ascertain, promises to be a success. The colonists came from the Southern States, from a climate not so very different from that of Espirito Santo, and those that I saw looked like men through whose hands a better future might be worked out for the Doce. We hope that their enterprise may be successful.*

Northward to the mouth of the Rio São Matheos stretches a sand-beach, broken only by one or two river-mouths, and back of which, between the shore and the tertiary bluffs, is a wide area of swamps and lagoons, — a region al-

* In the spring of 1868 there were on the Doce, according to Burton, four hundred Americans, who were doing well, and were "studying coffee." (Highlands of Brazil, Vol. I. p. 5.)

most impassable, and never yet mapped. Just north of the Doce and near the coast is a large lagoon called Monserras. During the dry season this is separated from the sea by the sand-beach, but when the rains come it opens for itself a channel to the sea, which channel remains open until the dry season returns.* A more waste and desert region than the shore between the Doce and São Matheos can scarcely be imagined; but it is the high-road, and must be followed in going to São Matheos. The weary sand-beach stretches ahead to the horizon, dancing in the hot air, and dimmed by the drifting spray from the ocean breakers, that pour their thundering, blinding surges on the desolate shore. A line of monotonous sand-heaps, like a great tumbling billow ready to burst on the low grounds behind, runs parallel with the beach, bare, or scantily covered by tufts of grass, dwarf palms, &c., — no shade, no water. The road is a strip out of Sahara. On one side is the sea, on the other a miasmatic, pathless swamp. Ordinarily the lagoons lie at some little distance from the shore, and are separated from it by a dense, impenetrable thicket; but at Pitanguinha there is a little lagoon by the shore where water may be obtained. The sands on these beaches are coarse, and do not pack so as to afford a good footing. Animals sink at every step, and the journey from the Doce to the São Matheos is excessively fatiguing, the traveller being, in addition, liable to suffer severely from thirst. About thirty miles north of the Doce is Barra Secca, where a little river, draining the marshes of the interior, empties into the sea. Just where it takes its rise, and what is its course, no one, even of the immediate vicinity, knew. It is usually set down as draining a lake called Tapada; but

* When I went to the Doce from São Matheos this bar was closed, but on my return, in the latter part of December, it was open, and dangerous to cross.

that is not possible, for the lake lies only a few hundred yards from the mouth of the river, and at a very much higher level. The tide ebbs and flows in the Secca and evidently runs up a long distance. The river comes from the south and, just before reaching the sea, flows along in the rear of the broad beach ridge, from behind which it escapes at the barra, and, cutting a channel across the beach, flows into the sea. The beach is constantly changing by the drifting of the sands, through the action of the wind and waves, so that the barra is as constantly shifting, never remaining long in one place. With a northeast wind it shifts to the south, with a southeast wind to the north, while an easterly storm sometimes closes it entirely. The river is so shallow that it may be forded at low tide, whence the name Barra Secca. At the time of my visit, in 1865, there were exposed in the banks of the stream at the barra strata of sandstone which were laid bare by the washing away of the beach-sands. These sandstones were formed, below high-water mark, by the cementing together of the sands of the lower part of the beach by the lime of shells, &c. They preserved the characteristic beach structure, and were full of shells; but of these sandstones I shall have more to say further on.

The lands behind the beach at Barra Secca are flat and sandy, and, though cultivated to a very slight extent, are of no real value. The lower grounds are damp, furnished with a soil, and are largely covered with a rather luxuriant forest-growth, back of which is the picturesque Lagôa Tapada, a large, irregular, shallow sheet of clear water, margined by grass-covered meadows and forest. Its waters drain off northward into the Mariricú.

On the beaches between the Rio Doce and the São Matheos the traveller sees, at frequent intervals, the shells and skeletons of sea-turtles, and at certain times of the year there is no sight more common than that of a flock of Urubús feeding upon the decaying carcass of a turtle, recently killed by some hunter for the sake of its flesh, fat, or ovarian eggs. Most abundant of the four species which occur on the coast is the Loggerhead Turtle, *Thalassochelys cauana* Fitz.

This species is very common on the Brazilian coast. One may frequently see it floating lazily about on the surface of the sea, inside the Bay of Rio, as well as outside the bar, but there are regions where it is especially abundant, and to certain beaches it resorts in great numbers, at particular seasons of the year, to deposit its eggs. Perhaps one of the most noteworthy of these is the beach between the Doce and São Matheos. In the month of November, 1865, I found the turtles laying in the vicinity of Barra Secca, and Mr. Copeland and I made a short stay at the place to capture some specimens for the Museum of Comparative Zoölogy. In the daytime the turtles remain out at sea, but in the night the females come on shore to lay their eggs. According to the statements of the fishermen, as well as to my own observations, they come in shore when the tide is low, and having gained the beach, creep to the upper part beyond high-tide level before they make their nests.

I have repeatedly watched the movements of these animals. They walk by means of their great flippers, assisted by their short hind legs, with a very slow, hitching motion, the body dragging on the sand. The flippers and feet make two irregular grooves in the sand, three or four feet apart, which look as if a great wagon with cogged wheels had been driven over the beach. These tracks are so prominent that one may see them even on a dark night, and they serve to tell the hunter whether the turtle has returned to the sea or not, and to guide him to her, or to her nest. While laboriously working her way up the beach, the turtle keeps her head stretched out, and from time to time snuffs and sighs as if fatigued, and now and then she rests. So intent is the animal upon the accomplishment of her mission to land, that one may mount on her back without alarming her ; but if too much disturbed, she hastily turns about and makes her way as fast as possible to the sea. Arrived at the top of the beach, sometimes just above high-water mark, sometimes a few feet higher up among the sand-hills, at others, even on the landward side of the sand-ridge, she stops and prepares to make her nest. This she accomplishes by means of her flat hind feet, and after this manner : She digs up the sand with one foot, and throws it to one side. Then she uses the other foot in the same manner, working with one foot after the other, alternately, like a machine, as Neu Wied has remarked.* The sand which would be likely to fall back again is moved out of the way by the foot, which moves forward before it is thrust down to deepen the hole, and scrapes the sand to one side. The whole is a very slow operation, occupying several minutes. During this time the animal remains with her head stretched forward, with very little motion of the body, and occasionally giving a hiss or breathing heavily. In this way I have seen a perpendicular hole dug in the sand a foot and a half or two feet deep, and a foot in diameter. The animal then remains quiet, and the deposition of the eggs soon begins. While one individual was laying, I caught the eggs in my hand, as they fell. They were laid two by two, or one by one, at an inter-

* Neu Wied has in his Journey a plate of a turtle laying. The figure of the turtle is very inaccurate, and the eggs are represented as of different sizes.

val of about half a minute, falling in a heap into the hole, to the number, if I remember rightly, of 143. A nest almost always contains over 100 eggs, and usually from 120 to 150. When the laying was going on, the animal appeared perfectly unconscious of the presence of persons about her. So soon as it was accomplished, she rested a moment, and then, with her hind feet, scraped the sand back into the hole until it was full. After this the body was raised a little, and sand was scraped underneath it. The body then descended, packing the sand tightly, and this operation was repeated several times. It was an exceedingly interesting sight to see this stupid reptile performing so strange an act, one to which she seemed to be prompted by something more than a blind instinct. The whole operation finished, she turned round and started for the sea, when we captured her, by taking hold of the shell behind, and upsetting her on her back, in which position it is impossible for a sea-turtle to turn over. In upsetting one of these animals two persons are usually required. As soon as the animal is alarmed, she thrusts her fore paddles into the sand and throws it behind her, so that, if one does not take the precaution to close his eyes, he is likely to be blinded.

The eggs are rather larger than those of a hen, round, and covered with a tough, white, parchment-like skin. The albuminous portion is clear, and does not become hard in boiling. The yolk is very large, deep yellow, and is the only part eaten. I found these eggs very palatable, though they tasted somewhat fishy. The Brazilians are very fond of them, and, while the turtles are laying in November, December, January, resort to the shore to collect them, filling in a short time the great panniers of their mules. They discover the nests by thrusting a long stick into the sand,

and then dig out the eggs with their hands. The nests are, I am told, sometimes despoiled by the Teiú lizard (*Teius monitor* Merr.). The Brazilians cook the eggs in various ways, but generally by boiling. The yolk is usually mixed with sugar and farinha. They are not so satisfying as the egg of the hen, and one may eat a dozen at a meal. They make very good omelets. The eggs which remain unmolested are hatched by the heat of the sun, in about twenty days, it is said, when the young turtles dig their way to the surface and escape to the sea, where the most of them fall a prey to sharks and other fish. The eggs are laid at such a depth below the surface as to insure a uniform temperature, for on the surface the sands are exposed during the day to a fierce, burning heat, while in the night they cool down very much, the diurnal variation of temperature amounting to thirty degrees or more. At the depth of a foot and a half, or two feet, the temperature is quite uniform, and would stand in December or January at about eighty degrees. I observed that the dogs at Barra Secca, when it was cold at night, scraped away the upper layer of sand near my tent, and lay down in the warm sand below, and I learned a useful lesson from them, so that, when benighted and unsheltered, I could find a warm bed on the sea-shore. The strength of the turtle is enormous. A stout stick placed in its bill was crushed like a straw. The flesh is dark red and coarse. We had it cooked in various ways, but did not find it very good, though we were obliged to use it for food. The animal is usually very fat, the fat being of a greenish-yellow color, and oil is tried from it for various purposes by the people of the coast. The ovaries were always full of eggs in all stages of development. These ovarian eggs are much esteemed by the inhabitants, and the turtles are killed in great numbers for their sake. All along the beaches one may see carcasses from which the plastron has been removed to allow the ovaries to be got at. Mr. Copeland and I prepared some six specimens of the species of turtle above described in the shape of skeletons, shells, skins, and alcoholic preparations of head and flippers, for the Museum of Comparative Zoölogy.

The food of this turtle appears to consist of fish, shellfish, sea-urchins, &c. I have observed growing on the carapaces of these animals a large barnacle, *Coronula*, like that which grows on the back of the whale on the same coast.

Prince Max zu Neu Wied, on his journey northward from the Doce, saw a living sea-turtle near Barra Secca, and watched her lay her eggs. He has referred this specimen to the Green Turtle, *Chelonia mydas*, but I do not think that his determination is satisfactory. The specimens which Mr. Copeland and I collected have been examined by Mr. J. A. Allen, of the Museum of Comparative Zoölogy, and have all proved to be Loggerheads. I do not think I saw a single *mydas* on the Brazilian coast; at all events, all the sea-turtles I examined were different from the green turtles brought into the New York market.

Eretmochelys imbricata Fitz. (*Tartaruga de Pente* of the Brazilians.) This species occurs quite abundantly on the Brazilian coast. It is taken at the Abrolhos together with the *Cauana*, and its thick scales are used to some extent for the manufacture of ornaments. One may occasionally find it in the markets of Pernambuco and Bahia, and at the latter city my friend, Dr. Anto. de Lacerda, had one which he kept captive in the pool of a fountain in his garden.

Sphargis coriacea Gray. (Leather-back Turtle.) Neu

Wied mentions hearing an immense sea-turtle with a leathery skin spoken of, and which he supposed might be *Testudo coriacea* Linn.; but he did not see a specimen. The fishermen described to me a similar turtle, and their description would tend to confirm Neu Wied's opinion. I did not see a specimen.

Chelonia mydas Schu. (Green Turtle.) According to Neu Wied and others, this species occurs on the Brazilian coast, but I have never seen it.

The number of sea-turtles destroyed every year on the beaches between the Doce and São Matheos is very large, and the destruction, if persisted in, must ultimately drive them from the coast.

Gerber's map of that part of Espirito Santo between Santa Cruz and the São Matheos is very inaccurate, because the materials from which his map was compiled are inaccurate, and little trust is to be placed in the position assigned to the lakes and minor streams. For this region the chart of Mouchez, though giving in considerable detail the hydrography and topography of the coast lands, is worse than nothing, for he seems to have disregarded all the previously published maps of the province.

The beach, backed by a high ridge of sand, runs northward, for perhaps two miles, to a place called As Pedras, where the same calcareous sandstone, as seen at Barra Secca, occurs, exposed for some distance along the beach over a considerable area. The beds dip seaward, and appear to be very thick at low-water mark; but they thin out before reaching high-water level. The arrangement of the materials in this sandstone is precisely like that of the beach, and this formation is only the lower part of a beach ridge which has been cemented by the lime of shells, &c., and then laid bare. The sandstone is exceedingly hard. Two sets of joints — one parallel with the beach line, the other at right angles to it — divide it into great blocks, which, in those spots where they have been undermined by the surf, lie upset and in confusion along the edge of the reef. Along these joints the rock is often harder than between them, so that when the surface of a block is exposed to the action of the sea, the edges wear less rapidly than the middle, and the cracks seen on a worn surface are oftentimes bordered by narrow ridges. These have evidently resulted from the penetration into the mass of water carrying lime in solution, after the joints had been formed, and the farther solidification of the rock on each side of the joint. The waves beat terribly against this reef, and it is badly broken up. Continuing northward along the reef, which sometimes forms a smooth pavement to the beach, at others obstructs it by broken masses of stone, we come to a rocky point where the shore bends in abruptly, and makes a little bay. The beach ridge ends as abruptly, and the shores of the cove have no ridge, but the reef rock continues straight on, and forms a line of rock, stretching out some distance across the bay, while at low tide one may see that the reef, only just covered by water, is continued across the bay to a point where the beach, furnished with a sand-ridge, follows once more its normal northward trend. This locality is very interesting, because it shows us that the sandstone is confined to the beach, and that, when the beach ridge and the sands behind have been swept away, by a storm or otherwise, the bared reef may stretch, like a wall or breakwater, across the coast indentation thus formed.

Still farther northward we find the beach bordered by a

fringing reef of this sort, — a sort of irregular flagging to its edge. Riding over these one sees the ordinary shells of the coast, with their colors still fresh, imbedded in the rock. There is a species of marine worm which constructs on these rocks, near low tide, great, rounded, flattened agglomerations of sand-tubes, which are sometimes more than a foot in height, and resemble immense sponges. The sand is very compactly cemented together, as might be inferred from their withstanding the continuous pounding of the Atlantic surf. Sometimes these masses are broken across, when they appear like fragments of some large Astræan. At one locality on the shore, opposite these rocks, there is a shallow lagoon of considerable extent, called Mariricú, which lies just behind the beach, and is separated from the sea only by the beach ridge, which latter rises to a height of but five feet above high-water mark. I found the surface of this lake to be, on the 18th November, 1865, about five feet and a half above the level of the sea at low water.* Since it is separated from the sea by only a narrow ridge of sand, there is a constant soaking of its waters through the beach at low tide. It has seemed to me probable that the waters of the lagôa percolating through the beach might have something to do with the solidification of the sands. The narrow beach which separates the lake from the sea is the only dry ground between the swamps and the ocean, and it is the road taken by the wild animals passing north and south along the shore. At As Azeites, a little settlement near the shore, and about three leagues south of São Matheos, there are several beach ridges, one inside of the other. Inside of these, draining the lagôas and swamps of the south, flows a little black-water stream called the Rio

* The tide here rises about six feet six inches.

Mariricú.* At As Azeites the swampy region is several miles wide, and a considerable part of it is overflowed; but the lagoons and streams are so masked by floating masses of water-plants (*balsas*), and a dense swamp vegetation, that the country is impassable, and consequently unknown to the inhabitants of the district itself. At Azeites the Mariricú is only a narrow, navigable channel leading through a wide, overflowed region, and bordered by balsas of grasses, arborescent arums, and trees. A narrow canal, several hundred feet long, is here cut through the floating vegetation to reach terra firma. On descending, the channel frequently expands into broad lagoons, diversified by balsas, and islands with clumps of trees, but the channel soon contracts, though the growth of floating grasses † on each side makes it appear very much narrower than it is in reality. In some places it is one hundred and fifty feet wide and ten to fifteen feet deep. The lands bordering it are, for a large part, only just above water, and are heavily wooded. In these forests one sees the Gamelleira (*Ficus*), and the Tucum palm (*Astrocaryum tucuma* Mart.), furnishing a valuable fibre, and the Ingá, while on the open grounds flourishes the Cashew, or Cajueira (*Anacardium occidentale*), some immense trees of which I saw growing at As Azeites. These lands are very excellent for the cultivation of rice. There are some stretches of higher sandy grounds, which, though not fertile, are more or less cultivated, producing mandioca, feijão, cotton, rice, and corn. The Mariricú empties into the São Matheos a few miles above its mouth. The tide, which goes

* In all probability during heavy freshets some of the shore lagôas may make outlets for themselves; but the only barras I saw between the Doce and São Matheos were those of Monserras and Secca.

† This is the resort of great numbers of alligators, capabaras, and of many species of water-birds, — ducks, parras, cranes, &c., — many of which breed there.

up the São Matheos some thirty-five miles, makes its influence felt up the Mariricú for some distance. Sometimes, during drouths, the salt water flows up the river and kills the floating grasses, and many of the fresh-water fish die.

The Rio São Matheos rises in the province of Minas, in the forest, south of the Colonia de Urucú; but I have no information as to the exact point, for the region of its head-waters is a forest inhabited by savages and quite unexplored. In a manuscript map kindly furnished me by my friend, Herr Robert Schlobach, Imperial engineer of the Mucury, the Rio São Matheos is represented as taking its rise a few miles south of Philadelphia. Its ancient name was Cricaré, or Quiricaré. It is formed by the union, at a distance of about sixty miles above its mouth, of two branches called, respectively, Braço do Norte and Braço do Sul. I made a horseback journey to the Fazenda do Capitão Grande, distant some forty-five miles from the city of São Matheos, and situated on the Braço do Norte, a few miles above its junction with the Braço do Sul, and, descending to the sea, mapped the river as far as São Matheos, below which I was prevented from continuing my work. At Capitão Grande the river, a stream some one hundred feet wide, shallow and swift, flows in the bottom of a valley, cut through the tertiary formation, which here has a thickness, above river level, of three hundred feet, more or less. On both sides of the valley the country is a plain, for the most part heavily timbered, especially on the slopes. At the fazenda, the soils on the slopes appear to be drift, and are exceedingly fertile. The coffee on the slopes was vigorous and healthy, without blight, and it was very heavily fruited with a berry of excellent aroma. I do not remember having seen anywhere better coffee-trees than those at Capitão Grande. The soil yields mandioca and the other products proper to the climate in abundance. In the narrow valley the climate is very hot, trying, and feverish, but the clearing away of the forests will change its character. On the Braço do Norte clays are occasionally exposed in the bluffs, together with a kind of coarse white sandstone such as would result from the hardening of the sandy clays of the tertiary. There are but few inhabitants on this branch of the river. Of the Braço do Sul I know nothing. The main river is a respectable little stream, one hundred and fifty to two hundred feet in width above São Matheos, and bordered by alluvial lands, which lie in the bottom of a valley cut through the tertiary beds, and vary much in width. The valley is bordered by bluffs, in which sandstones and clays of ordinary type are occasionally seen. The tertiary slopes are very fertile and largely cultivated, as are, to a considerable extent, the lands on the upper plain,* so that the country, with its numerous fazendas and cocoa-palm trees, wears a very pleasant aspect.

The cocoa palm (*Cocos nucifera*) is, according to Wal-

* The soil of the upper plain, when uncleared, usually bears a heavy forest growth. It is more or less sandy and clayey, and I should suppose would be difficult to work; but I am told that such is not the case. It is frequently covered with a thin layer of loose sand on the surface, owing to the washing away of the clayey portion by the surface water. It is especially fitted for the cultivation of mandioca, which flourishes well in a sandy soil, as well as for cotton, which also does well in a soil of that kind. Cane is planted to some extent. When cleared, these lands make excellent grazing grounds, but grass must be planted. On the Sertão flourish the *Nayá, Timburé,* and *Murí* palms. Northwest of São Matheos, on the north side of the river, are some extensive barren and swampy plains covered by a vegetation composed of shrubs, among which I observed a species of Vaccinium in fruit. The soil of the river-borders is Massapé, and very fertile; but these are generally very low and liable to be flooded. Near São Matheos are extensive swamps, which sometimes breed very malignant fevers.

lace and other good authorities, not a native of America, and the early explorers of Brazil do not speak of it. Its home appears to have been in the East Indies, but it has been introduced into America, probably by natural means, the impervious shell and thick husk fitting the fruit to bear long transportation by ocean currents. It is now found everywhere on tropical coasts, but in the East Indies it is more largely cultivated than elsewhere, sometimes even forming forests. It is a very valuable tree, furnishing food, oil, fibre for cordage, arrack, &c., &c.*

Von Tschudi says that Villa Viçosa is the southern limit of this palm in Brazil. This is not quite correct. South of São Matheos these trees are but rarely seen, but they, however, grow at Rio de Janeiro, and great numbers may be seen on the shores of Paquetá, while there are a few specimens on the islands of Cobras and Villeganhão. At the Barra do São Matheos they grow very well, and at the Fazenda do Campo Redondo, at a distance of several miles from the sea, there is a fine grove on the edge of a bluff. This palm, as has been frequently remarked, appears to flourish best on the sands of the sea-shore, and it is very rarely seen at any great distance inland.

A specimen was seen by Burton on the São Francisco at Brejo do Salgado, three hundred and fifty miles from the sea; and, according to the same observer, the cocoa palm occurs in occasional patches thence down the river. Burton speaks of a large grove twenty-eight miles to the southwest of Joazeiro, and he suggests that the saline character of the soil may make up for the want of sea air.

* See Hamilton, Description of Hindostan, Vol. II. p. 210. Meyen, Botanical Geography, p. 331. Transactions of Royal Asiatic Society of Great Britain, Vol. I. p. 546. Bennett's Travels in New South Wales, &c., Vol. II., Appendix, p. 295.

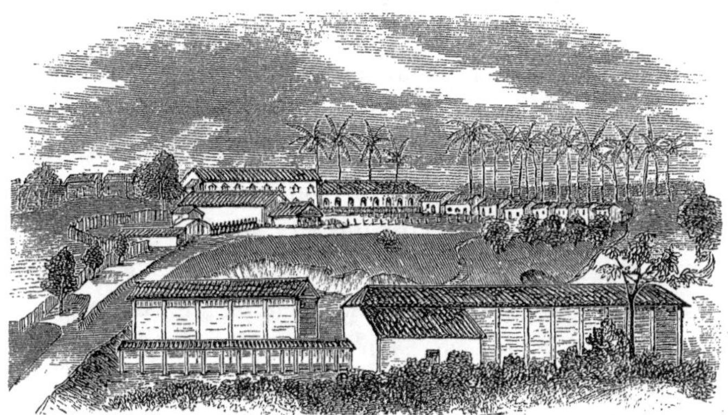

FAZENDA OF CAMPO REDONDO, SÃO MATHEOS.

Humboldt found this palm growing in the interior of Venezuela, and the Duke Paul von Würtemburg reports it as flourishing, at a distance from the sea, in the island of Cuba. It occurs, also, sometimes in the interior of India.

In Brazil I have seen the cocoa palm as far south as latitude 23° at Rio. It is grown abundantly in the provinces of Bahia, Sergipe, Alagôas, Pernambuco, and along the coast northward, but near the equator it is rarely seen. The northern limit appears to lie within latitude 28° north of the equator. Mayen says, on the authority of Humboldt, that in Venezuela it ascends to the height of 700 toises above the plain, or in the vicinity of 4,500 feet above the sea.

The products of the São Matheos region are principally mandioca meal (*farinha*), together with some sugar, cacáo, beans, &c., &c. The cacáo is principally planted on the lower grounds, where it yields well. The influence of the tide is felt at a distance of some twenty-five miles, or more, above the mouth of the river, and navigation for small vessels ends a few miles above the town. The height of the bluffs decreases on nearing the coast, and, at the city, they are not more than eighty to one hundred feet. São Matheos, like Nova Almeida and other Brazilian towns, is built partly on the edge of the bluff, partly at its foot, by the river-side, at the point where the bluffs leave the river and run southward to the Doce. It has about two thousand inhabitants, and is a place of some considerable importance, being reached by the coasting steamers and little schooners. The trade is principally in farinha, feijão, &c. On the opposite side of the river the bluffs run eastward for a few miles, and then trend off northward to Itahúnas, decreasing in height as they near the sea. This decrease in height, I am inclined to ascribe, in part, to the slope of the old sea-bottom, on which the tertiary beds were deposited.

Below the town the river grows wider, shallower, and is obstructed by sand-banks. The banks are low, with only occasional plantations, and are, for the most part, covered by a dense forest; but, lying between the tertiary lands and the sea on the northern side, at least, are quite extensive sandy plains with their accompanying sparse vegetation. The river-banks are generally muddy, and the aninga and other brackish-water swamp plants grow abundantly on them. Salt-water crabs and fish ascend to the town. The city of São Matheos cannot be more than seven or eight miles in a direct line from the sea, and is very incorrectly located on the maps. According to my observation, it lies about west-southwest of the barra. The river, descending from the city, makes a large bend southward, receiving the Mariricú; it then runs north a few degrees east, and, just above the barra bends round and enters the sea from the northwest. The maps show a Rio São Domingos entering the São Matheos, just above the villa, but I find no note of it in my journal. The Villa da Barra do São Matheos is situated on a ridge of sand, only a stone's throw from the sea, but it is nearly two miles from the mouth, because the river flows southward, behind the beach ridge for that distance, before escaping into the sea. On both sides of the river, but particularly on the southern side, there are extensive mangrove swamps, which furnish very interesting collecting grounds for the naturalist, for they are especially rich in crustaceans. Among the mangroves two genera are represented,—*Laguncularia* and *Avicennia.* The common red mangrove of the Brazilians is the *Laguncularia racemosa* Gaert., of which the wood is used for burning, and the leaf and bark for tanning. The second species is the *Avicennia tomentosa* Linné, used for the same purposes.

One league, or thereabouts, north of the Barra do São Matheos, is the mouth of the Rio Itahúnas, or Guaxindiba. This little stream, a black-water river, as has already been noted by Max. zu Neu-Wied,* rises in the Sertão, and reaches the coast at a point some three leagues north of the mouth; but a beach ridge, only a few hundred feet in width, prevents its reaching the sea, and causes it to bend abruptly to the southward, so that it flows along, through the low grounds,

* Prinz Max. zu Neu Wied, *Reise nach Brasilien,* Vol. I. p. 226, says that the *Peixeboi,* or manati, *Manatus americanus,* occurs in a large grass-grown lake south of the São Matheos and communicating with that river, and that it used to be captured by the inhabitants for the sake of its blubber, flesh, and ear-bones; and in the second volume of his *Beitrage zur Naturgeschichte von Brasilien,* page 602, he states that it occurred plentifully in the river and lake aforesaid, or in the vicinity of the Quartel Juparanãa, and was formerly frequently taken by the inhabitants. He adds, however, that he had never seen a specimen. While at São Matheos, and during a stay of three months in that part of the coast, I never heard the animal spoken of, and I do not believe that it is now to be found there. (See note on page 75.)

just behind the beach ridge, for two or three leagues. It is rather narrow, but very deep, and vessels of considerable size enter it. It is navigable for some distance into the interior, and furnishes an excellent water highway. Near its mouth a riacho * enters it from the southward, and this affords navigation nearly to the Barra do São Matheos. A canal is being opened to join this riacho with the São Matheos. The low lands between the bluffs and the coast are for the most part timbered; but they furnish some excellent pasturage. On the Itahúnas, a strip of sand margins the bluffs and appears to be the continuation of the sand plain on the São Matheos. The bluffs bordering the plain are only thirty or forty feet in height, but they grow higher as we go westward, and, at a distance of two leagues from Itahúnas, they are at least one hundred feet high. They have, as elsewhere, steep slopes, and, in part at least, are covered by a very fertile soil. On the Fazenda of Senhor Olindo Gomes dos Santos I saw most excellent crops, especially of mandioca and coffee, which were almost, if not quite, equal to anything on the São Matheos.

Off the shore, in front of the town of Itahúnas, are some ledges of rock washed by the waves. I could not examine them, but I thought them to be sandstones. The bluffs sweep round to the shore, just to the north of a little riacho, called Doce, if I mistake not, and a short distance to the south of the first point north of Itahúnas; they form along the shore, for a distance of several miles, a line of low cliffs, which, from their whiteness, have received the name of Os Lençôes. These cliffs are nowhere more than thirty to forty feet in height, this being the thickness of the formation exposed above the sea. The most prominent feature

* This term is applied to small estuaries. It means literally a small river.

in this line of cliffs, is a hard gray grit with an argillaceous cement, and in some places twenty feet in thickness. This rock is composed of materials precisely like those of the sandy clays of Lagôa Juparanãa, and elsewhere; but the rock is so compact and hard as to be used as a building-stone at Itahúnas, although I believe that the stone actually used is derived from quarries near the village. Associated with this rock are white and red clays, and beds of the common, coarse, red, lava-like sandstone.* The cliffs end at, or in the immediate vicinity of, the Riacho das Ostras, a little, black-water stream, of whose course I know nothing, but the bluffs continue along the shore much further. Riacho Novo is another black-water stream, which, before emptying into the sea, runs for several miles in a deep channel just behind the beach ridge. The tide enters this channel, and it is navigable for canoes for a considerable distance. The whole country between Itahúnas and the Mucury is wooded, the lower and wet grounds very luxuriantly, the higher and drier less so, the latter often supporting only a thick growth of small trees and bushes. Dunes of sand occur on the shore near the Mucury, and in one locality there are exposed on the sand-beach the stumps of trees, rooted in their soil and upright, something which points to an encroachment on the shore by the sea. I have observed dead trees, apparently mangroves, standing in the water off the shore below Caravellas. Has there been a recent depression of this part of the coast?

* I was delayed one morning at a little settlement just south of the Riacho das Ostras, and I undertook a careful examination of the beds for fossils, but I saw not the slightest trace of organic remains in them. I am sorry that I was unable to make notes on the arrangement of the materials in the cliff; but the place was like a furnace, and I was driven away by the sun, blinded and exhausted.

CHAPTER III.

PROVINCE OF MINAS GERAES. — THE MUCURY AND JEQUITINHONHA BASINS.

The Basin of the Mucury. — Porto Alegre. — Description of the River below Santa Clara. — Luxuriance of Forest Vegetation. — Santa Clara. — Minas Geraes a Land-locked Province. — Want of Roads. — The Philadelphia Road and the Mucury Colonies. — Difference in Topography and Soils between the Tertiary and Gneiss Lands west of Santa Clara. — Urucú, its Dutch Colony, Soils, Climate, &c. — Philadelphia and its German Colonies. — Great Fertility of the Mucury Basin. — Character of Country between Philadelphia and the Head-waters of the Mucury. — The Basin of the Jequitinhonha. — The Rio Pardo. — General Geological Structure of the Jequitinhonha-Pardo Basin. — The Head-waters of the Setubal, their Geological Features and Catinga Forests. — Geological Excursion from the Fazenda de Santa Barbara to Alto dos Bois. — Difficulty of geologizing in Brazil. — The Brazilian Campos. — The Chapadas between Itinga and Calháo. — The great Calháo-Arassuahy Valley. — Magnificent View over the Valley from the Chapada at Agua da Nova. — Calháo and the Geology of its Vicinity. — Description of the Country between Calháo and Sucuriú. — The Chapadas. — Minas Novas, its Geology, Gold-Mines, &c. — Occurrence of Gold in Drift. — Gold-Mines of the Arraial da Chapada; their former Richness; not yet worked out. — Decomposition of Clay Slates in the Minas Novas Region. — The Rio Arassuahy. — The Rio Jequitinhonha from its Confluence with the Arassuahy to the Sea described; its Geology, Vegetation, Commerce. — The Salto Grande.

THE Rio Mucury takes its rise in the province of Minas, about 150 miles west of Villa Viçosa, among the high lands which form the water-shed bounding the basin of the Jequitinhonha on the east. Its course for the first seventy-five miles, curves excepted, is approximately east-northeast, when it meets the Rio Preto, a stream rising in the same water-shed some forty miles to the north-northeast. This last river flows with a course almost parallel to the Mucury up to about the same meridian, when it receives the little Rio das Americanas coming from the north. It then bends abruptly to the south, to join the Mucury, the combined waters of the two streams flowing still southward until they reach the Rio Todos os Santos, a stream rising at a point south of the Mucury, and forty miles south of the source of the Rio Preto, and flowing also parallel with the Mucury. The Mucury soon bends gradually round to the east, and, making several broad curves, runs with a general southeast direction to Santa Clara, on the boundary between the provinces of Minas, Bahia, and Espirito Santo, a distance of thirty miles, in a straight line, but much more, following the course of the river. The Rio Urucú is another river flowing in a valley parallel with the Todos os Santos and Upper Mucury, but emptying into the Mucury proper on the right bank, some fifteen miles in a straight line below the mouth of the Todos os Santos. On the opposite or northern side the Mucury receives, about eight or ten miles farther down, the Rio Pampão, which comes from the north, and has apparently a course of about sixty miles; but its upper waters have never been explored.* Above Santa Clara the Mucury and its tributaries are swift and obstructed by rapids, but from Santa Clara the main river runs with a very tortuous course, as a *rio d'areia*, to the sea, into which it empties in latitude 18° 6′ S., and ten and a half miles south of the parallel of Santa Clara. Its waters come principally from the province of Minas, where it drains an irregular triangular

* Some of the lands of the Mucury company were situated on the Pampão. The only one who has visited them is my friend Mr. George Schieber, one of the surveying corps of the Mucury.

area, bounded as follows: by a line running due north along the Serra dos Aymorés, eighty miles, another line, 135 miles long, running a little east of northeast, and another on the south about 110 miles long, and running east-west along the water-shed dividing the basin of the Mucury from those of the Doce and São Matheos. This triangle is a right-angled one. From its southeast angle the area drained forms an irregular strip some six or more miles wide, along the middle of which flows the river. In the study of this river system, several points strike one as interesting. The parallelism of the Rios Preto, Mucury, Todos os Santos, and Urucú shows that they flow in parallel valleys, which are evidently determined by the trend of the foldings into which the gneiss is thrown. The coincidence in direction between the courses of the Rio das Americanas and that of the Rio Preto above the Mucury, and the latter to its confluence with the Todos os Santos, points to a valley running north-south, into which the rivers Preto, Todos os Santos, and Mucury empty as side tributaries. The Pampão flows in a similar valley. It is interesting to compare the basin of the Mucury with that of the Doce. In each, the greater part of the region drained lies west of the coast cordilheira, while east of the cordilheira the area dwindles down to a narrow strip bordering the river on each side. Between these two rivers are intercalated the São Matheos, Itahúnas, &c.

With these introductory remarks on the hydrography of the Mucury basin I propose to give a somewhat detailed description of the river basin in ascending the river to Santa Clara, and then traversing the basin thence westward to the head of the Mucury Pequeno, along the line of the Santa Clara, Philadelphia, and Minas road. The Rio Mucury, contrary to the general rule, enters the sea obliquely from the south. Its mouth is narrow and difficult to enter, being much less practicable than the São Matheos. On the left bank at the mouth, built on a sand-bank, and surrounded by mangrove swamps, is the miserable little village to which the ridiculous misnomer of Porto Alegre has been applied. The place is of importance only as the port of the Mucury district, which exports coffee, cotton, rosewood, &c. From Porto Alegre large quantities of salt are sent into the interior, where it finds its way into the very heart of the province of Minas Geraes. Dry goods are also imported through the same channel. The town is one of the most wretched I saw in Brazil. Its inhabitants are principally of Tupi origin. From Porto Alegre to Santa Clara, a distance of forty-five miles in a straight line, but at least fifty-five by the river, the stream is rather shallow, very tortuous, narrow, and affording navigation for *pranchas*, canoes, and a very small steamer; but the water is, for a considerable part of the year, so shallow, and the river is so obstructed by sand-banks, that it is navigable constantly only for canoes.* Just above Porto Alegre the tertiary bluffs appear, and sandstones and clays are exposed in them. At first these bluffs are not very high, in some places measuring only from eighty to one hundred feet; but, ascending the river, their elevation increases, and at Santa Clara† they are some 330 feet above sea level. At Santa Clara the whole thickness of the formation is not displayed in these bluffs; for, back from them the tertiary lands reach, in some places, an altitude of

* In the latter part of January, 1866, Mr. Copeland and I ascended the river in the little steamer. The water was so shallow that we were constantly running aground. A few leagues below Santa Clara we stuck fast, and the rest of the distance we had to make in the prancha we had been towing.

† Santa Clara itself is 327 palmos above sea level, which would make the level of the river below the rapids only a few palmos lower.

360 feet above sea level, according to actual measurement. The river valley is very narrow, and the sides have a steep slope. The alluvial lands are small in extent. Both they and the tertiary lands are covered by a very heavy and luxuriant forest, and the scenery on the river is of surpassing beauty; for here, as on the Doce, the trees crowd down to the water's edge, forming a dense wall of verdure. A host of species of beautiful-leaved and bright-flowing climbing plants hang a dense curtain from tree to tree, and sometimes depend in folds from the outstretched branches, like the drapery from the arm of an antique. The gneiss first makes its appearance below the tertiary rocks at a place called Dous Irmãos, some eight leagues, more or less, below Santa Clara. It has a northward dip. At Santa Clara navigation is made impossible by a series of rapids, and thence into the province of Minas the Mucury is rapid, and has many falls. At Santa Clara, which is only a collection of a few dwellings and warehouses, built on narrow alluvial flats on the right bank of the river, the valley is very narrow, and has steep banks. The rocks in the river are gneiss,* which is much veined with granite. This locality is very unhealthy, owing to the narrowness of the valley, and the great heat of the day, — which often gives way to damp fogs by night, — to the bad character of the river water, and to swamps in the vicinity on the top of the chapada.†

* The *Diccionario Geografico* says that iron ore exists within the district of Porto Alegre, but does not indicate the locality. Von Tschudi, *Reisen durch Süd-Amer.*, Vol. II. p. 338, says that he has found chrysolites in the river-sand at the Barra.

† I was attacked by fever here, and only escaped by removing to the high grounds of Minas. I owe a deep debt of gratitude to Signora Gazzinelli, who took a mother's care of me, and also to my faithful and generous companion. Mr. Schieber, whose kindness I never can forget.

Any one glancing at a map of Brazil will see that the rich and populous province of Minas Geraes is land-locked, and separated from the sea by serras and forests. The Serra da Mantiqueira and the Serra do Mar skirt it on the south, and on the east the coast mountains, collectively known under the name of the Serra dos Aymorés, clothed with forest, form its eastern boundary line. None of its rivers are navigable to the sea, though some of them are, for scores of miles, navigable in their upper courses; but all of them are obstructed by heavy falls or rapids in their descent from the plateau to the coast plains. Many of them, as is the case with the Doce, Mucury, Jequitinhonha, and São Francisco, are navigable in their lower courses, in some cases, even up to the confines of the province. From Rio a railroad to the foot of the Serra da Estrella, with a magnificent wagon-road, which, crossing this Serra and the Serra da Mantiqueira, connects with the Barbacena district, about 152 miles in a direct line from Rio. An excellent railroad crosses the Serra do Mar to the northwest of Rio, and enters the valley of the Parahyba, down which it extends many miles. It is to be extended northward into the province of Minas. But the greater part of Minas is destitute of wagon-roads, and the traffic is almost wholly carried on on the backs of mules. The coast forest- and mountain-belt bounding the province is almost entirely uninhabited and impassable. A very small quantity of salt and other articles of commerce finds its way in canoes up the river Doce, as already stated, and a larger quantity by the same means enters the province by the Jequitinhonha; but commerce with the sea-coast is carried on with great difficulty and at much expense. Thus, a bag of salt that costs two *milreis* at the sea-coast is worth eight or even eleven *milreis* by the time it has reached the interior of Minas. The Senator Theophilo Benedicto Ottoni,* some twenty-five years ago, conceived the project of opening a good wagon-road from Santa Clara to Minas Novas, through the broad forest region of the Mucury, and of colonizing that region. A company was organized for the purpose of accomplishing this object. Through agents in Europe, a considerable number of German, French, and Swiss colonists were secured, and two colonies were founded, one on the Rio Urucú, the other at Philadelphia, on the Todos os Santos, the Mucury colonies being founded in the year 1858. An excellent wagon-road, now out of repair, was constructed from Santa Clara to Philadelphia, and a mule-path was laid out to Minas Novas; but the colonists appear to have been, to a very large extent, of very poor quality. Through the misrepresentation of the agents of the company in Europe, the colonists were led to expect to find themselves, on their arrival, put into the possession of a house and cultivated farm. It was a bitter disappointment to them to be sent into the virgin forest. Nevertheless, extensive clearings were made, and the villages of Urucú and Philadelphia were built; but political opposition from the enemies of Ottoni was added to the difficulties the colonists had to contend with. The company failed; the colonists, disappointed in their hopes, deserted by wholesale,† and to-day the Mucury is dragging out a miserable existence, Philadelphia is in decay, and the road is out of repair. It was not because the lands of the Mucury

* To the Senator my companion and I are indebted for letters of introduction, which secured for us friends and assistance everywhere along our whole journey.

† I should state here that the worthless colonists were the first to leave. Those that I had the opportunity of meeting in the Mucury seemed to me to be of a good, industrious class, but they were crippled by the failure of the company.

were not fertile. They are exceedingly rich, while the climate is healthy and agreeable. It is not that the project of opening a road through the Mucury to Minas was unwisely planned, but it is owing to bad management on the part of the company, to the slanders of enemies, and to the bad character of a large part of the colonists themselves, that the enterprise has proved a failure.*

The Minas road, on leaving Santa Clara, runs for a few miles through a hilly region bordering the Mucury. This region is covered by the ordinary red drift soil, and is very fertile. Quite a number of German families still remain here. Beyond Barriado, where are a few settlers, one soon leaves the river valley and rises to the top of the tertiary chapada,† which, at a distance of eleven and three quarter leagues from Santa Clara, and near the Riacho das Pedras, is 1,226 feet above the level of the sea. It forms a plain like that of the Sertão ‡ below Santa Clara, and is well but not densely wooded. The chapada is covered by a layer of drift of a yellowish color, in which I saw, indiscriminately mingled with the clayey sand of which it is composed, rounded and angular fragments of quartz, sand-

* The story of the Mucury is a long and sad one, and I do not wish to enter into it here. Those who desire to read the history of the colony will find a very fair statement of the facts in Von Tschudi's *Reisen durch Süd-Amerika*, Vol. II. That of Dr. Ave Lallemant is prejudiced, unfair, and unreliable.

† *Chapada* means primarily a *plain*, but in Brazil the term is applied to elevated plains or small plateaus, usually consisting of horizontal deposits, and separated by deep valleys of erosion. The term chapadão is applied to chapadas of great extent, as the chapadão de Santa Maria in Minas.

‡ The term Sertão, plural Sertões, so often used in works on Brazil, simply means the interior of a country as opposed to the coast. It is applied, for instance, to the lands in the vicinity of the city of São Matheos. The word appears to have a somewhat indefinite signification in Brazil. The inhabitants of the Sertão are called Sertanejos.

stone, and gneiss. This soil, like that of the plains below Santa Clara, is weak, and much inferior to the gneiss soils. At Riacho das Pedras, a little stream flowing into the Mucury, the tertiary lands are left, and the road reaches a rolling gneiss country. The rock, wherever I saw it, was very coarse-grained and homogeneous, decomposed on the surface, and covered with drift clay, which is usually very fine in texture, and *very* red from the large percentage of ferric oxide. The hills were low and rounded, with a topography like that of the coffee region of the Parahyba do Sul at the Barra do Pirahy. Indeed, the two regions are precisely identical in soil, general topography, and climate. The country is covered by a dense virgin forest, far more luxuriant than that which clothes the tertiary plains. The country continues with the same general character to Urucú, where it becomes diversified by abrupt gneiss hills, many of which are bare and precipitous, and give to the scenery a very romantic and pleasing air. The soils of the Urucú are extremely fertile, and yield abundantly coffee, cotton, sugar-cane, mandioca, rice, etc. The climate is warm, but not so hot as on the coast, and a sea-breeze cools the air in the latter part of the day and evening. The climate appears to be healthy; and that is the universal testimony of even the discontented settlers. The hills of Urucú are all isolated masses, and form no well-defined mountain range, though they appear to be the remains of a range running about east-northeast, crossing the Urucú. Westward of the colony the country rises steadily in altitude. At a distance of eight leagues from Philadelphia the road crosses a pass in the Morro do Kupan at an elevation of 1,800 palmos. The Morro itself must be at least 3,000 feet in elevation above the level of the sea.

Philadelphia is a small village situated on the left bank of the Todos os Santos, about forty miles above its confluence with the Mucury, and twenty-eight and a half leagues west of Santa Clara. The Todos os Santos is, like the rest of the rivers of the Mucury basin, only a respectable brook, and of no especial importance in itself. Within a few miles of Philadelphia there empty into the Todos os Santos several little streams, among which are the Rios S. Jacintho, S. Antonio, and S. Benedicto, which flow through fertile, cultivated valleys, and are settled by German and Brazilian colonists. The ground on which the village stands is, according to the measurement of Herr Schlobach, engineer of the Mucury, 1,918 feet above the sea. Many of the neighboring hills are 300 to 400 feet high, so that the general elevation of the country would be considerably above 2,000 feet. So far as the quality of the soils in the vicinity is concerned, I can only reiterate what I have said in speaking of Urucú, and repeat my comparison between them and the soils of the coffee regions of the Rio Parahyba do Sul. In one word, I may say that the whole country, from the Riacho das Pedras to the head-waters of the Mucury, forms one of the most extensive and uniformly fertile agricultural regions in Brazil south of the Amazonas, and I cannot help expressing my firm belief that nature having so abundantly blessed the Mucury, a not far distant day will see it teeming with inhabitants, and the highway of a commerce with the interior of Minas.* The road from Sta. Clara to Philadelphia is well

* For an interesting and detailed description of Philadelphia and vicinity, *vide* Tschudi's *Reisen durch Süd-Amerika,* Vol. II. His sketch of the village is wretched, the hills to the south appearing like an Alpine mountain range. I cannot speak too strongly of the Mucury as an agricultural region, and I would call the attention of emigrants to it as one of the most fertile and healthy tracts I have seen anywhere in Brazil.

laid out with a very good grade, and with proper repair, might be made an excellent carriage-road. The soil is, however, very clayey, and the passage of the heavy ox-carts in wet weather, not to speak of the gullying by the rains, have cut it up fearfully. The bridges were in bad repair in 1866. The road from Philadelphia to Alahú is nothing but a miserable mule-path, badly laid out, and obstructed by bushes and fallen trees, and in wet weather most abominably muddy. Just before reaching Açude it passes directly over the top of the highest point, apparently for the purpose of giving the traveller a view of the surrounding country.

Westward of Philadelphia the country is more hilly. About one league from Philadelphia the path crosses a high hill, from which one has a magnificent view over the low swelling hills of the vicinity of Philadelphia, with the rugged mountains of Urucú in the background. A short distance west of the mill of Senhor José Maria, the gneiss becomes very micaceous, passing into mica slate, and is very full of quartz veins. In the latter occur large crystals of black tourmaline. Crossing the head-waters of the Mucury, near Poté, the country in the vicinity of Açude is much more hilly than usual, and many of the hills are of considerable height. The whole country is still most luxuriantly forest-clothed, and the soils are extraordinarily fertile, and in some places almost black. About twenty-seven to twenty-eight miles west, a few degrees north from Philadelphia, the water-shed dividing the basins of the Mucury and Jequitinhonha is passed, and one descends into the valley of the Rio Setubal. For the last league or so bordering the valley of the Setubal, the rock is chiefly mica slate, with much quartz in veins and layers. The soil is redder than usual, full of little flakes of mica, and boulders and fragments of quartz, angular and rounded, are found abundantly in it. Near the post, marking thirteen leagues from Philadelphia, the mica slate appeared to have a strike east-west. Dip vertical.

The Jequitinhonha, one of the most important rivers of Minas Geraes, takes its rise in the knot of the Cordilheira do Espinhaço, in the Serra Frio, three leagues west of the town of Serro, and about the same distance south-southwest of the Peak of Itambé, whence come the waters of two or more of its little tributaries.

The area drained by it forms an irregular triangle, of which one side, from its head-waters to Belmonte, is about 320 miles. An almost continuous range of Serras runs with a zigzag course along this line to the confines of the province. On the west it is bounded by the Serra do Espinhaço, the water-shed lying west of the Serra do Grão Mogor. This side of the triangle, which runs approximately north-northeast, is 165 miles in length. The remaining side, which marks a water-shed determined by a series of elevated plains or chapadas, and which runs almost east-west, is 210 miles long. The Jequitinhonha at first flows with a general northeast course for about 130 miles, when it receives from the northwest a small river called Itacambirussú, which rises in the Serra d'Itacambira, in the southern part of an oblong region west of the Serra do Grão Mogor, formed by a range of serras or highlands which leave the Serra do Grão Mogor, and bowing out westward join the Grão Mogor range again eighty miles to the north. The centre of this region appears to be flat, and is diversified with a large number of little lakes, which discharge their waters into the Itacambirussú. This river crosses the Grão Mogor range between the Serra do Grão Mogor proper and the Serra

Sobrado, and reaches the valley of the Jequitinhonha. It then runs with a southeast course for a few leagues, and empties into the river of that name. The Jequitinhonha then changes its course towards the east, and, some eight leagues farther down, receives the Rio Vacaria, a small stream which has its source in a number of lakes in the northern third of the serra-enclosed region west of Grão Mogor. After this it changes its course to the southeast, and in lat. 17° S., long. 1° 30′ E. of Rio unites with the Rio Arassuahy. The Arassuahy is a large stream which rises in the serras a few miles northeast of the Peak of Itambé, and flows parallel with, and on an average of fifteen to twenty miles, southeast of the Jequitinhonha. Its principal affluents are the Rio Soledade on the left, and the Itamarandiba, Fanado, Capivary, Agua Suja, Setubal, Gravatá and Calháo, which flow into it from the south, or right, in the above descending order, all, with the exception of the Capivary and Agua Suja, having their sources in the high lands separating the Jequitinhonha basin from that of the Doce and Mucury.

The Jequitinhonha, after having been increased by the waters of the Arassuahy, continues its course a few degrees east of northeast to the sea. It is obstructed by many dangerous rapids and cascades, and on the boundary line, in the very extreme northeast corner of Minas, there is a magnificent series of falls, which, in the aggregate, must have an altitude of 300 feet. Eight leagues farther down, at Caxoeirinha, it leaves the hills, and, reaching the coast tertiary plains, flows, a broad, beautiful stream, to the sea. Above Caxoeirinha, it is navigable of course only for canoes. The Rio Pardo is so closely connected with the Jequitinhonha in the general topographical features of the country through which it flows, that, before describing the topography and geology of the Jequitinhonha basin in detail, I will first give a sketch of its hydrography. This river, a much smaller stream than the Jequitinhonha, takes its rise in the Serra das Almas, to the north of the head-waters of the Vacaria, and flows with a course almost west-east, emptying into the sea a few miles north of the mouth of the Jequitinhonha, and receiving by a side canal, as will be hereafter described, some of the waters of that stream just before it reaches the sea. Its basin is triangular in shape, long from west to east, but narrow from northwest-south. It is bounded on the south by the water-shed of the Jequitinhonha, which runs west-east. The south side of the triangle coincides with this, and is 210 miles long. The west side, running along the Serra das Almas, is about sixty miles long, while the remaining side is about 195 miles in length. Gerber gives the area of the basin of the Pardo as 420 square leagues, and that of the Jequitinhonha as 2,200.* Hydrographically, above their lower courses the two rivers form separate systems, but topographically and geologically they are very closely united. The united basins are essentially eozoic and palæozoic. Gneiss, mica slate, sienite, clay slate, quartzite, and limestones form the bounding, mountain ranges, and the bottom rock of the region. Through the kindness of my friend, Dr. Anto. de Lacerda, of Bahia, I have in my possession a lithographed section across the country, from the Serra Congonha across the Grão Mogor, extending into the valley of the Jequitinhonha, a section constructed by the late Dr. Virgilio Helmreichen. According to this section, the serras of Congonha and Grão Mogor are composed of metamorphic slates, while the intervening country and the valley of the Jequitin-

* *Nocões Geographicas, &c., da Prov. de Minas Geraes,* por Henrique Gerber, p. 9, 1863.

honha is composed of primitive rock. I have never visited the Serra do Grão Mogor, and can therefore say nothing of it from personal observation.* At Calháo intelligent persons informed me that it is composed of slates. The serra is distinctly visible from near Minas Novas, though distant some thirty miles, so that its height may be inferred. The outlines of the hills are entirely different from those of the gneiss serras of the coast. Gold occurs in this serra, together with the ores of other metals, such as iron, which last is mined and smelted at a locality called Tropinha, two leagues to the south of the town of Grão Mogor. The region embraced between Minas Novas and Calháo is, according to my own observations, composed of clay slates, and this group of rocks undoubtedly extends considerably to the west and southwest of Minas Novas. From near the mouth of the Arassuahy to a little below Caxoeirinha the rocks are gneiss, mica slates, and the like. All these rocks have been folded, metamorphosed, and denuded.

During the tertiary, as I shall further on attempt to show, the plateau of Brazil was sunk so that the waters rose to a height of more than 3,000 feet above their present level, and flooded the great river basins of the whole country, this submergence being of almost continental extent. In the basins of the Jequitinhonha and Pardo, a great thickness of more or less arenaceous clays, sandstones, &c. was deposited, filling up the valley to a height in some places of fully 1,000 feet, converting it into an immense plain, whose level above the sea must be on an average quite 3,000 feet.

* Spix and Martius visited the Serra da Grão Mogor, which they describe as being only about 4,300 feet high. The prevailing formation of this region is quartzose slate (quarz-schiefer). Boulders of white quartz (sometimes fibrous(?)) are abundantly scattered over the surface, and contain asbestus. Gold and diamonds occur here.

These deposits I have called tertiary, because along the whole coast they are undisturbed, nowhere participating in the disturbance of the cretaceous, and because the drift sheet extends over them. I believe them not to be drift, because they were denuded by river action anterior to the formation of the drift sheet, which descends their slopes, and extends over the slate and gneiss hills left bare. Similar deposits were at the same time laid down in the valleys of the São Francisco, Parana, Parahyba do Sul. And indeed over the whole plateau to the westward, as we shall see further on. All these are older than the coast tertiaries.

In April, 1866, I entered the Jequitinhonha valley, on the Setubal, and after making a *détour* to Alto dos Bois, crossed the country to Calháo, from which place I found it practicable to visit Minas Novas, after which I returned to Calháo, and descended the Jequitinhonha to the sea. I propose now to give the results of my explorations of this region, following very nearly my line of travel.*

The tertiary clays are denuded away from the region of the head-waters of the Setubal, and the wider valley of this river is scooped out of these rocks, the river-bed being the solid gneiss or slate, or excavated in alluvial deposits laid down by the river. The Setubinho is a little river which flows from the southwest in a valley bounded, on the one side, by the hills of the water-shed between the Mucury and Jequitinhonha basins, and on the other by the tertiary plains, though the slopes on both sides are of the old metamorphic rocks, the tertiary beds merely capping the hills on

* The months of February and March, 1866, were exceedingly rainy over the Mucury region, and so was the month of April, which I spent in the basin of the Jequitinhonha. During this time I was obliged to travel over the worst possible roads in almost constant rains, so that my geological studies were made under a great disadvantage.

the northwest side. The slopes toward the Setubal, Seturna, and Setubinho are all very steep, and for a large part bare and excessively stony, the soil being full of rounded and angular fragments of quartz often of large size. Over large areas it is very barren, the only vegetation consisting of low, gnarly branched, sparsely scattered shrubs and trees.

One observes immediately on entering the valley of the Setubal from the Mucury that the forest thins out and disappears from the hillsides, though it extends down the wet valleys and over such areas as may have rich soils, but even there the forest has not the same luxuriance it had in the valleys of the Mucury, and there are many trees confined to each separate region. It is, however, on the open grounds that the change in the vegetation is most marked. On leaving the forest (*sahindo do matto* * as the Brazilians say) one of the first plants to attract one's attention is an arboraceous species of the order of Solanaceæ, called the *Boleiro.* This tree attains to a height of fifteen to twenty feet, and forms a conspicuous element in the landscape. Its leaves are light-green and curly, its flowers bluish-purple, and its fruit, which is of the size of a Baldwin apple, is edible ; but of the flora of the campos more anon.

The high and steep hill north of the Setubal is almost bare on the south and west ; but the northern side is covered by a stiff drift clay, and is clothed with a thick wood, densely filled with a luxuriant growth of a species of bamboo, the slender-stemmed *Taquara lisa* of the Brazilians. Thence to Corrego Grande the country is composed of highly micaceous and schistose gneiss, and is covered by a thick sheet of drift clay, in which are boulders and pebbles of

* The Brazilians speak of the plains as *fora,* outside, and of the forest as *dentro,* or inside.

quartz, and the country is wooded. At the Fazendas of Santa Barbara and Santo Antonio, and in the neighborhood, this forms a rich soil, which is very productive. Maize is largely cultivated in this region, and takes the place of mandioca for the making of farinha. Wheat grows well here, and I saw some most excellent sheaves at a farm-house near the Setubal, but the farmers complain that it has to be planted in clumps like rice, and weeded, which is very troublesome. At the Fazenda of Santa Barbara the country bordering the Rio Setubal is very hilly. The prevailing rock is mica slate or schistose gneiss, with a general strike of N. 80° E., and northward dip. I have recorded no southward dips.

Wishing to ascertain the character of the chapada west of the Setubal, Mr. Copeland and I made an excursion thither from Santa Barbara. The account of the journey I transcribe with few changes from my note-book, in order not only to give an idea of the country, but of the disadvantage under which the geologist labors in exploring in the rainy season.

For the last two months the rain had been constant, and it was still raining when we reached Santa Barbara. Our time was very limited, but to leave the Setubal without seeing the topography of the plains was not to be thought of, so, on the last day of March, we set out on muleback, and unencumbered by baggage, for Alto dos Bois, a point described as being so elevated as to overlook the plains, and enable one, if the weather were clear, to see across the valley of the Jequitinhonha, and discern the mountains of Grão Mogor. It was raining heavily. We crossed the Setubal by a bridge below Santa Barbara, finding the stream very much swollen and turbid, and the meadows bordering it inundated. We followed up the valley a short distance, passing through cornfields on the hillside, and crossing a high hill by a miserable path leading through a wood which was so tangled with bushes, *unha de gato*, *bamboos*, &c., that it was with difficulty that we could burst our way through it by main strength. We reached at last the valley of the little river Santa Rita. Thus far the country was of the same character as at Santa Barbara, and the surface was covered by the same drift paste and boulder deposit. For a distance of some two miles or more farther on, after a long and steep ascent, we reached the foot of the chapada, which presented a long, steep, even slope, as in the following ideal section.

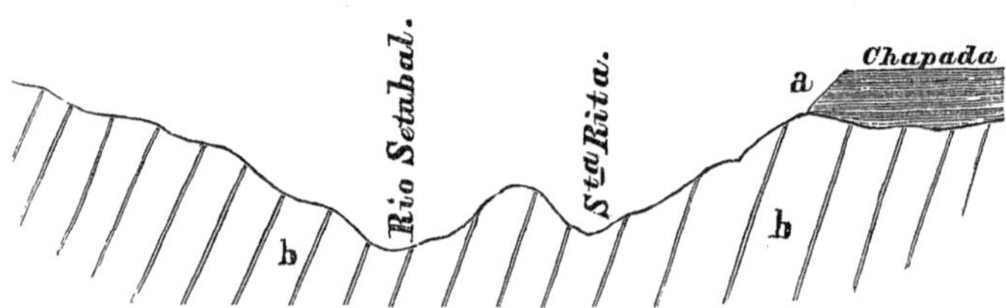

It was impossible to estimate satisfactorily the height of the chapada top from the Santa Rita valley, because of the wooded state of the country and the obscurity of the weather, but after having ascended and descended the same slope, I should judge that it is fully 800 feet, if not very much more. The lower part of the slope is covered with the ordinary drift paste, filled with boulders of quartz, gneiss, and mica slate. The quartz of this region is so crystallized as to break up into a coarse angular gravel. About half-way up I saw mica slate *in sitû* traversed by a thick quartz vein. Farther up the soil changed in character, and, when wet, was of an umber brown color, and quite free from stones. The road went straight up this slope, and being cut up into *pilões** by the feet of the mules, formed a sort of staircase of stiff, adhesive, and slippery clay, which made the ascent exceedingly wearisome. Arrived at the top of the chapada, we found ourselves on a wooded plain. The soil seemed very rich, and was of a dark brown color on the surface; but I observed that the material brought up from below by the *Saüba* ants was clayey and brick-red, like the ordinary drift paste of the gneiss regions. I saw a few quartz boulders lying on the surface. I looked carefully in the ascent for any cuttings which might enable me to ascertain the material of which the chapada was composed, but I saw none. The woods consist of large trees, but they are rather sparsely sown, and they belonged to the catinga class, losing their leaves in the dry season. There is an abundant undergrowth of bushes and Samambaia ferns (*Pteris caudata* (?). The beautiful Indaiá palm (*Attalea*) is very abundant here, and its plumy coronals add very much to the picturesqueness of the scenery. It was long after nightfall when we reached the open campos at As Trovoadas and crossed a very deep valley to the place called Batatal, where, jaded and wet, we spent the night. The next morning we visited the Alto, or the highest point of land in the vicinity. It had been raining; but, providentially, for a half-hour it cleared up, and we had an almost uninterrupted and most magnificent view of the country on every side.

The Alto is the name given to the highest point of the

* It is well known that mules travelling over a bad road follow in one another's footsteps, cutting up the road into a series of transverse muddy troughs, separated often by high ridges, over which the animal carefully steps.

swelling ridge between the valleys of the Fanado and Capivary. From this point the country descends with very long sweeping curves to the river valleys on both sides, the ridge itself growing gradually lower toward Minas Novas. The valley of the Capivary is very broad, and in profile rounded, not angular. The long and gentle curves of this landscape are very noteworthy, and are very different from those which result solely from water denudation or erosion. The Fanado valley is of the same character. East of the Capivary the land rises in a high ridge called the Caixão, which runs northward, presenting the same topographical features. This whole country is covered on the surface with red drift clay and pebbles, and this layer is twenty or more feet in depth. No solid rock is to be seen, but on the sharp descent into the valleys, and in certain gullies, especially at the foot of an isolated, dome-shaped mass which rises above the general level of the country at As Trovoadas, the rock, in a very decomposed state, is seen to be crystalline and metamorphic, and in beds highly inclined. At As Trovoadas, as nearly as I could make out, the strike was N. 60° W., and the dip 40° northward, but the observation was taken from decomposed rock, and may not be very reliable. The rock appears to be composed of feldspar and mica, with quartz in rounded grains, but when decomposed it is red, very soft, and crumbling. The soil in the gully, where the above observation was taken, is full of fragments of very limpid quartz and crystals of kyanite and black tourmaline.* This rock must be very deeply

* The *Diccionario Geografico* says (Vol. II. p. 98), that antimony occurs at Alto dos Bois, and I heard many persons on the Setubal say that gold had been found in the ridge separating the Setubal from the Capivary. Saint Hilaire also says that antimony is found at the Alto. But the *antimonio* of the Brazilians is only a pyrites.

decomposed, for the freshly cut gullies are excavated to a great depth, and the river valleys are evidently cut through soft material. Their shape is not such as usually results from river erosion, for in soft materials river valleys have sides either bluff or with angular slopes of about thirty degrees or more, and bluffs of that kind are to be found along the edge of the valley recently washed out by torrents.

The tertiary clays and decomposed metamorphic rock being of so nearly the same consistency, it is not at all wonderful that in their denudation they should both wear down together, and that the metamorphic hills should pass almost insensibly into the plains. I believe that the wide upper valley of the Capivary is one of erosion anterior to the drift, and that the present swelling outlines and long curved slopes of its bounding ridges are due to glacial action over a surface deeply decomposed.

The hillsides and slopes of this region are sparsely covered with coarse grass and small flowering plants. It was like a garden. Trees are few and scattered, and are noted for their gnarly branches and rough bark. A little palm (*Cocos flexuosa*) is not uncommon on the campos. Another conspicuous little tree of these campos is the Páo de Paina, which has a small stem covered by a thick growth of a woolly substance, used to fill pillows, &c. It is very abundant in the neighborhood of As Trovoadas. Sometimes the trees form clusters (*capões*) in low and wet places, and along the riverside, in the valley of the Capivary. In the dry season these campos are dried up, and vegetation is withered and appears dead. The lands, though rich, are worthless, except as furnishing pasturage for herds of cattle which abound on these campos.* The Ema, or American ostrich (*Rhea Americana*),

* According to Spix and Martius, from the Arraial do Rio Manso there

is found on the campos, but appears now to be rare in this vicinity. It occurs more abundantly farther north on the campos of the Rio Pardo, and on the campos of the São Francisco basin. It ranges from Ceará to Buenos Ayres. In Patagonia another species, *Rhea Darwini*, is found. There is also a species of deer (*Cervus campestris*), called by the Mineiros *Veado campeiro*, which is not uncommon on the plains. But after a half-hour's enjoyment of the magnificent view, down came the heavy rain with a strong chilling wind, and we left the Alto to return. The steep slope from the chapada to Santa Rita was so slippery and untrustworthy, that we were obliged to make the descent on foot in the deep mud, leading our mules.

From Santa Rita we crossed a steep, high ridge by a road terribly cut up, and so full of loose quartz stones as to afford a very insecure foothold for the mules, and at night reached the river Setubinho wet and sore, and there we spent the

stretches northward a low plateau, in a north-south direction, for several leagues, apparently tying in with the great chapada forming the dividing line between the Jequitinhonha and Arassuahy, opposite Minas Novas. "The vegetation of these uniform, elevated plains, which extends from Tejuco to Minas Novas, and gently decreases in height, shows a form which we had not before observed to a similar extent." Low crooked-branched and broad-leaved trees lift themselves here and there amongst a dense thicket of many kinds of bushes, which alternate now with bare rock-sheets, now with thirsty open fields, or, in the low places and beds of streams, with a somewhat higher and sappy wood (*Capão*). The bush they call here *Serrado*, or, when it is lower and without trees, *Carrasco*. The plants belonging to it do not all lose their leaves during the dry season, and in a note our authors add: "Particularly those of the genera *Sida, Ochna, Mimosa, Acacia, Qualea, Coccoloba, Kielmeyera, Laurus, Nycterosition, Arrazoa, Barnadesia, Albertinia, Anona, Banisteria, Malpighia, Aspidosperma*. The stemless palms, *Astrocaryum campestre* and *Diplothemiun campestre*, and the low *Cocos flexuosa*, one sees here." (Vol. II. pp. 473, 474.) A very interesting article in Danish, on the campos region of Brazil, from the pen of Eugen. Warming, is to be found in the *Tidsskrift for pop. Fræm. af Naturvidenskaben*, 3die Række, 5te Bind, 1ste Hefte, 1868.

night. The morning found the little river swollen to its utmost capacity, and running like a mill-race, turbid and yellow, and it was with much difficulty that we forded it. The Rio Setubal we found also in as swollen a state ; mules and horses gave out, and the weary travellers waded through the mud many a long mile, and arrived drenched at the Fazenda de Santa Barbara.

Leaving Santa Barbara for Calháo, I observed near a little brook at the Fazenda da Lagôa rounded quartz boulders, overspread by drift clay. Beyond this the road, after passing a low flat, ascends a gentle slope, on which the same are seen at a height of one hundred feet above the brook; and a short distance farther on and higher up the road is full of coarse gravel, intermixed with angular boulders. Mica slate is seen occasionally cropping out on the hillsides, and on the top of a chapada, before reaching the Fazenda of the Tenente Honorio Ottoni, I saw in the drift paste rounded pebbles. From the chapada one descends into the valley of one of the tributaries of the river Gravatá. The hills of the valley are mica slate and compact dark gray gneiss, with a strike of N. 40° W. and a dip apparently to the northeast, or vertical. Near the fazenda are numerous road cuttings through the hillside, showing the red drift clays to be many feet in depth, and containing an abundance of large fragments of gneiss, quartz, and mica slate. The river Gravatá, where the path to Calháo crosses it, runs in a deep, narrow valley bordered by gneiss slopes, rising to chapada plains above. I observed, in ascending the slope to leave the valley, that, for some five hundred feet, the slope was strewn with quartz boulders, which are very numerous and large. About half-way up the slope there is an exposure of a white or brownish schistose rock, much decayed, which seems to be wholly made up of very minute and rounded grains of limpid quartz, without visible cement in the specimens I examined. This rock had apparently a dip of 20° to the southward. The relation of this rock to the gneiss, or to the materials of the chapada, I did not make out; but I much suspect that the dip was only local. This chapada extends from the Gravatá to a little valley called Estrella, north of which is a plain of very wide extent, and perfectly level. I find in my diary a note that on the top of this plain I saw quartz boulders lying. Two little streams, the Agua da Nova and Diamantino, take their rise on the plain, in quite extensive, shallow, marshy lagoons, but they soon cut for themselves valleys down to the older rocks beneath. These plains are more or less thickly covered with bushes, gnarly-branched trees, and occasional thickets. I saw no Indaiás here, but a little crooked-stemmed palm, called Licurí, the bases of whose leaf-stalks were persistent for some distance down the stem, was quite common. Ferns are rare. Among the trees are several species which produce edible fruit, especially the Mangabeira, Bacuparí, Piquí, &c. Great numbers of cattle are pastured on these plains, and grazing is one of the principal occupations of this part of the country. The cattle are allowed to roam over the plains, and are taken care of by mounted vaqueiros, who dress from head to foot in leather, that they may be able to break through the thickets in their chase after the cattle.

The valley of the Agua da Nova not only deepens, but grows wider in descending, and opens out broadly, on leaving the chapada, into the great valley of the Calháo and Arassuahy. Running along the edge of the chapada, at the top of the slope, are occasional perpendicular bluffs, in which is exposed a thick horizontal bed of sandstone, which is seen forming similar bluffs on the opposite side of the valley. This bed forms the upper stratum of the chapada formation, and may be seen forming bluffs of the same sort along the valley of the Calháo. The sandstone is white, very compact, and rather fine-grained, but there are some beds which are coarse, containing pebbles, and with a hard, opaque, white cement, resembling that of the tertiary sandstone of Itahúnas. Veins of milky quartz traverse these rocks. They certainly bear an altered and old look, but they are here surface-layers, and have never been disturbed. The valleys of the Calháo, and of some of the little rivers west, coalesce several miles before reaching the Arassuahy, leaving the metamorphic rocks, over a very large area, denuded of the formation of the chapadas. This forms a great depression like a lake valley, some 800 or 1,000 feet below the plain, and which is surrounded on all sides by high, level-topped chapadas, which project in capes and promontories between the river valleys. The bottom of the depression is diversified by low, rounded, wooded hills. From the top of the sharp spur of the chapada, on the western side of the valley of the Agua da Nova, one has a magnificent view over this great valley. It was near the close of a clear afternoon that we rode out on the edge of this spur to descend, and suddenly, leaving the bushes of the plain, saw before us the beautiful valley. The level-topped chapadas beyond the Arassuahy extended like a wall to the north of the depression, blue in the far distance, while below us lay the billowy sea of foliage which clothed the bottom of the valley. Weeks of sore, weary forest wanderings, beneath a rainy sky, were forgotten, and the heart, homesick, tired, and often disappointed, gladdened as the eye revelled in the beauties of the landscape; but the sun was rapidly nearing the level horizon, and our camarada warned us that we must descend. So, turning our mules into the steep path, we soon passed into the thicket, and the landscape was lost to view. It was as when, after the curtain has dropped at the close of the last act of an opera, and the memory of the brilliant scenery and the rich music still lingers in the heart, one wakes to feel the sorrows and the realities of life again.

The country bordering the Calháo River, for several miles above its mouth, is composed of slates, which are seen exposed in the banks of some of the little brooks flowing into the Calháo.* These slates, on the right bank of the Arassuahy River at Calháo, are fine-grained and siliceous, and have the slaty structure well developed. They dip to the N. 70° W., at angles varying from 50° to 80°. The planes of cleavage dip to the south 20° W., but I have omitted to note the angle. The country forming the bottom of the valley is much more uneven than one would suppose when looking at it from the chapada, and some of the hills are several hundred feet high. They are everywhere covered by the characteristic red drift-clay on the surface, under which occurs usually a sheet of pebbles, as at Rio. This pebble-sheet is sometimes very thick, and being exposed on hillsides by the washing away of the clays, leaves them very barren. As a general thing the country is sparsely wooded, but ordinarily the forest (catinga) does not bear the

* At the head-waters of the Calháo Spix and Martius found the rock to be coarse-grained, whitish, unstratified, with little white mica, but with much black schorl, often in long prismatic crystals. They state that it is covered by a layer of gray or white pebbles of quartz containing *grisolitas* (crysoberyl) of a greenish-white, pale ochre, or citron-yellow color, and others of an olive, grass, or blue-green color (*Agoas marinhas*), precious garnets and white and bright blue topazes. (Vol. II. p. 502.)

same luxuriant look as that of the Mucury, and resembles a second growth. I have observed immense arborescent cactuses (*Cereus*) growing in the woods near Calháo.* The higher lands are apt to be dry, though the soil would otherwise be fertile, and during the dry season the trees lose their leaves. The river-borders, or *varzeas*, are very productive. Cotton seems to be the principal product, and it is of excellent quality. One thousand canoeloads were sent down the Jequitinhonha to the sea in one year, but a very considerable quantity of the cotton is manufactured at home into coarse cloths, &c. Large quantities of corn, beans, &c. are raised here. Calháo is a village of respectable size, situated on the right bank of the Arassuahy, at its junction with the Calháo, which is so small, and ordinarily so shallow, that the negrowomen wade across it to fill their water-jars in the Arassuahy. At Calháo the latter river is about the size of the Mucury below Santa Clara. Calháo derives its importance from being a sort of centre of the salt-trade with the coast *via* the Jequitinhonha.

At the point where the path from Calháo to Minas Novas crosses the Rio Setubal gray quartzites are exposed, with a strike of N. 65° E., dip 85° southeastward. At the passage of the Corrego de São João fine-grained siliceous gray schists are seen, strike N. 60° E., dip 88° to 90° southeastward; and at the passage of the Sucuriú the same rock is seen, and an observation gave strike N. 60° E., dip 80° N.

* Saint Hilaire says: "In general the cactuses in the province of Minas appear to belong to the catingas in the neighborhood of the Arassuahy and Jequitinhonha, for I have not met with a single species either in the Geraes, properly so called, or in the *carrascos*. (Vol. II. Part I. p. 103.) The same author calls attention to the number of Barrigudo trees (*Bombax*), and the absence of Melanostomaceous plants in the catingas.

At the villa of Sucuriú the same rock is seen, strike N. 30° E., dip 50° S. On the west side of the Corrego de São João there is an outlying chapada, the southern side of which presents a red and white cliff, in which are exposed horizontal white beds, which, I was informed, are composed of tabatinga, or clay over which is a thick bed of red drift-earth, such as is seen everywhere covering the country. The drift, as well as the sand and gravel of the streams in the vicinity of Sucuriú, contains gold, but in small quantities. I saw a few old abandoned workings. West of Sucuriú the road passes over a chapada, and descends into the valley of the Sucuriú, which is bordered by high slate hills, and then ascends to a chapada which, perfectly level and covered by *carrasco*, extends for a league to the Rio d'Agua Suja.* This chapada is precipitous along the edges, and is covered by a thick bed of red drift-clay, under which appears to lie a sheet of gravel, which in some places is cemented by oxide of iron, and forms a conglomerate. From the borders of this chapada one has the most extended views of the surrounding country, and in clear weather the higher points of the Serra do Grão Mogor are distinctly visible.

The valley or cañon of the Agua Suja in some places cuts through the whole chapada formation to the metamorphic rocks below, and is very deep and narrow. This river flows northward into the Arassuahy. After crossing a narrow

* The grass and bushes of the campos are infested by the carapato (*Ixodes ricinus*), a wood-tick which, swept off by contact with the garments, attaches itself by hundreds to the skin, and can only be detached by the application of tobacco or something of the kind. Strangers are apt to suffer severely from the irritation caused by these disgusting creatures. Even the Jabotí tortoise is attacked by them. When allowed to remain, the animal feeds on the juices of its prey until its body becomes as large as a castor-bean, to which in shape and color it bears a close resemblance.

chapada another cañon is reached, that of the Rio d'Agua Limpa or Mãe d'Agua, a little river flowing into the Agua Suja.* The banks of these valleys are wooded, and possess a fertile soil. At the ford slates of ordinary quality are seen. Strike N. 50° E., dip 70° southeastward. Between the Capivary and Minas Novas the country is very hilly and barren, the vegetation being of the character of the campos. In the numerous rain-gullies in the mule-paths the drift is cut through, and decomposed slates are exposed. They are as soft as the drift-clay, and were it not for the different tints of the laminæ and the quartz veins which traverse them, it would be difficult to recognize them as a metamorphic rock decomposed *in sitû*. The Ribeirão do Meio is a brook emptying into the Capivary, from the sands of which gold has been obtained. Spix and Martius have left us the following graphic picture of this part of the country: —

"The thick wood appeared to us a wide grave, for the dry season had stripped off all ornament of leaves and flowers; only once in a while thorny species of *smilax*, or cord-like twists of *cissus*, set with single leaves, climbed up, or the stately flower-panicles of Bromelias stretched themselves out from among the branches. . . . . Thorny acacias, many-branched *Andiræ* and *Copaiferæ*, and fig-trees rich in milk, appeared here in exceeding plenty; but what most pleased us were the giant stems of *Chorisiæ* (*Chorisia ventricosa*), which, contracted above and below, were swollen in the middle like huge casks, their cork-like bark being beset with stout shining brown spines. Here huge bunches of parasitic plants depended from the branches. Here myriads

* I give the names on the authority of my guide. According to Gerber's map, the eastern stream is the Agua Limpa, the other Agua Suja.

of ants have hung from the branches their nests full of Dædalian windings, and which, with a circumference of several feet, contrasted strangely in their black color with the bright gray of the leafless branches. The autumnal torpid wood echoed with the cry of many kinds of birds; especially croaking *araras* and *piriquitos*. Shy armadillos and anteaters (*Dasypus septemcinctus* and *Myrmicophaga tetradactyla*) met us, ... and sluggish sloths (*Bradypus tridactylus*) hung stupidly from the white branches of the embauba (*Cecropia peltata*), which here and there rose among the rest of the trees. Herds of howling monkeys were heard in the distance. The high dry grass was covered with crowded balls of little carapatos, which, when we accidentally disturbed them, scattered themselves with lightning-like rapidity over us, and excited a painful itching. Not infrequently a snake was heard in the thicket by the traveller riding hastily by." *

I passed through this same region in the wet season, when the trees were all in leaf, and the woods looking gay and pleasant. I saw scarcely any animals. I heard some guaribas howling; but neither armadillos, sloths, nor snakes of any kind were seen. It is a very mistaken idea, carefully spread abroad by our geographies and popular works and pictures, that one may everywhere expect to see in the Brazilian forests great boas wreathed about the trees, and all manner of birds and beasts in profusion. I have ridden day after day through the virgin forest without seeing or hearing anything worth shooting, and nothing more dangerous than a wasp!

In the year 1727 Sebastião Leme do Prado, with a band of Paulistas, travelling northward through the province of

* Spix and Martius, *Reise*, Vol. II. pp. 499, 500.

Minas, discovered gold in the river Bom Successo, and gave it the name which it bears He established here regular mining operations, and founded the city now known as Minas Novas, which grew to be a flourishing town. The precious metal was also discovered elsewhere in the vicinity, and in especial abundance on the hills bordering the Rio Capivary near the Arraial da Chapada, where it was very extensively mined. The gold was principally obtained from the sands and gravel of the river, and from the gravel sheet underlying the drift-clays on the slopes and tops of the hills. Very little gold was extracted from the veins of the quartz, some of which were known to be richly auriferous. The hills are dry, and water is to be found only in the rivers, which during the greater part of the year afford a good supply, so that the washing of the gravels on the high grounds was attended with much difficulty. Ditches, or *regas*, were dug round the hills to collect rain-water, which was brought into tanks, and in some of the washings all the water used was derived from this source ; and at Minas Novas and Chapada, washings said to be rich were pointed out to me as abandoned because of the scarcity of water, when just below, a hundred feet or more, tumbled a dashing stream. Notwithstanding the disadvantages under which these old miners labored, a large extent of ground was, as we shall see, worked over, and an immense quantity of gold was extracted, according to one authority 300 arrobas (=9,600 lbs. avoirdupois) being sent to Bahia alone. Many large nuggets were discovered in these mines. In the Lavra do Batatal a lump weighing 28 lbs. was found. In 1746 diamonds were discovered in the vicinity of Diamantina, and government prohibited the extraction of gold in order to encourage the search for diamonds. This prohibition put a stop to the gold-mining of Minas Novas,* and, though the prohibition has been removed, the blow has been fatal, for little gold-mining has since been carried on, and the present inhabitants content themselves with agricultural pursuits, or help to swell the number of miners who wash for diamonds on the Rio Jequitinhonha. The gold-mines are to-day practically abandoned, but the idea that they were worked out is very erroneous. At Minas Novas and Chapada the rocks are slates and quartzites, and resemble very closely those of the gold region of Nova Scotia. Indeed, it was the strong resemblance borne by the slates of Calháo and the vicinity to the Nova Scotian gold-bearing rocks that aroused my interest, and led me to turn out of my way to visit Minas Novas. These rocks evidently overlie the mica slates which flank the gneiss of the coast belt, and I believe they will prove to be Lower Silurian in age.† At Minas Novas their strike is N. 42° to 50° E., and their dip is vertical. They are traversed by great numbers of milky-quartz veins, some of which are well known to be auriferous. Some of these veins are of considerable dimensions. In an enormous gully cut out by the surface waters in the hillside above the cemetery on the Bom Successo at Minas Novas are several fine veins of corrugated quartz.‡ These veins run nearly vertically through the rock, and may be beds instead of veins. As they are exposed in the cliff they present the appearance of vertical fissures, in which cylindrical masses of quartz are piled in a single row, their

* The city is in decay, and is to-day of very little importance. The cotton raised in its vicinity has a very excellent reputation in Brazil; similar lands in Bahia and Pernambuco produce a good quality of cotton.

† Perhaps Quebec group.

‡ These appear to have precisely the same structure as the "Barrel quartz" of Nova Scotia.

ends projecting like logs. Some of these cylinders of quartz are two feet in diameter. In section they appear as represented by the accompanying woodcut. I spent some time

in an examination of this vein for gold, but could detect none. The large size of the quartz veins of the vicinity may be inferred from the dimensions of the quartz boulders scattered over the surface, some of which weigh many tons. I am not aware that any auriferous vein has been worked at or near Minas Novas, but at the Arraial da Chapada several were anciently more or less worked. A rich vein, according to universal testimony, crosses the *praça*, and it is well known that one miner followed it in secret until he undermined his neighbor's house, when his secret was let out. There one hears the terms "vein" and "gravel" gold, and I saw many beautiful specimens of crystallized gold in the hands of the inhabitants, some of which were taken directly from quartz veins, though it is true others were obtained from quartz boulders. There can be no doubt that rich auriferous lodes exist in the neighborhood, which have never been explored, and which one day must be developed, for all the gold which so richly abounds in the drift must have come from the underlying rocks.* In the Minas Novas region I have seen no signs of gneissose rocks, itacolumite, or itabarite associated with the auriferous slates.

The slates, &c., of the valley of the Jequitinhonha are decomposed to a great depth, and are as soft as earth, and can be easily worked with a spade. This decomposed rock, which is of a bright red color, preserves its lamination, and the quartz veins traverse it as in the solid rock. Excellent

* For an account of the gold-mines of other parts of Brazil, see further on.

opportunities for the examination of it are afforded at Minas Novas and elsewhere by the enormous gullies swept out in it on the hillsides by the mountain torrents. Some of these gullies are more than 100 feet deep, and show at the same time the very finest sections of the drift. Near the Arraial da Chapada the bright red cliffs of these ravines are very conspicuous elements in the landscape, and some parts of the country appear as if scored by a giant plough. Burton describes similar gullies in other parts of Minas and in São Paulo. In the latter country they have received the name of *vossorocas*. Burton supposes they were formed by the giving way of a hillside under the hydrostatic pressure caused by the soaking of the mass by water; and he says that the ground breaks away suddenly with the force of an eruption, the hollow in the hillside thus formed being afterward excavated to a greater size and depth by rains and streams, which sometimes gush out of the head of these gullies. The gullies which I saw did not strike me as having been formed in this way. I supposed that they had been hollowed out with more or less rapidity by the action of surface water, perhaps aided by springs, and without a regular land-slide. The surface of the undisturbed decomposed rock is always well marked, and has a regularly rounded contour like that of the gneiss, and is never irregular and jagged like a water-worn surface. The decomposed rock is immediately overlaid by a sheet of cascalho, or quartz pebbles, whose thickness varies from a few inches to eight or more feet. The pebbles are of all sizes, and are more or less rounded. I observed in several localities that there were large boulders lying in this gravel just above the rock. The cascalho is often so cemented by ferric oxide as to form a conglomerate, which requires to be broken up before it is washed for gold. Like the drift pebble sheet of the coast, it forms a concentric layer wrapped over the whole rock surface of the hills, and it is found lying on very high slopes and piled up in masses such as water never deposits.

It is in the cascalho that the greater part of the gold of Minas Novas and vicinity occurs. Over this gravel lies a mass of red drift-clay, varying very much in thickness, from a few inches to fifty feet or more. This is, like the drift-clays of the coast, a homogeneous mass, through which are scattered from time to time angular and rounded quartz boulders of large size. Over large tracts between Minas Novas and the Arraial da Chapada this sheet of clay is so thin that the cascalho bed lies on the surface, and the country is consequently stony and barren. The clay contains sometimes more or less gold. It is, however, to the cascalho sheets that the search for the precious metal has been principally confined. The gold occurs disseminated through the gravel in flattened grains, and occasional nuggets of considerable size, which are always in a crushed and battered state. The process of extraction was similar to that described by Mawe as employed at the mines of Jaraguá in São Paulo, and which I shall further unfold in the description of that province. It consisted in stripping off the clay sheet down to the gravel, which was broken up and washed on the spot in rude trenches to separate the pebbles, when the auriferous mud and sand were washed in the *bateia*, or wooden washing-pan. A great number of the washings were situated on the tops of hills, or slopes at some height above the water of the stream, and in these cases the washing was performed through the aid of rain-water. In several localities water was conducted to the washings from streams. Some of the old *regas*, or ditches, are still visible running for miles around the hills. The supply of rain-water was of course sufficient only during the rainy season, so that washing operations had to be suspended for the rest of the year. In the old washings, as in that above the cemetery at Minas Novas, or the Lavra da Santa Cruz, at the junction of the Rios Fanado and Bom Successo, the gravel lies in great piles. At the Arraial da Chapada the same thing is seen, but there the whole tops of hills have been deprived of their clay coating and washed over, so that to-day they are hoary with the quartz boulders that remain, the testimony of a departed industry. I was informed that the custom was with the miners, as a general thing, to wash the gravel on the spot. It seems wonderful that when the washing was near a river or stream the gravel was not sent down to this stream to be washed. To-day the washings, though owned by private individuals, who to some extent know their value, are unworked, the owners finding it more profitable to pursue agriculture or wash for diamonds in the Jequitinhonha. The abundance of gold over this region may be seen from the nuggets in the possession of the people, and which have been picked up on the hillsides or in rain-gullies. After rains one sees in the ravines the prints of the feet of those who regularly go in search of gold washed out by the surface waters, and in the streets of Minas Novas and Chapada little dams are built across the small rain-gullies by the children, to collect water to wash the soil for gold, which they collect in quills, and larger dams are built by the elder members of the population for the same purpose. No one who has been over the ground as I have, and has seen the irregular way in which the mining has been performed, and the immense area of drift which has yet been untouched, — drift rich in gold, as the occasional recent washings testify, — can doubt that the region is far from exhausted. It has only been forgotten. My friend Mr. J. S. Mills, of New York, an excellent geologist, who has discussed these observations with me, has suggested that the gold probably occurs in bands in the drift, the direction of which might be worked out by a careful topographical survey. Senator Theophilo Benedicto Ottoni, of Rio, about two years ago obtained from the Emperor a concession of the area of the Comarca of the Jequitinhonha, to explore it for gold and other minerals; and an attempt, which we hope may yet be successful, has been made to organize an American company for the purpose of thoroughly exploring and developing the gold-fields of Minas Novas and vicinity. With modern mining methods and appliances I have the fullest confidence that they would prove very remunerative. The system of washing by hose-pipes could be employed successfully in many localities.

Gold also occurs in the gravel and sands of the streams, these loose materials being derived in part from the drift, in part from the decomposed rock. Near the Arraial da Chapada is an outlier of the tertiary called the Serra do Macaco, which forms a very picturesque flat-topped mountain, with escarped sides, in which the horizontal layers of red and white clays are beautifully exhibited.

Now that we know that gold may occur in any formation, why may not the lower beds of this series be found to be auriferous in some places?

I have had no opportunity of making an examination of the gold of the Minas Novas region, and I know of no analyses ever having been made of it.

The sands of the Arassuahy above the Rio Setubal, or thereabouts, are rich in gold. I have never heard of their affording diamonds.

From Calháo I took passage in a canoe, and descended to the sea. On that voyage the following observations were made on the rivers Arassuahy and Jequitinhonha.

At Calháo the Arassuahy is about as large as the São Matheos. Its current is strong, and even during the dry season it contains much water. The country on both sides of the river below Calháo is, generally speaking, low and uneven up to the foot of the chapadas, while *vargens*, more or less wide, border the stream. These consist of alluvial deposits, and afford a rich soil. Their height above the average level of the river is about twenty feet. From Calháo to the mouth of the river the country is sparsely settled. The river-bed is much obstructed by ledges of slate, but there are no rapids, and canoe navigation is not very difficult. At the Pontal, at the mouth of the river, these slates appear, from the canoe, to lie very flat, and to be traversed by heavy veins of a crystalline rock, like granite, the outcrop of one of which crosses just above the Pontal. At this place, in the angle between the two rivers, is a little settlement, which the inhabitants hope may one day rival Calháo in its commerce in salt.

The traveller who has heard the Jequitinhonha constantly spoken of by the Mineiros as a "*majestoso rio*," feels much disappointed when he reaches it at its junction with the Arassuahy, for it is but little larger than the latter river. It is, however, much deeper.

Above the Arassuahy the Jequitinhonha flows in a wide cañon, separated from the valley of the Arassuahy by a long, narrow chapada, which extends from the western limits of the chapada formation well down into the angle formed by the union of the two rivers. The chapada forming the eastern boundary of the valley of the Calháo comes down into the corresponding angle on the other side of the Arassuahy, so that that river really escapes into the cañon of the Jequitinhonha through a cut across the chapadas. Below the mouth of the Calháo the river valley, comprised between the chapadas, is quite wide, uneven, and composed of mica slate, gneiss, &c. Some six or eight miles down the river there is a high hill, the Morro do Arião, which presents the smooth, black-stained, even rock surface so characteristic of the gneiss hills of the coast.

Fifteen or twenty miles below the Calháo the little river Piauhy enters the main river from the south. This stream takes its rise in the Serra do Chifre, a short distance to the north of the source of the Calháo, from which river it is separated by a strip of chapada called the Chapada do Piauhy, on the plains of which herds of cattle are pastured. The Piauhy is noted for its affording *grisolitas* (peridote or chrysoberyl), *pingoas d'agua* (white topaz or limpid quartz pebbles), and other valuable stones, like those found in the Rio das Americanas in the Mucury. The chrysoberyls are used in jewelry and by watchmakers, and at the time of my visit to Minas were selling for 11 $ 000 per pound, or about $ 5.50 American currency. The demand of late years for them has been very small. A few years ago, according to Senator Ottoni, several hundred-weight were extracted and exported, which drugged the market, and made it for a long time unprofitable to wash for them.

Two miles below the Barra do Piauhy, the mica slates dip to the N. 45° W., and at the Ilha do Cubango there are heavy vertical veins of granite, which extend in walls almost across the river, while, a couple of miles farther down, there are some high gneiss or mica-slate hills. The river is full of rocks, and the banks are rocky, though the banks of the river are generally low. At the Arraial d'Itinga the mica slates still show themselves, with a strike of N. 45° E., and vertical dip, and are traversed by granite veins.

The Arraial is a considerable little town, built on a ridge of quartz gravel bordering the river on the northern side, and which, being considerably higher than the river border, itself about twenty feet high, is not covered during the enchente. It derives its importance from its trade in salt, which is brought up the river from the sea, and is sent into the interior to the Sertão do Rio Pardo, together with merchandise, &c. On both sides, but at a considerable distance, the chapadas skirt the river, but they are rarely seen by the voyager by canoe, because of the intervening gneiss hills, which are sometimes 500 to 800 feet in height above the level of the river. Just below Itinga one has a distinct view of a chapada on the south side of the river, and in the cliffs at its top the characteristic white rock is seen. The height of the chapada top above the level of the river must be over 1,000 feet.

The rocks exposed in the river-banks between Itinga and the "Estreito" are gneiss, a compact variety. The hills have the ordinary topography of the gneiss regions of the coast, and often present bare, blackened cliffs and slopes. Back of the hills the flat tops of the chapadas are seen, and occasionally they accompany the river. The slopes of the chapadas invariably show gneiss almost to the top, where there are usually lines of white cliffs. The thick red bed at the top of the chapadas of Minas Novas I have not observed here, nor is it to be seen in the cliffs of the chapadas at the junction between the Arassuahy and Jequitinhonha. The hills and chapada slopes are thickly wooded, but the trees are all small. A small species of Barrigudo (Imbaré, *Bombax*, or *Chorisia*), with an enormously swollen trunk, is very abundant on the margin of the river. The course of the river is rapid, and its breadth is about equal to that of the Parahyba do Sul at São Fidelis. At the "Estreito" the river passes through a narrow gorge across a gneiss ridge. This chasm is in some places not more than 150 feet wide, and is a most romantic spot. The sides are bold, rounded masses of rock piled up one upon the other in picturesque confusion.

When the river is swollen, the "Estreito" is a fearful place to pass; the waters rush through with great fury, and below it are dangerous whirlpools, where canoes are frequently lost. Between the "Estreito" and the Pedra do Bode the river-banks are low and flat, and the country behind is often marshy and interspersed with shallow lagoons. In one of these I found an abundance of *Ampullarias*, but I could find no other shells. The Pedra do Bode, one of the noted landmarks on the river, is a gneiss hill on the north bank, presenting a smooth precipitous face to the river. It is of some considerable altitude, but is not so high as the chapada behind it. Thence to São Miguel the river is bordered by gneiss hills and chapada spurs, and back of these, on both sides of the river, are seen the level tops of the chapadas which accompany the river. The hills are often very abrupt, and present many bare surfaces. Some, which may not be wholly composed of gneiss, are very regular in their curves, and have steep slopes covered with a low vegetation. The Indaiá palm is very common on some of the hills, going to make up the greater part of some of the woods. A short distance above

São Miguel is the Caxoeira do Labyrintho, a series of rapids extending for more than a mile. In some states of the river these rapids are very dangerous, owing to the inclination of the river-bed and the numerous rocks which obstruct the river, and canoes are wrecked and lives lost here almost every year. In descending, the *pröeiros*, or oarsmen of the prow, row vigorously to give the canoe a good headway, so that it may obey the steering-oar, which must be handled very dexterously. São Miguel is a miserable hamlet on the right bank of the river at the mouth of the river São Miguel, which takes its rise in the same serra with the Rio das Americanas. It is important principally because of its commerce in salt. There are some large fazendas in the vicinity on both sides of the river, on which very large herds of cattle are raised. There are said to be some fertile lands here. Below São Miguel is the Caxoeira de Dorma, a series of rapids usually easily passed. On the right bank of the river, and some two or three miles below São Miguel, is a range of irregular gneiss hills, which have apparently a general north-south trend, and present a precipitous front to the west. A small stream springs from the top of one of these precipices, and hangs a white thread of water against the black wall of rock. Irregular gneiss hills occupy the right bank of the river for some four and a half leagues. On the left bank the great chapada stretches along, its sides descending with rounded, smooth slopes, often destitute of forest and green with low herbage. The stream here is full of rapids with a strong current, and is about 500 feet wide. The scenery on this portion of the river is exceedingly grand. Just above the valley of São Simão is a little village called Farrancho, inhabited by civilized *Machacalis*. Below this the chapada slopes advance to the river-side and border it for a league or more, forming the narrow valley of São Simão. The chapadas are of great elevation,—1,200 feet or more above the river; their sides descend with steep, smooth slopes to the river. The lower part of the slopes is thickly wooded, but toward the top the vegetation generally becomes low and scrubby; in some parts, however, they are wooded to the top.

The regularity of the slopes would justify one in assigning to the chapada formation here a great thickness. At the entrance of the valley a very siliceous gneiss is seen underlying the chapada. Leaving São Simão the chapadas recede from the river, and the country thence to the Salto Grande is gneiss. Immediately below São Simão are picturesque groups of hills,—the Serra da Vigia on the right, and the Serra das Panellas on the left,—and below these are the rapids of the Panellas. In descending we pass the Serra do Feijoal on the left, and other hills on both sides of the river, shoot the Caxoeira do Angelim, the Caxoeira da Farinha, and other rapids, and reach the eastern extremity of the fine Serra da Lua Cheia, which, coming from the southwest, breaks down near the river. Between the hills of the Feijoal and the Serra da Lua Cheia, the lands bordering the river are generally flat and low, so low that they are easily flooded by the enchente. The soils of these low lands, or vargens, are in general composed of a fine sand with but little admixture of clay. They appear to be very fertile, as the vegetation they support is very luxuriant. These flat lands are full of swamps and shallow lagoons, which are flooded during the enchente, and are left full of water when the freshet subsides. This water sometimes becomes putrid, from the decay of the vegetables abounding in the swamps, and fills the air with miasma, while the water entering the river poisons it. It was in the beginning of May that I descended the river. Between São Miguel, sezões, or fever and ague, were exceedingly prevalent, and I left all my canoe-men sick at Salto Grande. There was scarcely a house on the river where there were not cases of fever, and canoes on the up voyage were delayed at the Salto and elsewhere along the route because of the sickness of the crews. This general prevalence of fever among the canoe-men is principally attributable to their exceeding imprudence. They drink freely of the warm muddy river-water when overheated. They bathe in it under the hot sun, and go with dripping garments a great part of the time, spending frequently night after night under drenching rains with no other shelter than a woollen blanket. I had risen from a sick-bed to make the voyage, and was constantly exposed to the rain and cold ; but I avoided the river-water, and escaped, as did my fellow-traveller, a merchant from Calháo.*

The hills comprising the Serra da Lua Cheia are of considerable altitude, much broken up and very irregular in outline. In this serra are several conspicuous needles visible from a long distance, two of which are named respectively the Enchadão and Enchadinho. From this serra to the dangerous Caxoeira de Santa Anna, formerly called the Caxoeira do Inferno, the river is very rapid, full of islands, and there are some places difficult to pass.

The Caxoeira de Santa Anna is at all times so dangerous that the cargo is always carried round the rapids, and re-embarked below, the canoe descending empty. At the head

* I cannot let pass this opportunity of acknowledging the kindness of this gentleman, Senhor Baretto, a mulatto, who gave me my passage to the Salto from Calháo, and was of the greatest service to me.

of the rapids, which extend for nearly a mile, is a large island. Both channels are practicable in some states of the river, but with low water the northern is the only safe one. The Caxoeira consists not only of a series of rapids, but also of several *bancos*, or low cascades. Canoes constantly descend, — a most exciting feat. The ascent is accomplished only with the empty canoe and with great difficulty. Between this Caxoeira and the Salto Grande the river is very swift, full of rapids, and obstructed by rocks, while in some places it is very narrow, and bordered by a wide margin of rocks covered by the annual floods. Islands are numerous. At the town of Salto Grande, a wretched little place, on the right bank, a quarter of a mile above the Salto, or falls, and celebrated for its trade in salt, &c., the river is only eighty to one hundred feet in width, but on each side are low margins of bare gneiss * rock and sand-banks.

At the Salto the river reaches a point whence, within the distance of a mile or thereabouts, it descends some three hundred feet, more or less,† in a splendid series of cascades and rapids. At the head of the Caxoeira, when the river is not swollen, it is suddenly narrowed to forty or fifty feet, and plunges down a very steep incline into a gorge with perpendicular banks, making a wild and most romantic fall of about fifty feet. Below this are other falls, which, owing to the state of the weather, I was unable to visit. On each side of the rapids is a wide strip of rocky ledges, swept bare. When the enchente prevails, the stream swells too big for its channel, and pours in a terrible flood over the rocks

* This gneiss is composed of feldspar, quartz, and hornblende, and is well bedded. Strike, N. 30° W., dip vertical at the upper Porto do Salto.

† I am entirely unable, from the character of the country, to form a very reliable estimate of the total height of these falls, but I believe that 300 feet is much within the truth.

on each side, making a series of rapids to which those of Niagara are as nothing. The Salto Grande, during the floods, must be a sight worth a pilgrimage to see. The *Diccionario Geografico* says that the fall is twenty braças in height, and that the noise of the waters may be heard at a distance of four leagues, which is not very correct. The Salto consists of several falls and rapids, as above described. The Caxoeira is of course an effectual barrier to navigation, commerce requiring a transport of goods by mules around it, which have to be re-embarked above or below the falls. On the road from the village to the port below the country is seen to be covered as usual with drift-clay, in which are large boulders of the hornblendic gneiss, together with rounded and angular fragments of quartz. Below the Salto the river leaves the province of Minas Geraes, and enters that of Bahia; but to make my description of the river complete, I continue it here to the sea.

Between the Salto and the Caxoeirinha the river is narrow, with high gneiss banks. It is much obstructed by rocks and rapids; but this part of the river I am unable to describe in detail, because I was obliged to run the greater part of it, rapids and all, in the night. At the Caxoeirinha the river leaves the rocks, and becomes a *rio d'areia*. Up to this point the canoes bring from the sea very heavy loads of salt, &c., but here their cargoes have to be divided and rearranged. Here has sprung up a little settlement, which bears the same name as the rapids, but it is of no importance.

The river, on leaving the rocks, becomes immediately shallower, less rapid, and widens into a fine broad stream, comparable to the Doce, and from 800 to 1,000 feet in width or more. The lands also grow lower, and the river valley is cut through the coast tertiary band. The whole country is heavily wooded, but the vegetation did not bear to me the luxuriant air of that of the Doce. There are a few settlers along the river, and one or two large fazendas. At a place called Zinebra, a few leagues above the mouth of the river there is an old fazenda, with which is connected a good saw-mill. Below this a short distance, on the same or right bank of the river, an American colony has been established, and on the occasion of my visit I found two Southerners, Messrs. Ogden and Thompson, engaged in cutting a clearing in the forest. The locality they have chosen is a fertile one, but it seems to me doubtful whether, single-handed, they can ever succeed. Below Zinebra the tertiary lands leave the river, an isolated patch being found on the Po-assú, a channel on the north whence some of the waters of the Jequitinhonha escape into the Pardo. Thence to the sea, low alluvial lands, with a heavy forest growth and swamps, border the broad, beautiful river. It is, however, very shallow, and full of sand-bars. The river would be navigable for a little flat-bottomed river steamer, but it would have to be of very light draught. At the mouth the river becomes exceedingly broad and shallow, and is to such an extent obstructed by sand-bars, that the level of the river is always higher than that of the sea, and the salt water never enters, as is the case with most other rivers. So heavily does the surf beat on the bar, that vessels enter with great difficulty, and when once they have entered it often happens that weeks or even months may elapse before it may be safe to pass the bar again. Cargoes of corn laden at Belmonte have often to be relanded after lying in the hold of a vessel for weeks. Nor is this all. The sand-banks are constantly shifting, and a vessel at anchor may be heaped round by sand and detained for a long while. The result is, that the port is rarely ever resorted to by coasters. Belmonte is a little town situated on the alluvial border of the river, in a grove of cocoanut-trees, on the right bank, a short distance above the mouth. During the freshets it is liable to suffer from the eating away by the river of the bank on which the town stands. It is of scarcely any importance, doing very little trade, its inhabitants being principally fishermen. Cattle are raised on the plains of the vicinity, but there is small opportunity for agriculture.

From Caxoeirinha to the fazenda of Zinebra I saw next to nothing of the geology, owing to a part of the journey having been made in the night, and because of the prevalence of very heavy rains; but near Zinebra I saw a small exposure of shales, which appeared to be of the same character as those hereafter to be described in speaking of the Rio Pardo, but owing to the height of the river I could make nothing of them. The *Diccionario Geografico* says, in speaking of the river, that in 1840 beds of rose-colored marble were discovered. Through the kindness of Senhor Pirajá I have in my possession a specimen of this marble. It is exceedingly fine in texture, and of a delicate pink tint, compact and hard, and would take a fine polish. If it occurs in sufficient quantities, it would make a beautiful building-stone. For the present let us leave the Jequitinhonha. When treating of the geology of the Province of Bahia we shall have to return to it again.

CHAPTER IV.

THE ISLANDS AND CORAL REEFS OF THE ABROLHOS.

The Geology of the Abrolhos. — Trap-bed, Fossil Plants, &c. — Land Fauna and Flora; Spiders, Lizards, and Sea-Birds. — The Cemetery of the Frigate-Birds. — The Whale and Garoupa Fisheries. — Importance of these Fisheries. — The mythical Brazilian Reef. — The Coral Reefs and Consolidated Beaches confounded by Travellers and Writers. — The Author's Discovery of the Porto Seguran Coral Reef. — Coral-building Corals found almost wholly to the north of Cape Frio. — The Fringing Reef of Santa Barbara; its Structure and Life. — Corals found on the Reef. — Star-fishes, Ophiurans, &c. — Resemblance between the Echinoderms of the Abrolhos and West Indies. — The Chapeirões. — The Parcel dos Abrolhos; its Appearance; forms a serious Obstacle to Navigation. — Safe Canal west of the Islands. — The Parcel dos Parêdes. — The Recife do Lixo. — Its great Extent. — The Submerged Border and its Coral-Growth. — The Coral Fauna of Brazil. — The Millepores and their Stinging Properties. — The Reefs of Timbebas, Itacolumí, Porto Seguro, Santa Cruz, Camamú, Bahia, Maceió, and Pernambuco. — The Roccas.

THE islands of the Abrolhos * lie about midway between the cities of Rio and Bahia, a little south of the parallel of

* The general impression seems to be that the name is derived from the Portuguese words meaning, "Open your eyes," a name which would be exceedingly appropriate, for the islands, whitened by the dung of sea-birds, have a spectral look, and, in addition, the reefs with which they are surrounded are so dangerous, that, before the lighthouse was erected, it required much vigilance to enable vessels to pass them in safety, and they have been always justly dreaded. The author of the odd old Dutch *Reys-boeck van het rijcke Brasilien,* published in 1624, says that they are very *periculeus,* and adds: "Daerom als hy dese passagien passeren willen so nemen sy eerst met al haer volc het *Sacrament* ende wanneer sy die ghepasseert hebben bedrijven sy groote blijdtschapghelijck al by alle *Jurnalen* soo wel vande spaensche als van de onse te sien is!"

ISLAND OF SANTA BARBARA DOS ABROLHOS.

Caravellas, and at a distance of about forty miles from the mainland. The position of the lighthouse on the island of Santa Barbara is, according to Mouchez, lat. 17° 57′ 31″ S., long. 40° 58′ 58″ west from Paris. These islands are situated apparently near the middle of the submerged border of the continent, which here, over a very large area, lies at a depth of less than one hundred feet. They are four in number, with two little islets, and they are arranged in an irregular circle, three of them close together. All are rocky and rather high, Santa Barbara, the principal one, being 33.22 metres in height. The length of this island is about three quarters of a mile. Its outline is irregular, and it is very narrow. It is composed of beds of sandstone, shales, and trap, which dip approximately north-northwest, at an angle of from ten to fifteen degrees. Owing to this

Captain and crew took the Sacrament before passing them! The name, however, means rocks, and is so defined in Fonseca's Dictionary. There is a little group of reefs and islands lying on the western coast of Australia, in lat. 28° S., and known as Houtman's Abrolhos. These are, in great part at least, composed of coral

northward dip of the strata, the northern side of this island presents a steep slope to the sea, while on all other sides it is precipitous. The island is almost divided in two in the middle by a cove indenting it on the southern side.

THE ISLANDS OF THE ABROLHOS FROM THE SOUTH.

In the cliff below the lighthouse, the lowest beds seen are an arenaceous limestone (?), *a*, of the following diagram, —

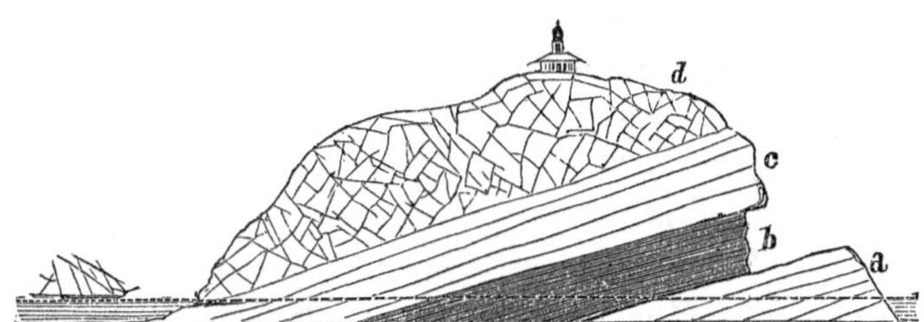

a rock so hard as to form a platform below the cliff. Over this is a hard, blue shale, *b*, with undeterminable organic markings, some of which appear to be the scales of Teleostian fishes. This is again overlaid by a thick bed of a yellowish sandstone, *c*, rather fine in texture, and sometimes more or less shaly, on the surfaces of some of the layers of which there are obscure impressions of plants. This sandstone is harder than the underlying shales, and so forms an overhanging cliff. The sandstones are overlaid by a bed of basaltic trap that occupies the greater part of the surface of the island, as is seen in the following little map, in which the darkly shaded portion represents the trap-bed.

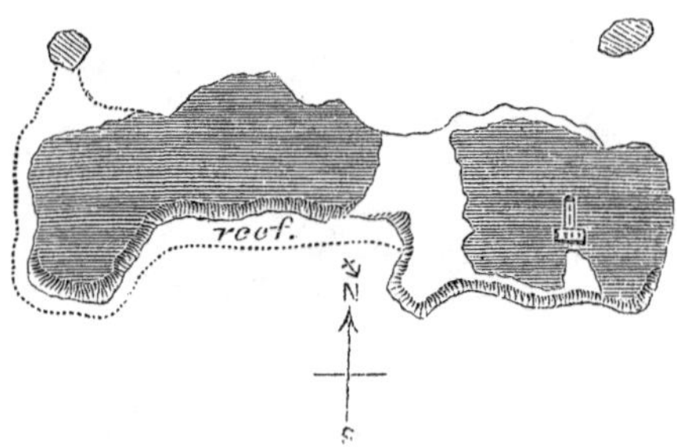

This trap-bed is divided by numerous joints into polyhedral masses of all sizes. On the upper surface of the bed these masses have lost, by decomposition, one concentric coat after another, until, in some cases, they have been rounded down to irregular spheres, like cannon-balls, and the greater part of the island is strewn with these boulders of decomposition. I have already called attention to similar boulders at Tijuca. The trap of Santa Barbara is traversed by but few veins, so far as I could see, the only mineral that I could find being chalcedony, incrusting cavities, and having the color and general appearance of Prenhite. Cracks and crevices are sometimes filled up with guano and phosphates from the dung of the sea-birds that frequent the islands; but I saw no regular deposit of guano. The surface of the rocks is sometimes covered in patches by an incrustation of a hard, brownish substance, which I have supposed to have been deposited by the surface waters, and to have been derived from the birds' dung. Darwin, who visited the islands on his celebrated voyage round the world, also speaks of it, and describes a similar substance as found on the island of Ascension, and on St. Paul's rocks.* The same geologist mentions having observed a columnar structure in the trap of Santa Barbara, but I find no note of it in my journal. The underlying stratified beds are somewhat altered, and indurated from their proximity to the trap.

In lithological characters the Abrolhos beds resemble the sandstones, &c., of the Rio São Francisco at Penêdo, to be described farther on, and which contain similar plant remains. They have been disturbed by the same upheaval, and I have little hesitation in referring both to the cretaceous.

If my identification be correct, it is interesting to observe the cretaceous rocks on the eastern border region of South America disturbed and associated with volcanic deposits, for along the eastern border region of North America the cretaceous and tertiary rocks have suffered no disturbance. I have seen no trace of secondary rocks on the coast opposite the Abrolhos and to the southward. On the Mucury and elsewhere the tertiary clays are everywhere seen to rest immediately on the gneiss; but the submerged border of the continent seems to be more or less overlaid by cretaceous strata, as in the eastern border region of North America.

The Ilha Redonda, lying just west of Santa Barbara, is composed of rocks of the same character as those of Santa Barbara, but I observed no basalt. Near the top of the cliff, on the eastern side of Redonda, there is seen a thick bed of a white or yellowish material which looks like chalk, and is easily cut with a knife when wet, but on drying it grows

* Darwin, Geological Observations, Part II. p. 33.

harder. It does not, however, effervesce with acids, and it appears to be an aluminous product of the decomposition of some rock. Mr. Henry Hughes, of Cornell University, examined a specimen of the rock, and reports it as containing quite a percentage of phosphoric acid, which has doubtless been brought into the rock by the percolation through it of rain-water from the dung-strewn surface of the ground above. The other two islands are composed of stratified rocks which appear to underlie conformably those of Santa Barbara and Redonda; but I was unable to visit them. On the shores of Santa Barbara I found fragments of pumice scattered about and much rolled by the waves. These have been observed elsewhere on the Brazilian coast, and it is somewhat puzzling to account for their origin. Darwin found pumice pebbles on the coast at Bahia Blanca, in the southern part of the Argentine Republic; but these, he says, had been brought down by the rivers flowing from the cordillera.*

The beaches of the Abrolhos Islands are formed in part of the *débris* of the rocks comprising the islands, but they consist largely, and in some places entirely, of coral and shell sand. It is very interesting to see how these materials are cemented together through the action of the sea water, and even the shingle is soldered into an exceedingly firm mass.†

* Darwin, Geological Observations, Part III. p. 4.

† I have seen the same in the island of St. Thomas, W. I. Darwin says: "On the shores of Quail Island in the Cape Verdes I found fragments of brick, bolts of iron, pebbles, and large fragments of basalt united by a scanty base of impure calcareous matter into a firm conglomerate. To show how exceedingly firm this recent conglomerate is, I may mention that I endeavored with a heavy geological hammer to knock out a thick bolt of iron which was embedded a little above low-water mark, but was quite unable to succeed." — Geological Observations, Part II. p. 21.

On the island of Santa Barbara was erected a few years ago an excellent lighthouse with a flash light, and the only human inhabitants of the island are the lighthouse-keeper and his assistants. A few goats were introduced on Santa Barbara some time since, which have multiplied until there is now a flock of several hundreds. These animals have almost deprived the island of vegetation, and can now only barely subsist. Redonda is covered with coarse grasses, with dwarf mimosas and a few ferns, &c., — a very meagre flora. The island Siriba boasts in addition a single tree of the same name, together with two dwarf cocoa-palms planted by the whalers from Caravellas. The land animals consist of little lizards of several species, which are extraordinarily abundant, more so than in any other locality I ever visited. An immense *Mygale*, the *Aranha caranguejeira* of the Brazilians, abounds in like manner, living under stones, but I did not observe that it made any nest. This huge spider preys on the lizards. It has been known to attack and kill young chickens on the island and suck their blood, and it is not improbable that it may destroy the young of the sea-birds, so common on the island.

Breeding-places for sea-birds are few along the Brazilian coast north of Cape Frio, and during certain seasons of the year several species resort to the Abrolhos in great numbers. Among these are the frigate-bird (*Tachypetes aquilina*), the *Piloto*, the *Grazina* (*Phaëton*), *Beneditos*, gulls, &c. Since the occupation of the island by men and goats, and the establishment of the lighthouse, these birds have resorted to the island of Santa Barbara less abundantly than formerly. One fact with reference to the frigate-bird is worth mentioning. At the southwest extremity of Santa Barbara is a little islet composed of a heap of large trap boulders of decomposition, and joined to the main island, as I shall have occasion hereafter to remark, by a fringing coral reef. This islet, whitened by the bird-dung, is called "*O Cemeterio*," or the cemetery. I was assured that to this sepulchral-looking spot the frigate-birds of the vicinity resorted on the approach of death, and that the place was strewn with their bones. At low water one day I visited the "cemetery," and I found such to be actually the case. There were remains of hundreds of these birds, some freshly dead, but the most of the skeletons were disarticulated and bleached. Nowhere else did I see a dead frigate, and it would seem that for generation after generation they had gone there to die. I do not know of a better station for an ornithologist desirous of studying the habits and embryology of the sea-birds of Brazil than the island of Santa Barbara. He can find as comfortable lodgings at the lighthouse as he could possibly desire, and he may at the right season of the year collect as many specimens, young and adult, of the birds frequenting the island as he may wish.

Before I go on to speak of the coral reefs of the Abrolhos and vicinity, a few remarks on the fisheries may not here be out of place. On the coast of Brazil are found several species of whales and smaller cetaceans, but these animals have not been carefully examined by competent naturalists, and I am unable to give as accurate an account of them as I could desire. These animals are captured at various stations, from Santa Catherina northward to Bahia. At present the two most important stations are Caravellas in the Abrolhos region and Bahia. I visited the Abrolhos during the whaling season, and during my cruise I saw several humpbacks (*Megaptera*), all apparently of the same species; but

I was not so fortunate as to see a fresh specimen brought in. I made the most diligent inquiries of the fishermen in relation to the different kinds they were accustomed to take, but they seem to confound the species, and I could obtain nothing very satisfactory from them.

From all that I could learn, three species are taken in the vicinity of the Abrolhos. The *noruega* is a humpback, which has the belly white and smooth, back very dark bluish, length fifty to fifty-five feet. This whale gives more oil than the *mystiça*, which the whalers said differed from the above in having the back black, and the belly and throat furrowed. Sometimes they have white spots on the sides.

The *caçeló* (cachelot) is distinguished by being wholly dark colored or black, and without spots or furrows. The fishery begins at Bahia, according to Castelnau,* about the 13th of June, and lasts until the 21st of September. At Caravellas I was assured that the whales always appeared later than at Bahia, and the fishery does not begin until the last week in June,† continuing through the month of September. This seems strange, since the whales, according to the fishermen, come from the south in June, and return in September, and one would naturally expect that they would arrive at the Abrolhos before they arrive at Bahia. The first whales appear in the Abrolhos waters at about the end of May, and they linger until October. The females often bring young calves with them, and appear to seek the shelter of the reefs. The head-quarters of the Abrolhos fishery is at Caravellas, or rather at the mouth of the river Caravellas, where are situated the armações, or trying-houses. In the year 1867 there were

* *Expédition dans l'Amérique du Sud,* Tome I. p. 150.

† So also Von Tschudi.

employed in this fishery seventeen launches. These vessels are large, well-made, pink-sterned, open boats, carrying one large square sail. The same build of launch is used at Bahia, and they are very good sailers. One of these launches costs, the hull alone, about 500 $ 000 ($ 250, more or less).* The crew consists of from fourteen to sixteen persons. Each launch takes in tow a whale-boat. These are of the ordinary build, and cost about 230 $ 000. The crew of the boat consists of seven men, — a harpooner and a steersman, the rest being rowers. The launch and boat usually belong to one person, who hires the crew for the season, or *safra*, furnishing them rations every ten days. When a large whale is captured the harpooner receives 120 $ 000, but if small only half that sum. The steersman receives half the prize money of the harpooner, the rowers each 24 $ 000 if the whale be large, and 12 $ 000 if small. The master of the launch receives 40 $ 000 if large, 20 $ 000 if small, and the crew 16 $ 000 if large, 8 $ 000 if small. The dead whale is towed in to land by the launch, aided by the boat if necessary. The distance is often great, and when the wind is adverse the whale often comes to land in a wretched condition, and frequently badly mangled by sharks, which abound in these waters. A small tug would be of much service in this fishery, not only to bring the whales promptly to the shore, but occasionally to tow the launches in case of a head wind. The whales are brought to the shore, beached in front of the trying-houses, and cut up. There are several of these trying-houses. The one that I visited was well constructed, and was fur-

* The reader will please bear in mind that the 1 $ 000 or mil-reis (not *mil-rei*, as foreigners will call it), has an approximate value of about fifty cents American currency.

nished with five cutting-tanks, which would accommodate the blubber of two large whales, together with ten tanks for oil having a capacity of about 15,000 gallons. There were twenty-six trying-pots.

The armações are hired by the owner of the launch capturing the whale, who furnishes the men necessary for the cutting up and trying out. The large females accompanied by young, *madrijos*,* are very fat, and are the most valuable prizes. There are killed every year and cut up at the Ponta da Balêa from thirty to ninety whales; but were the fishery pursued more vigorously, with proper economy of time and the use of a small steam-tug or two, the yield might be more than doubled. I learn through M. Bornand, of Villa Viçosa, that a company has been formed at that town for the prosecution of the fishery. A good-sized whale ought to afford 1,000 to 1,600 *canadas* of oil, the *canada* containing about ten bottles, or one and eight ninths gallons, the large whales giving much more. The oil, whose quality might, it seems to me, be improved by more care in the trying out, sells on the spot at from 1 $ 600 to 3 $ 000 per *canada*.

The whalebone is short, but sells well, but I have omitted to note the price it brings. The beach on which the whales are cut up is covered during the season by huge masses of rotting flesh, and is strewn with bones. There must be on the spot the bones of over 500 whales. These, with the flesh and the refuse from the trying-pots, would, properly and scientifically prepared, make an excellent manure, which, if judiciously applied, would go far towards rejuvenating the soils of the plantations of the vicinity, which are rapidly becoming exhausted. At present it seems ridiculous to hear the complaints of the planters, while

* *Madrijas?* I do not find the word in Fonseca.

hundreds of tons of the most valuable kind of manure are left to rot on the sands or are poured into the sea! The Abrolhos region is rarely visited by whalers, though I was informed that an American vessel some nine years ago spent a season on the ground, taking twenty whales.

The fishery at Bahia is carried on on a much larger scale than at Caravellas. Castelnau* estimated in 1850 that it gave occupation to 2,000 persons, and from 100 to 120 boats, giving a revenue of 200,000 francs. The same author estimated that on the whole coast of Brazil from 10,000 to 12,000 persons were engaged in this fishery, and that it produced a capital of 1,000,000 francs, but it seems to me that that estimate would be far too high for the present time. Castelnau speaks of the fact that whale-flesh is used as food by the lower classes in Bahia, and I saw it exposed for sale; but Dr. Antonio de Lacerda assured me that it was not healthy, and tended to produce elephantiasis. Castelnau states that, according to the fishermen, the whales enter the bay every morning, but always return to the open sea to spend the night, and I heard the same report. Whales are frequently taken very near the city, and one may sometimes enjoy the rare sight of sitting at a restaurant in the upper city and watching the chase and capture of a whale in the bay below!

The other fishery carried on in the waters of the Abrolhos is that of the *garoupa*, an excellent fish exceedingly abundant, and taken with the hook and line. The head-quarters of this fishery is at Porto Seguro, a town some seventy miles to the northward of the Abrolhos. At this town is owned a fleet of thirty-five or forty small vessels, each carrying from seven to ten men. The fishery really extends

* *Expédition dans l'Amérique du Sud. Histoire du Voyage.* Tome I. p. 152.

from Barra Secca northward to the Commandatuba, but the best grounds lie between lat. 17° and 18° S. The cruise is usually for twenty-five or thirty days. The fish taken are principally *garoupas*, but there are also several other kinds, such as the *meiro*, *vermelho*, &c. The fish are salted down in the hold, but, owing to the heat, they arrive almost invariably with a very strong and disagreeable odor. They are dried on shore and sent to Bahia. The yearly product of this fishery is from 160,000 to 200,000 arrobas (2,560 to 3,200 tons). The *garoupa* is a delicious fish, and with proper care might be prepared so as to be quite equal to the cod. The names of the fish taken in the vicinity of the Abrolhos, and which are used for food, are legion, and among them are some of the most delicious of marine fish. The Abrolhos Islands offer an excellent station for drying and curing fish, and there has been some talk of establishing there the head-quarters of a company to carry on this fishery. Immense quantities of codfish are now sent to Brazil, together with European sardines and canned fish from Portugal, and every venda is full of them. Some enterprising American should form a company for the development of this fishery. The Brazilian fish are as cheaply obtained as the Portuguese, they are nowhere to be excelled, and Brazil ought to be exporting her delicious fishes, canned, or otherwise prepared, to Europe, besides supplying her own market. The government would favor any undertaking of the kind proposed, and there are wealthy Brazilians who would aid in carrying it out.

In works on Brazil, from those of the old explorers to the present time, we find the uniform statement that a reef or consolidated beach, like that of Pernambuco or Barra Secca, extends around the greater part of the Brazilian coast.

There has been much confusion as to what this reef really was, some describing it as of coral, others as composed of sandstone, but in scientific works it is generally stated that no coral reefs exist on the coast of Brazil.

I am not sure who first expressed the opinion that the stone reef surrounded a large part of the coast, but I find it in Piso, whose first volume bears date 1648.* Since his time this general reef has been described over and over again almost in the same words, and it is even occasionally to be found laid down on maps.

Prince Max. zu Neu Wied has nothing to say concerning the true coral reefs, and, strangely enough, he does not describe the Porto Seguran or Santa Cruz consolidated beaches, notwithstanding he gives drawings of both. Von Martius,† however, observed coral banks at Camamú and near Ilhéos, and referred some of the corals to Lamarckian species.

Darwin, who just touched at the Albrohos, observed corals growing on the shore, but he did not see the reef. In his Geological Observations ‡ he says: "Round many intertropical islands, — for instance the Abrolhos on the coast of Brazil, surveyed by Captain Fitz Roy, and, as I am informed by Dr. Cumming, round the Philippines, — the bottom of the sea is entirely coated by irregular masses of coral, which, although often of large size, do not reach the surface and form proper reefs." Darwin speaks also of having received infor-

* Piso says: "Maximam Brasiliæ partem, nuno interrupts nunc continuato ducto tuetus. Ejus latitudo planissima est et quasi arte in superficii levigata ad viginti, subinde triginta passus et ultra se extendit. Tantæ vero altitudinis ut vix summo æstu inunditur." — *Hist. Nat. Brasiliæ.* Guilielmi Pisonis, M. D. de Med. Brasil. Liber primus. 1648.

† *Reise nach Brasilien,* Band II. Seite 684, 685.

‡ Part I. p. 58.

mation of the coral reef at Maceió, which further on I shall describe; and in another place in the same work, referring to the *Pilote du Brésil*, by Baron Roussin, a work I have never seen, he says: "A few miles south of the latter city [Pernambuco] the reef follows so closely every turn of the shore that I can hardly doubt that it is of coral." Dana also says: * "About Pernambuco, as I am informed by Mr. Titian R. Peal, there are some patches of growing corals, and they are said to extend along to 20° or 21° south latitude," which is not quite correct, as we shall see further on.

Staff-Commander Penn,† in treating of Cape São Roque, says that "the coast of Pititinga and the Cape is skirted by a reef which, between two and two and a half miles southward of the former, in front of two small villages, forms a curve with its outer edge and runs thence a mile from the shore, having two and three quarters fathoms of water inside of it." These reefs are represented on the map of Rio Grande do Norte by Almeida, and they appear to be coral reefs. Penn speaks of other reefs between the Punahú and the Touro which are of the same character. A little farther on ‡ he says: "The *recife*, a singular ridge of coral rock, borders the coast, generally at a distance of about a half to three miles, but in some places much farther off, and extends more or less from the northeast part of Brazil as far as Bahia. Traces of it may be found farther southward and along the north coast to Maranhão. The reef, which is about sixteen feet in breadth at the top, slopes to the seaward, is perpendicular on the shore, and said to be generally covered, but sometimes rises from distance to distance

* Coral Reefs and Islands, p. 108.

† South American Coast Pilot, Vol. I. p. 22.

‡ Op. cit. p. 25.

nearly three feet out of water. It is nearly always bordered by rocky banks, and forms a natural breakwater, having smooth water and shallow inside of it, with navigable channels for coasters, &c. It is broken occasionally, and forms by the openings entrances to the greater part of the ports, rivers, and creeks on the coast." Now such a description could never have been written by any intelligent seaman who had examined the coast. It is Piso's account of the *recife* told over again, and it is the more erroneous since it gives more detail.

Gardner not only mistook the structure of the Pernambucan stone reef or consolidated beach, but he describes the mythical coast reef in the same general terms. So no wonder that the whole structure and character of the coast reefs of Brazil have remained a puzzle to the geologist and the geographer, and that it has been a serious question as to what the Abrolhos reefs really were, one author declaring that they were formed of decomposed gneiss! The fact is, that the reefs of Brazil are of two kinds, the coral reefs and the consolidated beaches, which last are occasionally separated from the coast line, and sometimes run across the mouths of rivers, as at Porto Seguro, Pernambuco, &c., like narrow rock walls resembling artificial breakwaters. These, so far as I have observed or can learn, are never found at any great distance from the shore, neither are they continuous over any great distances. The Brazilian calls them *recifes.* That at Pernambuco, owing to the great trade with that port, has become famed, and many travellers have seen it and been puzzled over it.

From the Abrolhos northward to the shore of Maranhão, at very irregular and often very long intervals, are scattered true coral reefs, which lie in patches at a short distance from the shore, there being usually navigable channels between them and the mainland. It is very rare that one of these reefs is dry other than at very low tide, and the sea constantly breaks on its outer edge. These reefs are known by the Brazilians as *recifes.** Coral and coral-rock are called *pedra de cal*, or limestone. The whole confusion has evidently arisen in this way: A traveller has visited Pernambuco, and has seen the reef. He hears it called the *recife*, and is told that the coast of Brazil is bordered by *recifes.* On his way up or down the coast he sees from time to time the sea breaking against the coral reefs in a long line of surf. His pilot tells him that is the *recife.* He perhaps asks if it is made of *coral;* but this word in Brazil is almost exclusively applied to the precious red coral (*Corallium rubrum*), and the pilot says, "No, it is made of *pedra*, or *pedra de cal.*" Some of the coral reefs are laid bare at low water, but their great width is not visible from the deck of a ship sailing at a distance, and they look like walls. So the coral-reefs of Brazil have come to be confounded with the consolidated beaches; indeed, I should never have suspected the real character of the coral reef of Santa Cruz, close to which I passed in a steamer, had it not been that I had previously examined the coral reef at Porto Seguro. It seemed a low, narrow wall, and there was nothing that I could elicit from the pilot or captain that would have led me to suppose that it differed from the inner consolidated beaches at Porto Seguro or Santa Cruz.

I made my first acquaintance with the coral-reefs of Brazil while at Porto Seguro, in 1866. I had been for

* This word, as I shall show further on, is derived from the Arabic word *razif*, which means literally a pavement. Sometimes the form *arrecife* is used. Compare *reef*, Eng., *riff*, Germ., and *récif*, French. See *Recife* in Index.

several days collecting on the stone-reef or consolidated beach, before I heard of the outer reef. There was nothing that I could learn from the fishermen that could warrant me in considering it as anything else than a consolidated beach; but my studies of the latter class of reefs had satisfied me that the outer reef could not possibly be of that character, and when, on a spring tide, I visited it, in company with Mr. Copeland, I was not astonished to find it composed of coral. On that short visit I collected all the principal corals found on the coast, and made out quite satisfactorily the general structure of the reef, and of the chapeirões which surround it. I soon felt satisfied, from all that I could learn from the garoupa fishermen, that the Abrolhos *recifes* were true coral reefs, and my companion and I were on the point of visiting them when we received letters from Professor Agassiz, desiring us to come immediately to Rio, to return home with the expedition.

At Rio I found Mouchez's chart of the Abrolhos, on which is a note describing the reefs of the Abrolhos so clearly as to leave no doubt as to their being immense coral reefs. In order to settle the question I returned to Brazil the next summer, and went over the reef grounds of the Abrolhos as thoroughly as my time and the slim means I could command would allow.*

The coast of Brazil, north of Cape Frio, has quite a rich polyp-fauna, but very few of the madreporian polyps cross the southern tropic. The Bay of Rio offers only insignificant representatives of that order. All the specimens I could obtain were Astrangiæ, growing in small scattered

* I regret exceedingly that I have not yet been able to use the dredge on the Brazilian coast; but I hope that my studies of Nature in the tropics are only the preface to more thorough and detailed explorations in the future.

cells, on stones and dead shells in the shallow water off the Ilha do Governador. Professor Agassiz tells me that a fine species of Porites had been collected at Rio, and he also informs me that corals have been found at Desterro, in the bay of Santa Catharina,* a locality which, though extra-tropical, so far as latitude is concerned, is not really so in the character of its climate. Many species of Actinias are found in the harbor of Rio, together with one species of that curious locomotive halcyonoid, *Renilla* (*R. Danæ* Verrill). On the masonry of the new Custom-House docks at Rio I collected in abundance a slender, branching, tender, nodose halcyonoid undetermined.

As we go northward from Cape Frio the madreporians become quite common on the rocky shores, though the species are not numerous, and they are associated with species of *Millepora*, *Zoanthus*, and *Palythoa*, and various gorgonians. I have already called attention to the coral fauna of Guarapary and Victoria, and I have stated that I have no evidence of the existence of any banks of living corals or reefs south of the region of the Abrolhos. Here the conditions for the growth of coral reefs on a large scale are remarkably favorable. Over large areas the water covering the great submarine shelf, on which the islands are based, is much under one hundred feet in depth, and it is warm and pure. So it is not to be wondered at that very large coral reefs, both fringing and barrier, are found here.

When the tide goes out there is seen extending round about one half the circumference of the island of Santa Barbara a fringing reef, shown in the little map on page 177.† One may then walk out on its level surface as on a

* Collected by Mr. F. Müller.

† This reef is also represented in the cut on p. 175.

wharf, and from its ragged edge look straight down through the limpid green water and see the sides of the reef and the sea bottom covered with huge whitish coral-heads, together with a wealth of curious things not to be obtained without a dredge.

The surface of the reef, though flat, is somewhat irregular. It rises but a short distance above low-water mark, and it is overgrown with barnacles, shells, mussels, and serpula-tubes, together with large slimy patches of the common leather-colored *Palythoa.* The reef abounds in small pools, some shallow and sandy, others deep, rocky, and irregular. The former often contain scattered masses of corals, particularly *Siderastræa* and *Favia*, and they abound in small shells, crabs, *Ophiuræ*, &c.; but the deep pools are the richest in life. These are usually heavily draped on the sides with brilliantly tinted sea-weeds and corallines, the bare rock being gay with bryozoa and hydroids. The most common coral of these pools is *Siderastræa stellata* Verrill.* This is a coral growing in rounded or hemispherical masses with small cells. Professor Verrill states that it "differs from *S. radians* in having larger cells, which appear more open; thinner septa, and consequently wider intervening spaces; and four complete cycles of septa." This coral rarely ever forms masses more than six inches in diameter, though I have collected specimens 8–12 inches in length. Its color when alive varies much. Usually it is of a very pale pinkish tint, almost white, and it is not unfrequently blotched with deepened spots of the same color.

It is often seen growing in tide-pools in the reefs and rocks, with only just water enough to cover it. On the stone-

* Trans. Conn. Acad., 1868, p. 353.

reefs, as at Guarapary and Porto Seguro, it is often seen in the pools exposed for several hours to the full blaze of the sun, and of course liable to great and sudden changes of temperature. These pools are also likely to be very much freshened by heavy showers while the tide is down. Near Bahia I have seen it growing in tide-pools above sea level, and to which the waves had access only at high tide. It seems to stand exposure to the air with impunity; for on the reef at Porto Seguro I have observed it exposed to a hot sun for an hour or more during a spring tide. It is not confined to the tide-pools, but occurs also on the submerged border of the reef, where I have collected it in a depth of 3 – 4 feet at low tide. This species appears to range from Cape Frio northward beyond Pernambuco. Professor Verrill has separated under the name *var. conferta* what appears to be a variety, and which is distinguished by having in the central portion cells deformed through crowding. These cells are irregular, and deeper than the normal ones near the basal margin. Their septa and dividing walls are more elevated and convex, and sometimes adjacent cells are united by the breaking down of these walls. Usually with *Siderastræa* occur two species of *Favia,* — *F. gravida* Verr. and *F. conferta* Verr. The former — a solid, heavy coral sometimes flattened, and incrusting stones or dead coral, sometimes in round masses rarely more than three or four inches in width — is allied to *F. Ananas* and *F. Fragum* of the West Indies; but Professor Verrill shows that it has more spiny costæ than either of those species, while the septæ are narrower and sharper.

The other species, *F. conferta,* forms small hemispherical masses of about the same size as in the former species. It is interesting because of its affinity to *Goniastræa,* while it stands, according to Professor Verrill, in some respects intermediate between the genera *Favia* and *Mæandrina*. A hemispherical or almost globular coral with large cells, *Acanthastræa Braziliensis* Verrill, which is common on the border of the reefs below low-water mark, is rarely ever found in the tide-pools, though I have occasionally collected it from the deeper ones. On the edge of the reef it grows to a very large size. The color is a pale gray, when seen in the water. This is one of the corals most instrumental in building up the reefs. Occasionally an *Agaricia*, closely allied to, if not identical with, the West Indian *A. Agaricites* Edw. and Haime, is found in one of the little pools. It is a thin, spreading coral, attached by one side, and resembles the flat woody fungi growing on dead timber or the stumps of trees. This species often occurs almost at the water level. At Villa Velha, and elsewhere, it is found attached to *Mussæ*. It appears to extend along the whole coast between Victoria and Cape São Roque. The above are the principal madreporian corals found in the tide-pools. I have only rarely observed millepores growing in the pools, and these either in the deeper or the broad, sandy-bottomed ones on the reef. The only species was *M. Braziliensis* Verrill, a species easily recognized among the Brazilian milleporæ by the peculiar form of its branches, which Professor Verrill has described as "erect, angular, or flattened, or forming broad, convoluted, and folded rough plates, with acute edges and summits; the sides covered with sharp, irregular, angular, crest-shaped, and conical prominences varying much in size and elevation, often becoming continuous ridges, usually standing at right angles to the sides of the branches." Professor Verrill suggests that this may, after all, be only a variety of his *M.*

nitida, but I have never seen any intermediate forms. This *M. Braziliensis* sometimes grows to quite a large size, and it ranges along the whole coast from the Abrolhos to Pernambuco. On the submerged border of this reef occur the beautiful species of *Mussa* and *Symphyllia*, with which Professor Verrill has associated my name.*

While at Santa Barbara the weather was unfavorable for an examination of the reef below low-water mark, and my collections were principally made from the surface of the reef and from the tide-pools.

Of gorgonians I collected the same species as I found at Victoria, namely, *Hymenogorgia quercifolia*,† *Eunicea*

* *Mussa Harttii* Verrill is distinct from all others in its regular cells and its costæ, which are furnished with strong, sharp, and recurved spines. It grows in great abundance on the submerged border of the reefs in 3 – 6 feet water at low tide, forming beautiful hemispherical bouquets a foot or two in diameter. The color of the coral when alive is pale whitish. It is very fragile, and seems to prefer sheltered localities, and I do not believe that it grows on the outside border of the reefs, which is exposed to the heavy surf. It occurs, also, at Pernambuco, Porto Seguro, and elsewhere. With it is found another form, which, though closely resembling it, Professor Verrill considers to be generically distinct, and has called *Symphyllia Harttii*, giving the same specific name in case it should, after all, prove to be identical with the above.

† This *Hymenogorgia* is exceedingly abundant on the Brazilian coast, representing in the Brazilian polyp-fauna the *Rhipidogorgiæ* of the West Indian fauna. Professor Verrill describes it as follows: "It forms broad, fan-shaped fronds, often two feet high and a foot broad, consisting of broad, foliaceous branches, often resembling oak-leaves in form; but at other times large, oval, and irregularly incised or palmate. The branches of the axis are slender and rounded, and pass through the fronds like the midribs of leaves. The rather conspicuous, flat cells are scattered over the sides of the fronds." The color when alive varies from an ashen-gray to a light yellow or pink. The latter color often deepens in spots. It grows on the rocks and stone-reefs in clear water, and on the submerged border of the coral-reefs. It ranges downward to a depth of 5 – 6 feet or more. It is apt to fade in drying. On the fronds a small parasitic *Ovulum* (*O. gibbosum*) is often found. Professor Verrill has more recently restored both this and the West Indian *Rhipidogorgia flabellum* to the genus *Gorgonia*.

humilis,* and *Plexaurella dichotoma.*† The shallow, sandy-bottomed pools over the reef are more or less incrusted with irregular masses of a calcareous deposit, consisting of an agglutination of serpula-tubes, nullipores, bryozoa, &c., which usually lie in rather loose flakes on the surface, and are easily turned over by the hand; much of the reef itself is covered in the same way. The under sides of these masses, which are generally concave, are incrusted with beautiful bryozoa, and form hiding-places for great numbers of species of marine worms, chitons, little crustaceans, ophiurans, &c. Beneath this crust nestle in great abundance several interesting species of "brittle stars." The most common of these is the large *Ophiura cinerea* Lyman; almost quite as abundant are the pretty *Ophiothrix violacea* Müller and Troschel, and *Ophionereis reticulata* Lütken. The other species collected on the reefs were *Ophiomyxa flaccida* Lütken, *Ophiactis Krebsii* Lütken, and *Ophiolepis paucispina* Müller and Troschel.‡ All of the above species are, according to Professor Verrill, members also of the West Indian fauna, and it is interesting to observe the occurrence here of the huge West Indian starfish, *Oreaster gigas* Lütken, of which I collected two fine specimens at Santa Barbara, together with another West Indian species, *Linckia ornithopus* Lütken. The sea-urchin, common at Santa

* *Eunicea humilis* Edw. and Haime is a gorgonian easily recognized by its growing in low, densely branched clumps, with short thick branches, the color being usually lemon-yellow. It is a very common species along the coast.

† *Plexaurella dichotoma* Kölliker is another gorgonian, with few, large, round, blunt-tipped and thick-barked branches of a dark brown color when dry. It is as common as the preceding.

‡ In the collection of radiates made by Mr. Copeland and myself, on the Thayer Expedition, there are several additional species of Ophiurans.

Barbara and along the Brazilian coast, Professor Verrill considers to be the same as the West Indian species *Echinometra Michelini* Desor, and I am unable to detect any difference between the Abrolhos specimens and those which I have collected at St. Thomas. Many of the species of shells common on the Brazilian coast, south of Bahia, appear to be identical with West Indian forms, and one is astonished to find at Pernambuco, Bahia, the Abrolhos, and Victoria the large *Cassis Cameo*,* a shell so common in the West Indian waters. Professor Verrill has called attention to the number of species of Echinoderms in the Brazilian fauna that are identical with West Indian forms, in contrast with the almost complete distinctness of the polyp-faunæ of the two regions; and he has suggested that an explanation might be found in the fact that the Echinoderms remain longer in the swimming larval form than the polyps, and may be carried to greater distances by currents. There is no chance by which West Indian species could be carried south of Cape São Roque, owing to the equatorial current which sets along shore from that cape northwestward across the mouth of the Amazonas, whose fresh waters must have long presented a barrier to the migration southward of shallow water species.† It is interesting to observe that on the Brazilian coast, south of Pernambuco at least, we find no *Diademas* or *Tripneustes*, forms so exceedingly characteristic of the West Indian Echinoderm fauna.

The material composing the reef is an exceedingly hard, whitish limestone, ringing under the hammer, and, so far as I had an opportunity to examine it,— for the Brazilian reefs

* Formerly called *Cassis Madagascariensis*.

† Professor Verrill suggests, however, that the species found in both faunæ may have migrated *northward* from Brazil.

are never broken up by the surf, — showing no distinct trace of organic structure. The Santa Barbara reef extends around about one third of the island, and on the northwestern side it reaches across to the "Cemetery," so that when the tide is down that islet is joined to the main island by a broad, level platform of rock,* diversified by tide-pools, and forming an excellent collecting-ground for the naturalist. The reef, built up principally of *Acanthastræa*, *Siderastræa*, &c., has completed its growth on arriving at low-tide level, the upper surface being still farther added to by serpulæ, bryozoa, corallines, barnacles, &c., together with the coral-sand and *débris* of shells accumulating on the reef.

So far I have spoken only of fringing reefs, but there are other coral structures of greater interest in these waters. Corals grow over the bottom in small patches in the open sea, and, without spreading much, often rise to a height of forty to fifty or more feet, like towers, and sometimes attain the level of low water, forming what are called on the Brazilian coast *chapeirões*.† At the top these are usually very irregular, and sometimes spread out like mushrooms, or, as the fishermen say, like umbrellas. ‡ Some of these chapeirões are only a few feet in diameter. A few miles to the eastward of the Abrolhos is an area with a length of nine to ten, and in some places a breadth of four miles, over which these structures grow very abundantly, forming the well-known Parcel § dos Abrolhos, on which so many vessels have been wrecked.

* The reef is represented in the woodcut on page 175, but the cemetery has unfortunately been omitted. See also map on page 177.

† Singular *chapeirão*, pronounced *shăp-a-równ*[g], the *n*[g] representing a strong nasal. The word means literally *a big hat*. In the plural it is *chapeirões*, pronounced *shăp-a-rō'-en*[g]*s*.

‡ The Dutch used to call them "Jesuits."

§ The word *parcel* means *shoal* or *hidden rock*, plur. *parceis*.

I visited in my launch the northwestern part of this reef, where the chapeirões were sufficiently scattered to allow me to sail about among them.

Among these chapeirões I measured a depth of sixteen to twenty metres, and once, while becalmed, I found twenty metres alongside one chapeirão and three metres on top. The chapeirões, as a general thing, are rarely ever laid bare by the tide. They are here, as elsewhere, of all heights and dimensions; but in no case do they reach low-water level, nor, according to the testimony of the fishermen and whalers, are they ever in any part uncovered. They do not coalesce here to form large reefs as they do to the west of the islands. When the weather is clear and cloudless and the water calm, these chapeirões can be readily distinguished at a considerable distance. The surface of the sea appears to be flecked by shadows from a sky full of scattered cloudlets, producing a striking effect. The water, being shallow and clear, and with a sandy bottom, is of a very light greenish tinge, like that of the Niagara River at Buffalo. The general color of the chapeirões is brown, from their being incrusted with patches of *Palythoa*, and their position is marked by brownish spots on the surface of the sea. In the daytime a launch may sail in safety among them in calm weather, and a small vessel may traverse some of the chapeirão grounds without danger, but large ships are likely to find themselves in a labyrinth from which escape is not easy.* In windy weather the waves break over the chapeirões, but if there are white caps beside, and a cloudy sky, their position cannot be made out, and it is safest to keep well away from them. In stormy weather there is nothing to mark their position, and they

* See Chart of Abrolhos by Mouchez, note.

are very dangerous. Sometimes vessels striking heavily on small chapeirões break them off, and escape without receiving any serious injury, as has been remarked by Mouchez. At other times a vessel may run upon one of these structures and stick fast by the middle of the keel, to the amazement of the captain, who finds deep water all around, the vessel being perched on the chapeirão like a weather-cock on the top of a tower. Ordinarily in passing the Abrolhos vessels and steamships go outside of these reefs to the eastward in sight of the islands. It is not easy, however, to calculate one's distance from a point at sea, and especially from a light by night, and many vessels, notwithstanding the lighthouse, have been wrecked upon them. West of the islands there is deep water, and no chapeirões, and between the islands and the Parêdes there is a channel about eight miles in width, with plenty of water and no obstructions. The best way in passing close to the Abrolhos is to go to the westward of the islands, where one may run close to them with safety even in the night-time. There is then no danger whatever, and the sea is smoother. On the return voyage from Rio, September, 1867, the American steamship "South America" was, at the suggestion of the writer, taken through this channel. In case of necessity, good anchorage may be found close in by the island of Sta. Barbara on the northern or southern side, as the direction of the wind may determine.

Eight miles northwest of the islands, between them and the mainland, is the Parcel das Parêdes, literally, The Shoal of the Walls, an irregular area about seventeen miles long from north to south, and some nine miles in width, occupied by very extensive reefs and chapeirões. Mouchez has only given the general outline of the Parcel, which was all that was really necessary for his chart. The reefs within the Parcel are not drawn from an actual survey, and have no approach towards accuracy.

In the northern part of the Parcel the chapeirões so closely unite as to form an immense reef, which has grown upward to a little above the level of low water, and is quite uncovered at low tide. This reef, as are all the others, is exceedingly ragged in outline, full of indentations, and abounds in shallow pools. The fishermen describe two channels that enter the reef from the north, and almost separate it into three parts. My captain, Jacob Torgjusen, an intelligent Dane, says that the water in these channels is quite deep. The northeastern part of this reef is called the Recife do Lixo, because of the abundance of a shark-like ray called the *lixo*, which is furnished with large crushing teeth, and frequents the reef in search of shell-fish, on which it feeds.

RECIFE DO LIXO.

I spent one tide on the Recife do Lixo, during the full moon of the 13th of August, 1867, when the reef was uncovered, and examined it quite carefully.

The surface of the reef was remarkably even in height, and covered largely by calcareous sand, on which were patches of dead coral incrusted with millepores, barnacles, serpulæ, &c., with occasional living corals, such as *Siderastræa stellata* and *Favia*, and perhaps a *Porites solida*. The dead corals, nullipores, &c., usually forming incrusting masses over the sand, so loose as to be easily turned over, affording a rich harvest of Ophiurans, among which *Ophiura cinerea*, *Ophionereis reticulata*, and *Ophiothrix violacea* were especially abundant. In some situations sea-urchins, *Echinometra Michelini*, were very plentiful. *Volutæ* and *Cassis Cameo* may frequently be picked up. The reef is not very rich in shell-fish, but abounds in crustaceans.* A large naked

* On the Thayer expedition Mr. Copeland and I collected large numbers of crustaceans at all the principal localities on the coast between Rio and Bahia. On my second visit to Brazil I was too much engaged in geological studies, and in my examination of the reefs, to make extensive collections. The few crustaceans brought home from the second journey, my friend, Mr. S. I. Smith, of the Sheffield Scientific School of New Haven, has been kind enough to examine and describe; and since writing the above he has published a paper in the second volume of the Transactions of the Connecticut Academy of Arts and Sciences, entitled "Notice of the Crustacea collected by Professor C. F. Hartt on the Coast of Brazil in 1867, together with a List of the described Species of Brazilian Podophthalmia." Mr. Smith enumerates the following species as occurring on the reefs of the Abrolhos: *Milnia bicornuta* Stimp., *Mithraculus coronatus* Stimp., *Mithrax hispidus* Edwards, *Xantho denticulata* White, *Chlorodius Floridianus* Gibbes, *Panopeus politus* Smith, *Panopeus Harttii* Smith, *Eriphia gonagra* Edwards, *Goniopsis cruentatus* De Haan, *Dromidia Antillensis* Stimp., *Petrochirus granulatus* Stimp., *Calcinus sulcatus* Stimp., *Clibanarius Antillensis* Stimp., *Alpheus heterochelis* Say, *Gonodactylus chiragra* Latreille (?).

From other localities were described the following: *Callinectes Danæ* Smith, Pernambuco. *C. ornatus* Ordway, Caravellas. *C. larvatus* Ordway, Bahia. *Achelous spinimanus* De Haan, Bahia. *A. Ordwayi* Stimp., Bahia. *Uca cordata*, Bahia. *Cardiosoma quadratum* Saussure, Pernambuco. *Clibanarius vittatus* Stimp., Caravellas. *C. sclopetarius* Stimp., Caravellas. *Scyllarus æquinoxialis* Fabr., Bahia. *Panulirus echinatus* Smith, Pernambuco. *Palæmon Jamaicensis*

mollusk (*Aplysia*)* four to five inches long, grayish, and ornamented with dark rings and spots, may sometimes be found very abundantly on these reefs. It gives out a very copious deep purple fluid when handled. *Octopi* are very common in the cracks and crevices of the reef. The shallow pools are often very rich in life.

The uniform level of the surface of the reef laid dry is very remarkable. So very even and unencumbered is it that a loose coral a foot in diameter turned over, or a *cassis* lying on the surface, attracts attention at a long distance. The reef is so protected that the waves have no power to break off its edges and encumber its surface, as is the case with the coral reefs of the Pacific. Sand, resulting from

Olivier, Penêdo. *P. forceps*, Edwards, Pará. *P. ensiculus* Smith, Pará. *Peneus Brasiliensis* Latreille, Bahia. *Xiphopeneus Harttii* Smith, Caravellas.

The list of the Brazilian Podophthalmia is too long to be inserted in this volume. A very large number of the species examined by Mr. Smith have been identified with West Indian or Floridan forms. Mr. Smith suggests that one reason why my collection is so much richer in proportion in these forms than the Brazilian collections heretofore made, may be because my collections were made on the reefs and rocky parts, while the others were made at Rio, where there are no coral-reefs. The collecting-grounds at Rio are rocky as well as sandy. I suspect that the true reason is to be found in the fact that the crustacean fauna changes its character south of Cape Frio.

I called Mr. Smith's attention to the name *Cardiosoma Guanhumi*, in which the specific name appeared to have been derived from the word *Guayamú* or *Guâinumu*, the Tupí name for the species. The former is the way it was written for me by a Brazilian, but Fonseca gives the latter form. I feel quite sure that the name *Guayamú* was applied to several distinct species. In reference to *Uca una*, Mr. Smith has slightly misunderstood me. I do not remember the present vulgar name. Vça-una I found in Piso. It is Tupí, and means simply *black crab*, *vça*, or more properly Uçá, meaning *crab* and *una*, *black*.

* Mr. J. G. Anthony kindly informs me that the species, of which I furnished numerous specimens to the Museum of Comparative Zoölogy, is probably *A. Argo d'Orb.* Sander Rang describes a large species, *A. Braziliensis*, from the Bay of Rio. Hist. Nat. des Aplysies, p. 55, Pl. VIII.

the decay of coral, the breaking up of shells, &c., accumulates very slowly. The reef has grown as high as is possible, and is now dead, and at the lowest tide it is not more than two feet out of water. An ordinary tide would not uncover it completely.*

An irregular raised border, consisting principally of a growth of millepores, serpulæ, barnacles, &c., sometimes a foot, more or less, in height, separates this part of the reef, which is uncovered at low water, from that which is always submerged. It is here that the waves break at low water, and this favors the growth of these animals more than elsewhere. From the border this reef generally slopes off gently towards the edge under water, where it drops down perpendicularly into deep water, as at the islands. This submerged border at the Lixo has only a few feet of water on it at low tide, and one may usually wade out to its edge and collect. It is a perfect garden covered with growing corals of large size. Here grows *Acanthastræa* in large heads, more abundant on the edge of the reef. *Millepora nitida* Verrill forms pretty rosettes. This interesting species is thus described by Professor Verrill: † —

" Corallum forming low rounded clumps, four to six inches high, consisting of short, rapidly forking, rounded or slightly compressed branches, about .4 to .8 of an inch in diameter, which have remarkably smooth surfaces, and are obtuse, rounded, or even clavate at the ends. The larger pores are small, very distinct, round, evenly scattered over the surface, at the distance of .06 to .1 of an inch apart. The small pores are very minute, numerous, scat-

* The height of the reef is probably in part due to the recent uprise of the land.

† Trans. Conn. Acad., 1868, p. 362.

tered between the larger ones, and often show a tendency to arrange themselves in circles of six or eight. The tissue is, for the genus, very firm and compact." The color, when alive, is light pinkish. This species is abundant on the submerged border of the *Recife do Lixo*, growing in from three to four feet of water at low tide. Some of the rosettes I collected lay quite loose and without any attachment to the reef.

Among the millepores I obtained on the Brazilian coast Professor Verrill has distinguished three forms which so closely agree with *Millepora alcicornis* Linnæus, that he has separated them as varieties of that species. One of these, *var. cellulosa* Verrill, I found at Pernambuco, but I did not see it alive. Professor Verrill describes it as follows: "Corallum consisting of numerous, irregular, rather short branches, arising from a thick base, branches proliferous or digitate at the ends, the last division short, mostly compressed and acute at the tips. Some of the branches occasionally coalesce, so as to leave small openings. Cells numerous, crowded, rather large for the genus, each sunken in a distinct depression, the wall rising up into an acute ridge between them, texture rather open and coarsely porous."

Another form, which differs from the other in its "somewhat more porous texture, and the greater regularity and more scattered arrangements of the cells," together with its round and digitate branches and branchlets, having three to five short compressed divisions at the end, Professor Verrill has referred with doubt to the variety *digitata* of Esper. The third variety distinguished by having the branches in the same plane, and coalescing in such a way as to leave frequent openings, Professor Verrill has referred to the *M. fenestrata* Duch. and Mich.

Both of these last occur abundantly along the coast from Cape Frio northward, as far as I have examined. They seem to prefer the edge of the reef, where they form beautiful broad frills of a light yellowish-brown or pinkish color. In collecting these millepores, I was struck with their powerful stinging properties, and they burned me sometimes like hot iron, producing a sensation precisely like that caused by the *Physalia*, or our Northern jelly-fish, *Cyanea*. I was stung in the same way by the millepores of St. Thomas. The fishermen who were with me on the Brazilian reefs handled these corals with impunity, but they called them "sea-ginger," and told me that they were accustomed to play practical jokes on land-lubbers by persuading them to taste them. I presume that a thin and delicate skin makes me more sensitive than most persons to the stinging properties of these animals.*

The *Siderastræa* and *Faviæ*, already described in speaking of the Santa Barbara reef, are found on the Recife do Lixo, both in the pools and on the submerged border, and associated with them are a few forms which appear to be somewhat rare, as I could find but few specimens of them. Among these is the species of *Favia*, described by Professor Verrill under the name *F. leptophylla*, an interesting species which forms large hemispherical corals, easily recognized by "the very open, deep, rounded cells; few, thin, projecting septa; and thin distinct walls." Another and beautiful coral is *Heliastræa aperta* Verrill, which the following description, almost in Professor Verrill's own words, will serve to distinguish: The corallum is large, more or less regularly hemispherical, sometimes subspher-

* This stinging property of the millepores is in accordance with their acelephian structure, first announced by Professor Agassiz.

ical, and often a foot and a half in diameter. The texture is open and light, which character, together with the thinner and more acute septæ, serves to separate it from *H. cavernosa* Edw. and Haime, which it resembles in the large size and prominence of its cells. These in *H. aperta* are circular, large, moderately deep, with a broad central area, the margin projecting about .08 inch above the general surface. Septa in three complete cycles, narrow, thin, subequal, the summits considerably projecting, angular, acute, the inner edges nearly perpendicular, finely toothed, often with a distinct paliform tooth at the base. Columella well developed, of loose, open tissue. Costæ elevated and thin, rising obliquely upward to the summits of the septa, finely serrate. Walls very thin, inconspicuous. This species seems to be more abundant in the Bay of Bahia than in the Abrolhos region, and I have frequently seen it there in the heaps of corals brought from the Island of Itaparica to the city for burning into lime.

A very pretty *Pectinia* (*P. Braziliensis* Edw. and Haime) is another of these apparently rare forms, of which I have found only a single specimen, which was growing on the reef borders at the Lixo, in about two feet of water at low tide.

On the Recife do Lixo I collected a few specimens of a massive *Porites*, resembling *P. Guadaloupensis* Duch., which is very abundant on the Porto Seguro coral-reef. It is sometimes of a bright sulphur-color, though it varies very much in tint. Professor Verrill has described it as a new species under the name *P. solida*, and states that it differs from the West Indian forms in its larger and deeper cells, thicker walls, wider and more crispate septa, and more solid structure.

The beauty of the polyp growth on the submerged reef border is much enhanced by the great luxuriance of the gorgonians, which are the same as those already observed elsewhere, but on the Lixo reef I discovered a beautiful new species, described by Professor Verrill under the name of *Gorgonia gracilis*. It grows in little tufts, about six to eight inches high, with few slender and very delicate branches. The color is yellow or purple, and in rough water the species is apt to be overlooked by the collector, from its resemblance to a sea-weed. Towards the edge, as the water deepens, the reef grows more and more irregular. It is full of holes, and almost wholly composed of live and growing corals, which furnish a very insecure footing. On the edge and sides grow immense coral-heads, and the *Mussæ* are especially abundant. The outline of the border is exceedingly ragged.

The height of the perpendicular edge on the western side of the Porto Seguro and Lixo reefs varies very much, being in some places three or four feet, in others ten feet or more. I could not examine it on the eastern side, owing to the surf. I have introduced a diagram, showing the reef as seen in section, with the distribution of the different species of corals indicated. Just alongside the reef, at least on the western side, wherever I have examined it, — and the same holds good of the Porto Seguro reef, — the bottom slopes rapidly away from the reef-edge, and is composed of a soft, bluish, calcareous mud, washed from the top of the reef, which makes the reef appear much lower than it really is. A short distance away, in some places, a depth of seventy to eighty feet may be found. The diagram on page 211 represents a sketch of the edge of the reef of the Lixo, with soundings made by myself. The large reefs appear to

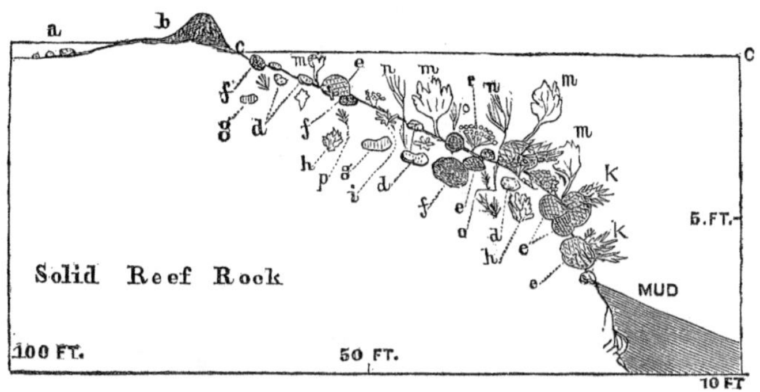

SECTION ACROSS BORDER OF LIXO REEF.

a. Tide-pool with *Siderastræa stellata* and *Favia gravida.*
b. Low dike-like border of serpula-tubes, barnacles, &c.
c c. Low-water level of spring tide.
d. *Siderastræa stellata.*
e. *Acanthastræa Braziliensis.*
f. *Heliastræa aperta.*
g. *Porites solida.*
h. *Millepora Braziliensis.*
i. *Millepora nitida.*
k. Varieties of *Millepora alcicornis.*
m. *Gorgonia (Hymenogorgia) quercifolia.*
n. *Plexaurella dichotoma.*
o. *Gorgonia gracilis.*
p. *Eunicea humilis.*
r. *Mussa Harttii.*

have been formed not only by the upward growth of large patches, but by the filling in and coalescence of chapeirões, which is a feature not hitherto spoken of in the growth of coral reefs. The ponds on the surface of the reef probably mark the intervals between the chapeirões where the filling in is nearly complete; though they may in some cases mark spots where the corals have been killed by the drift of sand.

I visited the eastern side of the Lixo, but the waves were breaking with too much force to allow me to see anything distinctly. The boatmen said that this reef drops down perpendicularly into deep water. I observed no sand-banks upon it. On the western side of the Lixo there are

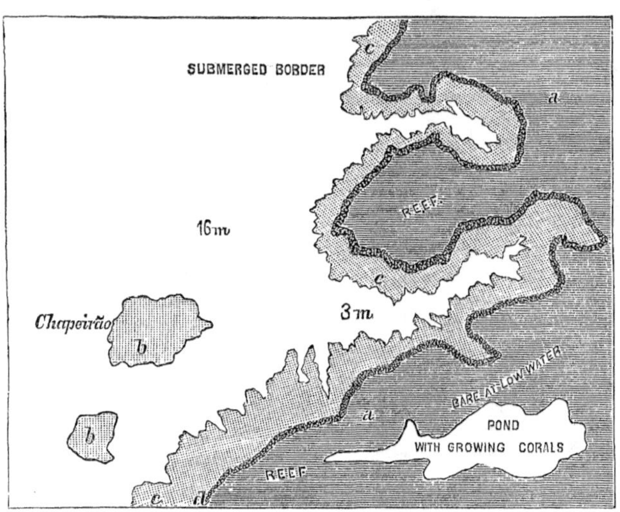

but few chapeirões, but the other sides are bordered by an abundance of them, and they stretch off southward, forming with two other reefs, called the Recife de Leste and Recife da Pedra Grande, the rest of the Parcel. The Recife da Pedra Grande was described by Jacob as being like a wall, quite straight, some three miles in length, and only two or three hundred feet wide. Southwest of the Parêdes are several other quite large reef-grounds. One of the reefs, Corôa Vermelha, has become converted into an island by the heaping up of sand in the centre. There are no reefs or chapeirões south of the Mucury. North of the Recife do Lixo are some small reef-patches, and about ten miles northward, a little to the west, is a very dangerous reef-ground some three miles long, and one and a half to two miles wide, called the Timbebas, between which and the reefs to the south is a wide open channel called the Itanhaem. This, as Mouchez says in his chart, is the most dreaded of all the reefs, because it is situated just beyond the limit of visibility of the coast, and there is nothing to mark its position, so that even the pilots of the coast keep well away from it. I once passed close to it in the steamer Santa Cruz, and saw the waves beating over it. Small coral patches occur close in shore along the coast northward as far as to Point Carumba, just off which is a reef-ground eight miles long and three to five miles wide. Mouchez says that the reef is uncovered at low tide in the western part, and that the rest of the ground is covered by chapeirões. These reefs are the celebrated Itacolumís. The next important reef-ground stretches across the mouth of the Bay of Porto Seguro. I found the reef uncovered over an area as far as I could see north and south while standing on the reef, with a width in some places of a mile; but it was exceedingly difficult to judge of distance on so plane and monotonous a surface. This reef is surrounded by chapeirões. The same line of reefs extends northward, passes the low point north of Porto Seguro, leaving a deep channel, allowing the passage of steamers, and is continued across the bay of Santa Cruz, as is represented in the sketch-map on p. 233.

In the summer I passed close to the northern part of the reef off Santa Cruz. It has, like the other principal reefs, finished its growth, and is in part converted into an island, on which I observed a few mangrove-trees growing. A schooner had struck on the outer side of this reef, and had afterwards been carried over and sunk erect, just behind the reef. Its masts were standing, showing a depth alongside of the reef of about thirty feet. Reef patches occur along the shore in the vicinity of Camamú. Quieppe Island is surrounded by them on all sides, and a little bay between Camamú and Boyapeba is full of chapeirões, while the entrance to the Barra Carvalhos, according to the "South American Coast Pilot," is similarly encumbered. Along the shores of Itaparica Island there are extensive coral-banks, from which coral is taken for the purpose of being burned into lime; and at low tide coral is largely collected from banks lying off Periperí, on the Bahia Railroad, to which place it is brought and burned. I saw large quantities of it at Bahia and Periperí. The corals were of the same species as are common on the Abrolhos reefs, but *Heliastræa* was more common. Limestone is very scarce on the Brazilian coast, and corals are largely used for making lime. They call the coral *pedra de cal** in Brazil, for this reason.

Still going northward, I am not aware that any coral-patches are to be found until we reach Maceió, but in the harbor of this town I examined quite an extensive one, which lies some distance, perhaps a mile, off the town, and at low tide is uncovered for a large irregular space. Its surface is flat, but irregular, composed of dead corals, and is full of holes. Walking out to the edge of the reef, where the sea was breaking heavily, I found it exceedingly rough. I was unable to see much, as the tide rose rapidly, flowing over the reef, and I had to wade back half a mile over the rough surface to my boat. I saw very few live corals in the ponds.† There were a very few little *Siderastrææ*, and

* It would not be worth stating here that the red coral does not occur on the Brazilian coast, if it were not that from my use of the term *coral* the report had been circulated that I had discovered a reef of red coral at the Abrolhos!

† Professor Agassiz tells me that he has fine millepores from Maceió.

an occasional dwarfed *Acanthastræa.* *Echinometræ* were abundant, and English sailors were spearing the large *Aplysiæ* for the sake of spilling their purple ink. In the hands of a sailor I saw a beautiful *Voluta* like *V. musica,* and I collected a specimen of *Linckia ornithopus.* My boatman said that on the outer edge of the reef large coral-heads grew, and that these were collected and brought on shore for burning into lime.

From the lighthouse at Maceió the reefs could be seen extending in an irregular line along the shore northward many miles. In the vicinity of Pernambuco are patches of growing coral, as Dana has remarked. I have specimens from these patches, but I was prevented from visiting them. I know of no one who has examined the reefs north of Pernambuco with a view to scientific results.

The *Roccas* are a very dangerous cluster of reefs in the latitude of Fernando de Noronha, noted for its annular shape. From the charts of the Roccas, together with what I have learned from those who have visited them, they must be true coral reefs.

Though there is a general resemblance between the Brazilian and West-Indian polyp faunæ in the representative species of *Siderastræa*, *Favia*, *Porites*, *Plexaurella*, &c., &c., yet one who has collected in the West Indies, as at St. Thomas, for instance, or Florida, is much struck with the absence of *Madrepora*, *Mæandrina*, *Diploria*, *Manicina*, *Cladocora*, *Oculina*, and other genera characteristic of the West-Indian fauna. The Brazilian reefs are built up by a very few species, among which *Acanthastræa Braziliensis* appears to be the most common, together with *Siderastrææ* and other massive forms; but the *Milleporæ* and *Mussæ* must contribute more or less to their growth.

CHAPTER V.

PROVINCE OF BAHIA. — COAST SOUTH OF SÃO SALVADOR.

The Tertiary Lands between the Rivers Mucury and Peruhype; their Vegetation, &c. — Colonia Leopoldina and its Coffee Plantations. — Villa Viçosa. — The Canal joining the Rivers Peruhype and Caravellas. — Formation of Beaches and Beach Ridges. — Coast between Caravellas and Porto Seguro. — Monte Pascoal. — Porto Seguro and its *Recife,* or Consolidated Beach. — Santa Cruz and its Reef. — Coast northward to the Jequitinhonha; the Lagoa do Braço, Campos, &c. — The Canal Po-assú and the Rio da Salsa. — Mangrove Swamps between the Jequitinhonha and Pardo. — Cannavieiras. — The Salt Trade of the Jequitinhonha. — Description of the lower Part of the Rio Pardo; Cacáo Plantations, &c. — Coast northward to Ilhéos. — Prince Neu Wied's Description of the Country between Ilhéos and Conquista, Possões, and Cachoeira; the Forests, Campos, Social Plants, &c. — Ilhéos. — Rio and Lagôa Itahype. — Dead Coral Banks. — Rio das Contas. — Bay of Camamú. — Turfa Deposits. — Villa de Camamú. — Coast northward to the Bahia de Todos or Santos. — The Bay of All Saints described. — Ilha Itaparica. — Rio Jaguaripe and Nazareth. — Rio Paraguassú. — Description of River below Cachoeira. — The Tram-road — Sant' Amaro and the Agricultural Institute.

THE Rio Peruhype has its source in the Serra dos Aimorés, a few miles north of Santa Clara, and flows in a deep, narrow valley, worn for the greater part of the distance through the coast tertiary belt. It is not more than fifty miles in length, and is consequently an insignificant stream. Between the Peruhype and the Mucury the lands are almost wholly tertiary, though between the mouths of these two rivers there is a strip of sand and marsh. I owe to the kindness of my friend Sr. Schlobach a profile of a railroad line surveyed by him to connect the Philadelphia

Road at Santa Clara with the Peruhype above Villa Viçosa, which profile is of much interest, since it shows that the surface of the tertiary beds slopes very regularly from the serra to the sea, — a slope which I believe to depend on the slope of the sea bottom on which the beds were deposited, though it is, in part at least, due to a thickening of the beds towards the serra. The country between the two rivers forms a great wooded plain; but the forests are by no means so luxuriant in their appearance as on the São Matheos or Doce. The soil is sandy, weak, and dry, though on the slopes of river valleys it is productive, and the bottoms of the valleys are covered by a very rank growth. This region is traversed by many little streams, all of which have cut for themselves deep valleys or cañons with very steep sides. The upland forests furnish many valuable woods, and rosewood is extensively cut both on the Peruhype and Mucury. In some parts the forest is very sparse of underbrush, but in others it is so matted and tangled with bamboos (*Taquaras*) and young *Airí* palms (*Astrocaryum Airí* Mart.) as to be almost impenetrable. There are large marshy areas on this plain, and these are in part flooded, some of them only during the rainy season, when they form shallow lagoons. The vegetation of the alluvial lands bordering the little streams is luxuriant beyond description. Here one finds an abundance of *Palmitos* (*Euterpe edulis* Mart.) and beautiful tree-ferns. The Bótocudos still hold the country, and I saw their deserted ranchos every few miles along the path. The forest abounds in game. Queixadas, coititús, antas, onças, monkeys, and birds of every description are very numerous.

Nowhere on the route did I see any good exposure of the tertiary beds, but in the descent into the river valleys

I repeatedly saw the drift, which here as elsewhere is composed of red clay full of fragments of quartz. All the streams are "black-water," and so is the Peruhype.* On this last river a colony of Germans, together with some natives of other countries, was established many years ago. It is called Colonia Leopoldina, and consists of a considerable number of fazendas situated on both sides of the river for several miles above São José, the head of navigation for steamers. Some of these fazendas are very large and valuable, as that of the late Sr. João Flach for instance. The cultivated lands are situated on the very edges and slopes of the chapadas, and they were formerly very productive.† Coffee is the staple product, and is noted in Brazil as being of very superior quality, passing by the name of Café de Caravellas. The trees are not allowed to grow to their full height, but are trimmed down, so that the picking is easily done by hand from the ground. Six feet or thereabouts appears to be the average height of the coffee-trees of this region. The trees, when trimmed in this way, are flat-topped with pendant branches, and a cafezal, or coffee plantation, at the colony is a very pretty sight. Mandioca, maize, cotton, and other Brazilian products are cultivated largely, and one riding through the fazendas sees here as elsewhere the orange, banana, lime, citron, pine-apple, &c. The mamão, a species of *Carica*, furnishing a large and savory fruit full of seeds, is frequently seen, though it is to be found almost everywhere along the coast, growing in corners of fields and gardens. The climate of the colony

* Von Tschudi says (*Reisen*, &c., Vol. II. p. 357) that the water of the Peruhype at Villa Viçosa is deep brown, and that, churned up by the paddle-wheels of a steamer, it looks like foaming porter.

† These soils become rapidly exhausted, and many of the old fazendas are very nearly worn out.

is very unhealthy, especially for foreigners, but it is far superior to that of the Mucury below Santa Clara. The Peruhype is navigable as far as São José for coasting steamers, which, however, do not enter the river at its mouth, but by means of a natural canal which, through a very extensive swampy tract, connects the Peruhype with the Caravellas River. The bar at the mouth of the river is very bad, and is rarely crossed by any except very small vessels. The entrance at Caravellas is good. Since the navigation of the Mucury is so difficult, and its bar so bad, it has been proposed to extend the Minas Road across the plains to the Peruhype below the colony, and ultimately make it a railroad. The only difficulty in the way of building such a road would be that of bridging the deep valleys of the streams, which would require very high viaducts or bridges. It is to be hoped that this project will one day be accomplished. Villa Viçosa, a town of some five hundred or one thousand inhabitants, is situated on the right bank of the Peruhype, at a distance of about four miles from the sea, and five or six below São José da Colonia Leopoldina. The lands of the vicinity are very productive in mandioca, and a large quantity of farinha is exported.* From the mouth of the Peruhype to that of the Caravellas the shore consists of a long sea-beach broken only by one river-mouth, — Barra Nova. From Villa Viçosa to the Villa de Caravellas a rather narrow tidal canal extends

* The water used for drinking at Villa Viçosa is very bad. Von Tschudi says that it is collected in shallow holes sunk in a sandy grass-grown plain southwest of the town. When first drawn its taste is very disagreeable, but it becomes potable after standing a day or two. All the towns along the coast, built on the sands of the sea-shore, as a general thing have bad water. This is the case at Caravellas. It is needless to say that the want of pure water is one cause of the prevalence of sickness in these places.

parallel with the coast. Westward of this, as Von Tschudi has remarked, is the *terra firma*, — the tertiary lands trending off northward at a greater or less distance from the canal. This stream, in which the tide ebbs and flows, and whose waters are salt, is sufficiently deep to allow the passage of the coast steamers. Between the canal and the sea the lands, which are but little above the water, consist in part of sandy flats covered with forest or cocoa-palm groves, but the greater part appears to be an immense mangrove swamp, similar to that which lies between the Jequitinhonha and Pardo to the north, and, like this last, it is intersected by a network of narrow canals, which have never been mapped. Gerber's map represents this area as a cluster of small islands, with the sea penetrating from the coast through numerous canals, which probably misled Von Tschudi, who also describes it as an archipelago. The so-called Rio Caravellas is only a narrow estuary, which penetrates into the interior for a distance of about twelve miles. From this estuary near Caravellas the canal extends southward to Viçosa, while at its head empty the Rio Caravellas and the Rio da Fabrica.

The water off this coast is very shallow, and along shore it is usually very turbid. Outside are the very extensive coral-banks, described in the preceding chapter, which not only break the force of the ocean waves, but give rise to a system of currents whose force and direction depend almost entirely on the winds. Owing to the protection of the shores from such powerful wave action as obtains almost everywhere else on the coast, we find here a sloping sand-beach, rising to but a few feet above high-water, and unaccompanied by a beach-ridge. The water along the coast is ordinarily as smooth and calm as that of an inland lake. Near the mouth of the Caravellas and a few miles to the southward, I observed dead mangroves standing in clumps in the water just outside the land. A short distance south of the Mucury, as already remarked, I observed dead trees still erect, and rooted below high-water mark. It would seem that there has been, within a few years, a slight sinking of this part of the coast. The mangrove swamps are sometimes formed in depressions caused by the wearing away of the lower lands by the sea; but this is very rare. They almost always occupy tracts protected from the waves by sand-beaches or otherwise, and filled up with sand-banks, or which, by the gradual rise of the land, have been brought so near the surface that the seeds of the mangrove take root. If one examines the sea bottom along the coast he will find that, especially in the vicinity of large rivers, such as the São Francisco, Pardo, Jequitinhonha, Doce, Parahyba do Sul, and, as a general rule, off flat lands everywhere, the bottom slopes away very gradually, and consists of beds of sand, and but rarely of mud, because such fine material is almost invariably swept out with the current into deep water. A beach must of course be formed within the limits of wave action. If these correspond with the edge of the land, then the beach will skirt it, but where, as is frequently the case, the water is so shallow that the waves break at a distance from the shore, there will shortly be formed, along this line, a ridge of sand, which will gradually increase in height until at last it will appear above water, forming a narrow strip parallel to the shore. This may become so high as to form a permanent barrier, enclosing behind it a lagoon of shallow water. Sometimes the formation of these beaches is due to the action of storms of extraordinary violence, which have caused the waves to disturb the bottom farther than usual from the shore. In the slow rise of a coast bordering shallow water the line of surf action would be gradually removed from the shore. If the slope of the bottom is uniform, and there are no storms, the effect may be to add slowly to the coast by a constant throwing up of sands by the waves ; and this is beautifully illustrated at the mouth of the Jequitinhonha by the plains stretching southward from Belmonte. These plains consist of a great number of parallel beaches, one lying in front of the other, and traceable for miles; but even here the growth of the coast has not been uniform, and occasionally an increased violence of the surf has thrown up a ridge a few yards or rods outside of that last formed, making a narrow ditch-like lagoon, like a river, running parallel with the shore. A lagoon of this kind runs behind the beach along the present shore line like a strip of insertion. If the water is very shallow, there will be no regular beach ridge, but there may be dunes if the sand is light enough to be raised by the wind ; but if the water is so deep that the whole force of the waves is expended on the shore line, breaking in a single line of surf, then we may expect to find a high ridge accompanying the beach, — a ridge which has owed its origin primarily, it may be, to some very heavy storm, and secondarily to the joint action of the winds and waves in piling up the sand. This is beautifully seen at the mouth of the Jequitinhonha. At Belmonte the sea is exceedingly shallow, consequently the shore is low and without a ridge. Going southward the water deepens, the surf is heavier, and a well-defined ridge begins, growing higher and higher the farther south we go.

Along the coast of Long Island, and of the Middle and Southern Atlantic States, we have these phenomena well exhibited. When between the rivers the shore is flat, such a beach may extend from the mouth of one to that of the other, while the lagoon behind may form a communication like a canal between the two, or the mouths of the two rivers may be united by a more or less wide strip of marsh and lagoon. These lagoons are liable to be filled up by sand and silt carried down by streams from the higher grounds. When the bottom is brought up to the level of low tide, or a little above, the seeds of the mangrove take root, and the shoal soon becomes covered with vegetation. Among the roots of the mangroves the silt of the water is deposited, and the sand-bank is overspread by a layer of soft mud, which may increase in thickness until the bank is covered only at high water. The mangrove in Brazil, Florida, and elsewhere is a very efficient agent in the silting up of marsh lands and swamps; but it is not alone in this work. It flourishes only in salt or brackish water, but after a time much of this swamp land becomes covered with fresh water, when arborescent arums, reeds, rushes, coarse grasses, and other aquatic plants tend all the more to choke up and stagnate the water, and often form rafts of vegetation, on which small trees grow, as is the case in the fresh-water lakes and lagoons of the vicinity of Barra Secca and the rivers São Matheos and Doce.

On the coast north of Rio the course of the rivers is at right angles with the coast. At the mouth of very many of them a branch runs off both north and south, parallel with and close to the shore, and in some cases uniting with a similar branch from a neighboring river, though usually an impassable swamp prevents a complete communication. In some cases these are only lagoons fed by the tide and rains; but occasionally it is a little river, which, having come down to the shore, is obliged to flow for several miles behind the beach ridge before it can find an exit to the sea in the mouth of a larger river. An example of this kind is the Itahúnas, already described. All the rivers flowing from the interior are white-water rivers, but the Itahúnas, probably having its source in the swamps of the chapadas east of the serras, is a black-water river. So also is the Mariricú, and so, in general, are all the little rivers which rise either in the fresh-water swamps along the shore or come from the chapadas eastward of the serras. Advantage has been taken, in numerous instances, of the streams and lagoons of the low lands to cut canals uniting settlements lying on two rivers; and a favorite project with some Brazilian statesmen has been that of opening a line of canals, extending along the coast from Santa Cruz, in the province of Espirito Santo, northward to Caravellas. I consider the project as impracticable, and that good roads over the plains of the chapadas would, in most cases, be far preferable.

The town of Caravellas is of small size. It owes its importance to its being the port of the surrounding country, and to its whale-fishery, described in the last chapter. The town is built on a sand-bank a few miles above the mouth of the river, and on the northern side. In the vicinity are large groves of cocoa-palms.

Contrary to the general rule, the river Caravellas just before reaching the coast makes a bend to the northward, and enters the sea very obliquely. The channel, narrow and marked by poles, continues for several miles northeastward beyond the Ponta da Balêa. There is another narrow channel running eastward, and still another southward, which last doubles sharply round the point on the southern side of the river. Between these channels are large sand-banks, which are, in some cases, visible even at high water.

The northward bend of the river at its mouth seems to be owing to the prevalence of a northward-setting current. The variation of the current through the tides is most probably the cause of the existence of the other two channels. Off this coast lie the coral reefs of the Parcel das Parêdes and the islands of the Abrolhos, already described.

The shore northward to Prado is low and flat, a long monotonous sand-beach, which is broken by the Barra Velha, the mouth of an inconsiderable stream, and, at a distance of about nine miles north of the Ponta da Balêa, by the Barra do Rio Itanhaem,* which is a small river like the Peruhype, arising in the Serra dos Aymorés. At its mouth this river bends abruptly southward, and flows for at least a couple of miles almost parallel with the sea-coast before it escapes into the ocean. According to Prince Neu-Wied *manati* have been captured in this river. Alcobaça, a small town of very little importance, is situated on the seaside between the river and the beach. Prince Neu-Wied says that the country about Alcobaça is healthy, but that the climate is unpleasant from the frequency of strong winds and storms.

Twelve miles farther north, the Rio Jucurucú empties into the sea. This stream is formed by the union of the two branches called respectively the Braço do Norte and Braço do Sul, which take their origin in the Cordilheira dos Aymorés. This river, like the Itanhaem, on reaching the coast is obliged to flow southward for two or three miles, behind a beach ridge, before escaping into the sea.

* According to Dr. José Candido da Costa, in his pamphlet entitled "A Comarca de Caravellas," a copy of which I owe to the kindness of Sr. Licino Lessa, the Itanhaem is narrower and shallower than the Peruhype. Near its mouth it forms a little basin. It is subject to freshets.

The bar of the river allows the entrance of schooners and small vessels, and the river is said to be navigable for small vessels (*sumacas*) for a distance of some six leagues. The valley of this stream is rich in lumber, and in 1857, according to Da Costa, almost all the inhabitants of the *municipio* of Prado were occupied in cutting it. The town of Prado, another unimportant place, is built between the river and the sea on the left bank. A few miles north of the Prado the tertiary bluffs, which from the Peruhype have skirted the coast at a greater or less distance in the interior, come down to the sea, and thence to Porto Seguro form a long stretch of picturesque perpendicular red cliffs, alternating with steep slopes covered with verdure, and occasional patches of sands or swampy ground.

From the sea the horizontal bright red and white bands of clays are distinctly seen. Near Porto Seguro, as well as elsewhere, I observed that the valleys of the rivers entering the sea did not have angular sloping sides, but their profile as seen in a cliff was as in the following sketch,

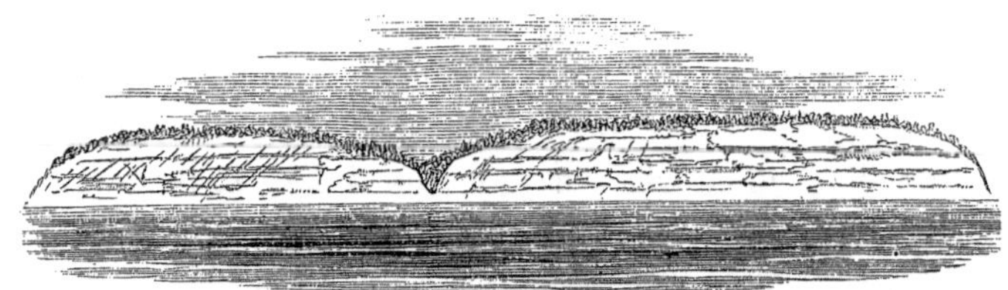

which is a kind of topography that we should expect to see in a glaciated surface. The bluffs would, I should judge, average two hundred feet in height.

About forty miles north of Caravellas, and a few miles inland, is the beautiful group of hills noted as being the first land of America seen by Cabral. The most conspicuous of these hills is an irregular conical peak which bears the name of Monte Pascoal.* Mouchez sets the hill down on his map as 536 metres in height, but I should judge that it was much higher. These hills are undoubtedly gneiss, but they lie far back in the virgin forest, in the as yet unexplored home of the savage Botocudo. The appearance of the group as seen from the sea is represented in the following woodcut. Along this coast several small rivers rising in the Serra dos Aymorés empty into the sea, but they are of very little importance, and are usually incorrectly laid down on maps.

MONTE PASCOAL FROM THE SEA.

* Monte Pascoal was the first point seen when Cabral discovered Brazil, on the 21st of April, A. D. 1500. Pedro Vaz de Caminha, one of the companions of Cabral, has described the events connected with the discovery in a letter written on the 1st of May, 1500. This writer speaks of the mountain as very high and *round,* with other serras to the south, together with flat land covered with large trees. " E neste dia a oras de vespora ouvemos vista de terra, a saber: primeiramente de huum gramde monte, muy alto e redondo, e doutras serras mais baixas aho Sul dele, e de terra chấa com gramdes arvoredos; aho quaal monte alto ho Capitam pos nome ho Monte Pascoal, e aa terra ha Terra da Vera Cruz." — *Corografia Brazilica,* Tome I. p. 13. A French translation of this letter is to be found in the *Art de Vérifier les Dates,* Tome XIII. p. 441. The original, which is of great interest, is preserved in the government archives at Rio. Cazal says, speaking of the Serra dos Aymorés: " Cuja porção mais alta he o Monte de Joam de Liam [João de Leão] e mais fóra o *Monte Pascoal* que se avista de muitas lequas ao mar." This João de Leão is a noted landmark, but I can give no idea of its height.

In front or to the east of Monte Pascoal lies the great reef-ground of the Itacolumís. This is separated from a low projecting sandy point called Ponta Corumba by a very narrow but deep channel. A few miles north of this point is the Rio Craminuan or Caxoeira, which is noteworthy from the fact that, contrary to the general rule, on reaching the coast, instead of immediately entering the sea, it flows northward nearly a mile close to the sea, but separated from it by a sand-bank, showing that the wash of the coast sands is here northward instead of southward, — a fact determined by the reefs lying off the shore. North of the Craminuan are the Rios Joassema, Frade,* and Taipé. The village of Trancozo is situated on the coast a few miles south of Porto Seguro, but it is of no importance whatever.

At Porto Seguro enters the Buranhaem River, a stream of moderate dimensions, which, according to the maps, rises to the southwest in the province of Minas, a few miles from the boundary line. Gerber represents this as a considerable stream flowing through a large lake called Gravatá, distant some thirty miles from the sea, and communicating with another considerable lake about half-way between this and the sea. At Porto Seguro I was informed that this was all very incorrect; but since I have never ascended the river, I cannot speak authoritatively. The stream is a black-water one,

* In speaking of the coast between Prado and Rio do Frade, Max. zu Neu Wied says: "Als ich im November dieses Jahres noch einmal diese Reise machte, fand ich bei starker Ebbe weite Bänke von Sand- und Kalk-Felsen, die sich tief in die See hinaus erstrecken, und wohl grossentheils durch Corallen-Thiere gebildet worden sind. Ihre Oberfläche ist in regelmässige parallele Risse getheilt; in den vom Wasser darin ausgewaschenen Löchern leben Krabben und andere See-Thiere; die Oberfläche dieser Felsbänke überzieht zum Theil eine grüne Byssus-artige Masse." — Prinz Max. zu Neu-Wied, *Reise nach Brasilien*, Erster Band, 297te Seite.

and flows in a deep narrow valley cut in the tertiary beds. The flat lands bordering it are said to be fertile, and favorably situated for agricultural purposes.* The forests are rich in valuable woods, Páo Brazil, Jacarandá, &c. Porto Seguro is quite a large and commercial town, situated on the left bank by the seaside. It really consists of two towns, one built by the river and seaside on a broad, flat, sandy, and marshy tract, and part on the top of the bluff on the northern side of the valley. The lower town is the business portion; the upper contains the ruins of ancient churches, monasteries, &c. Porto Seguro is noted as the head-quarters of the garoupa fishery of the Abrolhos. It has an excellent harbor, protected in front by the coral reefs, which break the force of the Atlantic waves, and by a reef like that of Pernambuco, consisting of a solidified beach. This last begins close to the shore a short distance north of the mouth of the river, runs like a wall of rock in front of the river, passing close to the point on the southern side, and continues on to the southward with occasional breaks for a distance of several miles. Its course is remarkably straight, and its height and width are very regular. It forms a more efficient breakwater to the harbor than the reef at Pernambuco does; but this is owing partially to the fact that the sea is broken by the outside coral reef which stretches across the mouth of the bay of Porto Seguro. The river escapes around the northern end of the reef. On the inner side it is overhanging, on the outer perpendicular and much undermined, as represented in the woodcut. As elsewhere its surface is diversified by ponds, in which several species of corals grow. The northern end of the reef is much shattered, cracked up, and dislocated,

* Lindley says that gold was found on one of the branches of this river.

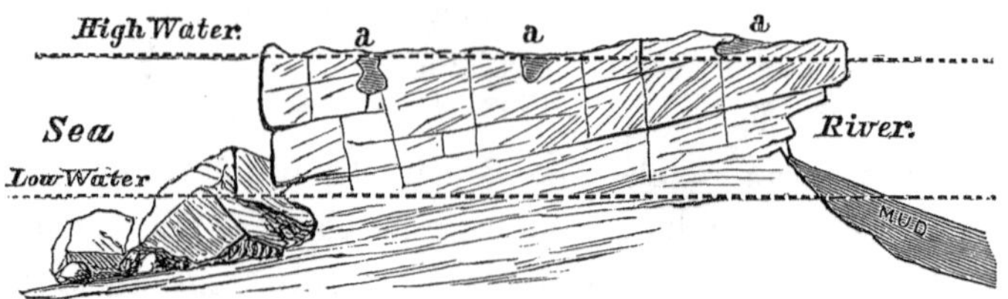

SECTION ACROSS STONE REEF AT PORTO SEGURO.

and it shows that the cementing of the beach-sands has taken place to a depth of many feet below low-water level. Beyond the reef, the rock slopes out with a smooth, rather even surface to a considerable distance, and I have waded out on it in some places a hundred feet or more. Its limits I was not able to determine, owing to the surf. On the outside the reef below low-water mark is covered with a growth of corals. Stony corals, *Hymenogorgiæ*, *Eunicia*, and the common polyps of the Abrolhos, are very abundant here. I found a single specimen of *Mussa Harttii* in a deep indentation at the northern extremity of the reef.

As at Pernambuco, Bahia, Barra Secca, and elsewhere, the rock is obliquely laminated, as in a sand-beach, the lamina dipping seawards at a small angle. It is composed of rather fine sand, with occasional small pebbles, compactly held together by a calcareous cement. It contains an abundance of recent shells, Venus, Cerithium, Chama, &c., &c., such as now live on the sea-beaches of the vicinity. On the inside the reef-rock is of little thickness, and the reef is flanked by a sloping bank of mud, on which a few mangroves have taken root. Oysters grow here on the rocks and mangroves, and a large species of *Littorina* is very common, being found even in the trees at a height of four or five feet above the level of high water. This same species occurs elsewhere, and I have observed it in great abundance higher than I could reach in the mangroves at Santa Cruz, a few miles north of Port Seguro. Crustaceans are abundant on the reef.

Little fiddlers (*Gelasimus palustris* Edwards)* are very common on the sandy beaches in some localities, boring holes in the sand. There are larger species of the same genus (*Gelasimus Maracoani* Latreille) with a nut-brown body, and one of the hands enormously developed, looking like a pair of broad-bladed shears.†

On the shore of the point on the southern side of the mouth of the river is a small patch of beach, which is only partially consolidated. This is separated from the reef by a narrow channel almost laid dry at low water. Outside the main reef, and opposite this, there are the remains of probably an older reef, which has elsewhere been almost entirely obliterated. Southward of the river the channel soon widens into quite a broad sheet of water, and the reef is left running at a distance of several hundred feet from the shore. The water inside the reef is shallow, and at low water one may wade about over a large part of the area and collect. The bottom is sandy, but on it grows an irregular crust, composed of millepores, corals, bryozoans, &c., which forms the nestling-place for holothurians, ophiurans, crustaceans, and a thousand interesting animals. A large naked mollusk, probably *Aplysia Argo* D'Orb, is very common here, and I have collected a dozen specimens in the same little pool. Inside the reef the water is deep enough to

* This species bears, in every respect, an exceedingly close resemblance to our northern *G. vocans* Dana. Mr. Sidney I. Smith considers the Brazilian species as identical with one occurring in the Gulf of Mexico, and even so far north as South Carolina.

† These fiddlers are called by the fishermen *Chama maré* (call tide), because of their congregating at low-water mark and waving their big claws.

admit vessels of ordinary tonnage and the coasting steamers. The reef is higher than that at Pernambuco, if I judge rightly, and less shattered; but, as above remarked, the Pernambucan reef is exposed to the full action of the sea. At high water the waves also break completely over the Porto Seguran reef. Prince Neu-Wied gives a sketch of Porto Seguro from the southern point. The width of the river is very much exaggerated.

The line of the tertiary slopes continues on without interruption to Santa Cruz; but the shore on leaving Porto Seguro soon separates itself from the chapadas, and forms a considerable point of land, of which the interior seems to be, for a considerable part at least, swampy. On the southern side the beach is backed by mangrove swamps, from which the water is drained by a little stream that flows across the beach. In passing very close to the point at low tide, I observed that the waves were breaking along a line at some distance from the shore, as if against the edge of a reef. It would appear that the point has been formed by the filling up of the channel behind a coral reef with sand. Santa Cruz, one of the most ancient towns of Brazil, and at the same time one of the most miserable, is built on the southern side of the mouth of the little river Santa Cruz, partly on the top of the bluff and partly on the sands at the base, between the bluffs and the sea. Though it has a fine harbor, and many natural advantages, it is of not the slightest importance, and is only a miserable little fishing village. It is situated in a shallow bay, about seven miles long, which is protected by coral reefs, that extend across it, offering anchorage for large ships.

The Rio Santa Cruz, anciently called João de Tiba, belongs to the same class of rivers as the Buranhaem, but it is a smaller river. It is said to rise in the Serra dos Aymorés, to have a course of about ten leagues, and to be navigable for a considerable distance for canoes. It is laid down incorrectly on maps. At Santa Cruz I was informed that its course was such as to cause it to approach the Jequitinhonha, from which, at Zinebra, it is separated by only a very short distance, and I was informed that it would admit of navigation up to that point by a small river steamer. Prince Neu-Wied says that the river has two branches, and that the head-waters of one of them lie so near the Jequitinhonha, that the report of a gun can be heard across the intervening space. The valley of the Santa Cruz is of the same character as that of the Buranhaem. It is fertile and richly wooded, furnishing building timber and some Jacarandá and Páo Brazil. It is noted for the abundance of *canna fistula* (*Cassia nigra*), a tree valuable for its medicinal properties.

The river on reaching the sea is prevented from flowing immediately into it by a *recife*, or consolidated beach, which, beginning on the shore just to the south of the village, continues in the trend of the beach, which is north a few degrees east, with an occasional break for a distance of about two miles, the river flowing behind it and escaping around its northern extremity. At low water the breakers show that the reef is continued under water with the same general trend northwards, tying in with a reef which, beginning at a point about a mile north of the river-mouth, fringes the beach for more than half a mile, as represented in the accompanying sketch-map of the bays of Santa Cruz and Cabral, in which not only the stone but coral reefs are shown. Mouchez has very incorrectly represented the harbor in his chart. Through the kindness of my friend

Hugh Wilson, Esq., superintendent of the Bahia Steam Navigation Company, I was enabled to revisit Santa Cruz in 1867, and to correct Mouchez's chart by drawings made from the top of the old church on the bluff, and these corrections I have introduced in the map. In front of the

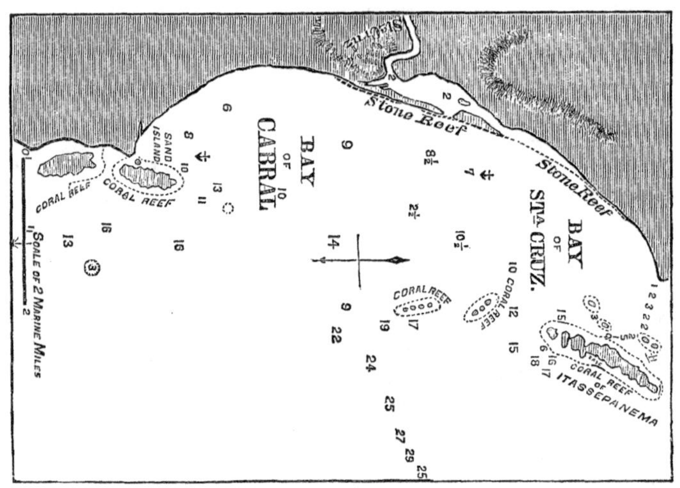

town the reef clings to the shore, though there is a channel behind it filled at high water. The reef is much shattered, and great blocks lie tumbled about in confusion, broken from it by the waves. For a distance of a mile or so north of the village the reef is backed by a narrow strip of mangroves. There is then a narrow break in the reef, forming a sort of bar, into which very small vessels may enter at high water. Thence to the barra the reef is backed, almost to the end, by a wider strip of mangroves. At the extremity the reef is double, the remains of an older reef being visible on the outer side. An example of a triple reef is represented in the chart of the mouth of the Rio Parahyba do Norte, published in the old work of Barlæus, in 1660.* I have never seen this reef, and know nothing of its present appearance. The bar of the Santa Cruz is good, and may be safely entered by coasting steamers or large vessels. At the time of my last visit I saw a brig lying behind the reef at the bar. There is sufficient water to enable steamers and vessels of considerable size to go up to the town. This harbor could be very much improved by stopping up the channels by which part of the river water escapes near the town as well as the break in the reef above described. This would cause the whole force of the stream to be expended on the proper channel of the river, and it would in this way be kept from being filled up by mud and sand banks. The country in the vicinity of Santa Cruz, embracing the region of the Jequitinhonha, is so naturally rich, that as the coast becomes settled it must one day be developed, and one cannot doubt that Santa Cruz will ultimately become the port of the Jequitinhonha, and a place of much importance. From Santa Cruz the bluffs run northward, following close to the shore, with occasionally some sandy tracts in front, but they gradually trend off from the coast and cross the Jequitinhonha between Zinebra and the Po-assú.

A few miles north of the Santa Cruz is the Rio Sant' Antonio, a little black-water river of no account. About a mile and a half north of this river I saw exposed at low water, in a little bay, an area of several acres of dead coral, principally *Acanthastræa*, &c., *in sitû*, but much worn by the waves. These corals must have grown when the sea bottom over this region stood at a considerably lower level than at present, for I have seen them nowhere living at a

* For a reduced copy of a part of the chart of Barlæus, as well as a recent sketch map after Almeida, see chapter on the Province of Parahyba.

level where they would be likely to be exposed at low water. When they were growing the sands forming the low grounds bordering the coast here could not have had so wide an extension seawards. This reef patch has evidently been killed by the joint rise of the coast and the encroachment of the shore sands. I collected on the shore here a great number of specimens of the pancake sea-urchin (*Encope emarginatus*), so common in the Bahia de Todos or Santos. Neu Wied speaks of the great abundance of this echinoid on the shore near Sant' Antonio. He refers it to *Echinus pentaporus*, or as he writes it, *Eschinus pentaporus*.

Thence northward the coast is bordered by a narrow strip of plain, which widens as the bluffs recede from the coast, and finally opens out into broad, sandy campos, which extend from the Mugiquisaba River, a small black-water stream, to Belmonte. It is interesting to observe how this whole plain has been added to the coast, by the throwing up by the sea of the sands of the Jequitinhonha, and moreover that this growth has taken place to the south of the river rather than to the north, because of the sweeping of the sands southwards by the oblique beat of the waves, and perhaps by the drift of a southward flowing current. In going northward from the Mugiquisaba to the Jequitinhonha the water gradually shallows. The result is that the surf beats more heavily on the southern part of this coast, and we find a high beach-ridge developed, while, near the mouth of the Jequitinhonha, the water growing shallow, the bottom is disturbed farther out by wave action, and the sands are constantly being driven in shore to widen the plain, and of course extraordinarily heavy storms are here likely to throw up beaches outside the present shore line. The plain consists of a large number of parallel beaches, one lying in front of the other, sometimes united, at others separated by miniature valleys, occasionally only a few feet in width, but often continuous for a considerable distance. Many of these beaches have their slopes almost as perfect as if it were but yesterday that they were swept by the waves. In the depressions water accumulates, sometimes forming shallow lagoons. The present beach on going southward from the barra separates itself from the plain, and at the same time increases in height as the surf action becomes heavier. Behind this beach is a narrow river-like lagoon, called the Lagôa do Braço. This is at first narrow and shallow, but to the south it deepens and widens as the beach-ridge becomes separated from the plain. Its waters are fresh and clear, and very rich in fish. *Cambôas*, or fish-weirs, are common along its shores, which are muddy, and to a very large extent support a luxuriant growth of mangroves and guaxúma-bushes. In making a voyage to Porto Seguro my companion and I became separated from our baggage troop, which was following the beach, and, taking a road on the western side of the Lagôa, were unable to cross it anywhere, because of the swampy nature of the banks and the depth of the water. Prince Neu-Wied says that about one half of the way between Mugiquisaba and the Jequitinhonha is the barra where a dried-up arm of the latter river once emptied into the sea. By the dried-up arm he probably means the Lagôa do Braço, which, as he travelled along the sea-beach, he probably did not examine, as he says nothing about it. I saw no outlet to the Lagôa. When swollen it may break through the beach ridge, as is the case with other seaside lagoons. It appears to be drained principally by the slow percolation of its waters through the beach into the sea.

The plain is to a large extent open and very sparsely covered with coarse grass, bromeliaceous plants, cactuses, &c., with here and there clumps of trees. The Aricurí palm is very common on this plain, together with the pretty dwarf Gurirí. This latter, all along the whole coast, is found growing on and just back of the sea-beaches, and is one of the marked elements of the beach flora. The Piassaba palm (*Attalea funifera* Mart.) flourishes on the Mugiquisaba, as remarked by Neu-Wied. I have not seen it south of this point. Cactuses, with procumbent prismatic stems, form large patches on the sands. Another very characteristic plant of the coast sands is the *Ipomœa littoralis*, a convolvulaceous plant, with long, thick, cord-like creeping branches, pink blossoms, and large thick oval leaves. This plant grows sometimes on the beach almost within reach of the waves, and its prostrate stems are often buried by the sands.

The right bank of the river, for some eight or ten miles above its mouth, is very low and liable to be flooded. A very large part consists of mangrove swamps. The Po-assú * is a very narrow ditch-like canal which runs through the alluvial grounds, leading off part of the waters of the Jequitinhonha into a small black-water river called the Salsa, which flows into the Pardo. So much higher do the waters of the Jequitinhonha stand above those of the Pardo, that, though the tide rises and falls in the latter river where the Po-assú enters it, there is a constant flow of water from the Jequitinhonha into the Pardo. The Po-assú is so narrow, tortuous, and filled up with trees, that

* This name is Tupí. *Pó-açú* means *left hand,* so that it might mean left-hand channel. *Yg-apó* means a swamp. I am inclined to think that the name was first *Yg-apó-açú,* or *the great swamp,* which has since passed to the canal.

its current is measurably slackened, else it would sweep out for itself a broader channel and draw off a larger quantity of water from the Jequitinhonha. What the consequence of this would be I shall state further on. The soil of the grounds bordering the Po-assú is massapé, and supports a very vigorous and thick forest growth.* From the Salsa, after having received the Po-assú, side arms stretch out eastward into the flat grounds lying between it and the sea. The whole area embraced between the Pardo and Jequitinhonha, the cross-stream and the sea, is one vast swamp, comparable to that which lies between the Peruhype and the Caravellas; the Po-assú being comparable to the Braço de Viçosa, both the Po-assú and Braço being channels by which the waters of a river with an inefficient mouth-opening are enabled to escape into the sea through the mouth of another river. There are some quite extensive sandy tracts in this area and along the coast. These are planted with groves of cocoa-palms, from the fruit of which cocoa-oil is to some extent manufactured. A large tract lying northward of the Salsa, and embraced between it and the Pardo, is of the same character. The Pardo is a large river, and has a bar on which the waves beat fearfully, but it usually admits at high water of the passage of steamers under the proper pilotage. The Pardo does not bring down so much sediment as the Jequitinhonha, its

* In the forest of the Po-assú the *Quitára* (Desmoncus), a trailing palm, is exceedingly common, interwoven with the trees. I saw it in fruit early in May. The fruit is round, about the size of a small cherry, bright red, and in clusters like grapes. This is a common species found in the Catinga woods on the low grounds along the coast, sometimes in such great abundance as to be a nuisance, for it hangs its pendant leaves, terminated with hooks, over the mule-path. The name Quitara is the one given it by my guide. On the Rio Negro the Desmoncus macroacanthus is called *Jacitára.*

channel is deeper, and the tide runs up the river for quite a number of miles.

Just before entering the sea the river makes a bend southward, and the barra is really cut through a strip of sand-beach which forms the shore for a long distance north and south. A wide channel, or riacho, leaves the river on the south just before it cuts through the sand, and runs through the mangrove swamps for a considerable distance southward. In journeying from Cannavieiras* to Belmonte I followed this route. My companion and I took canoe up this riacho as far as was possible, then coming to a sandy tract, we chose the sea-shore for some distance, when we embarked in a canoe on another riacho which led southward to the Jequitinhonha, a dismal, but to the naturalist exceedingly interesting journey. There was a perfect network of channels leading through the dense mangrove growth, and for some distance little land was to be seen. The mangroves standing in the mud or dirty water, with their bared basal roots, their frequent aerial roots hanging ropelike from the branches, and just before reaching the soil forking like tripods, their dense green foliage meeting overhead, and their curious cigar-shaped seeds pendent in the tree or occasionally sticking upright in the mud, were objects of wonder and admiration ; but among their roots innumerable fuzzy and muddy-legged *Guayamú* crabs glared at us with expressionless eyes, and then hustled away into their holes in the black mud. Beautiful orange Aratús, with white fingers, lay like rich fruit fallen from the trees; but they, too, took fright and ran nimbly off to their holes, or hid in crevices under the roots, while myriads of fiddlers marched away, waving their

* This is the usual orthography, but one meets with *Canavieira*, *Canavieiras*, or even *Canasvieiras*.

big arms, and took refuge in the sandy spots. Parrots* screamed among the branches; but, *peste!* mosquitoes and all the horrible blood-sucking race of *Maroims, Piums,* &c., swarm; and, how ardent soever the naturalist may be, his patience can ill stand the plague of flies that falls upon him in these swamps. The air of the swamps is very unhealthy, and the vicinity of Belmonte and Cannavieiras is feverish. The town of Cannavieiras is situated on the left bank of the river, just above the mouth. It is built on an old beach, an island surrounded by a channel, which, leaving the river some distance above the town, enters it again just before the river reaches the sea. The town is quite a considerable place, numbering, perhaps, two thousand inhabitants. It derives its chief importance from the fact that it is the port of the Jequitinhonha and Pardo, and is one of the stations of a line of steamers. Canoes descending the Jequitinhonha do not go to Belmonte, but come here by way of the Po-assú and Salsa, to exchange the cotton, corn, and other articles of export of Minas and the lower river for salt, dry goods, &c. The salt trade is very large. In descending the Jequitinhonha, we passed every day several large canoes on the up voyage, the most of them carrying salt.

It has been proposed to straighten and widen the Po-assú, so as to facilitate the navigation between the two great rivers. At present the waters of the Pardo during the annual freshets wear away the unstable land on which Cannavieiras stands. Were the Po-assú widened and a free passage opened for the waters of the Jequitinhonha, the result would prove disastrous to Cannavieiras as well as to Belmonte. At pres-

* *Psittacus ochrocephalus* Linn. builds its nest in the mangroves. See Neu Wied.

ent the struggle between river and sea is only sufficient to keep the bar of the Jequitinhonha open. Draw off any considerable part of the water of that river, and it is very doubtful whether its bar would not become a permanent obstruction to navigation, while the sands of the Jequitinhonha, thrown into the Pardo, would probably shoal the water and render its bar worse. Santa Cruz is the natural port of the Jequitinhonha, and it would certainly seem that the commerce of the Jequitinhonha would be vastly improved by using the river Santo Cruz so far as it is navigable by steamers, and then building a good wagon-road thence to the Jequitinhonha.

The Pardo is joined with the Poxim, a small river said to take its rise in a large lake, and whose barra is a few miles north of that of the Pardo, by a narrow arm running parallel with and close to the shore, and called the Patipe. Cannavieiras itself is situated on an island formed by a channel called the Rio Sipó which leaves the Pardo and joins the Patipe. The Poxim, just before it unites with this last river, subdivides and enters the sea by two mouths, one of which is called the Barra do Patipe, and the other on the north the Barra do Poxim. One or two channels continuing northward from the Poxim like the Patipe, running just inside the coast line, empty into the sea a couple of leagues to the north at the Barra de Commandatuba. This whole coast has been very indifferently mapped inside the coast line. The Jequitinhonha and Pardo are often shown as entering the sea by the same mouth.

The map of Senhor Henrique Gerber, so excellent for the provinces of Minas Geraes, Rio de Janeiro, and Espirito Santo, is exceedingly defective so far as the province of

Bahia is concerned; but it must be remembered that his map does not pretend to be a map of Bahia.

This part of the coast is better shown on the map of the *Tenente Manoel Ernesto de Souza França*, published in the report of the President of Bahia in 1866; but even this only gives the general features, and is not based on a careful survey.

According to Prince Neu-Wied's edition of Arrowsmith's Map of the Brazilian Coast, no connection is shown as existing between the Jequitinhonha and Pardo, and the latter river is represented as dividing into three a long distance from the coast, two of which unite before entering the sea, while the Poxim and Commandatuba, though joined together, have no connection with the Pardo.

The *Diccionario Geographico* declares that the Rio Pardo divides into two streams, one of which enters the sea under the name of the Rio Pardo, the other emptying into the Bay of Ilhéos under the name of Cachoeira or Patipe, which is all simply ridiculous.

Mr. Copeland and I ascended the Pardo to the head of canoe navigation, a journey of about three days, and the following notes were made on the return voyage.

At the Caxoeirinha do Rio Pardo the river reaches the low country and becomes a *rio de areia.* Here navigation is obstructed by a series of rapids caused by the river falling over ledges of slate conglomerate, of which the dip was found to be 45° to the S. 10° W., a dip corresponding with that of no other strata I have studied on the Brazilian coast. The material is a very highly altered conglomerate, composed of pebbles of milky quartz, granite with quartz in lamellæ, &c., imbedded in a slaty mass. The rock is exceedingly hard, and appears to resist decomposition. The stratification is not very distinct. On the water-worn surfaces the pebbles stand out very prominently, but a fracture passes straight through pebbles as well as cement. At the Caxoeirinha these rocks are overlaid by tertiary beds, which form plains elevated three hundred feet, more or less, above the river, and which descend with very steep slopes to the stream. The alluvial deposits along the river reach a height of about twenty-five feet above the river level. During the enchente these are sometimes at least overflowed. The lands here are very fertile, and the whole country appears to be densely wooded, but uninhabited.

Slate conglomerate and sandstone continue to show themselves in the river-bank for a mile or more below the Caxoeirinha, when to these rocks succeeds a fine-grained bluish slaty rock, an altered shale, in which I could find no trace of fossils.

A few rods above the fazenda of Sisterio there are some ledges of a fine-grained, very hard, bluish, compact, altered sandstone, with occasional bands of grit and conglomerate, and often very beautifully obliquely laminated. In this rock I observed remains of plants, and one surface of rock had the impressions of several stems of a thick-noded equisetaceous plant. One stem showed three nodes and another two, the length of the internodes being about three inches. Owing to the hardness of the rock, I was able to bring away only the impression of a single node. This plant resembles in its swollen nodes a plant not uncommon in the upper Devonian rocks of St. John, New Brunswick, and called *Asterophyllites ? scutigera* Dawson. These beds appear to me to be palæozoic.

Below this point the river valley becomes wider and the alluvial lands more extensive, though the river itself is not so very wide, as it averages for a considerable distance only about three hundred feet. The country is abundantly forest-clothed, and the river-banks, which are very fertile, are more or less cultivated, several large fazendas being scattered along the river. Some of these are very picturesquely situted, and are surrounded by orange, banana, jack and cocoa trees. Cacáo (*Theobroma Cacáo*) is quite extensively cultivated here. This plant is a native of the Amazonas, where it flourishes almost without culture;* the *cacaoeiros*, when once they have begun to bear fruit, requiring little care. It needs a warm damp climate and a rich alluvial soil, and appears to suffer nothing from an occasional freshet. South of the Amazonas it is cultivated but rarely, though even in the province of Bahia there are some quite large plantations, especially at Ilhéos on the Pardo, and at Valença. I saw no cacáo in the province of Espirito Santo. A little is cultivated in the province of Rio de Janeiro. In the Amazonian region the fruit, when ripe, is collected twice a year and dried in the sun; the seeds are then separated from the shells, and are employed principally for the manufacture of chocolate or other preparations for beverages. They furnish a thick yellow fatty substance which is sometimes extracted and used for various purposes.

Coffee does not produce well here. The trees grow very rank and high, with spreading branches; but the berries ripen very unequally, and not unfrequently have to be gathered six times during the year. This is owing to the peculiar climate of the region. In Rio de Janeiro, Minas, and elsewhere, there are well-defined wet and dry seasons during the year, and these appear to be necessary for the successful culture of coffee; but here on the Pardo a very large quantity of

* See Bates, The Naturalist on the Amazons, pp. 87 and 162.

rain falls, and is distributed throughout the entire year, making the climate very damp, so that while it is especially adapted for the culture of cacáo it does not do for coffee. On the higher chapada lands, however, where the soil is sandy and drier, coffee and cotton may be cultivated. Alluvial, flat, damp lands are nowhere proper for coffee, which flourishes best on hillsides.

The jack (*Artocarpus Braziliensis* Gom.) is cultivated in Brazil, particularly in the province of Bahia and to the north, though I have seen it at São Matheos, and occasionally as far south as Rio. It also occurs in Minas Geraes, but in the province of Bahia it becomes of considerable importance. The timber is valuable for building purposes, being very durable and strong. The fruit is immense, being sometimes a foot and a half in the longer diameter. It consists of a stringy, mucilaginous, sweet, and nutritious pulp, in which are imbedded large seeds, which, when cooked, are edible and nourishing, and are largely used for food. In some parts a kind of farinha is prepared from them, but its use is by no means general. The *Fruita pao* or bread-fruit, *Artocarpus incisa,* is also cultivated in Brazil, and may be seen very frequently in the province of Bahia.

The forests on the banks of the Pardo remind one of the Doce in their luxuriance. The trees by the river-side are loaded with parasites and interlaced with pendent rope-like *cipós.* Ferns are very numerous, and one species, with beautiful fringed pinnæ, climbs up the trees to a height of forty feet or more. One sees here a species of grass called *Capim da Colonia,* (an exotic ?) which has been introduced on the Pardo, and has spread within the last few years quite extensively over the river-banks. At the Estreito there is an isolated patch of tertiary, but below that the lands bordering the river are all low. A short distance below the Furado, a channel cut across a bend in the river, the lands become lower and less heavily wooded, and aninga and Guaxuma bushes appear on the banks, — an infallible sign of the approach to salt or brackish water. There is a fern common to the mouths of the rivers on the coast which appears to prefer this kind of water. It has a tall, erect, narrow frond, with stiff, long, narrow pinnules bent upward toward the stem.* It grows abundantly in the vicinity of the Furado, on the muddy banks reached by the brackish water.

From Cannavieiras to the Rio Poxim runs a canal, separated from the sea by a beach-ridge, and thence northward this same canal extends to the Commandatuba, and is said to afford water communication to the latter river from the Pardo. The Commandatuba flows into this channel, which is prolonged just behind the beach for some distance northward before it opens out into the sea. At the Barra da Commandatuba a colony has been established, and there is a little village of thirty or more houses there. The shore between the Commandatuba and the Una is low and flat, but a short distance inland one sees the tertiary slopes, which stretch along northward from the Jequitinhonha and Pardo.

The Una is a little river which, according to Prince Neu Wied, is so dry at ebb tide as to be easily forded. There are some excellent lands on this river, and the forests are rich in jacarandá and other valuable woods. This river is noteworthy for bending northward just before reaching the sea, and flowing a little distance behind the beach-ridge.

Opposite the mouth of the Pardo several high hills, lying

* This must be a species well known, but in the absence of specimens I cannot give its name.

some twenty miles inland west of Commandatuba, are visible at sea, and appear to be gneiss, but when off the Una a heavy mountain range, with outlines like those of the gneiss hills north and south, is seen stretching off northward, and traceable beyond Ilhéos, probably tying in with the mountains of Camamú. This range is called the Serra de Itaráca.* Some of the hills in this range must be three thousand feet in height or more. I insert here an outline sketch of this coast, to show the character of the topography. I was

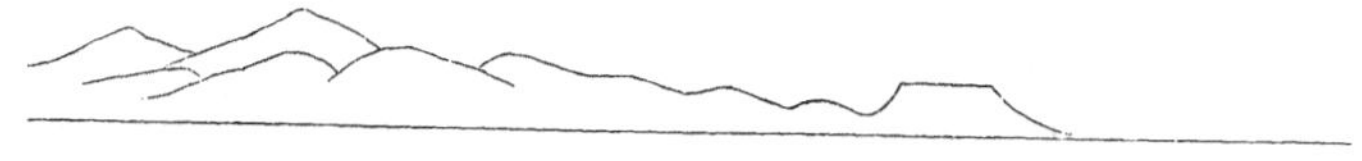

much struck with the outline of the table-topped hill on the right, which appears to be composed of soft horizontal rocks. May it not be an outlier of the great tertiary sheet that covers so large an area in the interior, and which in some places must have extended down to the coast before the deposition of the coast tertiary beds? I would suggest a comparison between it and the hills of Monte Alegre on the Amazonas.

From the Una northwards the tertiary plains descend to the coast and extend, with high, steep, wooded, rarely perpendicular slopes, seaward to Ilhéos. A few insignificant little streams empty into the sea along this coast. As a general thing, a strip of flat sandy or marshy ground runs along under the bluffs. The Piassaba palm (*Attalea funifera*) grows very abundantly on the slopes and the plain above, as Prince Neu-Wied has remarked. This author says that he did not see it anywhere north of Ilhéos. It now covers large tracts on the Rio Paraguassú, below

* Prince Neu-Wied says that gold is found here.

Caxoeira, and along the Bahia and São Francisco, and its fibre is quite an important article of commerce. Prince Neu Wied speaks of finding rounded fragments of pumice * on the shore near this place, and states his belief that they were drifted to the Brazilian coast from the island of Ascension.† He speaks of finding them on the shore near Porto Seguro, and I have picked up large quantities on the island of Santa Barbara dos Abrolhos.

The little village of Olivença is situated on the top of a hill by the shore, about nine miles south of Ilhéos. Its inhabitants, civilized Indians, employ themselves in making straw hats, baskets, &c., which they send to Ilhéos by way of the beach, on mule-back. They plant only enough to support themselves.

As above remarked, in speaking of the Rio Pardo, it was believed that that river divided, part of its waters flowing into the sea at Cannavieiras, while another part, under the name of the Rio Cachoeira, flowed into the bay of São Jorge dos Ilhéos. There is a road, or rather trail, through the forest following this river into Minas, and Prince Neu Wied travelled over it, crossing the head-waters of the Cachoeira. He says nothing about that river being only an offset from the Pardo, and in his map of the coast between the twelfth and fifteenth parallels, after Arrowsmith, he represents the Rio Cachoeira as taking its source very near the Pardo, at some considerable distance below the Villa da Vareda, and this is probably correct. Just above the mouth of the river a small river called the Rio do Fundão empties into it.

* *Reise*, Vol. II. p. 85.

† This hardly seems probable, since there is a southward-flowing current setting along the Brazilian coast south of Cape St. Roque.

In the year 1817 Prince Max. zu Neu-Wied made a journey into the interior from Ilhéos to the borders of Minas, and then went by land through the interior to the Bay of Todos os Santos. His voyage is full of interest, but I extract from it only a few points which bear upon the geography, natural history, &c. of the country.

His picture of the forest on the river near Ilhéos is so grand, and so true to nature, that I translate a portion of it: —

"Life and luxuriant plant growth is spread everywhere. Nowhere is there a little spot without plants. On all stems species of *Passiflora*, *Caladium*, *Dracontium*, *Piper*, *Begonia*, and *Epidendrum*, besides many ferns, lichens, and mosses of different kinds, bloom, climb, luxuriate, and attach themselves. The forest is made up of the genera *Cocos*, *Melastoma*, *Bignonia*, *Rhexia*, *Mimosa*, *Ingá*, *Bombax*, *Ilex*, *Laurus*, *Myrthus*, *Eugenia*, *Jacaranda*, *Jatropha*, *Visinia*, *Lecythis*, *Ficus*, and a thousand other, for the most part, unknown species of trees, whose fallen flowers one sees lying on the ground, and can hardly guess from which giant tree they came. Others covered with blossoms shine afar, white, bright yellow, bright red, rose-red, violet, sky-blue, &c., and in swampy places there rise, thickly crowded together, the great elliptical leaves of the *Heliconias* on long stems often ten to twelve feet high, and noteworthy for their bright red or fiery flowers. On the highest trunks, high up in the forks of the branches, grow immense tufts of *Bromelias*, with large clusters [*Blumenkolben und Trauben*] of light cinnabar-red flowers, or of some other color; from these fall great bundles of roots like cords, which hang down to the earth and form a new obstruction for travellers. Such bromelia-clumps fill all the trees, and when with the lapse of years they die, and are dislodged by the wind, they fall with a crash. A thousand llianas, from the slightest thread to the thickness of a man's thigh, and of hard, tough wood (*Bauhinia*, *Banisteria*, *Paulinia*, &c.), entwine the trunks, climb even to the highest top of the tree, where they blossom and bear fruit where no human eye can see them. Many of them are so wonderfully formed that one cannot look upon them without amazement, as, for instance, certain species of *Bauhinia*. From many of these the trunk around which they had wound decays, and here stands a giant coiled serpent, whose origin makes itself easily understood in this way." *

In penetrating into the interior the country becomes gradually higher, and at the same time drier, while the forest becomes less and less luxuriant, and finally passes, on the higher ground, into a catinga, which begins, on the road to Minas, at Porto da Canôa, on the Rio Cachoeira. Bromeliaceous plants become more abundant, with several species of *Solanaceæ*, *Mimosas*, and the stinging Cansanção (*Jatropha urens*). In the valleys, however, the forest is still dense and thick. In the catinga grows a cactus (*Cereus*) with immense stems reaching a height of fifty or sixty feet, with a diameter of two feet. On the western side of the Ribeirão da Issara is a range of hills, called the Serra da Sussuarana, which are not very high, but covered with masses of loose rocks and stones, with a thick growth of catinga. The country onward to the Giboya, a little stream flowing southward into the Pardo, is covered with catinga. Here, as elsewhere, the low grounds of the valleys are filled with high forests, but the woods on the slopes and high lands grow lower, and come under the class of catinga. The

* *Reise*, Band II. Seite 106.

Prince says that the Giboya flows over *Granit-tafeln* (gneiss). I should judge from his description that the country between the Giboya and Ilhéos was of the same character. A short distance to the westward of the Giboya is a range of mountains, "whose hills, of a considerable height, have a rounded outline, and are strewed over with masses of rocks and granite blocks in which especially very large pieces of white quartz occur. The whole vicinity is overgrown with very thick forest or catinga. These mountains bear the name of Serra do Mundo Novo. The first mountain is the highest; it rises, it is true, with gentle slopes, but it requires a full hour to ascend it. Thence onward hills and valleys alternate, until one at last descends into a considerable depression. The Rio Pardo flows to the left, in a deep valley parallel with the road. . . . . So soon as we had left behind us the fatiguing Serra we found the wood still more changed into catinga, for in the depression itself it was only 40 – 60 feet in height, filled with many bromelia and cactus clumps, hung with moss tufts (*Tillandsia*), and intermingled with many kinds of trees which reached only an inconsiderable height. Here is found the *Páo de Leite* (in all probability a Ficus), which is feared because of its corrosive milky juice, but nowhere was seen the beneficent, nourishing milk of the *Palo de Vaca*, which Humboldt has described; this milk would in our situation be a great comfort. Farther on we found the cork-like Barrigudo-tree (*Bombax*), which here grew to only a small height, many species of *Mimosa*, of Bignonia, and so forth, and between them rock masses and granite blocks. All this shows that one has gradually ascended from the wet, dark region of the coast forests to higher and drier country." This description is very interesting, because it shows that here we have a region which has not been so much affected by decomposition as the coast belt, and it shows us how the forest zone narrows down as we go northward.

Beyond the Barra da Vareda one enters a catinga wood, and gradually ascends, the hills being gently rounded, and, as the Prince remarks, announcing the open plains and ridges which make up so large a part of the interior of Brazil. "The wood has in many places lagôas grown up with swamp reeds, in others extensive naked places which have been burned over so as to produce grass for the cattle. Such places become covered immediately with high ferns (*Pteris caudata*), whose horizontally placed fronds wear an agreeable look. With the end of the wood one reaches pleasant green fields, which, despite the dry climate, appeared to have the fresh green of our European meadows."

Going westward the country becomes more and more open, and there are extensive open flat tracts of great extent, covered with a sparse catinga vegetation, immense candelebra-like cactuses, and ant-hills, and diversified by shallow lagôas. From Tamburil to the boundary of Minas Geraes one passes through a monotonous and somewhat hilly country, cut through by deep ravines, and covered by catinga. So soon as one has climbed the ridges which uniformly command one another, and throughout are covered in the same manner with catinga or carrasco, small narrow fields, grown up with many rush-like grasses, are reached.

In some places near Ressaque the Prince found mica slates with *staurotides* in single crystals, together with hornblende. This succession of gneiss and mica slate corresponds to what I observed on entering the basin of the Jequitinhonha, but I have seen no staurotides in Brazil.

The carrasco is a more or less thick, usually matted,

growth of bushes, with stiff, gnarly stems, which grow to a height of ten or twelve feet, and this is the character of the vegetation of a large part of the campos of the interior, particularly of the wide elevated plains.

The country grows flatter and flatter in going into the interior, and at the same time the bushes grow lower until the *campos geraes* are reached, where, "far as the eye can reach, open wooded plains, or gently rounded hills and ridges, are spread out, covered with dried grass or scattered bushes."

"In the valleys which intersect these wide, naked ridges and plains, one finds the banks of the rivers and brooks bordered by woods. Here also single clumps of bushes are found, hidden here and there in the deep places, particularly as one approaches the borders of Minas Geraes, and this kind of wood is in part one of the peculiar characteristics of these open places. One often believes that he has a continuous plain before him, when he comes suddenly upon a narrow, steep-sided valley, hears deep below him the murmur of a stream, and looks down upon the forest-trees which, variously colored with numerous flowers, line its banks. Constant winds prevail here during the cold season, with, for the most part, a cloudy sky, and in the dry months a burning, oppressive heat, whereby the grass is dried up, the region is glowing hot, and there is a scarcity of potable water."

In the winter these high lands are quite cool, and hail not unfrequently falls. Prince Neu Wied describes a large tract over which the vegetation was dead and leafless, and he was informed that it had been killed by frost, though it may be that the effect had been produced by excessive dryness. To one accustomed to the climate of the coast, that of these high regions is apt to be very disagreeable. The Prince found the temperature early on a foggy morning 14° Reaumur = 63.5° Fahr., and at noon on a dry sunshiny day $19\frac{1}{2}$° Reaumur, = 75.87°, and this was just at the end of the rainy season, in February. On the 22d January, at Catolé, he found in the shade between two and three o'clock, P. M., a temperature of $24\frac{1}{2}$° Reaumur, = 87.12° Fahr., and he states that it sometimes reached 30° Reaumur, = 95.50° Fahr., in the shade.

These campos stretch westward to the Serra das Almas, and southward to the valley of the Jequitinhonha,* and are only sparsely inhabited, principally by herders of cattle. Agriculture is confined to the bottoms of valleys and moist places.

The *Ema*, or American ostrich (*Rhea Americana*), occurs in abundance on the campos of the basin of the Pardo, together with the celebrated *seriema* (*Palamedea cristata* Linn., *Dicholophus cristatus* Illiger), a large bird the size of a crane, very swift of foot, and noted for its shrill voice like that of the peacock, whence the English name *crested screamer*, sometimes applied to it.†

With reference to the climate of this part of the coast, Neu-Wied says that generally the months of February, March, April, and May are the rainy months. The four months following are the cold season, while the hottest weather is in October, November, December, and January. Our author says that he never observed a lower temperature than 13° Réaumur, nor a much higher one than 30° Réumur.

* See description of the campos of the Comarca da Jequitinhonha in Chap. III. of this work.

† See *Eug. Warming, Skildringer af Naturen i det tropiske Brasilien. V. Camposdyrene, Tidsskrift for Pop. Fremst. af Nat.*, Tredie Række, Femte Binds, Tredie Hefte, 1868, p. 231. This is a very interesting article on the fauna of the campos region.

On the 5th March, one of the hottest days of the voyage, Neu-Wied observed a temperature of 28.50° Réaumur, and in the twilight of the same day 15° Réaumur, and an hour later, when the dew had begun to fall, 14° Réaumur, which observations are interesting as showing the diurnal varieties of temperature.

Arraial da Conquista, or Victoria, is the name of a little village lying a few leagues to the north of the Barra do Vareda, and noted for its cotton and large herds of cattle, which are sent to Bahia to be sold.

On the voyage to Bahia, near a place called Urubú, lying between the Rios Caxoeira and Contas, Neu-Wied describes brooks whose waters were salt, discolored, and whitish, and he speaks of others near the valley of the Contas. Thus far a great part of the country between the Arraial da Conquista and the Rio das Contas has been hilly and more or less wooded.

Between the Arraial da Conquista and Os Possões (Poções) the country is very uneven and covered with low woods. Neu-Wied calls attention to a very interesting fact stated by Humboldt,* that the number of species of social plants in

* Alex. de Humboldt, *De Distributione Geographica Plantarum*, 1817, pp. 51, 52 :—

" Rarissimæ autem sunt plantæ sociatæ (*Plantes sociales, gesellige pflanzen*) in plaga æquinoctiali. Difficiliter enim, ex genere arborum Silvis Orinocensibus nomen ponas, quippe in quibus magnus specierum numerus æque commixtus sit. Neque in locis planis sub zona torrida Novi Orbis, plantas sociatas fere alias ullas enumeres præter Rhizophoram Manglen, Sesuvium Portulacastrum, Croton argenteum, Bambusam Guaduam, atque propter capita fluvii Amazonum et in calidis Provinciæ Jean de Bracamoras, amœnissima nemora Bouguinvillea et Godoya repleta. Augentur vero stirpes catervatim nascentes quo magis per Mexicanum imperium versus Cancrum procedis, vel per cacumina Andium te tollis, ubi altitudine 1800 haxapodarum reperies Escalloniam myr tilloidem, Brathim juniperinam et multijugas Molinæ species."

the tropics is very small, and says that not unfrequently we find large tracts taken possession of by the samambaia (*Mertensia dichotoma* or *Pteris caudata?*) to the exclusion of all other shrubs. This is the case all along the coast. It is apt to spring up in abandoned dry fields, and bury them with a thick mantle of foliage. I observed it growing at Itabapuana, São Matheos, and in numerous other places on the coast, where it was so abundant as to be a nuisance, and I have noted it as being very common in Minas. It seems to have the same habit as our *Pteris aquilina*, which in the same way takes possession of fields and drives everything else out. The rapid growth of the samambaia in Brazil is often aided by fires set in the bushes and grass over dry places, which deplete the flora and give the fern a more open field to grow in. The most remarkable social plants of Eastern Brazil are the mangroves, *Conocarpus* and *Avicennia*, in addition to which and the *Pteris caudata* are some species of *Rhexia*, *Cecropia*, *Bignonia*, together with the *Ubá*, *Taquarassú*, some grasses, a bamboo, and the dwarf palm of the coast, *Gurirí*. The Piassaba and Carnahuba palms would also appear to merit being included under this head.

Between the Arraial da Conquista and Os Possões, Neu Wied describes a locality with high, gently rounded hills covered with the samambaia, and he states that sometimes such tracts are burned over so as to produce a growth of grass. The whole country here is exceedingly dry, and during the hot season the vegetation is withered and scorched. Water then fails, and cattle die if not removed. In this dry region is found a beautiful *Bignonia*, eight to ten feet high, with large bright citron-yellow blossoms, and a *Cassia*, together with the *Licurí* palm, a species which I found growing over the taboleiras at Alagoinhas on the Bahia and São Francisco Railroad. The soil is of a reddish-yellow color. The deep valleys are filled with dense forest. Everywhere one sees the round yellow hills of the white ant scattered about. Carapatos * are exceedingly numerous, incrusting twigs so as to make them fairly red, and worrying the traveller by day and night. Between Os Possões and Urubú the country is of the same desolate character. In this region many cattle are pastured, and in some localities a little cotton is planted. Of the journey from Urubú to the Fazenda da Cachoeira Neu-Wied draws a striking picture of the country in these words: † "I followed the way through an inhospitable deserted wilderness, in which, crowded together, mountain after mountain rose behind one another; all lay before us, monotonous, covered with thick-woven brushwood, rough and wild, and mingled with projecting rock masses. Some of these mountains are naked and consist of variously formed masses of rock, as a general thing gently rounded above; in the places bare of wood the soil shows itself as a red-yellow clay. Bushes of finely-plumed thorny mimosas, mingled here and there with beautiful flowering plants, amongst which I will mention only one splendid plant, a new species of *Ipomœa*, with large, brilliant fiery blossoms, made on both sides a border to the way."

"The rock masses of the strangest forms, often like towers or pulpits, standing singly out above the bushes, are everywhere in these mountains inhabited by a little cavia," called the Mocó, — *Cœlogenys rupestris*. "In these dry rocky

* Bates says that there are two species on the Amazonas. See Naturalist on the Amazons, p. 173.

† *Reise*, Vol. II. p. 236.

woods there reigned a heat beyond belief; not a breeze stirred, and the rays of the sun were reflected from every side; only the proud *Araras* [macaws] in our neighborhood appeared to enjoy it. They flew screaming about, while the most of the other birds took their *siesta* on the shady branches."

The mouth of the river at Ilhéos forms a very good harbor, and is entered by the coasting steamers. Where the valley of the river opens out into the sea are two isolated gneiss hills, standing one about a quarter of a mile east of the other. The western of these hills, once an island, has been joined to the bluff on the north by a broad strip of sand, on which is built the villa of São Jorge dos Ilhéos. This causes the river to make an abrupt bend southward. The eastern hill, also formerly an island, has been joined to the low lands on the south by a strip of sand, which compels the river to turn towards the north, when it enters the sea between the two hills. The mouth of the river is shallow, but is usually entered without much difficulty. The waves beat very heavily on this coast, and it is interesting to see how they are striving to throw up a spit across the river-mouth on the northern side. At high tide the water flows through a channel cut across the sand-beach, uniting the eastern island with the shore; but the waves tend to increase the height of this beach, and the river is obliged to escape around the western side of this hill. To the northeast are the reefs Sororoca, Itapitinga, and Itaipins, which I have not examined, and north of these are the islets Ilha Grande and Ilha Pequena, distant about two miles north, a few degrees east of the barra.

The town of Ilhéos is about as large as Caravellas, but is of much more importance. The banks of the Rio Caxoeira and of its tributaries near Ilhéos are thickly settled, and there are many large fazendas for the cultivation of cacáo, sugar, &c., which products are exported to Bahia.

The rocks of the hill on the western side of the barra consist of a well-bedded gneiss like that of Bahia, but much disturbed and broken up. The approximate strike is north a few degrees east.

A few miles to the north of Ilhéos the little river Itahype empties into the sea. The banks of the river are largely settled. The river itself is very narrow, but deep, and altogether is only about twenty-eight miles long. On the northern side is a little lagôa communicating with the Itahype by a narrow canal. Neu-Wied says that sea-shells are found on the banks of this lake, and that kettle-formed holes, like those hollowed by the action of the sea, are to be seen in the rocks bordering it.

Spix and Martius found coral banks in the lake, showing that it was formerly a bay which has been cut off from the sea by the throwing up of a beach across its mouth. "These banks show themselves in many places in the lake at a depth of from six to twelve feet," and furnish material for the manufacture of lime. The corals observed by Spix and Martius were referred by them to the old species *Madrepora cavernosa*, *hexagona*, and *astroites*, Lam.*

On the sea-shore near the Serra Grande, south of the Rio de Contas, Spix and Martius found "banks, five to six feet high, of a soft, coal-black substance which soiled the finger when pressed, and which, carefully examined, seemed to be made up of coal and quartz grains." This appears to be some recent formation.

* Von Martius, *Reise*, Band II. Seite 684.

From Ilhéos northward to the Rio de Contas the coast lands, as a general thing, are about two hundred feet in height, and level topped, with abrupt, steep slopes to the sea, and there can be no doubt that they are tertiary. The appearance of the coast just south of the mouth of the Contas is represented in the following sketch. I am told that the

VIEW OF THE COAST SOUTH OF THE RIO DE CONTAS.

soils of these plains are sandy, but are in part quite fertile. On the slopes cacáo is planted.

The Rio de Contas, or Jussiape, is a considerable river, which, according to the *Diccionario Geographico*, rises eight leagues to the northwest of the Villa de Rio de Contas, or at a distance of some one hundred miles from the sea and to the west of the Serra de Sincorá.* The course of the main stream is almost east-west. It is navigable only for a distance of some four leagues above its mouth.

The lands on the southern side at the mouth of the river are moderately high, and two or more rocky islands and a reef of rocks project northward, so that the channel of the river is bent toward the north, as is the case at Ilhéos. There are other rocky points on the same side of the river, in which a rock like that of Bahia is exposed. The town is situated in a little cove, just inside the bar, on the southern side. I found the rock in the hill and point west of the town to be

* According to Almeida, it takes its rise in a lake forty or fifty miles north of the Villa do Rio de Contas. Burton says that this name should be Rio *das* Contas. I follow Cazal, Almeida, and the common usage of Brazilian writers.

of the same character as that of Bahia, and exposed with a strike of N. 45° E., and a vertical dip. The town is a small one, and of so little importance that the coasting steamers rarely stop there. Its commerce consists principally in farinha, of which something like 50,000 sacks are exported every year; cacáo, yearly exportation more than 4,000 arrobas; coffee, 800 to 1,000 arrobas; together with a little rice and sugar, and large quantities of woods, especially *jacarandá, cedro, putumujú, vinhatico.* Along the river above the town are many fazendas. The population is said to be increasing. On the opposite side of the river the lands bordering the coast are flat, sandy along shore, but inside they are largely overgrown with mangroves. The beach extends southward in a sharp point, which tends to close up the river off the point just east of the town; but the current keeps it open. The bar is not difficult, and there is good anchorage for vessels off the town. The Rio de Contas appeared to me to be about the size of the Mucury. The tertiary bluffs trend off northward of the river, gradually leaving the coast. A narrow channel coming from the north and flowing through the low lands, parallel with the coast, empties into the river opposite the town. I have seen in some Brazilian work — a *Roteiro*, if I rightly remember — a statement that there had been found on the banks of this river the bones of some immense extinct quadruped. Spix and Martius also refer to them. (See Chapter VII. of this work.)*

* Cazal speaks of the existence, in different part of Brazil, of the bones of an immense extinct animal, which he suggests might be the behemoth. I translate a few sentences from his note on it. He says: "Morse gives to this quadruped the name of Mammoth, and says that the Indians of North America pretend that its species still exists in the woods which are to the north of the great lakes. . . . . This beast must have been of a slow march, not proper

In 1866 I touched at Camamú, but unfortunately during bad weather, so that I was unable to see much. I have never revisited the spot. I regret much that I am unable to give a more precise and detailed description of the bay on which it is situated, for it is of great interest, both to the geologist and the zoölogist, because on the borders of the bay are the *turba* deposits which have attracted so much attention, while off the mouth of the bay are extensive coral reefs. This bay, and the streams which flow into it, have been most erroneously represented on the maps and charts. Arrowsmith represents it as a deep wide bay opening broadly to the sea, half as large as the Bahia de Todos os Santos, and sown with little islands; and Mouchez's chart gives one really no idea of it. My friend, the Rev. Mr. Nicolay of Bahia, kindly furnished me with a copy of a recent map by Sr. José Nascimento, which is here subjoined, not only because of its value as a contribution to the hydrography of the coast, but because Mr. Nicolay has indicated on it the localities where the turba occurs, as well as several other points of interest.

The shores of the bay and its arms are for the most

for a hunting or carnivorous animal, and with a belly so capacious that only vegetables could suffice to nourish it.

"Among the many skeletons which have been encountered in the different provinces of the New World, perhaps none may help to form a better idea of the animal than the skeleton which was discovered, at the close of the past century in the *Termo da Villa de Rio de Contas*, by the persons engaged in clearing out a hollow in the rock (*caldeirão de pedra*), in order to make a tank for the cattle. . . . . This skeleton, considerably injured, occupied a space of more than thirty paces in length; the ribs were a palm and half broad; the shin-bones were of the length of a man of medium stature; the tusks were almost a *braça* in length; a molar tooth, without the root, weighed four pounds; in order to remove the lower jaw, the strength of four men was necessary." — *Corografia Brazilica*, p. 67, note. See *D'Archiac, Paléontologie Stratigraphique*, p. 231.

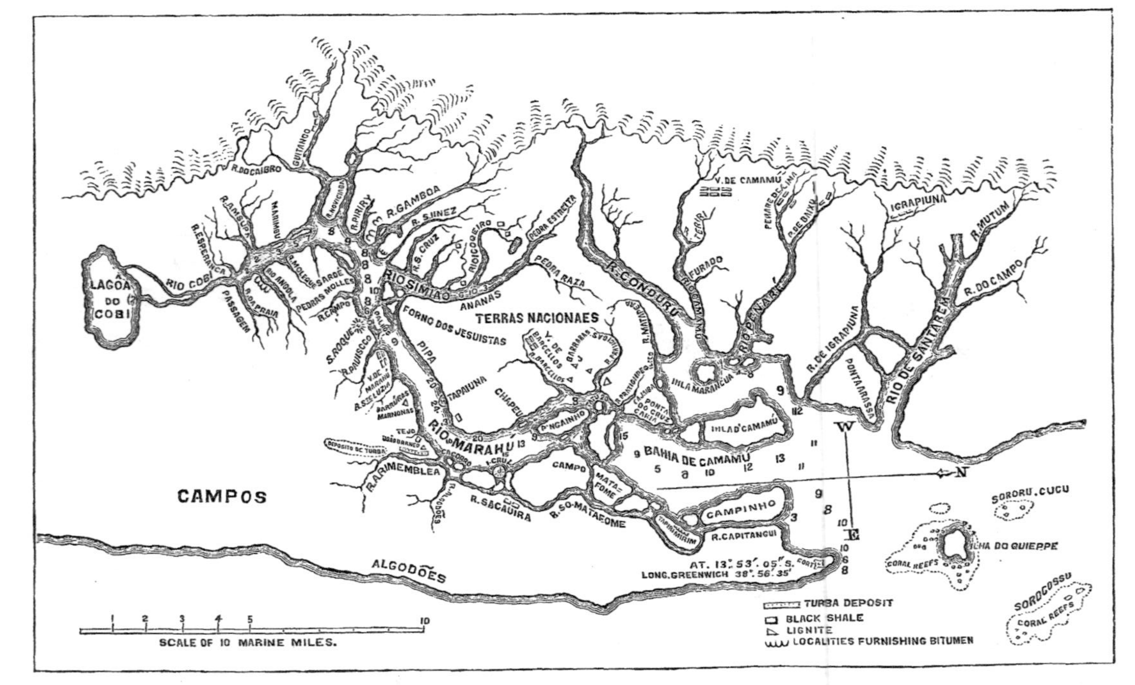

LAGOA DO COBI
RIO COBI
R. ESPERANÇA
PASSAGEM
R. DO CAIBRO
GUITANGO
R. GAMBOA
R. PIRIRY
R. S. INEZ
R. S. CRUZ
RIO SIMIÃO
PEDRAS MOLLES
SARDE
R. CAMPO
S. ROQUE
FORNO DOS JESUISTAS
ANANAS
TERRAS NACIONAES
PEDRA ESTREITA
PEDRA RAZA
R. CONDURU
V. DE CAMAMU
FURADO
RIO PENAR
IGRAPIUNA
R. MUTUM
R. DO CAMPO
R. DE IGRAPIUNA
PONTARASSA
RIO DE SANTAREM
PIPA
TAPAIUNA
CHAPEU
RIO MARAHU
INLA MARANCUA
INLA D' CAMAMU
BAHIA DE CAMAMU
CAMPO
MATA FOME
CAMPINHO
R. CAPITANGUI
R. SACAUIRA
R. SO-MATAFOME
R. ARIMEMBLEA
CAMPOS
ALGODÕES
AT. 13°. 53'. 05". S.
LONG. GREENWICH 38°. 56'. 35"
W
N
E
SORORU. CUCU
ILHA DO QUIEPPE
CORAL REEFS
SOROCOSSU
CORAL REEFS
TURBA DEPOSIT
BLACK SHALE
LIGNITE
LOCALITIES FURNISHING BITUMEN
1 2 3 4 5 10
SCALE OF 10 MARINE MILES.

part flat, and largely of recent origin, and there are extensive mangrove swamps bordering them. From over this region the tertiary clays have been almost entirely denuded. Gneiss occurs at the town, and just to the westward is the considerable range of hills called the Serra do Condorú, which is a continuation northward of the coast serras. It is in the low grounds of the Marahú, a broad river-like arm extending off for some distance to the south of the bay, that the turba deposits are found. The specimens of turba which I have seen were of a very light material, grayish or brownish in color, and felty in texture. The material burned readily when ignited in a candle, affording an abundant smoky flame, leaving the mass, however, of the same dimensions and form as before. The material appears to be merely a mud impregnated with bitumen; and as it appears to exist in large quantities, it would be very valuable for gas-making or the manufacture of kerosene. Prof. Arthur M. Edwards of New York, the microscopist, informs me that a Mr. Southworth brought home with him some specimens of carbonized wood, and Prof. Edwards * has some fragments of leaves from the locality, but they are unfortunately too badly preserved for identification. Mr. Nicolay personally examined the turba deposits, and has kindly furnished me with the following observations, which throw some little light on the mode of occurrence of the bituminous layers, though their age is still left in obscurity: —

* Since the above was written, I learn that some of the turba distilled in New York yielded of first quality, one hundred gallons light, clear oil to the ton; second quality, seventy-five gallons. Prof. Edwards says, in speaking of the specimens he examined: "I consider the turba a sand impregnated with bitumen, but it has evidently not been deposited under water, because it contains no diatoms, and the vegetable remains in it are wood, a few leaves, and fibres like fine roots." An English company has recently been formed to work this deposit for the above-mentioned purposes.

"The Camamú series does not appear in any way connected with that of Bahia or with those of the South, unless, as has been reported, turba is found on the Rio de Contas.

"This series may be designated as follows: —

"In a basin of gneissose rocks are bituminous schists, sands, and marls containing fossils, (fresh water ?) and it is presumable above that the turba.

"There are also sandstones, —principally red concretions, —not apparently connected with the bituminous strata, but lying nearer the sea, and possibly outliers of the Bahia series.

"The following section is from the pit sunk by Sr. João da Costa, filho: —

"20 feet	0 in.	Argillaceous and arenaceous schists.
3 "	0 "	Bituminous clay.
4 "	0 "	Ferruginous and arenaceous sandstone.
0 "	9 "	Schist with lignite.
0 "	9 "	Bituminous stratum.
1 "	6 "	Micaceous schists.
1 "	6 "	Schist with lignite and bitumen below.
2 "	6 "	Schistose strata.
2 "	0 "	Bituminous strata, some quite pure.
12 "	0 "	Schistose rocks.
15 "	0 "	Bituminous strata.
45 "	0 "	Gneissose rocks.
108 feet."		

Mr. Nicolay states that the bituminous strata vary much, from pure bitumen to an arenaceous kind interstratified with bituminous shales; "but it is to be observed that, at the depth of eighteen feet, two veins of *pedra molle*, an imperfect turba occur. These are in many cases present on the surface, and the connection should be ascertained, but except in this instance *pedra molle* and turba only appear above the limestone, i. e. so far as is known, which forms the bottom of the estuaries which unite to form the Camamú Bay."

The town of Camamú is only a small one, but is of some little importance, exporting to Bahia coffee, farinha, rum, rice, cacáo, and woods for building purposes. Spix and Martius speak of the occurrence of coral banks in the Bay of Camamú, and they refer some of the corals to the old Lamarckian species, *Madr. Uva*, *M. Astroides*, and *M. Acropora*. The Ilha de Quieppe lies just off the mouth of the bay, and is surrounded by coral reefs. The dangerous reef of Sorocosú (*Sororocossú ?*) lies a short distance to the northwest. About ten miles north of the mouth of the Bay of Camamú lies the large island called Boyapeba. Mouchez, in a note to his chart says that the pilots represent the gulf lying between this island and the Bay of Camamú as being full of corals, probably chapeirões. The island of Boyapeba is about five miles in diameter, moderately high, with a few prominent hills, and separated from the mainland by a narrow channel. On the southeastern side of this island is the dangerous point Castelhanos, on which the French ship Béarn was wrecked a few years ago. The village of Boyapeba, near the northern extremity of the island, is noted for its little commerce with Bahia in piassaba, rice, and mangrove bark for tanning.

Separated from this island on the south by narrow channels, and from the mainland by a channel which unites them with the Barra do Rio Una, is a much larger island, called Tinharé, some ten or fifteen miles long and five or six wide. Wedged in between these two islands and the mainland is the little island Tupiassú. The northern portion of the island of Tinharé appears to be high; while the rest, together with the island of Tupiassú and the mainland opposite, is low and in great part swampy. Mouchez represents doubtfully the island of Tinharé as crossed by two or three channels. The channel west of this island is very narrow but deep. Into it flows from the west the Rio Jequié, a small stream of little importance. The northeastern extremity of the island of Tinharé is prolonged northward into a sharp promontory called the Morro de São Paulo, which, according to Dr. Anto. de Lacerda, is composed of gneiss.* West of this is a deep bay some two miles wide, and four or five deep, into which the Rio Tinharé empties from the south, and the Rio Una from the west. This bay is remarkable for a long sharp sand-spit, the continuation of a sand-beach, which projects southeastward into the deep curve west of the Morro de São Paulo. In this bay are several small islands. Mouchez's chart appears to be very inaccurate. Prince Neu-Wied, after leaving Ignez, reached a river, which he calls the Jequeriçá, which he descended for some distance, when he was arrested and carried across the country to Aldêa, a little place near Nazareth. In his edition of Arrowsmith's map he has laid down his route, which he represents as following the Una instead of the Jequiriçá, which flows into the sea some ten miles north of the Una. One cannot doubt from his description that it was the Jequiriçá which he descended, and that, owing to that river having been represented as emptying into the bay of the Una, he laid down his itinerary incorrectly. The town of Valença is situated at the mouth of the Una, and is noted for its large cotton-factory, and its commerce in woods and coffee. From the Una to the Jequiriçá runs a

* Pissis also says that it is gneiss. *Mém. de l'Institut de France,* Tome X. p. 357.

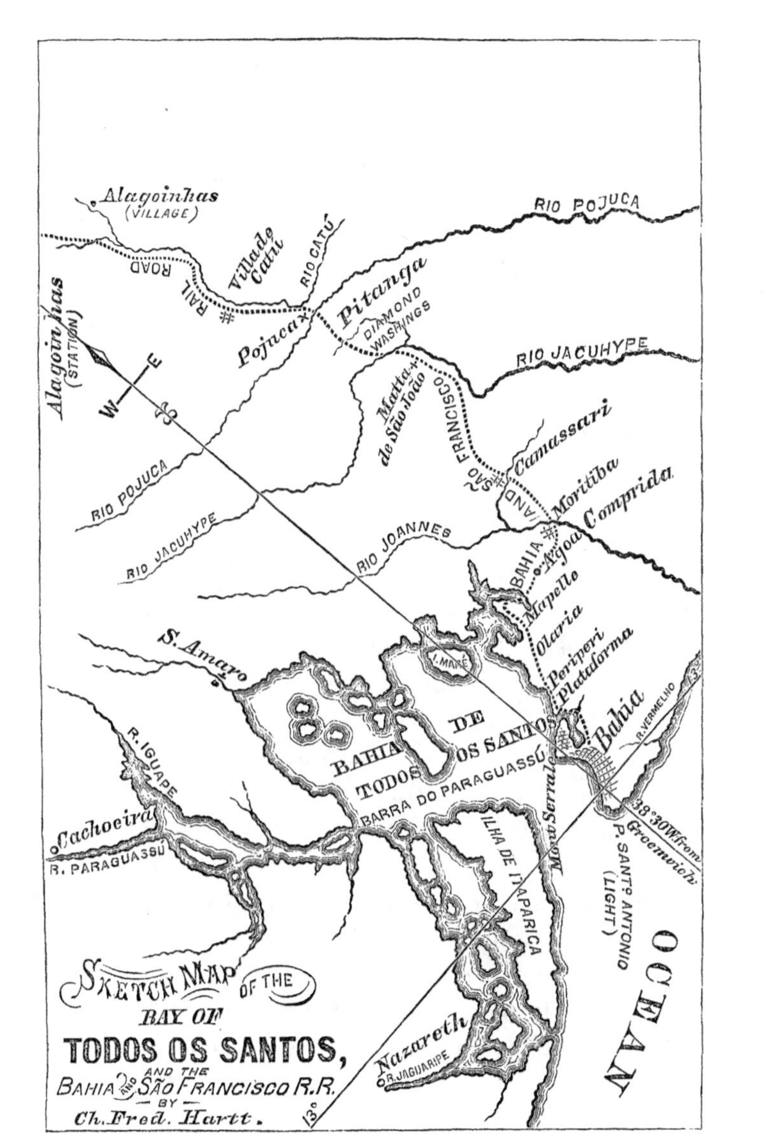

SKETCH MAP OF THE
BAY OF
TODOS OS SANTOS,
AND THE
BAHIA AND SÃO FRANCISCO R.R.
BY
Ch. Fred. Hartt.
Alagoinhas (VILLAGE)
Alagoinhas (STATION)
RAIL ROAD
Villa de Catú
RIO CATÚ
RIO POJUCA
Pojuca
Pitanga
DIAMOND WASHINGS
RIO JACUHYPE
Matta de São João
SÃO FRANCISCO
Camassari
Moritiba
Lagoa Comprida
RIO JOANNES
BAHIA
Agoa
Mapello
Olaria
Periperi
Plataforma
S. Amaro
I. MARÉ
BAHIA DE TODOS OS SANTOS
BARRA DO PARAGUASSÚ
Bahia
R. VERMELHO
38°30'W from Greenwich
P. SANTº ANTONIO (LIGHT)
OCEAN
Mont Serrat
ILHA DE ITAPARICA
R. IGUAPE
Cachoeira
R. PARAGUASSÚ
Nazareth
R. JAGUARIPE
13°
W
E

long sand-beach, behind which the lands are at no great distance moderately high and hilly. North of the Jequiriçá the beach is interrupted by a few rocky points.

The Bahia de Todos os Santos, leaving out the islands, is a quadrilateral figure, with unequal sides, as follows: * A line from the Ponta de Sant. Antonio to the Ponta Garcia runs approximately southwest, and measures a trifle over eighteen marine miles; a line running along the western side of the bay would run nearly north-northeast, and measure thirty-three miles; the northeastern side of the figure representing the width of the bay, which is remarkably uniform throughout, would run about southeast, length fifteen miles; while the other side of the figure, bringing us back to the point of departure, would be approximately, south-southwest, sixteen miles. A large island, of a sigmoid shape, called Itaparica, and about four or five miles wide in the widest part, lies in the bay with a general northeast trend, its axis lying several miles to the west of that of the bay, and its southern end distant about two miles northeast of the Ponta Garcia. About three miles northeast of the Ilha Itaparica is a considerable island called the Ilha dos Frades, which lies just off a point projecting southward from the head of the bay. This point is deeply indented by numerous little estuaries, and several large islands lie off its west side. The Ilha Itaparica and the Ilha dos Frades divide the bay into two strips, a long and narrow one to the west, a shorter and wider one to the east. With this general description of the bay, let me now enter into a little more detail. Between Ponta Garcia and the Ponta da Caixa de Pregos of the Ilha Itaparica, the river Jaguaripe finds an outlet to the sea, although its waters are free to enter the channel west of Itaparica.

* These measurements I take from the best maps and charts in my possession.

This so-called river is an estuary, which at the mouth is some two miles wide, but narrows rapidly down, being at Nazareth, a town some ten to fifteen miles up the river, only a few hundred feet in width. Steamers ascend to Nazareth, but navigation ends a few miles above that town. The river rises about fifty miles west of Cachoeira, and has a southeastern course. All its branches below Nazareth appear to have the estuary character, opening broadly into the river. I have never ascended the river, and can therefore give no definite information with respect to the geological character and agricultural capabilities of the country bordering it. Nazareth and Jaguaripe, a town about a league above the mouth, are noted for their extensive manufactories of tiles and earthenware, which are sent to Bahia. Opposite this last town a canal runs northeastward, joining the Jaguaripe with the estuary of the Jacoruna, separating a large tract from the mainland called the Ilha de Santa Anna.

If the coast should rise fifteen or twenty feet, the island of Itaparica would be joined to the mainland at its southern extremity; for at low tide the channel now separating it from the mainland at the junction of the canal with the estuary of the Jaguaripe is only some six feet deep at low water, but it deepens gradually on going northward. The west side of the channel is bordered by a line of islands called the Ilhas de Parajuhia, which extend almost to the mouth of the Paraguassú. It is interesting to compare the eastern exposed side of Itaparica with its long stretches of sea-beaches uniting rocky points with the sheltered western side, which is exceedingly irregular, and cut up by bays and estuaries. Of the geology of Itaparica I know nothing, except that I am informed by Dr. de Lacerda and Mr. Nicolay that the high hills are gneiss. In Dr. de Lacerda's cabinet at

Bahia is a pebble from Itaparica full of what appear to be tertiary shells; but it bears so strongly the appearance of a European formation, that I more than suspect that it has found its way across the Atlantic in ballast.* Darwin speaks of deposits of tertiary shells, and Elie de Beaumont says that M. Pissis found tertiary deposits resembling the European *mollasse* in the bay of Bahia. I have seen none, and it is singular that, if they exist, Mr. Nicolay, who has explored the bay most industriously, should not have found them. In all probability both Darwin and Pissis mistook (most pardonably) the recent consolidated beaches for tertiary.† Coral banks exist along the shores of the island, and the corals are collected and burned for lime. The species are the common ones of the Abrolhos region, though Heliastræa appears to be more abundant than at Bahia.

The Rio Paraguassú is the largest and most important stream that waters the province of Bahia. It rises in the Serra da Chapada, in the diamond district west of the Bahia de Todos os Santos, and, obstructed by many falls, reaches at last, a few miles below the city of Cachoeira, the head of an estuary which empties into the Bahia de Todos os Santos on the western side and to the northwest of the northern extremity of the Ilha de Itaparica. Through the kindness of my friends Dr. Antonio de Lacerda and Mr. Hugh Wilson, I was

* One must be on his guard on the coast of Brazil against collecting specimens from rocks brought as ballast not only from other parts of the Brazilian coast, but also from other countries. At Maceió I saw palæozoic rocks which I was told came from North America. So the Irish carboniferous limestone brought to St. John, New Brunswick, finds its way all along the coast of that province.

† I think it is Darwin who speaks of tertiary fossils as occurring at the head of the bay. I know of no reason why fossils should not exist as the tertiary rock of that locality.

enabled to ascend the river, or rather estuary, as far as the city of Cachoeira, on the occasion of the *fête* of the opening of the new steam tramroad just commenced by an English company, and intended to unite the city of Cachoeira with the Villa do Urubú on the São Francisco, and the following observations were made at that time. The entrance of the Paraguassú is quite narrow, and bordered by high lands. The water is deep. Some of the hills at the entrance may be five hundred feet high. Bluffs on the southern side showed that they were composed of horizontal beds of a soft, light yellowish brown sandstone. I believe this to be an extension of the tertiary formation of the coast.

On entering the narrow mouth the river immediately widens into a lake-like expansion, in which is a long, narrow island. High bluffs on the shore of the island and on the left bank of the river opposite show red sandstone with white streaks, and a beautiful oblique lamination. About a league from the mouth, just above the Barra do Rio Camurugipe, the river narrows abruptly, and is here more than one hundred feet deep. The land on each side of the river is three to four hundred feet high, flat-topped, with steep slopes toward the river, and occasional bluffs which show red sandstone in horizontal strata, as below. The soil is red, but thin, and the vegetation is very scanty, consisting of piassabas and low bushes, with a few cocoa-palms (*Cocos nucifera*) and Dendês* (*Elæis Guineensis* L.). At one place near the battery, called the Fortinho, the sandstone beds appeared to have a slight dip northward, but this is evidently local, as the whole formation is quite undisturbed and horizontal.

* This palm, an African species, goes by the name of Caiauhé on the Amazonas. The fruit gives an orange-colored oil, much used in Brazil for culinary purposes. It is also employed in the manufacture of fine soaps.

Above the fort the river opens out into another broad expanse, and a beautiful view is presented. The whole country is an elevated plain, with steep slopes to the estuary and its branches. These slopes have the same character as below, and are covered largely by piassaba-palms. An occasional sugar-factory may be seen on the river. From this lake-like expansion of the river a broad arm extends off to the southwestward for some six miles in a broad valley, into the head of which empties the Rio Capanema, and on which is situated the village of the same name. This arm is called the Rio de Capanema, and it is said to receive the waters of a lake. Maragogipe is a prettily situated and flourishing village, built at the base of the chapada, at the junction of the Capanema with the Paraguassú.* The slopes of the chapadas are here more or less cultivated with corn, &c., and there are large groves of cocoa-palms on the island in front of the town.

From the northern part of the expanse of the Paraguassú extends northwards another arm called the Iguape,† into which a little river empties. The valley of the Iguape is broad and exceedingly fertile, and there are very many extensive sugar-plantations situated in it.‡ The tertiary chapadas grow higher as we ascend, and on the Iguape they must be at least seven hundred feet in elevation. Passing the Iguape the river narrows very suddenly, and thence to the city of Cachoeira it is only a few hundred feet wide. More

* Cazal, *Corografia*, Tom. II. p. 125, says that in the vicinity is found *antimonio* and *bôlo-armenio*.

† This name is rather uncommon, being found only a few times on the Brazilian coast. It appears to be another form of *Ygarapê*, canoe-path or channel, a name applied to the side channels of the Amazonas. (See Index.)

‡ Cazal says that the soils of the valley of Iguape are the best known for the culture of cane, being of black *massapé* and strong. *Cor. Braz.*, Tom. II. p. 125.

properly speaking, the mouth of the Paraguassú is here, and the irregular sheet of water below, to the bay, is a tidal estuary. At the point where the river flows into the estuary, rock shows itself in the river-banks, and, according to Senhor Przewodowski, who accompanied me on the voyage, is the *coração de negro* rock, like that of Bahia. The strike is N. 40° E. Dip vertical. The country bordering the river consists of rounded hills, back of which are the elevated chapadas. Bricks, tiles, and pottery are largely manufactured on the river, from the clays of the alluvial river-banks. The water of the river is very turbid and brownish.

The stream narrows as we ascend, and is so shallow that two of the little steamers with a part of the excursionists from Bahia got aground. There are a few sugar fazendas on the river below Cachoeira. Above Cachoeira the river is obstructed by falls and rocks. The town of Cachoeira, a respectably sized village, is noted more particularly for its export of sugar and cigars.

The country bordering the Bahia de Todos os Santos is in general terms very productive, and it has long borne the name of the "*Reconcavo*." Sugar is the principal product, though mandioca, &c., are cultivated. The tertiary deposits extend across from the Paraguassú to Sant. Amaro, a flourishing city on the river Serigi, some three leagues above the mouth, according to my friend Dr. Brunet, the Director of the Agricultural Institute of Bahia. Dr. Brunet informs me that the lands of the vicinity of Sant. Amaro are very fertile. Sugar and farinha are the principal products. Here is situated the Agricultural Institute above spoken of. Eastward the tertiary beds extend to the Bahia and São Francisco Railroad.

It is interesting to compare the Bahia de Todos os Santos

with the Bahia do Rio de Janeiro. The latter is also divided almost into two parts by the Ilha do Governador and the islands to the northeast; but the most interesting point is to observe the difference in the character of the rivers emptying into the two bays. Those of Rio are all small, and at their mouths they are bordered by flat grounds and mangrove swamps, owing to the rapid building out of the land; for the lands about the bay are mountainous, and the streams, swollen by heavy rains, bring down an immense quantity of sediment into the bay. The lands bordering the Bahia de Todos os Santos are high, and stretches of sands and mangrove swamps are rare; while the rivers, bringing down less sediment, are not so contracted at their mouths, but open out broadly into the bay.

CHAPTER VI.

THE SÃO FRANCISCO BASIN.

The Explorations of Halfeld, Liais, St. John, Allen, Ward, Burton, &c. — General Shape of the Basin. — Its uniform Width. — The São Francisco Valley hollowed out of a Series of horizontal Beds of Limestone and Sandstone. — The Chapadas. — The so-called "Serra" separating the São Francisco from the Tocantins Basin an irregular Strip or Table-land of Sandstone. — The Serras of Araripe and Dous Irmãos. — Table-topped Hills in the Valley of the São Francisco; Outliers of the Chapadas. — Doubts about the Age of the Sandstones and Limestones. — Limestones of the Rio das Velhas. — Remains of Extinct Quadrupeds in Brazil, spoken of by Cazal, Spix and Martius, &c. — Claussen's Discoveries in the Caves at Curvelo. — Dr. Lund's exhaustive Researches at Lagôa Santa. — Caves described; their Number, Extent, Stalactites and Deposits of Bones in Saltpetre Earth. — Immense Quantities of small Bones brought in by Owls, &c. — Large Number of Fossil Animals discovered by Lund. — Former Existence of Megatheria, Mylodons, Mastodons, immense Armadillos and Cats, Horses, &c., in Brazil. — Remains of a Race of Man of high Antiquity. — Reinhardt's Generalizations. — The Rio de São Francisco of the Sixteenth Class among the Rivers of the World, but third in Rank in Brazil. — General Description of the Stream. — Its Affluents, the Rios Pará, Paraopeba, and Das Velhas. — The Rio das Velhas alone capable of being made navigable for Steamers. — The São Francisco navigable with but few Interruptions for two hundred and sixty-four Leagues below the Rio das Velhas. — Cost of removing Obstructions. — Proposed Railway from Joazeiro tò Piranhas. — Fertility of Low Lands of São Francisco Valley. — Liais's Picture of the Campos.

Before we take up the geology of the interior and western portion of the province of Bahia, which latter embraces a considerable part of the basin of the São Francisco, I propose to give, in a few words, a sketch of the geology and hydrography of the whole basin; and this is the more needed, since further on we shall have to visit the lower part of the great river, and it will be necessary to consider some questions bearing on its navigation, &c.

No river in Brazil has been so carefully studied and mapped as the São Francisco and its tributary the Rio das Velhas. Halfeld explored the main river from the confluence of the two streams to the sea, and published a few years ago a magnificent chart of the river in atlas form, accompanied by a detailed description of every league. He also made an estimate of the cost of removing the obstacles to navigation, so as to render it a great interior water highway. But, as Burton and Liais have remarked, the chart of the river is rather a detailed plan than a scientifically accurate map, as it wants the meridians and parallels.

M. Emmanuel Liais, the author of *L'Espace Céleste*, made a most elaborate survey of the Rio das Velhas and Upper São Francisco. In his report* on this survey the obstructions, such as rocks, rapids, &c., are shown in diagram and described, and estimates are given of the probable expense of removing them. Almost all the well-known explorers of Brazil have visited some part of the São Francisco basin. Among the later are Messrs. St. John and Allen of the Thayer Expedition, the former a geologist, the latter an ornithologist, but a good geological observer. Mr. St. John made an exceedingly careful examination of the valley of the Rio das Velhas, and of the São Francisco as far as the Barra do Rio Grande, whence he crossed over into the basin of the Parnahyba, and continued his explora-

* Hydrographie du Haut San-Francisco et du Rio das Velhas, ou resultats au point de vue hydrographique d'un voyage effectué dans la province de Minas Geraes, par Emm. Liais. Ouvrage publié par ordre du gouvernement impérial du Brésil, et accompagné de cartes levées par l'auteur avec la collaboration de MM. Edw. José de Moraes et Ladislas de Souza Mello Netto. 1865.

tions across the provinces of Piauhy and Maranhão to the city of São Luiz. Mr. St. John had the good fortune to examine the geology of a route along which the geological features of the country, though somewhat monotonous, are more exposed to view than along almost any other that could have been assigned him. He did his work thoroughly, and his report will be of great interest when published, and we hope that it will not be long delayed. Mr. Allen, suffering from ill health, parted with his companion and went on to Chique-Chique, where he took mules and crossed the province of Bahia to Cachoeira. Mr. Allen has made a valuable report on that part of the country which he visited alone, and he has been kind enough to give me some notes on the geology and physical geography of the province of Bahia, which I shall insert in a subsequent chapter.

Finally we have Burton, who in 1867 explored the Rio das Velhas and São Francisco to the sea, and he has given us an account of his journey in the second volume of his "Highlands of Brazil." This book is like a series of sharply drawn photographs of nature and life along his route. It is exceedingly full of facts of every kind; but his peculiar style, and wholly unique geological language, render his geological observations in many cases quite valueless.

The basin of the São Francisco is a peculiar one. It is long, of very equal width, the lower half making a strong curve, the convexity of which faces the northwest.*

The head of the basin, bounded by high serras of eozoic and older Palæozoic metamorphic rock, narrows rapidly to-

* Liais calls attention to the very uniform width of the São Francisco valley, and says that it varies from fifty to eighty leagues. See his paper, *Le Rio San-Francisco au Brésil*, *Bull. de la Soc. de Géographie*, Fifth Série, II. p. 389.

ward the south, and at the apex, near Ouro Preto, are some of the highest elevations in Brazil. It is separated from the coast river basins by the ridges of the great metamorphic coast belt and the line of chapadas running along a part of its summit, the whole collectively called the Serra do Espinhaço by Baron von Eschwege. This metamorphic belt flattens down between the two provinces of Bahia and Pernambuco, and the basin bends round to the east and crosses it, opening out to the sea.

The whole length of the basin is not far from twelve hundred miles, and its greatest width not more than two hundred and forty. It is separated from the Paraná basin by metamorphic serras, which become lower as one goes northward, finally dipping under a sheet of sandstone corresponding to that of the chapadas along the Serra do Espinhaço, and which run northward to the province of Piauhy, forming a table-land, flat topped, without mountains, and of a varying width, which sends out broad spurs of chapadas between the affluents of both the São Francisco and the Tocantins basins.

Mr. Ward tells me that the valley of the São Francisco, along the western flank of the Grão Mogor, is skirted by high chapadas composed of sandstones and limestones, having precisely the same character as those of the opposite side of the valley; the chapadas sending out spurs into the valley. These chapadas are, when dry, covered by the ordinary campos vegetation, but magnificent groves of Burití palms are found in the damp shallow valleys. Mr. Ward estimates the height of these elevated plains at 2,500 - 3,000 feet. They break down abruptly on reaching the São Francisco valley. Mr. Ward describes the flat-topped hills of the valley as outliers of the sandstones and limestones, and he, I think, rightly ascribes their lower height to their having suffered denudation.

On nearing the limits of Piauhy the basin reaches the metamorphic belt, and, being prevented from extending northward by the sandstone-crowned metamorphic ridge of the Dous Irmãos, bends round as above described and flows to the east. The chapadas I have mentioned continue on, with more or less breaks, along the northern side of the basin forming the serra of Araripe. The upper part of the basin is occupied by horizontal limestone deposits, while farther down are sandstones, shales, &c. The lower half of the valley, almost to the sea, is cut down to the bottom metamorphic rocks. The basin was primarily determined by the ancient denudation of the metamorphic rocks; but as it stands to-day it is worn in the great sandstone sheet, which I shall in the course of the succeeding chapters attempt to show covered the whole plateau of Brazil. It is the fashion among map-drawers to throw in a range of mountains separating two great river basins, especially if there is known to be any high ground between them, and this mountain range is carefully drawn along the summit line of the water-shed. Brazil is usually represented as traversed in all directions by mountain chains, drawn as if they were all alike narrow ridges. But rivers may take their rise on elevated plains, and the water-shed may be only an insignificant bulging. It is so with the great divide between the La Plata and Amazonian systems. In the case of the Tocantins and São Francisco the streams traverse an elevated plateau of sandstone, which forms on top a plain. The branches taking their rise on the divide, and flowing in opposite directions, have cut for themselves valleys that widen out towards their respective main rivers, so that the high lands separating the two basins, instead of being a narrow mountain chain, consist of an elevated plain sending out jagged spurs between the valleys of the tributary streams. Outliers of these chapadas form isolated table-topped hills and ridges bordering the main stream of the São Francisco as far down as the Great Falls.* These high lands, usually drawn on maps as narrow ridges, appear like ranges of upheaval, and the maps are only calculated to mislead the geologist and physical geographer.

The stratigraphical relationship of the limestones has yet to be worked out, but they appear to underlie the sandstone of the Tocantins-São Francisco divide, for both Gardner and Ward speak of them as appearing in the lower part of the slopes in descending from the chapadas into the São Francisco valley.† I shall leave the detailed description of these deposits to Mr. St. John, to whose report it properly belongs, and confine myself only to a few points well determined by other geologists, and which I need to state here in order to complete that general sketch of Brazilian geology and physical geography which I am attempting to give.

According to Reinhardt, the limestone of the Rio das Velhas is dark gray in color, fine-grained and crystalline. It splits into thin slabs, and is so sonorous that plates of it were used formerly as bells to the churches. Lund, Burmeister, Reinhardt, St. John, and in fact all the geologists who have examined the limestone, testify that it is with-

* I think that I may safely say that all the great north-south ranges of high land in Brazil, north of the latitude of Diamantina, except the Grão Mogor, usually described and mapped as mountain chains, are ranges of chapadas or narrow plateaus resulting from denudation.

† Mr. St. John tells me that they underlie the sandstones of the São Francisco valley.

out fossils, and it has been supposed, if I rightly understand Lund, to be very old, and probably Palæozoic. Reinhardt would make it Devonian. I see no good reason for coming to such a conclusion. The want of fossils is no criterion of the age of a formation; it is no proof that a rock is old simply because it contains no fossils; nor would the apparently metamorphosed and somewhat crystalline condition of the rock necessarily show that it was very ancient. The metamorphism of rocks is largely due to the action of water; and in the decomposition of the rocks of Brazil we see what a powerful agent meteoric water is in working changes in the rocks. We know that the limestones of the Rio das Velhas are plentifully soaked with water by the heavy rains, and the metamorphism of the limestone is doubtless due to this cause. Compact limestones without fossil remains are not by any means uncommon all over the globe, and they may be of any age. These limestones were deposited at the head of the São Francisco basin, in a bay sheltered on the east, south, and west by high lands.

In these limestones are the celebrated bone caverns, of which I will give some account before I describe the great river and its navigation.

It appears to have been Cazal who, in 1817, first called attention to the existence of bones and skeletons of giant extinct quadrupeds in Brazil;* but similar remains had been found even as far back as 1602, or earlier, by the first explorers in Bolivia and on the Pampas. Spix and Martius have described many localities in the provinces of Minas Geraes and Bahia where mammalian remains were found; and Eschwege, St. Hilaire, and other travellers spoke of the occurrence of those remains, not only in the

* See p. 261.

deposits in shallow hollows in the rock, but also in the saltpetre caverns of Minas Geraes; but no one of these travellers made a systematic examination of any of the localities, and we are indebted almost entirely to the distinguished Dane, Dr. P. W. Lund, for what we know of the fossil fauna of the bone caverns of Brazil. Lund had been travelling many years in Brazil in company with Riedel, the botanist, and was on his return with him to Rio when he accidentally heard of Claussen, another Dane, who was residing near Curvelo, in the valley of the Rio das Velhas. Claussen had been examining the saltpetre caves of the vicinity, and collecting bones from them, and Dr. Reinhardt* tells us that when Lund visited him he was trying to study them out with the aid of Buckland's *Reliquæ Diluvianæ.* This was in 1834. Lund saw that here was an immense field to explore; and as soon as he could honorably withdraw from Riedel, he returned to Cachoeira do Campo to examine some caverns there; but he soon rejoined Claussen, and worked with him for some time. The two, however, seem not to have agreed well together, and in 1835 Lund withdrew to Lagôa Santa, and he has remained there ever since devoting his time to an exhaustive examination of the bone caverns of the vicinity, reaping, as we shall see, a rich harvest for science. Burton tells us that the distinguished geologist is confined to Brazil by consumptive tendencies, and is bedridden by rheumatism.

The region in which Lagôa Santa is situated is composed

* For most of the facts in this account of Lund's researches in the bone caves of Brazil I am indebted to the very interesting paper of Professor J. Reinhardt, in Lütken's popular *Tidsskrift,* entitled *De Brasilianske Knoglehuler og de i dem forekommende Dyrelevninger.*

of beds of limestone and shales, the limestone lying below and the shales above, the whole being covered by a bed of red earth, which Reinhardt describes as resulting simply from the decomposition of the shales, but which Lund, if I rightly understand him, believes to be the same red clay which covers the whole country, and which Professor Agassiz and I would refer to the drift. These beds are, as already remarked, horizontal, and are traversed by narrow, often ramifying channels, caused by the widening of the joints of the limestone by the penetration of surface waters which sometimes form therein subterranean streams; for in the limestone region the streams sometimes disappear and pursue an underground course, often for a long distance. Professor Reinhardt gives a ground-plan of the principal ramifications of one of the most noted caverns, called the Lapa Vermelha, situated about a mile from Lagôa Santa, and this plan I have reproduced below.* According to

GROUND VIEW OF THE LAPA VERMELHA.

* It was drawn by Lund's former assistant, the late Peter Andreas Brandt, and gives only the larger galleries of the cavern. Besides these, there are innumerable smaller ones, some of which are only mere cracks where the joints have been widened by the water.

Reinhardt, this cavern extends over two thousand feet into the rock, growing narrower and narrower, until it becomes only a mere canal. The floors are usually horizontal. Sometimes the caverns are only narrow cracks, at others they are wide arched galleries, which not infrequently open into large halls. The walls and roof are smooth and without sharp corners. Reinhardt is inclined to believe that the excavation of the caverns has been partially due to the surface waters, which have in soaking through the rock dissolved away the surface of the walls. That this has been the case to a considerable extent is proved, as Reinhardt has remarked, by the projection from the smooth limestone wall of very thin sheets of clay, which would certainly have been worn away if the whole hollowing out had been performed by running water. One thing seems quite certain, that these caverns were excavated before the valleys of the region in which they occur. The roof and sides of the caves are often covered by very large and beautiful stalactites of a great variety of forms. A stalagmitic crust sometimes covers the earthen floor, and in some caves there are large pillars. These stalactites, formed by the exceedingly slow deposition of calcareous matter by the water trickling through the rock, since the time when the clay of the floor was deposited, bear some testimony as to the great age of the bones therein buried.

The earth covering the floor is a yellowish-red clay, which is, according to Lund, like the superficial soil of the country. All authors describe it in very much the same way. It is however very clear, from its mode of occurrence, that it is not drift, and that it is a deposit introduced into the caves by the action of the surface waters; but precisely how may be a question. The earth often contains fragments of quartz and other rocks. Reinhardt is of the opinion that it has been introduced from above by the water flowing into the caves through the overlying red earth and decomposed shales. It appears, in some cases, to have once filled the caves from floor to roof, and to have been subsequently more or less completely washed out. This red earth is strongly impregnated with saltpetre, and its extraction is so profitable, that the Brazilians have removed it entirely from many of the caves.* Reinhardt says that a small cartful sometimes produces as much as two stones, or an *arroba*, of the salt. Bones of extinct animals occur buried in this clay in almost all the caverns, but in such small quantities, in the majority of them, that they do not reward the pains of the collector. Lund told Burmeister that he had examined at least one thousand caves; out of these only sixty contained bones in any quantity, and but half that number really paid for working. The number of caves is astonishing, and Burmeister tells us that almost every bank has its cavern. They are not confined to the immediate vicinity of Lagôa Santa, but are found in great numbers throughout the limestone region. The skeletons found in them are usually disarticulated. The bones are often much broken and almost invariably scattered about, so that the discovery of an entire skeleton is hardly to be thought of. Besides this, the earth in which the bones occur is much cemented together, and has to be broken up to allow of their extraction. The bones are not all of the same age, and a large proportion of those in some caves belong to now existing animals. But in other caverns there are found remains of extinct animals of high antiquity.

* See note on saltpetre, near end of Chapter VII.

Some of the caves contain immense quantities of small bones belonging to bats and small animals of existing species. Near Caxoeira do Campo is a cave about 120 feet long, 30 to 40 feet high, and 6 to 9 feet broad. Over a part of the bottom lay quite a thick bed of earth filled with small bones. Lund carried out half a cubic foot of this earth, and counted all the half-underjaws he found in it. Of small opossums (*pungrotter*) there were 400, and about 2,000 of different kinds of mice, besides bats, porcupines (*pigrotter*), and small birds. Another interesting instance is related by Reinhardt. Lund had the whole of the clay brought from a cave at the fazenda of Escravania, which was only 24 feet deep. This earth filled 6,552 firkins. Lund determined the number of half-underjaws found in a certain measure, and calculated that in the whole mass there were the remains of not less than 6,881,500 individuals of cavias, opossums, porcupines, and mice! Beside these there were immense quantities of bones of small birds, lizards, frogs, &c. And all these bones had been brought into the cave by owls! Now owls are unsocial birds, and we cannot resist the conclusion of Reinhardt, that the deposit must have been gathering for many thousands of years. While these bones belong to the present geological epoch, those buried in the red clay below the stalactite accumulations belong to a more ancient time, and are for the greater part of extinct forms; and it is from this source that the bones of the Megatherium and other giant animals are derived.

Of these animals there have been discovered by Messrs. Lund and Claussen 115 species of mammals, belonging to 58 different genera, distributed as follows: — *

* See D'Archiac, *Géologie et Paléontologie*, p. 722, from which the table on the next page is taken.

	Genera	Species.
Quadrumana	4	6
Cheiroptera	3	7
Carnivora	9	18
Rodentia	15	32
Edentata	13	28
Pachydermata	9	10
Ruminantia	4	7
Marsupialia	1	7
	58	115

With bones of extinct animals occur those of now living species, as, for instance, *Cervus rufus*, *Cervus simplicicornis*, *Sciurus æstuans*, *Echimys Cayennensis*, *Myrmecophaga tetradactyla*, *Lepus Brasiliensis*, *Felis concolor*, and *Felis mitis*.* Among the extinct quadrupeds may be mentioned the Mastodon, whose remains have been only rarely found in caverns, but more often in pits and holes. There were bones of species of Macrauchenia, Toxodon, Chlamydotherium, and of the gigantic Glyptodon (Hoplophorus), Mylodon, and Megatherium. Among the carnivora were wildcats and jaguars, and a species of Smilodon (*S. neogæus*),—an immense cat-like animal with enormous knife-like canine teeth in the upper jaw, allied to the fossil European species. Of monkeys Lund found but few, and they belonged to the genera Callithrix, Hapale, Ateles, and Protopithecus, the latter being an extinct genus. In six or seven of the holes Lund found stone implements and remains of man so buried with the remains of the extinct fauna, as to leave no doubt that man was contemporaneous with it in Brazil as in Europe. In the Sumidouro cavern they were found mingled

* Reinhardt, *op. cit.*, p. 315.

with bones of the extinct cavern jaguar (*Felis protopanther*) an immense Capibara (*Hydrochærus sulcidens*), together with remains of llamas and horses, which last certainly existed in Brazil, as in North America, long before the conquest. According to Reinhardt, the race of men whose remains Lund has found appear to have been well built, but slender. The same writer states that a skull he examined was dolichocephalic and somewhat prognathous. It was of medium size and ridged with a very prominent cheekbone, a small forehead, and eyes wide apart. The walls of the skull were extraordinarily thick.

Reinhardt* has come to some interesting conclusions with reference to the history of the cave fauna, and I translate them in full. They are: —

"1. That Brazil, in the post-pliocene time, was inhabited by a very rich mammalian fauna, of which the present may be said to be a fraction or stunted remainder, since many genera, nay, even large systematic groups, such as families and suborders, have disappeared, and only very few have come down to our day.

"2. That the Brazilian mammalian fauna, in the whole post-pliocene time, had the same peculiar stamp which in the present distinguishes the South American fauna in comparison with that of the Old World, while the extinct genera belong to families and groups which still to-day particularly characterize South America. Only two of these genera, one extinct, the Mastodon, the other still existing, the horse, belong to families which are entirely confined to the Eastern hemisphere, and form exceptions to the rule.

"3. That the mammalian orders were far from being richer in genera formerly than now. The Ruminants, Pa-

* Lütken's *Tidsskrift*, 3die R., 4de Bind, 4de Hefte, p. 351.

chyderms, Elephants, and the Carnivora have suffered the greatest loss. Some orders, as the Cheiroptera and Monkeys, number perhaps to-day more genera than formerly.

"4. That in South America the post-pliocene mammalian fauna was more distinct from the present fauna, and was more especially rich in peculiar and now extinct genera than was the case with the corresponding fauna in the Old World.

"5. That the poverty in large animals, one may almost say the dwarfish character, which in our day the South American mammalian fauna, in comparison with the mammals of the Eastern hemisphere, was far from obtaining, or rather did not obtain at all in the prehistoric fauna. The post-pliocene Mastodons, Macraucheniæ, and Toxodons, those giant armadillos and sloths, could well compete with the Elephants, Rhinoceroses, and Hippopotami which at the same time lived in Europe."

Liais shows that the Rio de São Francisco is, so far as length is concerned, to be counted as belonging to the sixteenth * class among the rivers of the world, since its length is about 2,900 kilometres, or a little more than 1,802 miles, and he tells us that in Europe there is only one longer river, namely, the Volga. In America it is surpassed by only the Amazonas, the Mississippi, the combined Paraná and La Plata, the St. Lawrence, and the McKenzie, while in South America it occupies the third rank. The São Francisco takes its source in the highlands between lat. 20° and 21° S., and flows almost due north to its confluence with the Rio das Velhas, in lat. 17° 11′ 54″ S., and long. 1° 43′ 35″ west of Rio. It receives two considerable affluents on the right bank before reaching the Rio das Velhas, — the Pará, which unites with it in about lat. 19° 10′ S.,

* Burton says seventeenth or eighteenth.

and the Paraopeba, a much larger stream, which enters in about 18° 49′ S. Both of these streams rise in the same highlands with the São Francisco, and flow northward, inclining toward the west, entering the main stream very obliquely. The Rio das Velhas is the main branch of the São Francisco. It takes its source in the Serra da Mãe dos Homens, near Ouro Preto, and runs almost parallel to the São Francisco, from which it is separated by a little chain of limestone hills called the Serra do Espirito Santo. Between Sabará and its mouth the river has to descend nearly 263 metres, but it makes so many turns that the descent per metre is very much lessened, not only through the increased distance, but through the friction of the river against its banks. Liais makes the descent of the river 0.394^{m} per kilometre, while the velocity of the current varies from 0.30^{m} to 1^{m}. The river is some 80^{m} in width, and, were a few obstacles removed, it would be navigable by steam from its mouth to Sabará, 120 leagues. To remove these obstacles in the way of navigation, Liais calculated that an expenditure of £ 260,000 would be required. Burton thinks that it could be done for £ 55,000. The Rio das Velhas flows in a narrow valley, cut through the limestone, and bordered by bluffs like an Iowan stream; and Liais's map shows it doubling sharply about narrow ridges, sometimes isolated, at other times having the character of spurs to the main line of bluffs. Here and there older rocks form high ridges and peaks, but these are rare. Among these is the Serra da Piedade, eastward of Sabará, — a mass composed principally of iron ore. It is 1,774 metres in height.* (Buril.)

* For a graphic account of a visit to the Piedade, see Herr Eug. Warming's *En Udflugt til Brasiliens Bjerge*, Lütken's *Tidsskrift*, &c., 1ste Bind, 1ste Hefte.

The low lands bordering the Rio das Velhas are alluvial, rich, healthy, and suitable for cultivation with the plough. The country back from the river is wavy campos land, fit only for grazing.

On the main São Francisco also, and its affluents, there is much valuable land. The two rivers are quite well settled, and from one end to the other there is seen a succession of fazendas, hamlets, and not a few considerable towns.

Liais has drawn a beautiful picture of the scenery of the campos and of the Rio das Velhas, and rather than spoil it by a translation, I beg the reader to allow me to give it in his own words: * —

"La présence d'un épais tapis de graminées sur toute la surface du sol donne, au premier abord, l'idée d'une grande uniformité d'aspect. Cependant il n'en est pas ainsi, et les paysages des Campos sont des plus variés. Des bouquets d'arbres dans lesquels les feuillages les plus divers s'allient aux fleurs de toutes couleurs portées par les guirlandes des lianes ou par de superbes orchidées ou broméliacées parasites, rompent la monotonie du tapis de verdure, et l'on se croirait dans un parc admirablement cultivé. D'autres fois, sur le bord de petits ruisseaux, croissent des groupes de gigantesques Mauritia vinifera, palmiers précieux de ces régions. Leur tronc élevé, surmonté d'un magnifique parasol formé par de vastes feuilles en éventail, produit un effet des plus pittoresques, lorsque surtout une immense prairie est parsemée çà et là de ces végétaux gracieux. D'autres fois, et ce fait s'observe surtout dans les régions les plus sèches, des arbustes tortueux couvrent tout le terrain, et dans ces parties des Campos se font remarquer les belles

* *Bull. de la Soc. Géog.*, 5 Série, XI. pp. 396, 397.

fleurs des Carïocar, des Cochlospermum, des Vochysia. Enfin, souvent, au milieu d'une vaste plaine, on voit surgir une de ces curieuses chaînes de montagnes de grès rougeâtre ou verdâtre, à sommet coupé en table, et si abondantes dans tout le Brésil, où M. de Castelnau les a déjà signalées. Les flancs arides de ces collines, parfaitement alignées et qui se prolongent sur plusieurs lieues de longueur en gardant le même niveau et prêsentant l'aspect d'un toit, sont couverts par des Melocactus et par de magnifiques Kielmeyera, dont les grandes fleurs roses rappellent celles des camélias. Lorsqu'on monte sur ces collines, qui parfois atteignent jusqu'à 500 mètres au dessus du niveau de la région environnante, un admirable panorama se déroule sous les yeux du spectateur. Je me rappelle en particulier un magnifique tableau de ce genre que j'ai aperçu en gravissant les flancs de la serra de Curumatahy. Le regard embrassait toute la largeur de la vallée du Rio das Velhas. Son fond offrait l'apparence d'une immense plaine, d'où on voyait sortir comme des îlots les serras du Paraúna, de Buenos-Ayres, da Garça et du Bicudo. La rivière, accompagnée sur ses deux rives d'un cordon de grands arbres, dessinait son cour au fond de la vallée par une ligne d'une verdure fraîche qui tranchait sur la teinte rougeâtre des graminées desséchées et éclairées par les feux du soleil couchant. De belles teintes violettes couvraient les flancs des montagnes rapprochées, et dans le lointain, à une énorme distance, une chaîne de montagnes bleu pâle se montrait à l'horizon. C'était la serra da Mata da Corda avec ses dômes dioritiques, qui limite à l'ouest le bassin du San-Francisco." *

* A similar panorama is to be seen from the edge of the chapadas bordering the Calháo-Arassuahy valley.

The Upper Rio de São Francisco flows with a more direct course, and its current is consequently more swift than that of the Rio das Velhas. It is also much impeded by rapids and falls.* Notwithstanding that its general level is higher than that of the Rio das Velhas, its banks are very unhealthy, and terrible fevers, called carnadeiras, from time to time drive away the population from the vicinity, so that Nature has made the Rio das Velhas more fit to sustain a population and be a water highway than the Upper São Francisco. From the mouth of the Rio das Velhas the São Francisco would be navigable for steamboats, with some interruptions on account of obstructions which might be removed, as far down as the Villa da Bôa Vista, a distance of about 264 leagues. From that point to the Porto das Piranhas, a little over 70 leagues, the river is not navigable. From the Porto to the sea steamers already ply. To remove the obstructions from the main river, and make it navigable for steamers, Mr. Halfeld estimates the probable cost at about £108,900. A canal has been proposed to unite Bôa Vista and Porto das Piranhas! This is certainly not advisable when a railway could be constructed at vastly less cost. Burton estimates that an expenditure of £203,000 would be sufficient to open the Rio das Velhas and São Francisco, and build a railway around the obstructions of the Paulo Affonso to the Porto das Piranhas.

The opening of steam navigation in 1867 below Porto das Piranhas has given an immense impetus to the trade of the whole country adjacent to the Lower São Francisco, and Burton says that its effects are visible even in the neighbor-

* The worst of these is a series of rapids and falls called the Pirapora, which forms an obstacle that it would cost enormously to remove.

ing provinces of Piauhy and Ceará. But what a future is in store for the great São Franciscan valley when it shall receive the gift of a steamboat and shall hear the scream of the locomotive! When any one stops to consider how much Nature has done towards furnishing Brazil with a great interior water highway, it seems wonderful that it should not long since have been improved.*

The railways of the province of Rio have already been described in the "Journey in Brazil." Government is pushing the Dom Pedro II. line northward into Minas, with the view of continuing it over the Mantiqueira and across the highlands into the valley of the das Velhas. A tram-road has been commenced from Cachoeira, on the bay of Bahia, to Urubú, on the great river, and there is hope for the São Francisco, even if the Pernambuco and São Francisco and the Bahia and São Francisco railways fail, as they probably will, in reaching the river.

* The difficulty seems to be not Paulo Offonso, nor Sobradinho, nor Pirapora, but *politics*, and the jealousies of those who have had anything to do with the matter. A steamer some time ago was built in sections, and started on its overland journey to the São Francisco; but I cannot learn that it has yet reached its destination. Burton says that a M. Dumont brought to Rio from Bordeaux two small steamers, which were to be transported in sections to the Rio das Velhas, and commence running in 1869, so that it is probable that steam navigation has been by this time opened on that river. I have tried in vain through my Brazilian correspondence to inform myself on this as well as other matters relating to Brazil, but it seems wellnigh as difficult to keep one's self posted in the progress of affairs in the interior of Brazil as it is to obtain news from the heart of China.

CHAPTER VII.

THE PROVINCE OF BAHIA, — INTERIOR.

Journeys of Spix and Martius, Nicolay and Lacerda, Allen, and other Explorers. — Geological and Physical Features of Country between Malhada and Cachoeira, described by Von Martius. — Sandstones. — Remains of Mastodons found near Villa do Rio de Contas. — Immense Copper Boulder from Cachoeira. — Rev. Mr. Nicolay's Report of Journey from Cachoeira to the Chapada Diamantina. — Occurrence of Diamonds in Sandstones. — Limestones. — Sterile Plains. — Diamantiferous Sands of the Chapada. — The Diamond Mines of Sincorá and Lencôes. — Annual Yield of the Provinces in Diamonds. — Mr. Allen's Report of a Journey from Chique-Chique, viâ Jacobina, to Cachoeira. — Country between Chique-Chique and Jacobina an immense Limestone Plain. — The Chapada at Jacobina a detached flat-topped Mass of Sandstone. — Gneiss Hills. — " Lake Plain," east of Jacobina. — Knobs. — Potholes, probably of Glacial Origin. — Eastern Sandstone Plain. — Climate, Vegetation, &c. of Route. — Difference in Topography between Gneiss Regions of Bahia and the Mucury described and accounted for — Former greater Extension of Forests. — Von Martius's Description of the Country between Cachoeira and Joazeiro. — Country near Feira da Conceição. — Serra do Rio Peixe. — Rio Itapicurú. — Want of Rain at Queimados. — Serra de Tiuba. — Tanques and Fossil Bones near Coche d'Agua, Barriga Molle, and Neighborhood. — Monte Santo. — The Great Meteorolite of Bemdêgo. — Rock Inscriptions. — Villa Nova da Rainha. — Joazeiro to be the Terminus of Bahia and São Francisco Railroad. — Rio de Salitre. — Salt Licks. — Mr. Allen's Note on the Salt of the São Francisco Valley. — Saltpetre. — Geology of Country between Carunhanha and Urubú. — Change in Geological Structure, Climate, Vegetation, &c., below Urubú.

The interior of the province of Bahia, notwithstanding its rich diamond-mines, is almost a *terra incognita* to the geologist and geographer. It forms, however, so important a part of the empire, that I have deemed it worth while to collect the most important facts that bear on its geology and physical geography, and with these before us, I think that we shall be able to come to some trustworthy conclusion as to its general structure. Though Spix and Martius explored the province while geology was yet in its infancy, they made many interesting observations. The Rev. Mr. Nicolay, a few years ago, visited the diamond district in company with Dr. de Lacerda, and he has kindly furnished me with some notes on the route he followed. Mr. J. A. Allen, ornithologist on the Thayer Expedition, crossed the province from Chique Chique, on the São Francisco, to Cachoeira, and I am indebted to him for a very interesting sketch of the country he traversed. That part of the São Francisco valley comprised in the province has been examined by Von Martius, St. John, and others, and finally by Burton, so that we know its general geological features. These observers furnish us with three complete sections across the country between the São Francisco and the sea, and Mr. Nicolay gives another incomplete one. In examining this material, we shall take up these sections in their order, going from north to south, and we will first follow Von Martius in his journey from Malhada to the coast.

This little town is situated on the Rio São Francisco, opposite the mouth of the Rio Carunhanha, in the province of Bahia, at the extreme northern angle of Minas Geraes. Von Martius says that the vicinity "is composed of limestone, which the burning of the woods not infrequently changes on the surface into a white chalk-like crust. This rock formation we left, on the third day's ride, between the fazendas Curralinho and Pé da Serra, where we observed granite, and on it here and there layers of a porous iron sandstone, in part weathered to iron ochre." He describes this whole region as being covered by a *catinga* growth with *Cerei* and *Cnidoscoli*, &c. The water is bad and slimy. The population is principally engaged in raising cattle and horses. Leaving this part of the country, high banks of red granite, some bare, others covered thickly with *cacti*, were met with. Near the Serra dos Montes are rounded hills and mountains, composed of diorite, and destitute both of soil and vegetation. The rock forming the Serra dos Montes Altos is gneiss and granite. The soil is in many places highly impregnated with saltpetre, but this salt is extracted to no important extent. In the Serra de Caytelé quartzose shale or quartz rock, like that found so extensively through Minas, abounds. It is flesh-red in color, almost horizontal, and frequently traversed by heavy veins of white quartz. East of Caytelé is a level high land of this same quartz rock. On leaving this, one descends to reach a hilly country composed of gneiss, covered with catinga forests.

The mountains in the vicinity of the fazenda of Joazeiro are of granite and gneiss granite. The road thence to the Villa do Rio de Contas "rises gradually, and leads finally into a valley shut in on both sides by high mountains. The Serra da Villa Velha rises at least 1,200 feet above the villa. The base of the mountain is composed of mica slate, on which rests red quartzites (*Quarzschiefer*), and over these white rocks of the same kind." The strike, according to our author, is from N. N. W. to S. S. O., with a westerly dip, which is higher in the upper beds than in the lower. The foot of the mountain is covered with light vegetation, which resembles the flora of Serro Frio; on the top it resembles that of Tejuco. Quartz rock, thinly laminated and elastic, was observed. Gold occurs in veins in the rock, and also in the sands and gravels of the Brumado and other streams, where it is found in grains and nuggets. Spix and Martius speak of one nugget having been found weighing eight pounds. Two leagues north of the villa are other gold deposits. The great sandstone formation is rightly described as extending northeastward under the names Morro das Almas, Serra de Catulé, Serra da Chapada, &c., to Jacobina.

Spix and Martius describe the top of the Morro Redondo as flat, and speak of the occurrence there of a hard, white sand rock, on which were drawings in red paint, supposed to have been made by the Indians. This rock rests upon granite, which in some places contains augite. North of this is the Serra de Tiuba, and between it and the São Francisco they found green pistacite in the granite.

Over the quartz rock lies a red sandstone, concerning which Von Martius shall speak in his own words: —

"The third formation, which we met here, is that of the so-called red *Todtliegende*, or older sandstone. It occupies the highest point of the mountain, as even at Brumadinho, and shows, without distinct stratification, here and there a thickness of several hundred feet. This rock is here composed of grayish white quartz grains, in which pieces of reddish quartz sandstone and of red Grauwacke slate are imbedded, and it is not infrequently intermixed with much silver white mica. To this formation or to one of the overlying clays belong probably certain nodules of clay ironstone which are hollow inside, and contain a very fine red powder, which, according to the results of an examination made by my honored colleague, Hofr. Vogel, is composed of iron oxide, argillaceous and siliceous earth, with some lime and magnesia, and is used by the inhabitants as a tonic. The highest mountain of this district, the Serra de Itaubira, probably presents the same formation of the red *Todtliegende* on its conical head. We saw it northwestward from the Morro Redondo, rising high into the blue ether, and are of the opinion that it is at least 5,000 feet high."

From the vicinity of the river Sant' Antonio a beautiful variety of alabaster is obtained, and, according to Spix and Martius, it occurs in large quantity. It is sent to Bahia to be made into images and ornaments. Dr. Lacerda kindly presented me with a fine specimen of this mineral.

Bones and teeth of the Mastodon* occur apparently in plenty in the vicinity of the Rio de Sant' Antonio, near the Villa do Rio de Contas, and near the old fazenda de Bom Jesus de Meira, eight leagues from the Villa, buried in the soil.

Between the Villa and the Rio de Contas are quartzites, where mica slates passing into granulite make their appearance lying in granite. The granulite "holds here and there masses (*Knauern*) of a very hard coarse-grained cellular gray quartz. On this formation we observed layers of a light green, somewhat porous, very compact sandstone, which appears exactly like that which in Germany is here and there interstratified with the *Quadersandstein*. Parched woods, leafless in the dry season, stretch out in immeasurable extension over the hilly or even mountainous land; large tracts are covered with bushes of the Arirí palm, (*Cocos schizophylla* Mart.), ... and here and there a lighter clump of the Aricuri palm (*Cocos coronata* Mart.)."

The base of the Serra das Lages is composed of clay and mica slates, chiefly greenish-gray in color, some approach-

* See note to p. 261.

ing chlorite slate, and containing octahedral crystals of iron. Higher up quartzites appear, and on the top the vegetation resembles that of Minas. Near the fazenda of Lages, on the top, are heavy deposits of iron ore, in the form of magnetic iron, specular iron, and brown iron-stone; "the last furnishes not infrequently considerable quantities of stilposederite (phosphate of iron)." The prevailing strike of the quartz rock is from north to south *im Stunde* 22, 23, and 24, the dip of the strata at high angles from 40° – 60° towards the east.

In the Serra de Sincorá the quartz rocks (*Quarzschiefer*) have a strike of N. S. *im Stunde* 22, 23, and 24, and dip with high angles toward the east. It forms the division between the high and low lands of the province of Bahia; east of it obtains a changeable wet climate, while to the west there is a dry climate.

Leaving the Serra of Sincorá one meets with granite, hornblende, and clay slates and diorite. These are overlaid with layers of clayey sand of an ochre-yellow color, which in some places is even ten feet thick. "Near Carabato there overlies the granite an older sandstone (*Graues Todtliegendes*), which is composed of fine-grained quartz, feldspar, and mica, and approaches feldspar porphyry. In this are imbedded rounded masses of quartz."

At Olho d'Agua great blocks of white quartz are exposed.

"The soil, which already at Olho d'Agua began to be hilly and mountainous, continues with similar irregularity, and covered with *catinga*, until finally in the vicinity of the Fazenda do Rio Secco, which we reached at the end of the fifth day, the road sank gradually between some high, bare, granite mountains, where the traveller reaches a plain, which, covered merely with dried shrubs a few feet high, presents a more free aspect. . . . . At Rio Secco there rested on the granite, which when bedded showed a strike of W.W.W.–S. S. O., and a westerly dip (*Einschiessen*) under a high angle, a fine-grained hornblende rock and iron-stone." The country between Villa da Pedra Branca and Cachoeira Spix and Martius found to be composed of gneiss and granite, with occasional beds of hornblende rock and mica slate.

In the last century there was found near Cachoeira a huge mass of native copper, which was carried to Lisbon. Spix and Martius* visited the locality where this mass was found, and could discover nothing that would justify them in believing that the copper was derived from the rocks of the vicinity, which consist of gneiss. They afterwards saw the specimen in the museum at Lisbon and examined it. It bears the following inscription: —

"Maria I et Petro III imperantibus, cuprum nativum mineræ ferri mixtum ponderis libr. MMDCXVI in Bahiensi Præfectura prope oppidum Cachoeira detectum et in Principis Museo P. MDCCLXXXII." According to Vandelli, in the *Memorias da Academia Real das Sciencias de Lisbôa*, Vol. I. p. 261, the outside of the mass is of a hardened dark yellow color. A portion of the surface was analyzed, and gave ninety-seven per cent of pure copper, with no trace

* According to Von Martius, *Reise*, Band II. Seite 746, copper occurs at the following places in Brazil: —

Riberão de São Domingos, near Pé do Morro, in the Comarca do Serra Frio, Minas Geraes, where it is found in greenstone.

Primeiros Campos, in the Serra Curaça, Province da Bahia, chloride (*Saltzsaures*) and sulphuret, occurring in granite.

Arraial do Pinheiro, Cattas Altas da Itaperava and Inficionado in Minas Geraes.

of either gold or silver. Vadelli says that a second and smaller piece was found near it. Spix and Martius appear to have considered the mass as a meteorolite, but I have seen fragments of amygdaloidal trap from the vicinity of Cachoeira, and I am inclined to consider it an erratic, derived from this trap.

Mr. Allen gives me the following note: "The country between Malhada and the coast, in all its leading geological features, as given by Von Martius, bears a most striking resemblance to that traversed by myself some 100 to 200 miles to the north of this line. Some minor features, as the occurrence here and there of clays and slates, etc., I noticed at only one or two points, and only as insignificant patches."

"At Chique Chique I observed very small patches of magnetic iron ore, at times apparently in place and resting on the limstones, but commonly occurring as detached patches and irregular fragments of large size. It is undoubtedly similar to that spoken of by Burton as occurring in large quantities a little below Chique Chique."

Mr. Nicolay says that the country rises toward the Chapada Diamantina by a series of terraces, and he estimates the height of the chapada at 3,000 feet above the sea, which would coincide with Mr. Allen's estimate of the height of the chapada at Jacobina, and of my own estimate of the height of the chapada at Minas Novas. At the "chapada," says Mr. Nicolay, there are "shales, sandstones, and conglomerates. The sandstones vary much in quality, both as to composition and hardness, but are all evidently the direct products of primitive rocks. Upon these sandstones there is (or was) a stratum of quartzite, in many places still very distinct, in which are found crystals of magnetic and other pyrites, and among the sands created by the disintegration of this rock, as marked by these crystals, diamonds are usually found.

"The superposition of the harder upon the softer strata is the cause of the presence of those caverns called *gruna*, which frequently perforate the hills, and in which many diamonds are found. They are all formed by the percolation of water through the rock, and the disintegration of the softer strata; but in the larger number of cases not a cavern but a ruin is formed, and the surface presents a wild confusion of enormous blocks or slabs of conglomerate sixty to seventy feet square, and from ten to fifteen feet thick, for the larger examples. As yet I am not aware that any fossils have been found in this district. The chapada forms the eastern limit of the barrier of the great river São Francisco, and I can trace it from the sources of the Paraguassú into Goyaz.

"The chapada is separated from the next division by the valley of the river São José on the south, a tributary on the left bank of the Paraguassú. The next range, which may be called the limestone range, that rock being developed in magnificent cliffs, especially on the eastern side, and presenting numerous caverns, is distant about twelve miles.

"I am not aware that this limestone has been more than casually examined by Vivian near Joazeiro, by myself at Mocambo, and by Cato at Rio Una, an affluent of the Paraguassú, right bank, nor have I heard that any fossils have been found in it. It is very distinctly bedded.* Im-

* Mr. Allen, who has carefully read Mr. Nicolay's report, says: —

"The limestones mentioned by Nicolay greatly puzzle me. They seem to occur on tributaries of the Paraguassú, and hence must be *east* of the chapada. If so, beds of limestones must occur on both sides of the divide which sep-

mediately to the east of this occurs a belt of violent dislocation, say twelve leagues in breadth, or more, presenting irregular hills of primitive rock [gneiss. C. F. H.] with valleys between them, *not having usually any outlet*,* and for the most part covered with a forest of ancient growth.

" Here the road is strewed with large quartz pebbles, and boulders of all sizes, qualities, and colors. The *brejos*, or hollows between the hills, are sometimes lakes, more often swamps, and some occasionally are quite dry. This is the Serra do Mocambo, Serra da Calderão da Onça, and Serra da Saude, and is marked on the south of the river Paraguassú by the Mato dos Macacos.

" To these hills succeeds a zone of *taboleiros* or a *taboleiro* or table land, where gneissose rocks are often exposed on the surface, which is nearly level, but varied by occasional small lakes or ponds, and riachos or watercourses, having no final issues for their waters, and often dry during a part of the year. This is crossed by the deep cut, formed by the Rio Paraguassú, which, like other rivers to the north, presents, to the extent of its *enchente*, or overflow, sometimes a mile in width, a belt of verdure; all the rest is arid, a region of cacti and prickly and aromatic plants.

" Upon this surface, however, at long intervals, appear isolated masses of primitive rock at Bahú, about 150 feet high [above the plain], and a range of similar rocks or hills, known as the Serra das Pedras Brancas, from which some outliers, singularly rounded on the surface, are presented at Pedra Redonda. This crosses the Taboleiro at about

arates the valley of the São Francisco from the sea. If so, it is a patch belonging undoubtedly to the limestones developed so extensively in the São Francisco valley. In his gneiss taboleiro I recognize my 'lake plain.' (See report of my journey.)"

* This is one of the features of a glaciated surface. — C. F. H.

four leagues from its eastern extremity, but without entirely breaking its continuity. Beyond this the Serra da Boquerão, also of primitive rock, for the most part bare, and immediately to the east the Serra Mangabeira, where I expect on further examination to find sandstone; beyond this is undulating wooded ground for six leagues, to another taboleiro of the same geological character as the other, but presenting a superior vegetation, and which is again bounded to the east by a chain of nearly continuous elevations, which forms the main buttress of the system, the west limit of the Lagoa do Rio Paraguassú, and to the east of which, only so far as I *know*, excepting at the chapada, sandstones are developed."

"Throughout the entire district the bottom rocks are gneissose, varying occasionally to porphyry and granite on the one hand, and hornblende and quartz rock on the other, occasionally presenting micaschist."

Mr. Nicolay further remarks that "not only near the limestone ranges, but on the edge of the great taboleiro, saline streams are found."

From Mr. Nicolay's report, as well as from what he has stated to me in conversation, there can be no doubt that the diamonds of the interior of Bahia occur in a sandstone bed, forming part of the great sheet which once overspread the whole country, tying in with the sandstones and clays of the Jequitinhonha basin; and this sandstone, as we shall see from Mr. Allen's report, is found also at Jacobina, at which place, in 1755, diamonds were first discovered in the province of Bahia.

I saw specimens of the diamantiferous rock from the chapada in the hands of Mr. Nicolay. It was not itacolumite, but it seemed to me to bear a very close resemblance to the sandstone bed overlying the clays in the Jequitinhonha basin. It also bore a remarkable resemblance to the tertiary sandstones on the Bahia Railroad near Pitanga, where diamonds also occur. The diamantiferous sands I saw in the possession of Dr. de Lacerda at Bahia appeared to have resulted from the disintegration of the chapada sandstones.*

It is much to be regretted that the diamond mines of the Chapada Diamantina have never been critically examined, for I feel convinced that from their study the mystery of the origin of the diamond is to be solved.

The metamorphic basis of the province of Bahia presents a long and, on the whole, a gentle slope towards the sea, and a shorter and equally gentle incline towards the São Francisco valley. Along the highest part of the province there runs an irregular strip of sandstones, occasionally sending off spurs in various directions, and not infrequently forming isolated patches. These sandstones, lying nearly if not quite horizontally, form a series of chapadas or tablelands, and flat-topped hills of greater or less extent, and with an elevation of about 3,000 feet. On the eastern side of this line of chapadas is the diamond district, embracing

* Specimens of diamantiferous sands sent from Bahia to M. Damour were found to contain the following minerals: hyaline quartz, jasper and silex, itacolumite, disthene or cyanite, zircon or hyacinth, feldspar, red garnet, magnesian garnet, mica, tourmaline (green and black), hyalotourmaline (*feijão*), talc, wavellite (*caboclo*), yttric phosphate, titaniferous yttric phosphate, diaspore, rutile, brookite, anatase, hydrated titanic acid, tantalite, baierine or columbite, titaniferous ferric oxide, stannic oxide, mercuric sulphide, and gold. (*Bullétin de la Société Géologique de Paris*, 2de Série, séance du 7 Avril, 1856, p. 542.) Another paper on the diamantiferous sands of Bahia, by the same author, is to be found in the *Bulletin de la Société Philomathique*, 5 Fevrier, 1853. I have been able to consult neither of these papers, and I quote through Burton.

the head-waters of the Paraguassú and Itapicurú, forming an irregular area nearly 150 miles in length from north to south.

The Serra or Chapada do Sincorá lies many leagues to the southeastward of the Serra da Chapada proper, of which it is a spur or outlier, and it appears to cross the Paraguassú with a northeast trend, but the maps vary to so great a degree that I can form no satisfactory conclusion as to its extent. The only description I can find of the serra is contained in a letter from the geologist Helmreichen.* According to him the serra "bears the same raw and unhospitable character to the eye as that of the Grão Mogor; extensive campos form the country between its western slope and the Serra da Chapada, while the country from its eastern slope toward the coast is covered by thick woods." He says that there is a close analogy between this serra and the Grão Mogor in geological structure, and that probably it is composed of itacolumite. "The first discovery of diamonds was here made on the banks of the Macujé, and the Commercio (the chief place), distant ninety miles from Bahia, is on the Macujé on the lands belonging to the Fazenda de São João. Diamonds were found in the serra of Sincorá over an extent of twenty leagues. The washings on the west side of this serra have up to the present turned out to be poor. Considerable quantities of diamonds were, however, washed from the Macujé itself, and from the points where the Paraguassú and Andarahy cut through the serra. On the Andarahy the principal washings are confined to the brooks of the vicinity, which flow into it on its right banks.

* Quoted by Von Tschudi, *Reisen durch Süd America,* Zweiter Band, 154te Seite. Helmreichen did not himself visit the Serra do Sincorá, but he obtained his information from a traveller in whom he put confidence.

Here there are many snakes, much fever and ague, and many diamonds." A very rich deposit has been discovered within the last few years at Sincorá, and the city has grown to a very large size.

The city of Lençôes, which is the government head-quarters of the diamond district, is situated about thirty miles to the north a little east of Macujé or Santa Isabel do Paraguassú, and is a large and very important place, and in the vicinity great quantities of diamonds are washed.* Castelnau says that along the course of the river of Lençôes there are pot-holes, some of which are of the depth of twenty-five with width of one or two *braços!* In these *caldeirões*, as they are called, a considerable number of diamonds have been found. The same author says that these pot-holes are found also in the chapada and are always rich.

Diamonds also occur at a locality not far southeast of Chique Chique, at a locality called the Corrego de Santo Ignacio, visited by Burton, who describes the vicinity as composed of itacolumite.(?) He says that there occurs here as at the chapada "a bouldery, not pebbly conglomerate, which resembles that of the Scottish Old Red," † so that it would seem that the hills among which the Santo Ignacio

* "Les plus beaux diamants de la chapada viennent dos Lençoës (les draps), lieu situé à vingt lieues de Santa-Isabel, cette bourgade tire son nom d'un gros ruisseau, enclavé dans une gorge profonde ; il se precipite de sommets élevés sur de large dalles, et après y avoir parcouru environ trois cents mètres, il se jette en formant des cascades dans le rio São José. Tout à l'entour de ces mines, des montagnes entierès, des blocs énormes composés en grande partie de cailloux roulés et cimentés par une pâte ferrugineuse et presque noire, témoignent de grandes révolutions géologiques. En général, les pierres ont des formes très régulières, et celles qui présentent la cristallisation en octaèdre forment la grande exception." Castelnau, *Histoire du Voyage*, Tome deuxième, p. 343 (note).

† Burton, Highlands of Brazil, Vol. II. p. 336.

diamond diggings are located belong to the same formation as that of the chapada.* Burton describes the Serra do Pintor as table-topped.

I regret that I have been unable to obtain precise information concerning the annual yield of the diamond-mines of Bahia as well as to the size,† quality, and relative value of the stones. According to the report of the President of the province, published in 1866, there were exported through the custom-house as follows: during the year 1862–63, diamonds to the value of 1.647 :450$000; during 1863–64, 1.476 :900$000; and during 1864–65, 1.381 :500$000; which figures show a decrease during these three years, but they cannot be considered as giving even an approximate idea of the annual yield, for only a very small proportion of the diamonds actually exported pass through the custom-house, so great are the facilities for smuggling. So nearly

* Mr. Allen, in looking over my MS. has kindly added the following note: —

"The Serra do Assuruá I did not visit, but I saw it at a distance and inferred it to be of sandstone. As it overlies the limestone, it seems to be evidently a part of the sandstone formation noticed by me to the eastward, as already mentioned. I am now fully convinced of the truthfulness of your generalization in respect to the former great extent and subsequent denudation of the sandstones. The occurrence of gold and diamonds in the above-named serra, as also at Jacobina, was repeatedly spoken of to me by many trustworthy persons."

In 1858 the government conceded to the Companhia Metallurgica do Assuruá the right to mine gold and other metals within the space of four leagues. *Oliveira, Exploração de Mineraes*, published as an *Annexo* in a government report for 1866. My copy has no title-page.

† The diamonds of the chapada are often of considerable size, and Burton says: "The Chapada of Bahia also produced a stone weighing 76½ carats, and, when cut into a drop-shaped brilliant, it proved to possess extraordinary play and lustre. It was bought by Mr. Arthur Lyon of Bahia for 30 contos [$15,000], and it is now, I am told, in the possession of Mr. E. T. Dresden." (Burton, Highlands of Brazil, Vol. II. p. 153.)

as I am able to ascertain, the annual production in diamonds of the province cannot fall far short of three millions of dollars.

I am indebted to Mr. Allen for the following: —

Notes on the Geological Character of the Country between Chique-Chique, on the Rio de São Francisco, and Bahia, Brazil. By J. A. Allen.

Chique-Chique is a small village, situated on the Rio de São Francisco, about fifty miles below the mouth of the Rio Grande; it is a little north of the parallel of Bahia. My journey thence to the latter place was by the route usually taken by mule-trains in passing from the Villa da Barra do Rio Grande and Chique-Chique to the coast, namely, by way of Engenho Velho, Jacaré, Jacobina, Arraial do Riacho do Jacuhipe, Villa da Feira da Sta. Anna, and Cachoeira. As I found it necessary to perform the journey over this unsettled and poorly watered district in company with the large eastward-bound tropas, I was obliged to pass on hurriedly, and had not time to explore the country adjacent to my route, or for a satisfactory examination of many of the interesting localities immediately upon it. The following is a summary of such geological observations as I was able to make, the geology of the country being to me at the time a matter of secondary interest.

The country between the São Francisco at Chique-Chique * and the coast at Bahia presents three natural regions, which are

* Mr. Allen says: "This name is always written by the inhabitants of the village as above; never, so far as I observed, Xique-Xique. The place takes its name from the abundance of a low branching form of *Cereus*, called by this name that grows here." Burton uses Chique-Chique for the town and Xique-Xique for the cactus; a distinction, it seems to me, without a difference, since both are pronounced alike. The vicinity of Chique-Chique appears to be a perfect paradise of cactuses. The name is applied to several species of the plant. See Burton, Highlands of Brazil, Chap. XXII.

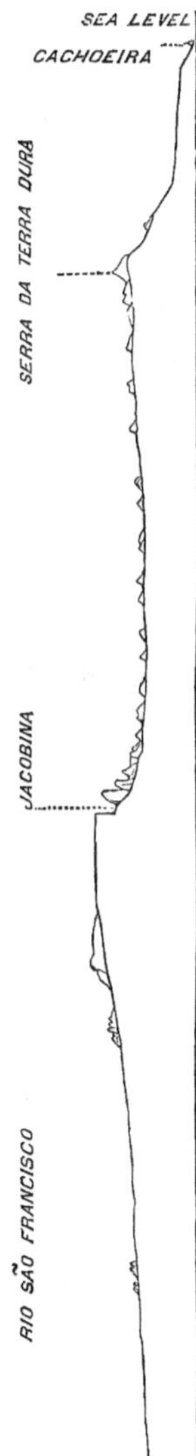

plateaus, differing widely from each other in geological and other characteristics. The first or western extends from the river above named to the vicinity of Jacobina, a distance of rather more than two hundred miles. It is a vast limestone plain rising almost imperceptibly from the level of the São Francisco River to the summit of the divide which separates the waters flowing westward and northward into this noble stream from those that reach the sea by other channels. Here and there large exposures of the underlying rock occur, commonly scarcely rising above the general level, but occasionally forming irregular, pinnacled hummocks or low serras.* The first so-called "serra" met with occurred at Sta. Euzebia; thence eastward they were frequent. At Sta. Euzebia these elevations rarely attained a height of fifty to one hundred feet, above the general level of the country; farther east, and particularly in the Volta da Serra, they rise much higher than this, and in some cases probably reach an altitude of nearly a thousand feet above the plain. The limestone for many leagues to the eastward of Sta. Euzebia is very compact, dark blue, and distinctly stratified, resembling lithologically some that I had previously seen on the lower portion of the Rio das Velhas and at Lagôa Santa, as also

* Burton says that the limestone at Chique-Chique is largely quarried for burning, and he suggests that it would make a good hydraulic cement. According to him, "Chique-Chique annually sends up and down stream between the Villa da Barra and Joazeiro fifteen hundred to two thousand alqueires." Mr. Allen remarks that "the weathering of the limestone often leaves the surface studded with acicular or small attenuated points."

on the São Francisco at Urubú. Further eastward this stratified variety passed into an earthy, light-colored, unstratified kind, which was frequently greatly decomposed at the surface, becoming soft, white, and chalky sometimes to the depth of several feet.* This gave to the distant, nearly verdureless hills in many places the appearance of being covered with snow. Bands of the stratified rock, which was sometimes quite shaly, alternated with those that were unstratified; in some instances the passage of the one into the other was easily traced. The strike of the limestone strata varied from E. and W. in the western part of the plateau to W. N. W. and E. S. E. † in the eastern. The dip was at first nearly vertical or somewhat to the southward, but afterwards an inclination to the northward was observed. Large caverns were reported to be of frequent occurrence throughout this limestone district, from some of which, I was told, very large bones had been taken. Many of the caverns are doubtless very rich in palæontological treasures, but want of time and other circumstances would not allow me to visit them, though I greatly desired to do so. The extent northwards of the limestone I had no means of determining. Its southern limit I once saw at a locality known as Olhos d'Agua, about seventy-five miles west of Jacobina. At this point we turned southward a few miles from our usual course to visit a spring of water, and found an extensive exposure of a compact quartzose sandstone, which was horizontally stratified and overlaid the limestone unconformably, the dip of the latter beneath it being at a considerable angle. The sandstone here stretched away to the southward for miles like a vast level floor, its surface covered only with detached angular blocks of the same rock, and supporting a few cacti. A distant low, even serra seen in the same

* Von Martius thought that this chalky crust resulted from the burning over of the surface; but the heat would not be great enough to produce such an effect.

† This strike is very remarkable, for it usually varies in Brazil from N. N. E. to E. N. E. — C. F. H.

direction, and called Serra das Pedras d'Agua, was doubtless of the same formation. At Jacobina a similar horizontal sandstone was observed, which gave rise to a beautiful level grassy plain, called the Taboleira de Jacobina, and which occupied the summit of the water-shed.

Near the Volta da Serra two large hills of hornstone, or chert ("pedra de fogo" of the Brazilians), were crossed in the few leagues intervening between this considerable serra and the Jacobina taboleira. One of them attained the estimated height of several hundred feet and was several miles across, while the other was about half these dimensions.

In respect to the relative age of the limestone of the western plateau I obtained but few data. It must, however, be much older than the sandstones already referred to. The compact stratified portion has a very striking lithological resemblance, as previously observed, to limestones seen on the lower part of the Rio das Velhas, which rested conformably upon very old clay slates. No fossils were seen in it, but nodular concretions were frequent at a few places. The Taboleiro de Jacobina is the most elevated part of the region under consideration. On several sides the country slopes gradually, but to the eastward the descent to the Jacobina valley is abrupt, through a narrow precipitous defile, called the "Tombador" (literally the "tumble down"). The contrast of the scenery here with that of the taboleiro, which the traveller may have left but an hour or two before, is very great. Almost vertical walls of rock, nearly a thousand feet in height, quite surround the head of the valley, while lower down are isolated, dome-shaped peaks within these enclosing walls.*

These peaks are composed of compact gneiss, though a few of the smaller appear granitic, but the upper portion of the walls is quartzose sandstone. The valley seems to have been formed by a rent in the sandstone, which was subsequently eroded to its

* This scenery must bear the closest resemblance to that of parts of the Arassuahy-Calhão valley. — C. F. H.

present size and form. From the village of Jacobina the level summits of the distant valley walls to the westward are conspicuous features in the landscape.

Leaving Jacobina and winding among the hills for a few leagues we soon enter upon the second or middle plateau, which extends thence eastward to the Serra da Terra Dura, a point midway between Jacobina and Cachoeira, at the head of Bahia Bay. Throughout this distance of nearly two hundred miles I found generally only different varieties of gneiss, usually very compact, and sometimes so granitic in structure as only here and there to present well-defined stratification. At one or two points hornblende rock was noticed, of which there was a considerable exposure on the Riacho de Jacuhipe, near the arraial of that name. The dip was always eastward, and usually very great; the strike varied somewhat at different localities. From Jacobina nearly to the Serra da Terra Dura it was generally N. N. W. and S. S. E.; sometimes N. W. and S. E., and at one or two points N. and S. East of the Arraial do Riacho de Jacuhipe it was nearly E. and W., as in the Morro da Lagôa do Boi and Morro do Curral Velho, which have this trend. In the Serra da Terra Dura, but a few miles further east, it varied at different points from N. N. E. and S. S. W. to N. N. W. and S. S. E., averaging about N. and S.

Over this whole region there is an almost entire absence of loose materials on the surface. Vast exposures of nearly bare rock, sometimes of acres in extent, occur almost constantly, and nowhere is there more than a very thin stratum of soil. Slight knolls and shallow basins alternate, which rarely differ more than twenty or thirty feet in elevation.* In the rainy season many of these basins become filled with water, forming shallow lagoas, varying in area from less than one to more than fifty acres, from the most of which the water evaporates in the dry season. They are filled with rank aquatic vegetation, the dead parts of which, accumulating year by year, already form large deposits of partially decomposed vegetable matter. So numerous were these lagoas for more than fifty miles that it seemed natural to speak of this region in my notes as the "Lake Plain." Almost everywhere the elevations are evenly rounded, indicating that the rocky crust has been exposed to severe and probably long-continued abrasion. But the absence of abraded materials seemed most remarkable. Very rarely were even loose boulders observed, though a few such were repeatedly noticed. At frequent intervals there were singular holes in the rocks, usually nearly filled with water, to which the inhabitants give the name of "caldeirãos." † These "caldeirãos" are of frequent occurrence, but I was unable to learn whether all

* The country just below the falls of Paulo Affonso, at Piranhas, for instance, though composed of gneiss, is worn down in the same way almost to a plain; but while it agrees with Mr. Allen's "lake plain" in the thinness of the soil, the part I saw was abundantly strewn with loose rocks. See reference to *Piranhas*, in index.

† The term *caldeirão* has the same derivation as our English word *caldron* (*chaudière*), and it has the same signification. It is applied to true pot-holes, but sometimes to rock basins in which water collects; but these last are more frequently called *poços*, and when excavated they form *tanques*. Mammalian bones (Mastodon, &c.) are not infrequently found in the caldeirãos of the lake plain. — C. F. H.

were of a similar character. Nearly all of the considerable number examined proved to be genuine pot-holes, and some of them were of great size. The largest one I measured was elliptical in outline, eighteen feet long, nine or ten in width, and twenty-seven deep, with smoothly worn sides. Beneath the water that partially filled it there must have been many feet of materials that for ages have been falling into it, so that its whole depth must be much greater than my measurements indicate.* Near the Serra da Terra Dura the country becomes somewhat diversified by the presence of abruptly rising points or knobs that at intervals dot its surface, as

* Mr. Allen tells me that these pot-holes often occur out on the plain, far away from any high land, and that they are sometimes found excavated in the summits of slight bulgings in the plain, or even on the top of a hill, as in the case of the Morro do Caldeirão. These holes must have been excavated by falling water. There is only one suggestion that I can make as to their origin, and that is that they were formed by glacial waterfalls, in the same way as the pot-holes found over the glaciated regions of North America, as, for instance, in New Brunswick and Nova Scotia, where I have had an opportunity of examining them. It is well known that glacial waterfalls, notwithstanding the constant movement of the ice, are very often stationary, and in the Alps they hollow out enormous pot-holes in the rocks. The lake plain is noted for the small amount of decomposition which has taken place over it, owing, I believe, largely to the fact that it has never been covered by the virgin forest, having always been dry. — C. F. H.

illustrated in the accompanying sketch, taken from an eminence known as the Morro do Caldeirão-assú, there being a large pot-hole near its summit. This elevation reached the height of about one hundred feet, and was the most considerable elevation crossed in this part of the journey. It overlooked the plain for many miles in every direction, and the knobs rising from it looked not unlike small rocky islands in the sea. Those examined were composed chiefly of quartz, the great hardness of which may have prevented their being so much abraded as the softer strata enclosing them. At its eastern border the middle plateau becomes more broken, and merges gradually into the Serra da Terra Dura. These hills are all composed of gneiss, varying somewhat in character, but generally very hard and compact, with a rather small amount of mica. The average trend of the strata has already been given.

From the Serra da Terra Dura a considerable descent is made in reaching the third, eastern, or coast plateau. This is characterized, so far as observed by myself, by the general absence of rock exposures, and a deep superficial deposit of compact sand, probably detritus from the naked abraded plains to the westward. Its extent and general features are too well known to require a detailed description here.*

The three plateaus described above are separated from each other by low mountain chains and belts of broken country. The sandstone serra of the Tombador and the gneiss range of Jacobina divide the western from the middle one, while the Serra da Terra Dura separates the latter from the eastern. In some respects these several districts are somewhat alike, but geologically they widely differ. The first, as previously observed, is a vast limestone plain, two hundred miles in extent, rising gradually to

* This plain, which Mr. Allen has represented on his profile, he describes as continuing down to Cachoeira, and it is undoubtedly composed of the tertiary sandstones which extend over so large an area at the head of the bay of Todos os Santos. — C. F. H.

the eastward; the second, of equal breadth, is gneissose, and apparently everywhere of nearly uniform altitude; the third is narrower, somewhat lower, and sandy. The whole region between the São Francisco and the sea is covered usually with a low forest, or catinga, except a narrow belt along the coast, where a moister atmosphere cherishes a more luxuriant growth. The country everywhere wears a barren aspect, the vegetation is dwarfed and scanty and the aridity of the climate is excessive. The greatest aridity and the highest temperature obtain in the limestone district, where for nine months of the year little or no rain falls and all the herbaceous vegetation annually withers. Cacti in great variety, including some of gigantic proportions, with various species of bromeliaceæ, are leading forms in the vegetation. With the exception of a few species, the trees are leafless throughout the long dry season, and the streams either become dry or form mere chains of brackish pools. The convolvuli and other vines that overrun the catinga, though dead at the time of my journey, indicate an excessive luxuriance of foliage and flowers during the short rainy period.

The middle or gneiss district differs but little in its climate and vegetation from the preceding; it is, however, less arid, and cacti are proportionately less frequent, though still a leading feature of the vegetation. The eastern or sandy plateau is also quite arid in its western part, but gradually becomes moister towards the sea, where the vegetation exhibits the ordinary luxuriance characteristic of the Brazilian coast. In the vicinity of Jacobina, however, where the great altitude of the land arrests the currents of damp air from the sea and condenses their moisture, mists and light rain-falls occur at frequent intervals throughout the year, and the forests are not only of larger size, but their verdure is perennial. In the Serra da Terra Dura a nobler forest growth is also seen, resulting from causes similar to those existing at Jacobina.

The population of the middle and western plateaus is extremely sparse. The settlements consist of but a few families each, and occur only at long intervals. Nothing approaching the character of a village is met with between the São Francisco and Jacobina. The middle district is more thickly settled, little hamlets being more or less frequent, and there are a few small villages. The eastern is comparatively well settled and largely under cultivation; towards the coast the soil is very productive.

A journey across the limestone plain is always tedious and difficult. Extra animals must be taken to transport food for both men and beasts, and in the dry season water must in like manner be provided for use in crossing the long stretches where none can be obtained. During the rainy season the swollen streams and noxious exhalations from the temporary pools render the journey equally troublesome and far more dangerous.

The topography of the arid, gneiss country of the interior of Bahia, Sergipe and Alagôas, with its great plains and bare surface, is in striking contrast with that of the forest-clothed gneiss region of the coast of these provinces and of those to the south, where the surface of the gneiss never forms plains, but is always hilly and ridgy, and covered by a thick bed of drift-clay. This difference in topographical features has resulted, at least in so far as the last surface-moulding is concerned, from the different climatic influences to which these provinces have been subjected. Over both of these parts of the country an extraordinary amount of erosion has taken place. One might at first be inclined to consider that the amount of denudation had been greater over the gneiss plains than in the hill-roughened basin of the Mucury, since on the former the upturned gneiss strata are planed down to a more even level; but it seems to me that this feature furnishes no criterion. The peculiar topography of the wooded gneiss region is owing to the prevalence of a very moist climate, giving rise to numerous streams, which have furrowed the surface with an intricate system of water-courses, such as we do not find over the gneiss plains of the north, for the grand physical features of the northern country in question are such as produce a dry climate, and preclude the possibility of that unequal erosion such as is produced by the flow of surface waters fed by heavy periodical or constant rains. Decomposition must have played its part before the drift as well as afterward, for else how could those feldspathic sandy clays of the tertiary, spread over the coast plains and the great Amazonian valley, have been formed? I cannot resist the conclusion that the present forest belt was wooded before the drift, and that this forest and its attendant peculiarities of climate, were the causes of decomposition as at present, which decomposition has aided immensely in rounding down the hills and producing topographical features which received their finishing touches from the glaciers.

In the interior of Bahia, behind the arid region, where, as in the Chapada, or the Serra de Tiuba, the country throws up barriers to the air-drift and causes the condensation of moisture, we find a different kind of topography, and the surface is deeply furrowed. There are forests, and decomposition has taken place to a greater or less extent.

The limits of the forests, of the belt of decomposition, and of the area over which copious rains fall, coincide very remarkably, and show a dependence upon each other, but the forest belt has a smaller area than that of decomposition or of the rains. The wooded belt seems to have narrowed greatly within comparatively recent times, losing its foothold in the west, where immense regions, now campos, over which the climate and soil would normally be proper for the growth of forests, have dried up, the climate has become hot, less rain now falls, and the forest cannot regain its lost place. Doubtless there are many natural physical causes to be taken into consideration in studying the distribution of the forest, catinga, and campos floræ; but there is one agency which has been at work in Brazil, whose effects we can hardly overestimate, and that is the burning over of wood and campos lands by man.* The very physical features of the highlands of Brazil determine a difference of luxuriance in the floræ of different regions, and there are, as I have already shown, regions where for ages the climate has been such that forests could scarcely have had any noteworthy extension, so that there must have always been in Brazil, naturally, virgin forests, catingas, campos, and barrens. On the coast, where the forest is dense and moist, and the climate is wet, forest fires are next to impossible, and one never sees a scorched and dead wood, such as covers so large an area in the province of New Brunswick, for instance. But in the interior, where the catinga forests drop their leaves, and are as dead for several months in the dry season, fires are easily kindled and the wood killed; and fires set in open fields or campos, for the purpose of producing a new crop of grass, may spread to the neighboring catingas. It is the opinion of many writers that a large part of the catinga and campos region of the Brazilian highlands was once covered by forests, and that their present bare appearance and the character of their floræ is in very great measure due to frequent and extensive burning over of the country. Every year the Brazilian campos lands are systematically and almost entirely burned over, for the purpose of producing a new crop of grass. This burning of

* For a very interesting article on the effect of the burning over of the campos, see *Tidsskrift for Pop. Frem. af Naturvid.*, — *Camposfloraen og Camposbrœndene*, by Eug. Warming.

course has destroyed all those trees and shrubs and plants of all kinds that cannot bear the scorching, and has wrought a great alteration in the character of the whole flora of the region ; the climate also has suffered a change, for with the destruction of the woods and forests it becomes hotter, the unprotected earth is like a furnace, streams run dry a few days after a shower, and the springs disappear. The wholesale and careless destruction of the forests on the Brazilian coast, unless put a stop to, will in the end work a sure ruin to the country. Brazil owes her climate and fitness for agricultural purposes to her forests, and it is absolutely necessary that they should be preserved over a very large part of the country, especially on the coast. The climate of the Bahia has already suffered from the destruction of the forests of the Reconcavo, and the burning over of the plains. But I fear that Brazil will learn this fact only when it is too late.

This whole subject of the former wooding of the Brazilian campos is the same as that relating to the North American prairies, which many suppose to have once been wooded. Dana has shown that the existence of forests depends upon moisture, and any climatic change which may lessen the amount of moisture over a region may cause the thinning out and final disappearance of its forest; and it is more than probable that some such influence, beside that of forest clearing and burning, has been at work in Brazil.

Taking the Estrada de Capoeirassú on their journey from Cachoeira to Joazeiro, Spix and Martius ascended the steep slope of the same name as the road to a height of about seven hundred feet, when they reached the top of the dry plateau.

The rock in the vicinity of Cachoeira is gneiss, reddish or yellowish in color, with a N. – S. or N. E. – S. W. strike and westerly dip. In some parts specular and magnetic iron were observed to take the place of mica in the rock.

Two leagues from Cachoeira the country becomes barren and uninhabited. Here is a little place called Feira de Conceição. "The plain, as a general thing elevated from six to seven hundred feet above the sea, forms here and there shallow hollows, in which during the rainy season brackish water, often unfit for the use of cattle, collects. In other places one sees in several directions rows of hills with gently sloping sides. The only rock we found was gneiss, gneiss granite, or granular granite, for the most part of a reddish or yellowish color, though sometimes also blackish or white. This rock lies entirely bare over a large extent of surface, or is covered by a thin coat of a heavy red clay, which appears to originate from the decomposition of the same rock. Besides this, fragments of granite and fine granite lie scattered about. In the low lying and wet places one finds little woods very much like the capões of Minas Geraes; the higher plains and the hills are in some cases bare of all vegetation, in others clothed with single cactus stems and plants or with thick bushes and low trees. All these plants belong to the catinga group, for they shed their leaves in the dry season, and for the greater part clothe themselves only on the entrance of the rainy season. Only in the low wet places do the leaves remain on during the whole year. . . . . The wood is never wholly without sap during the leafless season."

The trees leaf out with marvellous rapidity, and a short rain suffices in two or three days to clothe a wood with spring-like verdure. Von Martius speaks of the roots of the

Imbuzeiro (*Spondias tuberosa* Arr.) which extend underground near the surface, and are full of swellings as large even as a child's head, hollow and full of water, — a provision against the drought. During the dry season these hot plains are almost destitute of life. The same kind of country extends on to the Rio do Peixe, and during the dry season it is almost wholly without water. When Spix and Martius reached the Rio do Peixe, they found it only a string of brackish-water pools.

Crossing the Rio do Peixe our travellers passed over a range called the Serra do Rio do Peixe, which is described as being composed of gneiss and granite, and strewn with gigantic and isolated blocks of gneiss. In some places hornblende rock was observed.

The Rio Itapicurú was, like the Rio do Peixe, dried up, forming only a string of pools. Over all this region this is the state of the streams during the dry season; but a week's rain fills their dry beds and converts them into torrents, which if the rain does not continue, soon become dry again, for the surface water runs speedily off from the bare rocks and exceedingly scanty soil. Such a country is of course barren and unfit for culture. Rain falls abundantly enough on the sea-coast, but the air soon parts with its moisture, or becomes so heated that but little or none is condensed in the interior. The Arraial das Queimådas lies only about one hundred and thirty miles from Bahia. But very little rain falls there, and Von Martius relates that the inhabitants assured him that in some parts in their vicinity it had not rained for three years. A little cotton and maize are cultivated there. One league N. W. from this place red gneiss was found, strike N. N. E. At Bebedor, one league farther, white granite, with a N. W.-S. E. strike, was seen. Approaching the Serra de Tiuba light green *pistacite* became more and more abundant in the granite, at first in grains, taking the place of mica, afterwards in bands through the rock. Thin layers of a slate-like hornblende-stone were also observed in the rock.

The Serra de Tiuba, where our travellers passed it, is about twelve hundred feet high above its base, and is composed of reddish granite passing into sienite. It is wooded to its summit, probably owing to the condensation of the moisture in passing over the serra, producing a damper climate; and I am strengthened in this opinion, since Von Martius says that the trees increase in height in going up the serra.

West of the serra a more level country succeeds, and continues to Villa Nova da Rainha, southwest and north of which are mountains which show in their valleys high catinga woods and a comparatively thick covering of fertile soil. From Villa Nova Spix and Martius made a rapid journey to Monte Santo, to visit the great meteorolite of Bemdêgo.

The way ascends gradually, with occasional patches of catinga, to the fazenda called Coche d'Agua, on the west side of the Serra de Itauba. The hills at the base of the mountains are, like the serra itself, of gneiss-granite. "The rock is for the most part bare, but here and there a thickness of from four to five feet of reddish clay lies in the shallow hollows which the *Sertanejos* excavate to a depth of several feet in order to make tanks for the keeping of the rain-water. In these hollows numerous bones of ancient animals are found, for the most part in a broken state, and so scattered that one hardly dare hope to find a complete skeleton. The bones in a recognizable condition which we had the opportunity to gather are the under jaw, a vertebra, and a part of a shoulder-blade of a mastodon."

At the fazenda called Barriga Molle similar bones were found, and at Mundo Novo and Pedra Vermelha Spix and Martius found in a *tanque* the head of a femur. Other localities for these fossil remains mentioned by Von Martius are the Fazenda de São Gonzalo and Caldeirões, and at the Fazenda Cançanção, near the Monte Santo.

Monte Santo is an isolated hill of mica slate which has a N. - S. strike. The height of the mountain is, according to a barometrical measurement, about seventeen hundred and sixty feet above the sea. The rock is said to contain disthene. In the vicinity are several serras characterized by Von Martius as resembling one another in their "rounded-off, long-drawn-out ridges, without steep sides, gaps, or rugged cliffs," the whole being covered with catinga vegetation.

At a place called Bemdêgo, near Monte Santo, Spix and Martius examined an enormous block of meteoric iron, already visited by Mornay * in 1811, and Von Martius gives a long description of it in his *Reise*, to which the reader is referred. It was discovered in the year 1784 by a man who was searching for a lost cow. Coming under the notice of the Governor, an attempt was made to carry it off, under the impression that it was silver; but the cart broke down, and Spix and Martius found it long afterwards lying in a brook nearly buried with sand. They give the greatest length at eighty Paris inches, the greatest breadth $43\frac{1}{2}''$, and the greatest height $34\frac{1}{2}''$.† The specific weight was

* Phil. Transactions, 1816, p. 270.

† Mornay, *loc. cit.*, gives its dimensions as 7 × 4 × 2 feet, the cubic contents at 28 feet, and the weight of the whole mass at 14,000 pounds. See Dana's System of Mineralogy, p. 16. Mornay's paper is accompanied by a plate, showing the shape of the meteorite. Von Martius also figures it in his atlas.

7.731. They estimated the weight of the whole block at 17,300 Paris pounds. Fragments of the mass were cut off and carried to Europe, where they were examined by Fickentscher. Wollaston had already made a chemical analysis of it. The latter found it to contain, iron, 96.1; nickel, 3.9; while Fickentscher obtained

Iron	96.10
Nickel	5.71
Mixture of { Iron, Nickel, Silicon, Carbon }	.46
	98.07

The prepared surface showed the Widmannstadtian figures.

In the vicinity of Bemdêgo were found some strange characters painted on a rock, apparently by Indians, and Von Martius gives a sketch of them in the atlas accompanying his work.

Burton * gives a long list of localities in Brazil in which inscriptions occur either engraved in or painted upon stone, and he figures a number of the hieroglyphics he observed on the banks of the São Francisco, a short distance above the rapids of Itaparica. He gives also copies of certain other glyphs, observed by Mr. C. H. Williams, of Bahia, on the Rio Panema, one of the influents of the Lower São Francisco. Characters of this sort, which his Majesty the Emperor of Brazil has supposed to be the work of maroon negroes or *Quilombeiros*, but which Burton refers to the Indians, appear to be very common in numerous localities on the Lower São Francisco and in the northwestern

* Highlands of the Brazil, Vol. II. Chap. XXVII.

part of the province of Bahia.* Von Martius speaks of having found, in the immediate vicinity of the locality he describes, fragments of Indian pottery, showing that there had anciently been an Indian encampment on the spot. He afterward found rock sculpturings on the banks of the Japurá.

Near the Villa Nova da Rainha is the Serra do Gado Brabo, which is described as a granite hill covered in places with a layer of red clay affording gold.

From Villa Nova towards Joazeiro the road leads for some six leagues through catinga. The rock for the whole distance appears to be gneiss, often with pistacite, and the country is quite even. Granite blocks lie strewn about over the surface, with loose pieces (*Fündlinge*) of verdegris-colored quartz, fibrolite, schorl, and common opal.†

From Joazeiro, Spix and Martius made an excursion to the Rio de Salitre to visit a locality where salt was extracted. This river flows into the São Francisco from the south, about a league to the west of Joazeiro. According to Von Martius, the rock in the vicinity of the town is granite. Going west-southwest towards the Rio de Salitre this rock was soon left, and whitish-yellow dolomite succeeded. Burton says that at its mouth the Rio de Salitre has tall banks, white with

* Mr. Wallace, in his Travels on the Amazon and Rio Negro, describes similar picture-writings as occurring at Monta Alegre, Serpa, at several localities on the Rio Negro, and on the Uaupés.

† The Bahia and São Francisco Railroad is to terminate at Joazeiro. According to the survey of Vivian, Joazeiro is set down as 936 feet above the sea. Halfeld gives the height of the river at this point at 1383 palmas (998 feet).

From a MS. map, furnished me by Mr. Nicolay, showing the line of Vivian's survey, I take the following heights of places along the line: —

Alagoinhas, 300 feet. Agoa Fria, 763 feet. Coité, 1,145 feet. Faz. da Sta. Luzia, 1,106 feet. Queimadas, 888 feet. Faz. da Arueira, 1,997 feet.

the finest limestone. Passing the limestone band, Von Martius describes a band of finely bedded mica slate composed of crystalline quartz grains and white or bright brown mica. A clay slate of a dark green color, highly laminated, and containing crystals of magnetic iron ore or pale flesh red or bluish, with chlorite, occupies a large area along the river. This rock sometimes passes into mica slate.

Associated with it are greenstone and gray limestone, with garnet, and folia of chlorite, and mica slate. At the Fazenda Aldea, between the hills of the last-named formation and the river, Von Martius describes a flat piece of ground, about 60,000 square feet in area, over which the soil is highly impregnated with salt. There are many other similar localities. The soil is alluvial, and deposited by the river. It is ochre yellow in color, and contains more or less of pebbles and vegetable material. After a rain or freshet has covered this soil, and the sun has dried it, an efflorescence of salt appears on the surface. It is not pure, and it appears to contain sulphate of lime, chloride of lime, chloride of magnesia, and saltpetre.

The basin of the São Francisco, from the Rio Verde northward to the Rio de Salitre, is extensively covered by saline deposits, and some of the streams, as, for instance, the Rio Verde,* a river navigable for some distance for canoes, are brackish. Von Martius says: "To the west the mountains withdraw themselves still farther from the stream, and the country consists of a uniform, dry plain, grown up with grass and low bushes. Here one sees, especially in the low places, and particularly after rain, white crusts of salt weather out, and the places where it makes its appearance most abundant-

* Burton remarks that the fish seemed to be attracted in swarms by the brackish streams.

ly (lagoas, salines) are the salt-mines of the inhabitants." These mines lie sometimes at a great distance from the river. The salt is collected by scraping up the crust of salt from the earth, mixing it with water to separate the earthy impurities, and then allowing the salt to crystallize out by the heat of the sun.

Burton says that sometimes the liquor is "strained in *bangués* (coppers or hides) evaporated over the fire and allowed to crystallize." Salt made in this way must of course vary very much in quality, and there is every gradation from almost pure salt to a useless dirty variety bitter with magnesian salts. The origin of the salt of the São Francisco valley is unknown. So far as we know, no deposits of rock-salt * occur. The amount of salt manufactured on the São Francisco is insufficient to meet the demand, and much sea salt finds its way overland by way of Joazeiro.

Mr. Allen has been kind enough to give me the following note on the salt of the São Francisco valley.

"A saline efflorescence occurs at innumerable localities in the drier portions of the Brazilian plateau, as in other arid districts, but chiefly along the banks of the streams. At Jacaré, situated about midway between Chique-Chique and Jacobina, the efflorescence arising in the dry months from the annually overflowed banks or bottom lands of the Riacho do Jacaré is scraped up at intervals, of course with more or less earth, and the whole leached. The lye thus obtained is placed in small troughs to be evaporated by the sun, by which means a small quantity of impure common salt is ob-

* According to the Engineer Nesbitt, rock-salt occurs below Chasuta, on the Rio Huallaga, one of the great tributaries of the Amazonas, and he says that the banks of the river for more than a league are pure rock-salt! (Brazil and Brazilians, p. 578.)

tained. At times the amount of other accompanying saline compounds, as sulphate of magnesia, &c., is so great as to render it quite unfit for use. Though the river bottoms, or the overflowed portions of their valleys, afford the principal sources of native salt, the borders of the half-dried lagoons not unfrequently abound with a similar efflorescence. Such salt licks occur as far south at least as Januaria, where they are numerous. From this fact the city is more commonly known in the neighborhood as *Salgado* than by the name of Januaria.

"The banks of the lower portion of the Rio das Velhas are also remarkable for a similar thick incrustation of what appeared to be nearly pure sulphate of magnesia. This incrustation is often of considerable thickness, appearing not unlike thick hoar-frost. To the presence of these saline impurities in the waters of the streams of many portions of Minas is attributed the great prevalence of the disease known as goitre, that occurs there with such frequency.

"Most of the streams of Bahia are brackish, at least in the dry season."

"At Jacaré the apparatus I saw in use in the manufacture of salt was extremely rude, a section of a hollow tree serving for the leaching-tub, and small logs hollowed out for evaporating-vessels.

Saltpetre, as above remarked, occurs with the salt over a large area in the provinces of Bahia and Minas Geraes; but in the limestone region of the São Francisco valley, where caves are abundant, it is found, as we have already seen, mixed with the earth in the bottom of the caves. Von Martius says that fifteen leagues up the Rio do Salitre there are extensive caverns excavated in limestone and filled with a black earth containing $\frac{75}{100}$ of saltpetre. The process of extraction is very simple. The earth is lixiviated with hot or cold water, and the lye is reduced in strength by evaporation until the saltpetre crystallizes out.* When salt occurs mixed with the saltpetre the lye obtained by the lixiviation of the earth is first evaporated down sufficiently to allow the salt to crystallize, after which the saltpetre is obtained by further evaporation.

Saltpetre is quite largely extracted in some parts of the São Francisco valley, and on the Rio das Velhas. Burton says that on the Upper Rio das Velhas it sells for 10$000 (about $5.00) per arroba.

From Carunhanha to Urubú the São Francisco flows through a flat country, bordered here and there, at a greater or less distance from the river, by isolated hills and ridges. Just below the Rio Carunhanha is the Serra da Lapa, composed of limestone † of a bluish color, and horizontally stratified. Below Urubú the river valley, according to Burton, becomes more contracted and is bounded by "Serras," which on one side or the other accompany it at a short distance. These serras are masses of horizontally stratified sandstone, with which the valley was doubtless at one time filled, but which has suffered very extensive denudation. Burton says that below Urubú, with the change in the geological structure of the valley there is ushered in a change in the climate and the vegetation, which Mr. Allen assures me is very marked. The sandstones in many localities are described by Burton, Halfeld, and others as itacolumite. The sandstone lying near the surface of the

* Burton, Vol. II. p. 291.

† Mr. Allen says: "I well remember the serra of horizontal blue limestone at and below Urubú mentioned, as you state, by Martius and Burton. The limestone for many miles east of Chique-Chique was lithologically of the same character, but the strata there were inclined. See my report."

tertiary chapadas of the Jequitinhonha basin also resemble itacolumite.*

Below the Barra do Rio Grande the river valley is described by Burton as broadening and forming a dead flat, which in places, as at Chique-Chique, is covered with patches of blowing sand, reminding one of an African desert.

A short distance below Chique-Chique, near Tapera da Cima, are heavy deposits of magnetic iron ore, which Burton compares with the Itabirite and Jacutinga of Gongo Socco and vicinity. The relation of these deposits to the other rocks he does not give, but Halfeld states that they have a north south course.

At Pilão Arcado Burton speaks of finding a conglomerate underlaid by soft green shale, traversed by quartz veins. Here gneiss makes its appearance. The Serra do Tombador, near the Ilha Grande do Zabelé, Burton describes as composed of magnetic iron ore resting on limestone. The river for a great part of the way between Chique-Chique and Joazeiro is bounded by conical hills and ridges. Below Joazeiro outliers of the great horizontal sandstones, &c., appear constantly on both sides of the river, and they continue, as we shall see further on, even below the fall of Paulo Affonso. The bottom rocks are gneiss, slates of various kinds, limestone, &c., but no competent geologist has been over the country, and its structure has yet to be worked out.

* Mr. Allen gives me the following note: "Respecting the sandstone occurring below Urubú, Burton's observations and my own also agree. I mention them in my note as quartzites and quartzose sandstone. I could see no difference between them and those observed 200 miles to the eastward at Olhos d'Agua and in the Jacobina Taboleiro. Near Jacobina I observed them also disintegrated, forming beds of white quartz sand on the declivities of some of the hills." The term "itacolumite" is very loosely used by travellers, and is applied to compact schistose sandstone as well as to the true itacolumite.

CHAPTER VIII.

PROVINCE OF BAHIA, GEOLOGY OF THE VICINITY OF SÃO SALVADOR AND THE BAHIA AND SÃO FRANCISCO RAILROAD.

Topography of the Vicinity of São Salvador da Bahia. — The Upper and Lower Cities. — The Population, &c. — The Harbor. — The Commerce of the City and Province. — The Climate, &c. — The Bahia Steam Navigation Company. — The Bahia and São Francisco Railroad. — The Paraguassú Steam Tram-road. — The Gneiss of Bahia. — Decomposition. — Drift Deposits — Consolidation of Beaches. — Stone Reef at Rio Vermelho. — Blown Sands covering the Drift of the Hills. — Mr. Allport's Description of the Cretaceous Beds of Monserrate and Plataforma. — Fossil Fishes, Crocodiles, &c. — Description of several species of Fossil Mollusks. — Cretaceous Beds of Plataforma and Vicinity. — Prof. Marsh's Notice of the Reptilian Remains — Fossil Fishes at Agua Comprida. — Gneiss at the Rio Johannes. — Taboleiros and Sand Plains of Camassarí. — Peculiarities of the Topography of the Tertiary Hills. — Tabatinga Clay. — Sand Plains and Taboleiros of the Imbuçahy. — Peat-Bog. — Drift. — Diamond-washings at Pitanga. — Cretaceous Strata at Pojuca. — Piassabas. — Campos of Alagoinhas. — Tertiary Hills. — Character of Vegetation.

THE sea-coast line going southward doubles sharply northward on itself on reaching the mouth of the Bahia de Todos os Santos, forming a sharp peninsula or cape directed southward and terminating in the point St. Antonio. This point, which is composed of gneiss, is about four miles long, and has a mean elevation of about two hundred feet. On the esaward side the land is hilly, the hills being often hemispherical or hemi-elliptical, and presenting very remarkably regular outlines. On the coast, as in the case with the Morro do Conselho, and the hills in the vicinity of Rio Vermelho, these hills are bare of trees, so that their form is beautifully seen. On the western side, which is occupied by the city of Bahia, this cape is very even in height, and is bounded by a steep, in some cases precipitous slope,* which, with a somewhat zigzag course, is continued many miles to the northward, apparently marking the line of a fracture.

The city of São Salvador da Bahia, usually known as Bahia, stretches along the edge of the bluff for several miles. The *cidade alta* is irregularly but well built for a South American town, and there are some beautiful residences, especially in the southern part of the city. There are many churches, some of which are very fine, together with schools of various kinds. Besides these there is a medical college, a public library of several thousand volumes, a large theatre, a public garden, and a museum; but the latter is no credit to the city.

Below the bluff is the *cidade baixa*, which occupies a very narrow strip along the bay. This is the business portion, and though it for the most part consists of but one or two streets, it is closely built up with warehouses and stores. Here are the Alfandega, or custom-house, the markets, the marine arsenal, the consular offices, banks, several hotels, &c. It is a hot and busy place. The city stretches along the shore for several miles to the northeast, forming the suburbs of Monserrate and Itapagipe, the latter known as the head-quarters of the Bahia Steam Navigation Company. The two cities are united by very highly

* "In 1671, as a result of heavy rains, there occurred a destructive slide from the summit of the bluff, which precipitated a large quantity of earth upon the lower town, destroying houses, burying thirty persons alive, and filling up half of the Praia. . . . . Similar slides have frequently occurred in the history of Bahia, notwithstanding the expenditure of immense sums in endeavoring to prevent them." — Kidder's Brazil, Vol. II. p. 39.

inclined streets, passable with great difficulty by carriages or teams.*

The whole city numbers from 160,000 † to 180,000 ‡ inhabitants, principally of Portuguese and negro descent, though there are many foreigners, as at Rio. Quite a number of Englishmen are engaged in business there, but Americans are few. The society is almost thoroughly Europeanized, and there is much real culture among the people.

The Bay of Bahia forms one of the best harbors on the Brazilian coast, and, next to that of Rio, it is the most resorted to. It is, however, so wide, and the entrance is so open, that the ocean-swell rolls in, preventing shipping lying at the quays, which is also the case at Rio. In ordinary weather the shipping is protected by the high land on which the city stands, but southwest storms cause a heavy swell. The city owes its importance chiefly to its harbor, which eminently fits it to be a port for foreign trade, while it makes the city the centre of the trade of the coast for a long distance north and south, as well as of the interior. The products of the São Francisco and of Sergipe find their way for the most part to Bahia. The borders of the bay itself, or the Reconcavo, so called, are highly cultivated, and produce much sugar, tobacco, piassaba fibre, &c. The principal product of the interior is cotton, which is cultivated to a

* A street with a moderate grade to connect the two towns was in process of building in 1867. By this time it is probably completed. Dapper, in his *America*, published in 1673, gives a very curious and interesting copperplate engraving of the city. He represents the upper city as built on a plain, surrounded behind by a narrow crescent-shaped lake, and with high mountains in the background. Two inclined planes for carriages elevated by machinery are seen uniting the upper and lower towns.

† Pompéo, *Geographia*, 1864.

‡ Dr. Candido Mendes de Almeida, *Atlas*, 1868.

considerable extent in the less arid regions. Cattle are raised in great quantities, and large herds are sent into the province from the country west and north. Perhaps no better idea of the commerce of Bahia could be afforded than by giving the official tables of the exports during the year 1864–65*: —

Articles.	Official Values.
Agoardente	372:813$120
Cotton	1.303:277$553
Sugar	6.316:627$583
Cacáu	173:225$356
Coffee	1.614:063$450
Cigars	45:839$000
Hides	356:008$300
Diamonds	1.381:500$000
Other Articles	81:029$049
Tobacco	2.060:833$745
Woods	237:266$997
Piassaba	141:437$653

The cacáu comes principally from Ilhéos, and the coffee from the Colonia Leopoldina. Large quantities of good cigars are made in Bahia, and are sold at a very low price. At Cachoeira, or rather at São Felix, opposite, is a very large manufactory of cigars (*charutos*) and cigaritos. The latter are much esteemed in Brazil. The hides come from the interior of the province and from the São Francisco.

I have in preceding pages called attention to the whale-fishery of the port.

The following table will show the value of national pro-

* Relatorio of the President, the Commendador Manoel Pinto de Souza Dantas, Bahia, 1866, 2, Quadro No. 11.

ducts exported to other provinces during the latter half of the year 1865:—

Ports.	Values.
Alagôas	115:052$949
Ceará	26:062$318
Espirito Santo	9:545$877
Maranhão	27:812$877
Pernambuco	1.048:050$020
Pará	60:008$970
Parahyba	1:485$500
Rio de Janeiro	368:492$986
Rio Grande do Sul	212:865$511
Sergipe	150:930$837
	2,020:307$845

The climate of Bahia is hot, but not unhealthy.* In the lower city, where the streets are narrow and protected from the sea breeze by the high ground behind, it is very warm and uncomfortable, but I have never suffered so much in the lower town from the heat as I have in early summer in New York. In the upper city the climate is exceedingly pleasant and healthy for a tropical city. Bahia has suffered at times from yellow-fever,† but it has been for many years remarkably healthy.

Bahia is connected with Europe by two lines of steamships, and with New York by the Brazilian mail steamers. It is the head-quarters of the Bahia Steam Navigation Com-

* For several years past Dr. Antonio de Lacerda has kept a journal of meteorological observations for M. Arago, but I do not know that they have yet been published. The climate of Bahia appears to be more moist than that of Rio, and rains are more frequent, being distributed through the entire year.

† Pompéo, writing in 1864, says (*Geographia*, p. 449): "Until 1849 Bahia was sufficiently healthy; but since that time the yellow-fever became almost endemic, attacking with preference the Europeans." Pompéo is not quite accurate here.

pany, under the able direction of my esteemed friend, Mr. Hugh Wilson. This company has at present a fleet of sixteen steamers. Of these one runs regularly between the city and Maceió, in the north, calling at Penêdo, and at various ports in the province of Sergipe. Regular steam navigation has been established on the São Francisco, below the Porto das Piranhas. Several steamers of the company are engaged in running between Bahia, Cachoeira, St. Amaro, and various other points in the Reconcavo, and there is a regular line between Bahia and Caravellas, or the Colonia.

A railroad has been commenced to Joazeiro, but after extending ninety miles in a northerly direction from Bahia, it ends in a desert sand-plain at Alagoinhas, as we shall see, and there is no prospect of its ending anywhere else for some years to come.

From Cachoeira to Urubú, on the São Francisco, a steam tram-road has been projected. It was formally opened in July, 1867, but I have not learned what progress it has made up to this time.* The road will run up the Paraguassú valley, through the diamond region, with side branches to Feira de Sta. Anna and Lençôes.

This railroad has a future before it, as it is to run through a much more important tract of country than the Bahia and São Francisco Railroad.

The gneiss composing the point on which Bahia is built is composed of a very compact, sometimes trap-like variety of gneiss, not infrequently without mica and with very indistinct planes of stratification, as has already been observed by Allport † and Darwin. The latter has given the following description of the Bahian gneiss : ‡ —

* Major James informs me that it is proposed to make this a railroad.

† Quarterly Journal Geological Society, Vol. XVI. Part 3, p. 263.

‡ Geological Observations, p. 140.

"The prevailing rock is gneiss, often passing, by the disappearance of the quartz and mica, and by the feldspar losing its red color, into a brilliant gray primitive greenstone. Not unfrequently quartz and hornblende are arranged in layers in almost amorphous feldspar. There is some fine-grained sienitic granite, orbicularly marked by ferruginous lines, and weathering into vertical, cylindrical holes almost touching each other. In the gneiss, concretions of granular feldspar, and others of garnets with mica, occur. The gneiss is traversed by numerous dikes composed of black, finely crystallized, hornblendic rock, containing a little glassy feldspar and sometimes mica, and varying in thickness from mere threads to ten feet; these threads, which are often curvilinear, could sometimes be traced running into the larger dikes. One of these dikes was remarkable from having been in two or three places laterally disjointed, with unbroken gneiss interposed between the broken ends, and in one part with a portion of the gneiss driven, apparently whilst in a softened state, into its side or wall. In several neighboring places, the gneiss included angular, well-defined, sometimes bent masses of hornblende rock, quite like, except in being more perfectly crystallized, that forming the dikes, and, at least in one instance, containing (as determined by Professor Miller) augite as well as hornblende.

". . . . The folia of the gneiss within a few miles round Bahia generally strike irregularly, and are often curvilinear, dipping in all directions at various angles; but where best defined, they extended most frequently in a N. E. by N. (or East 50° N.) and S. W. by S. line, corresponding nearly with the coast line northwards of the bay. I may add that Mr. Gardner found in several parts of the province of Ceara, which lies between four hundred and five hundred miles north of Bahia, gneiss with the folia extending E. 45° N.; and in Guyana, according to Sir R. Schomburgk, the same rock strikes E. 57° N. Again, Humboldt describes the gneiss granite over an immense area in Venezuela and even in Colombia, as striking E. 50° N., and dipping to the N. W.

at an angle of fifty degrees. Hence all the observations hitherto made tend to show that the gneissic rocks over the whole of this part of the continent have their folia extending generally within almost a point of the compass of the same direction."

On the surface the gneiss is in some places decomposed to a great depth, and so soft as to be easily removed by a spade or mattock. It is well exposed in the numerous cuttings on the streets in different parts of the city and vicinity. On the Ladeira do Bom Gosto, on the southern side ascending the hill towards the cemetery, is a very heavy excavation, in which the decomposed gneiss *in sitû* is beautifully shown, and where, at the time of my visit, workmen were cutting it away with hoes. The decomposition consists of a loss of alkalies on the part of the feldspar, reducing it to the state of kaolin. The iron of the rock is oxidized and stains it a deep brick-red, though the color is rarely uniform, but likely to be in streaks, portions of the rock being white. Between the lighthouse at the barra and the first little point above to the westward, the rock is the dark gneiss above described; it is well bedded, but the stratification is much disturbed. The strike here is N. 60° E. Dip N. W. 35°. Nearer the lighthouse the rocks become much distorted; there are here several large veins traversing the rock; one trap vein about nine feet wide has a N. S. direction, while a granite vein I observed runs N. E., S. W.

The lighthouse is built on a rock which is joined to the mainland by a narrow and low isthmus. The road which skirts the shore crosses this isthmus, making a considerable excavation necessary, while the road branching off to go to the lighthouse runs through a similar cutting. The greatest thickness thus exposed is twelve feet. The upper part, *c c c*, consists of the ordinary red earth one finds everywhere spread over the surface of the hills at Bahia. In it are seen occasional loose irregular masses of quartz and Bahian gneiss, the latter decomposed. The under part, *b b*, consists of an irregular sheet of boulders of Bahian gneiss more or less decomposed, sometimes a foot and a half in diameter, in some cases angular, in others rounded, together with others of that peculiar friable quartz so common in the quartz veins of the Bahian peninsula, also either angular or rounded milky quartz pebbles, of all sizes, the

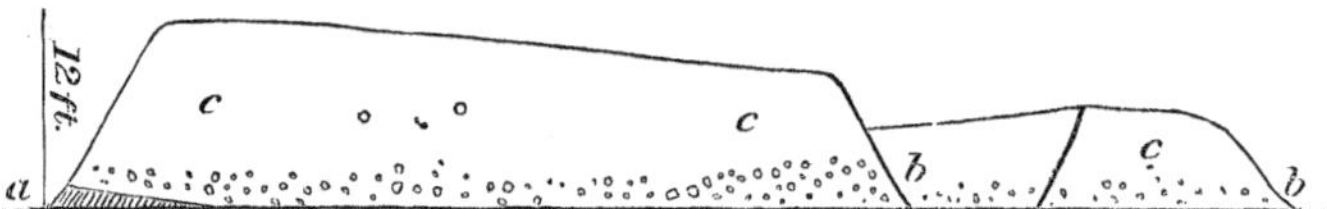

whole being thrown together, without any arrangement, in a confused mass. The greatest thickness of the boulder bed exposed is about four feet, but it thins out at *a* to only six inches, and is seen to lie immediately upon the decomposed gneiss.

This is the most remarkable gravel deposit which I have seen on the coast. According to my estimation, the height of the bottom of this sheet above sea level is twenty to twenty-five feet.

On the eastern side of the lighthouse the gneiss is much broken up and faulted. There are here some interesting granite veins, in which the materials are seen to be very coarsely crystallized on the side of the fissure, while the crystals of the middle of the vein are very small. Near this place I found gneiss of a light color, full of garnets, almost vertical, and with a strike of N. 25° E.

Going up the beach and passing the first little point, one observes a sheet of a soft dark brownish or yellowish sandstone lying against the hillside, and extending from near low-tide mark to a height of some fifteen to twenty feet. This sand has been evidently washed or blown up from the shore over the hillside. The grains of sand in this rock are very fine and uniform, and the whole is loosely cemented by oxide of iron.

I had observed that the red drift coating in no place came down to the surface of the sea, but that it was invariably worn away up to about that level to which the waves must have reached at the time when the late rise of the land began. Below that line the hillsides were bare and the rock surface broken and irregular. I found that the sands just described overlapped the drift and in part protected it, but the waves had washed away a portion of it, so that the drift is seen underlying the sands and extending to below half-tide mark, showing, I think, conclusively that the land was more elevated when the drift was deposited than it is at present.

The seaward sides of several gneiss hills are bared and much broken up by the waves, which beat heavily on this coast. It is interesting to observe the exceedingly irregular surfaces developed by this action, and how the harder rocks become prominent while the softer are washed away. This gneiss is of the same general character as that at the Barra, though it varies much in color and general appearance.

At the lighthouse, and especially to the eastward, the beach is partially solidified by the cementing together of its materials by the lime derived from shells. The shingle of the beach and the sand and gravel packed away in cracks in the rocks become cemented solidly together. Where the beach is composed of sand, it is converted into a very hard sandstone up to a level considerably above half-tide, and in this sandstone the structure and irregular bedding of the beach are beautifully preserved. Very commonly, after this solidification has taken place to some depth, the newly made rock is washed bare, and broken up by the waves. In this case it forms detached masses, sometimes an extensive sheet of sandstone, which appears to crop out on the beach with a seaward dip of a few degrees, whatever the trend of the beach may be. Masses of this kind we find at intervals on the shore between Bom Fim and Bahia, and on the coast thence to Rio Vermelho. The sandstone of this reef or solidified beach often contains layers of coarse materials as well as an abundance of shells, the latter usually broken, but sometimes perfectly preserved, and with their colors quite fresh.

It is very interesting to see how this consolidation goes on. On a long sand beach one may trace it, sometimes, in every stage of progress. It frequently happens that, owing probably first to a heavy storm, a ridge is thrown up behind the beach. This opposes a barrier to the sea for a long time thereafter, and may increase in height and extent from the drift of the sand by the wind, which is very likely to take place, especially when the shores have a northerly trend, on account of the prevalence of northeast winds. Along this line shells are thrown up in great quantities and buried in the sands. These shells become dissolved by the water soaking through the sands, whether salt or fresh, and the carbonate of lime thus derived is deposited as a cement to the materials of the beach to a height somewhat above half-tide, rarely ever to high-tide mark. The copious rains must tend notably to this result, and the waters from marsh lands soaking through the beach must also assist in the solidifying process. Sometimes, after the solidification of the lower part of the beach, the loose parts are swept away by a storm, it may be, sometimes, by the bursting of the barrier by the freshet of a river, which has inundated the low grounds behind. In this case the beach is left standing like a wall running in the water parallel to the coast. Such a reef is seen at Rio Vermelho, where, after skirting the shore for some distance, it projects partially across the mouth of a little bay, like a wall or breakwater of rock.

The reef at Rio Vermelho illustrates very well the general character of these consolidated beaches. It is composed of layers of calcareous sandstone and conglomerate, often somewhat irregular, dipping seaward, the dip being only a few degrees, or about that of an ordinary sand beach. The height of the reef is very uniform. In the finished and isolated reefs, as that of Pernambuco and the one under consideration, the recent rise of the land has brought this level somewhat above that of the sea.

The solidified portion is seen to be but a sheet of varying thickness lying on the surface of the beach. On the inner

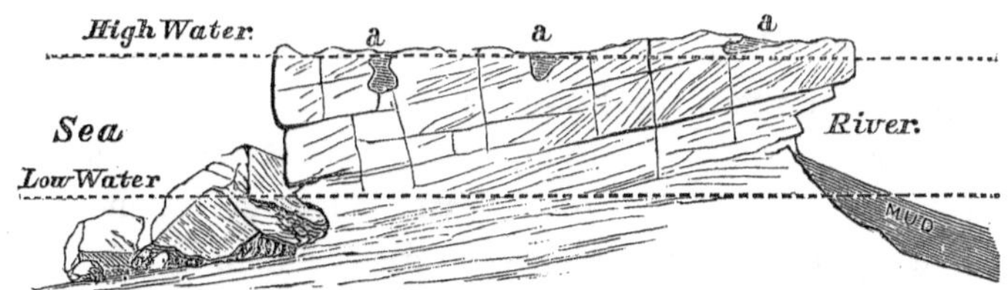

side it is quite thin, and from the action of water behind it is undermined and broken off, until, at last, it forms a low perpendicular wall, undermined below and sometimes projecting several feet. Usually this side of the reef is flanked by a slope of sand or mud, and sometimes by large oblong blocks of the sandstone.

The surface of the reef is, broadly speaking, horizontal, but it is marked by longitudinal ridges and much worn away, sometimes honeycombed and exceedingly rough, with large, shallow or deep, irregular pools of water, the homes of several species of corals, &c. The whole mass is divided by joints into great blocks. On the seaward edge the reef is often worn away by the waves and undermined, presenting always a perpendicular wall to the sea. The upper bed almost always projects a little, and great blocks broken from it lie in front, which afford some protection to the reef. One often finds a depth of twelve to fifteen feet or more at low tide along this side of the reef. In most cases corals grow on the faces of these reefs below low water.

The reef rock is quarried quite largely at Bahia, as well as at Pernambuco, for building purposes, and in both cities it has been used for flagging. In the sidewalks of the upper town at Bahia one may see it in large blocks which are full of shells of species now common on the coast.

Going northward beyond the Morro do Conselho, the gneiss hills recede more and more from the coast, and low lands come in between them and the sea. These I have examined for a short distance north of the Morro do Conselho. The coast is fringed with patches and strips of sands which, in part forming dry plains with a scanty vegetation, in part swampy, do not rise to a greater height, as a general rule, than eight to ten feet above the sea. There are many patches of this kind near Bahia, and some are of considerable extent. On the sea-coast they occupy several little bays between the hills. As we go northward beyond the Morro do Conselho these sands become wider in extent and occupy large tracts. The sands on this coast are very fine, and are easily raised and carried by the winds, so that extensive sand-hills are formed, and the shores of Itapuan, a few leagues north of Bahia, are white as snow.

In an excursion made among the dome-shaped gneiss hills to the east of the city of Bahia, and bordering the sands, I observed that the light, white sand was blown up on the seaward side of some of the hills over the rich red drift, so that one half of a hill is snow white and the other brick red. These hills are cultivated in places, and it was very interesting to see a red field set in a framework of white sands, the drift soil being turned up in the process of tilling.

The city of Bahia, as already described, is built in part on a low strip of land, which, beginning some two miles above the lighthouse on the bay side, accompanies the shore for a mile or two farther on, when it sweeps round westward, separating itself from the gneiss ridge, forming a long, irregular tongue of land occupied by the suburbs Jequetaia, Bom Fim, and Itapagipe.

From Jequetaia to Bom Fim this tongue of land is flat and composed of recent sands. At Bom Fim and Monserrate the land is higher and more irregular, the height being eighty to one hundred feet. It is composed of cretaceous strata. Allport says: * —

"The rocky cliff forming the southwest of the hill, on which the fort of Monserrat is built, presents to view several alternations of conglomerate, sandstone, and shale. Towards the northeast these beds pass into a gritty shale of a bluish-gray color, and full of pebbles; the latter gradually disappear, and the upper strata, as far as the seaward exposed portion extends, consists of beds of shale, alternating with bands of sandstone, both of which contain the same species of fossil shells. The entire series of these

* Quarterly Journal Geological Society, Vol. XVI. Part 3, p. 263.

deposits is covered with the usual red loam, and have the general inclination to the northwest.

"The seaward exposed position of the cliff of Monserrat, about thirty feet in height, consists chiefly of conglomerate, with irregular wedge-shaped bands of shale and clay, and also bands of sandstone. The conglomerate is composed of more or less rounded pebbles of gneiss, granite, quartz, and other crystalline rocks, and occasionally of sandstone, the whole forming an extremely hard rock. The pebbles vary in size from the finest gravel to large boulders.

"In the shale, near the base of the cliff, were found the fossils about to be noticed, consisting chiefly of scales and other portions of fish, bones and teeth of saurians, together with lignite, a few *Mollusca* and some *Entomostraca*.

"Two miles from the above hill, in a northeast direction, is the Plataforma, another hill of the same formation, but loftier; the conglomerates and shales have here the same lithological character, and in the latter are found several fossils similar to those found at Monserrat.

"The geological position of the above formation is undetermined, as they have not been traced in connection with other deposits; but a probable inference may, perhaps, be made from an examination of the fossil remains.

"With regard to the fish remains, Sir P. Egerton, Bart., F. G. S., to whom the specimens have been submitted, states that "the scales are those of *Lepidotus*. The species appear to be a new one. The nearest approach to it is an undescribed species from the lithographic stone of Pappenheim [middle oölite].

"Numerous fish-bones were found associated with the scales; and probably the greater portion belong to *Lepidotus* also. But these and the crocodilian teeth and bones, which are also common in these clays from Monserrat and Plataforma, have not yet been systematically examined. Professor Owen, on a cursory view of the large vertebra, figured in Pl. XVII.,* suggested that it would be a dorsal vertebra of a Dinosaurian reptile allied to the *Megalosaurus*."

At Monserrat Mr. Allport collected a Melania described by Morris as *Melania terebriformis*, and species of Unio, Paludina, Neritina (?), together with seven species of cyprids, enumerated and in part described by Jones, namely, *Cypris* (?) *concultata*, *Candona candida* Müll, *Cypris* (?) *Monserratensis*, *Cypris* (?) *Allportiana*, *Cypris* sp. non descript.

I was unable to visit the locality at the fort, but I examined the beds quite carefully near the Pedra Furada. On the shore, south of this last-named locality, are exposed thick beds of sandstone, shale, conglomerate, and limestone, the continuation of the same series as Allport describes. This limestone is of a compact texture, and mottled with brown, gray, and green. It contains some sand, together with little pebbles of gneiss and quartz, generally angular. Freshly broken, this rock shows the fossils very indistinctly, but on the weathered surfaces they stand out in fine relief. At the same time the surface becomes granular, as if it were composed of coarse, round, or oval grains of sand. These are, however, calcareous, and the structure may be oölitic, though the grains look as though they had been rounded mechanically.

Associated with this rock is a fine-textured argillaceous, light slate-colored limestone, in which there is no trace of this oölitic (?) structure, but instead one finds occasionally

* Quarterly Journal, *loc. cit.*

the cone-in-cone structure beautifully developed, — a somewhat unusual circumstance in rocks of this age. In these rocks, which, at the most, form a bed but two feet thick, several species of shells occur in immense numbers. Of these *Melania terebriformis* and *Paludina* (*Vivipara*) *Lacerdæ* sp. nov. are the most abundant. Species of *Unio*, *Planorbis*, teeth and bones of crocodiles, and bones of Dinosaurian reptiles are also to be found here, together with scales and bones of Lepidotus and other fishes.

This limestone with its fossil shells resembles very closely specimens of fresh-water fossiliferous limestones from the Weald of England.

There are also thick beds of shale, in places black and finely laminated, but for the great part not well laminated and very soft, of a very light color, and full of little flakes of mica. In this shale are to be found layers abounding in entomostracan remains, of which the most interesting is an estherian with its valves marked with concentric ridges like an Astarte, and apparently new. Fish scales and skeletons are not uncommon.

There are some heavy beds of sandstone in this section, which are seen near the Pedra Furada. This sandstone is fine-textured, soft, and of a light, greenish-gray tint. It is a rare thing that it affords fossil remains, and these are usually carbonized stems of plants. Von Martius speaks of rich beds of *Blätter-Kohl* near the mouth of Itapagipe and of a brown coal in the sandstone near the city of Bahia. This last is said to have been worked for a short time by the direction of the government, by Feldner, a German, early in this century. I saw nothing to indicate the existence of coal or lignite at Bahia, and I do not believe that any deposit of the slightest value exists there.*

* Gardner says that Dr. Parigot found the Bahian coal to be lignite.

The Monserrate fossils clearly indicate a fresh-water origin for the beds in which they occur. The following are the mollusks thus far obtained from this locality: —

Melania terebriformis Morris, Geological Journal, Vol. XVI. p. 266, Pl. 16, Figs. 3 *a*, 3 *b*, 3 *c*.

"Shell subulate, consisting of 7–8 flattened whorls, marked with numerous oblique, somewhat prominent, rounded ribs, which are in some specimens stronger towards the anterior part of the shell. In some individuals the posterior part of each whorl is slightly raised, making the suture more distinct. The last whorl is somewhat constricted. The aperture is ovate. The lip of the columella is somewhat thickened and reflexed." Length 18 mm. Mr. Morris gives a figure of a smooth variety, with a less cylindrical shell. The top of the spire of the common forms of *M. terebriformis* is rarely perfect in adult specimens. The younger shells are sharp-pointed, and might be mistaken as belonging to another species.

The test is usually well preserved, and the shells are sometimes hollow.

Locality, Monserrate, and vicinity, Bahia. Age, Lower Cretaceous. Collectors, Allport, Nicolay, Williams, Lacerda, and C. F. H.

Melania Nicolayana, sp. nov.

Shell minute, subulate, nearly cylindrical, with six or more whorls which are flatter and more oblique than in *M. terebriformis*. Sutures distinct. Surface smooth. Length 2½ mm.

This form occurs associated with the preceding, but it is apparently a very rare shell. I dedicate it to the Rev. Mr. Nicolay, to whom I am under many obligations.

Locality, same as preceding. Collector, C. F. H.

Vivipara (*Paludina*) *Lacerdæ*, sp. nov. Geological Journal, Vol. XVI. Plate 14, Fig. 2.

Shell about 20 mm. in length, ovate conical, usually with four very ventricose whorls separated by deep sutures, subumbilicate, smooth, shell rather thick.

This species is figured by Morris as above, and is referred to by him in a note to Mr. Allport's paper as having "a smooth shell subumbilicate, and showing four ventricose whorls, deeply sutured." This shell appears sometimes to have had a truncated spine like our modern *V. excisa*. It is exceedingly common in the limestones at Monserrate, and is occasionally found in the shales. I have never seen it at Plataforma. I take much pleasure in associating with the species the name of my esteemed friend Dr. Antonio de Lacerda.

Collectors, Allport, Nicolay, Lacerda, C. F. H.

Vivipara (Paludina) Williamsii, sp. nov.

Shell smaller than that of *V. Lacerdæ*, from which it is also distinguished by being more conical and having whorls much less ventricose and shouldered. Length 9 – 10 mm.

Quite common with the preceding species at Monserrate.

Collectors, Allport, Williams, Nicolay, C. F. H.

Planorbis Monserratensis, sp. nov.

Shell minute, flat above, concave below, whorls two and a half in number, flattened from above, rounded, increasing rapidly in width, and apparently flaring a little near the mouth, which is oblong, wider than high. Surface smooth.

This pretty little shell is quite common associated with the Vivipara, Melania, &c. at Monserrate. In the conglomerate at Plataforma, in which the reptilian bones occur, there is a form which much resembles this, but the spiral seems more open and the shell is oblique in outline, apparently from the rapid widening of the body whorl. This last may be a species of *Valvata*. The width of the shell of the species just described is about 2 mm. Collector, C. F. H.

Unio (Anodon ?) Totium-Sanctorum, sp. nov.

Shell small, ovate-elongate, compressed, wider behind than before, the ends and lower margin forming a very regular curve, which is somewhat straightened below. Hinge line much shorter than shell. Umbo quite prominent, strongly and broadly flattened by a wide and shallow but well-marked depression which runs downward but obliquely backward across the valves, giving to the shell the appearance of having two rather prominent but wide ridges extending from each side of the umbo, one obliquely forwards widening and growing lower toward the margin, the other much more obliquely backwards with the same characters. The shell is rather thick, and is marked on the outside by numerous fine, concentric lines or wrinkles.

Collectors, Allport, Nicolay, and C. F. H.

During my last visit to Bahia I undertook a careful and systematic examination of the rocks exposed in the cuttings along the Bahia and São Francisco Railroad,* and I walked over the greater part of the line to Alagoinhas, extending my observations for several miles beyond among the taboleiros.

The observations made on this excursion I give in detail, taking them with little change from my field-book.

The Bahia and São Francisco Railroad, leaving the station at Jequetaia, follows the low sandy ground skirting the gneiss hills, and runs along the northern side of the bay of Itapagipe. At a distance of about two miles from the station at Jequetaia (Calçada), on passing a little gully, there are seen exposed in a cutting beds of a loose-grained sandstone and a shale, the latter being of a greenish-gray tint, and very finely laminated, but too much decomposed on the surface to offer anything of much interest. These beds

* I have to thank Mr. Mowry, the superintendent of the railroad, for a pass over the line, and Messrs. Tiplady, Turner, and Orecchi for many favors received during the excursion.

have a strike of about northeast, as near as I could make out, and a dip of a very few degrees to the northwest.

Going on towards the long bridge, we meet no beds well exposed in the railway cuttings, but in a quite deep one only a short distance from the bridge we find thick beds of finely laminated, dark-colored shale, much decomposed, in which are fossil fishes like those at Monserrate, together with a great abundance of cyprids.

Crossing the long bridge, the land on the opposite side becomes much higher and more irregular, and there are some heavy cuttings, in which, as well as on the sea-shore at low tide, the rocks are very well exposed. These consist of alternate beds of conglomerate and dark shale, with occasional layers of sandstone, which have an average strike of N. 60° E. and a dip of 30° ± N. W.

The surface of the beds of shale, exposed on the shore between tides, is very much burrowed into by a marine worm, which excavates a tube that enters the shale perpendicularly to the depth of an inch or more, and then, bending round rather abruptly, comes to the surface again. A little crab also burrows into the shale in places, making deep tubes, whose sides are corrugated after a manner that makes them resemble the empty moulds of fossil stems of plants or corals, and they are well calculated to puzzle an observer, who did not know how they were formed.

The decomposed shale has been used in the works of the Bahia Steam Navigation Company at Itapagipe as a fire clay, and Mr. Ford, the chief engineer, — to whom, by the way, I am indebted for many kindnesses, — assured me that it answered well for that purpose.

In the cutting at Plataforma, which is a very heavy one, there is seen a thick bed of shale with occasional bands of sandstone, affording fossil fish, and cyprids, over which lies a heavy bed of conglomerate. This is composed of fragments of the principal rocks of the neighborhood, presenting the same appearance as that of the beds whence they were derived. The pebbles of this conglomerate, mostly of quartz and gneiss, are of all sizes, even up to eighteen or twenty inches in diameter. They are but slightly rounded, and are more or less angular. It appears to be a deposit rapidly accumulated, and from its hardness it now forms quite a high ridge. This conglomerate is a massive bed, showing in itself few traces of stratification. It has afforded a few reptilian bones. The same bed makes its appearance on the shore at low tide, where it may be examined. All these beds are well displayed on the shore, the whole distance from Plataforma to the little bay of Periperí, and, as the strike of the beds is tangental to the curve of the shore between these two places, and their dip is away from the shore, they may be examined both in ascending and descending series, and over a considerable area. Above water-mark disintegration makes their examination very unsatisfactory. Below that line it has not obtained to so large an extent, and, save a thin coating of half-decomposed material on the surface, the rocks are nearly in their natural state.

About a half-mile from Plataforma, and where the railroad passes close to the water's edge, there is exposed in one place a section like that on the opposite page, of which the height is about ten feet.

In many of the other beds, especially in those which are finer in character, reptilian and other remains are quite abundant; but, owing to the compactness of the rock, they are difficult to extract. These fossils consist in spines of fish,

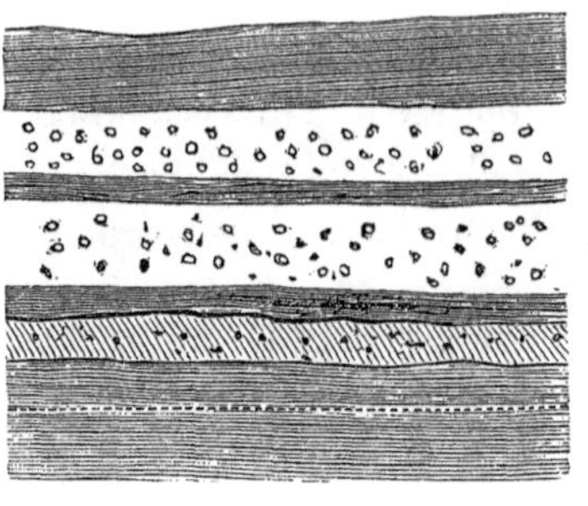

with occasionally a vertebra, bones of dinosaurians, and a few teeth, both of reptiles and fishes. Of these reptiles' teeth the most interesting are those of *crocodiles*. These vertebrate remains are especially abundant in a calcareous conglomerate, forming a bed a foot or two in thickness, and which I have designated in the above section as a bone-bed.

There are a few beds of a coarse arenaceous shale, which afford some very perfect specimens of a Teliostean fish, apparently different from the species found at Monserrate.* It is worthy of remark that this shale, as well as some of the other rocks of this series, are bituminous as well as calcareous.

The reptilian remains from Plataforma and Monserrate I placed in the hands of Professor O. C. Marsh of Yale College for description. The following with a few unimportant omissions is a notice of them published by him in the American Journal of Science and Arts, Vol. XLVII., May, 1869 : —

"The most interesting of the reptilian remains collected by Pro-

* Several fragments of the jaws of fishes have been found. One of these, the right ramus of the lower jaw of a little *Pisodus*, had the minute shot-like teeth preserved. All the fish remains from this locality are in the hands of Professor Agassiz. I regret that his illness prevents him from furnishing descriptions of them for this work.

fessor Hartt in the Bahia deposit is the tooth of a large Crocodilian, from the arenaceous shale near Plataforma station, on the Bahia railroad. This specimen is in an excellent state of preservation, and indicates a species new to science. It is larger, more slender, and more pointed than the teeth of existing crocodiles, resembling most nearly those of some extinct American species. It is conical in form, round at the base, and slightly compressed at the apex. The crown is two inches and three lines in length, along the outer side, and ten lines in diameter at the base. One edge is somewhat more convex than the other, and this is also true of one of the sides, and hence the tooth appears slightly curved in two directions. On either edge of the crown there is a sharp ridge, most prominent near the apex, over which it passes, but gradually disappearing before reaching the base, resembling in this respect the teeth of *Thoracosaurus*, from which, however, this specimen differs in being longer, and less curved than the teeth of that genus usually are. The sides of the crown are covered with fine, interrupted, undulating striæ, which appear to be different from the dental sculpture of the Crocodilia hitherto described. These striæ are most distinct near the middle of the tooth, becoming much more delicate at the base, and nearly obliterated at the apex.

"In size and general appearance, this specimen resembles somewhat the teeth of *Crocodilus antiquus* Leidy, from the Miocene of Virginia, but differs from that species in being less tapering, and in having the ridge on the edges extend farther downward. It resembles still more closely the teeth of a new species of crocodile discovered by the writer at Squankum, N. J., in the tertiary green-sand, which will soon be more fully described under the name *Thecocampsa Squankensis* Marsh. Both species have essentially the same proportions, and similar dental striæ, but the cutting ridge of the New Jersey specimens is more prominent, and extends nearly or quite to the base of the crown. The two species were apparently about the same size, both being considerably larger than existing Crocodilians.

"Other parts of the skeleton of the Brazilian species would perhaps show generic characters to distinguish it from the modern procœlian crocodiles, but in the absence of these, it may for the present be placed in the same genus. Its form, cutting edges, and especially its peculiar striæ, readily distinguish it from any species with which it is liable to be confounded, and it may appropriately be named *Crocodilus Harttii*, in honor of its discoverer, whose recent researches have thrown so much light on the geology of Brazil.

"Several specimens of reptilian teeth collected by Mr. Allport at Montserrate, a locality in the same deposit about two miles southwest of Plataforma station, evidently belong to this species, as the illustrations accompanying his paper (Plate XVI., figures 1, 2, 3, and 5) clearly indicate. The explanation of the plate refers to the specimens as, 'Teeth of crocodile with delicately wrinkled surface,' but no further description is given.

"In the same paper Mr. Allport has given figures of several Crocodilian teeth from the localities at Plataforma and Montserrate, which are quite different from those above described. These are represented in Plate XV., figure 5, and Plate XVI., figures 4, 6, 7, and 8, and are referred to on page 268 as, 'Teeth of crocodile with strong continuous striæ, and coarse riblets.' These specimens, taken in connection with some imperfect remains in the collection made by Professor Hartt, indicate the existence in this deposit of a second and smaller species of Crocodile, probably allied to the modern gavials. The teeth are not so large as those of *Crocodilus Harttii*, and are more tapering and more curved. They also differ widely in the striæ and lateral folds. These specimens may provisionally be referred to the genus *Thoracosaurus*, and, as the species is evidently new, it may be called *T. Bahiensis*.

"An interesting fossil, found by Professor Hartt at Plataforma station, is a fragment of a bone, evidently reptilian, but the exact affinities of which it is difficult to determine from this specimen alone. It resembles in some respects the extremity of an ulna, but after a careful comparison the writer is inclined to consider it the proximal end of a rib. It is much flattened at the articular extremity, and tapers gradually to the broken end, which is somewhat triangular in outline. Its length is about four inches, the transverse diameter of the perfect end two and a half inches, and of the other, one and a quarter inches. The larger extremity is divided into two articular facets lying oblique to each other, the smaller one being elevated about half an inch above the other, and covering rather more than a third of the entire terminal surface. In form and general proportions this specimen is not unlike the upper end of a right dorsal rib of some of the amphicœlian crocodiles, especially a rib in which the head and tubercle have so closely approached each other that their articular surfaces are nearly confluent. The size and other characters of the specimen, however, seem to exclude it from that order, and it probably belonged to a Dinosaurian reptile, possibly the same as a large vertebra from Monserrate, which Mr. Allport figured in his paper in Plate xvii., and which Professor Owen suggested might prove to be allied to *Megalosaurus*.

"The only other specimen in this collection that need be particularly mentioned here is a small flat bone, about two inches in length, with one articular extremity partially preserved. This appears to resemble most nearly the fibula of a tortoise, and probably should be referred to that group of reptiles."

I have not yet observed at this point any of the species of *Melania* and *Vivipara* of Monserrate. In some of the beds, however, a minute gasteropod is very abundant, and a unio-like shell also occurs, though somewhat rarely. These, with the estherians, point to a fresh-water origin for these beds, while the alternations of shales with sandstones and conglomerates show that there were intervals of quiet deposition of fine material with times when currents spread coarse material over their surface.

This locality, to a patient and painstaking collector, would yield a fine harvest.

From Periperí to Matto de São João I went over the railroad several times both by cars and with a trolly; but I have had an opportunity of examining a few points between these two stations, and with the aid of the information of my friend, Mr. Tiplady, the engineer of the first section of the railroad, and to whom I am much indebted for aid and hospitality, I am able to give the following observations.

At the tunnel at Periperí the rocks consist of shale and sandstone, with a slight northward dip, and contain but few fossils.

Between Olaria and Mapelle stations there are heavy beds of shale and sandstone, the latter a rather soft, bluish kind, which is quarried for building purposes along the railway. It is not very durable, as it weathers very rapidly. Fossils are very rare in it. The tunnel at Mapelle is through shale, the dip being northward and slight.

At a place called Cotigipe, between Mapelle and the next station, Agua Comprida, thick beds of shale and conglomerate are cut through.

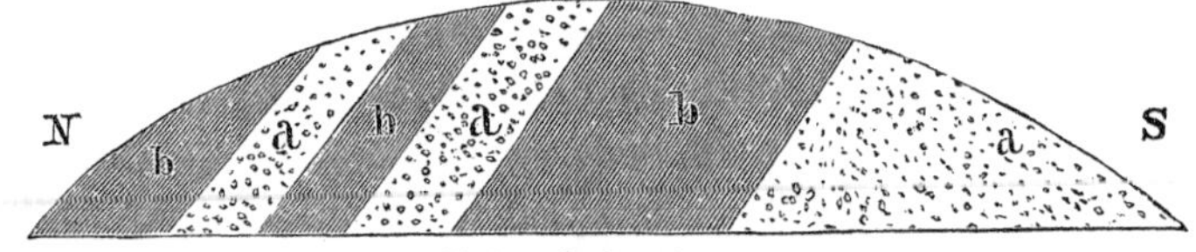

a. Shale. *b.* Conglomerate.

In the shale of bed, *a*, of the above diagram, which varies much in character from a thin, soft, black, almost papyraceous shale to a coarser greenish variety, fish remains and perfect fishes are quite common, but they appear to be of the same species that is found at Monserrate; together with these occur bones and teeth of reptiles, and an abundance of cyprids. I have collected nothing from the conglomerates. The dip of the beds is northeast, at a small angle. Thus far the country has been irregular, the hills being rather angular in their outline, and presenting altogether different topographical features from those of the gneiss regions. The cretaceous hills here are generally round topped, with long slopes.

Passing Agua Comprida, at Sapucaia, there are several cuttings through shale, and here we find the conglomerate in the hills on the west side of the road. Farther on no more of this rock is seen on the line. At Moritiba are found beds of sandstone of the same series, but their dip, according to Mr. Tiplady, is south. There are here some horizontal beds of pinkish sands of a much later formation.

Cutting No. 82, not far from Moritiba, is through a heavy bed of red clayey sand, sometimes used for moulding in the railroad foundry at Periperí. This bed is quite horizontal, and appears to belong to the same series as the sands to be described farther on.

Just before reaching the Rio Johannes there is a cutting through decomposed gneiss. Mr. Tiplady informed me that in the river-bed, a few rods above the railroad, the gneiss is exposed, while the rocks below the railroad are sandstones. In this vicinity the hills are rounded and more or less dome-shaped, yet not more than two to three hundred feet high; but in going on towards Parafuso they become lower, and the surface is rolling. At Parafuso there is a long, low cutting, in which I made the following section.

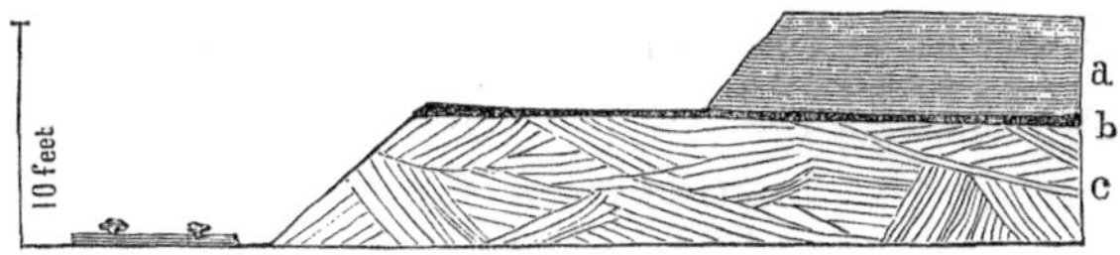

The lower bed consists of a soft, yellowish sandstone, with beautiful oblique lamination. I have looked long and carefully in this and other similar localities for fossils in these sandstones, but without success. Over this bed, which is quite horizontal, lies a thin sheet of hard, red iron-stone, and over this a bed of clay and pebbles.

In cuttings 8, 9, and 10 of the Third Section, which are through long and low banks, are seen similar beds of light pink-colored sand, slightly consolidated, and showing beautiful false bedding.

At Camassarí I spent several hours in an examination of the vicinity. The station stands on a plain composed of the sands just described, over the surface of which is spread a sheet of clayey gray or white sand, often containing pebbles and broken pieces of rock. In the vicinity are a few low hills, composed of beds of arenaceous pinkish and white-mottled clays, and sandstones which vary much in character, being sometimes soft and pinkish in color, while at others they are cemented by ferric oxide, dark red, and very hard. These beds are horizontal and very thick, and belong to the great coast tertiary formation. In some places they lie in an unbroken sheet, except by rivers, forming very extensive elevated plains, like the sertões and chapadas of São Matheos, the Mucury, and elsewhere along the coast; but sometimes, as on the Bahia Railroad, they are denuded in such a way as to form deep and wide valleys, in which stand isolated hills that rise abruptly from, or rather pierce through, the thick beds of sands and clays formed from their *débris*, and which occupy the bottom of the valley. In the denudation of these horizontal tertiary beds the tendency is to form, at first, a mass with a flat top, and sides steep and abrupt, as represented in the following diagram.

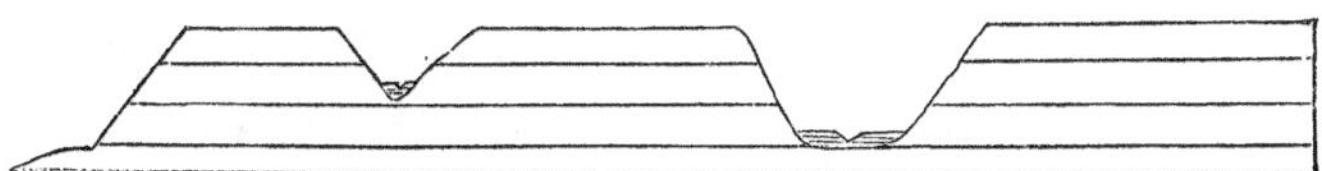

Such is the character of the slopes that border valleys. These are rarely perpendicular, and are almost always covered by vegetation. When, however, the sea or a river washes the foot of a slope, it is always perpendicular; as is the case with the red clay cliffs, which at intervals gird the shore of Brazil, from the Amazonas to Rio. As the valleys widen and approach one another, the hills preserve the same topographical features, and we may find all gradations between the broad chapada, the narrow hill with its flat top, and the roof-shaped or conical hill. These detached hills are called *oiteiros*, or *taboleiros*. At Camassarí the topography of these hills may be easily examined. The valley here is quite wide, and there are a great many small scattered hills, that rise like islands from the plain. I observed that most of the hills had rather a long smooth slope to the east, and an abrupt one to the west. The hills are covered thickly with fragments of rocks. This seems a little singular, because it is not owing to the dip of the beds, for, as already said, they are horizontal. I have questioned whether it might not be in some way due to glacial action. The whole hill, in every case, was once covered with a sheet of clay and fragments of rock, as I shall describe more fully hereafter. The finer materials of this superficial coating have been washed from the steeper slope, leaving the coarser masses behind, while on the long slope these last still lie buried below the surface. The sands occupying the valleys are not deposited horizontally, but form a series of sand-banks sloping more or less gradually toward the middle of the valley. The surface is not a plain, but rolling. The following is the outline of the country across the hills from Camassarí station, going west for a little more than a mile. It is intended to show the topographical features of the older tertiary hills and sand plains.

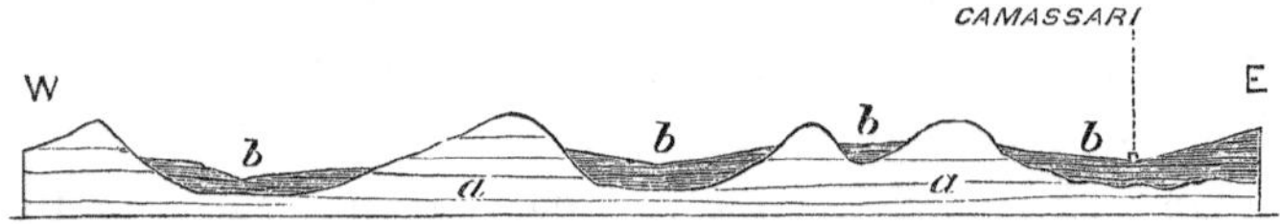

a. Tertiary clays and sandstones. *b.* Sands.

While at Camassarí station a man brought in a mule-load of "Tabatinga," * or pure milk-white feldspathic clay, which he said came from not far off, probably from a bed interstratified with the sands. The material is exceedingly fine and chalky, and becomes a pure white clay on wetting. It is used instead of lime for whitewashing walls. This Tabatinga clay is also found in the tertiary beds of the coast, where it is often pinkish or yellowish in color. I had observed the same material, as a cement to the superficial sands and gravel, in the cuttings in the vicinity of Camassarí.

Leaving Camassarí and crossing the taboleiros † one soon enters the hills, which are one hundred and fifty feet or more in height, composed of the tertiary sandstone and clays above described, and have some of their sides strown with large blocks of the red sandstone. A long cutting through a heavy bed of quartz gravel, of which the cement appears to be sand and clay, is soon reached, and one then goes on to another sand plain, which is more extensive than

* This word is of Tupí origin, and is derived from the two words, *Tauá*, clay, and *tinga*, white. Tabatinga is a corruption.

† This word is also applied to the plains.

that first described. Another bed of gravel is seen at a place called Embira Branca, and farther on, when fairly out on the taboleiros, at a place called As Pedras, there is what appears to be an old beach, composed of rounded quartz pebbles, filled in with snow-white sand. The beach is irregular in outline, and presents many projecting points. The surface of the sand, which is almost level, abuts against this gravel bank, and lies in the indentations in its outline, like a sheet of water. Just here this gravel is naked, but not far distant it is covered by a dark grayish soil, similar to that which covers the taboleiros. It appears to have been washed off here.

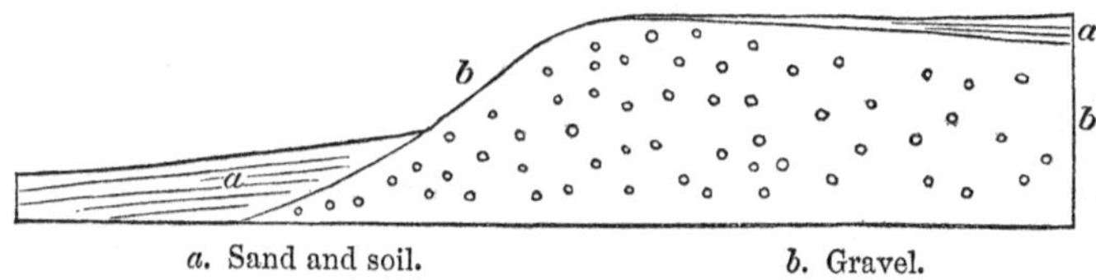

a. Sand and soil. *b*. Gravel.

A little farther on we find a cutting in which a cap of gravel is seen overlying the sand.

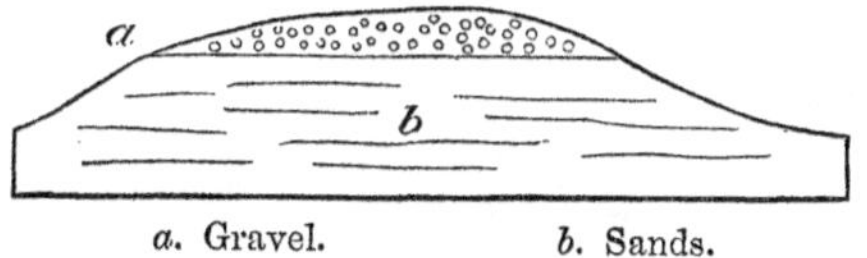

a. Gravel. *b*. Sands.

Thence to the river Imbuçahí the land slopes very gradually, and for miles on each side of the railroad is remarkbly even, being interrupted only by the isolated tertiary hills that show themselves above the surface of the sands. From the Imbuçahí to Feira Velha the sands rise with a very gentle slope. On this side they are bounded by tertiary hills and chapadas, some of which are roof-shaped ridges.

Both at Camassarí and on the Imbuçahí the sands are barren, and form slightly rolling plains, supporting only a scanty vegetation, which consists of several species of grasses, and a multitude of small flowering plants. Trees are very few. In the lower grounds, where moisture gathers, and where there are in some places streams, a long coarse grass grows most luxuriantly. A quarter of a mile south of the Imbuçahí is such a grass-covered area, and here excavations by the side of the railroad show that a bed of peat has accumulated, which is two feet thick in some places.

Leaving the taboleiros we enter a valley among the hills, which are at first low, but become higher farther on. They belong to the tertiary series, are often very irregular in shape, and about 350 feet in relative height. As we go up the valley it narrows more and more. The bottom is occupied by beds of sands and clays, which form a series of undulations whose outlines are in contrast with those of the bordering hills, as exhibited in the following diagram.

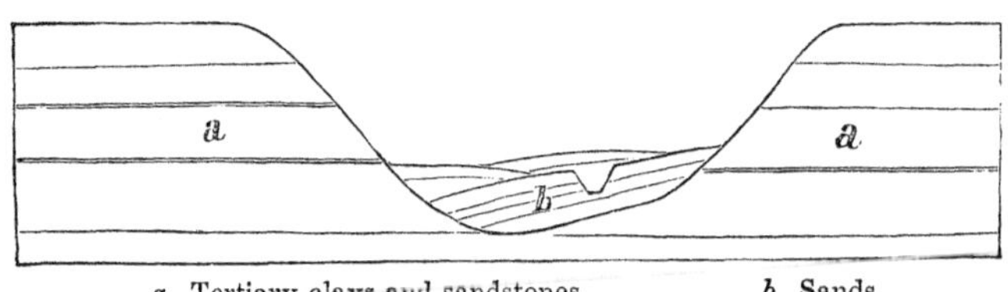

a. Tertiary clays and sandstones. *b.* Sands.

In order to illustrate the structure of these lower grounds, I will describe what is seen in a number of railway excavations in this valley.

Two cuttings towards Bahia of a cutting called Jacumerim, the road just pares away the southern side of a low hill, about twenty-five to thirty feet high, as seen in the following sketch.

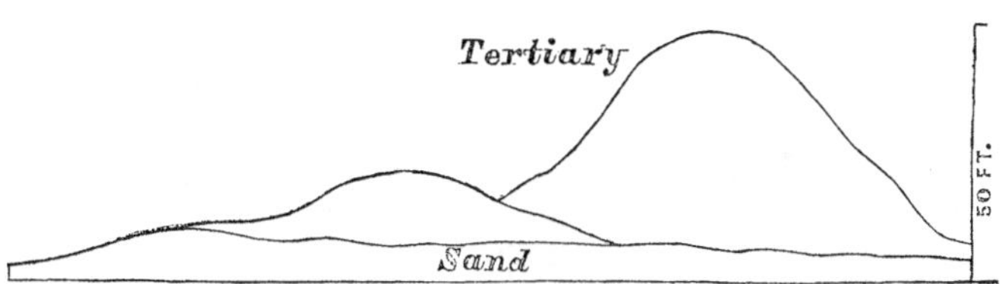

This hill is composed of sand, which is exposed for a depth of twenty feet. The same material forms the lower ground alongside. The hills in the background are composed of the tertiary clays and sands. The most interesting feature exhibited in this section is a sheet of fragments of red sandstone from the tertiary hills extending under the soil over the surface of the sands, not only on the low ground, but also over the whole hill.

The following cut represents a section across this same hill, but at right angles to the first. The boulders of sand-

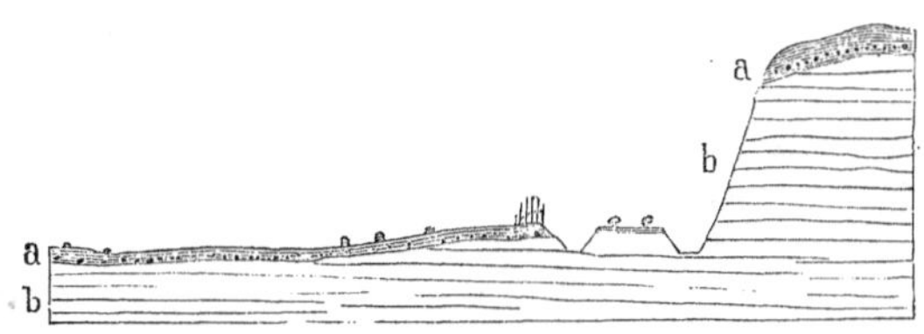

a. Soil containing pebbles and boulders of sandstone.
b. Stratified sands.

stone are here seen scattered over the low flat by the side of the railroad. These boulders must have come from the adjacent tertiary hills.

Going up the road we soon come to another cutting in which a sheet of gravel overlying a bed of yellow sands is exposed, as seen in diagram on preceding page. This gravel consists of large, well-rounded pebbles of quartz or decomposed syenite, with a white paste. In the next cutting, Jacumerim, we see the following section.

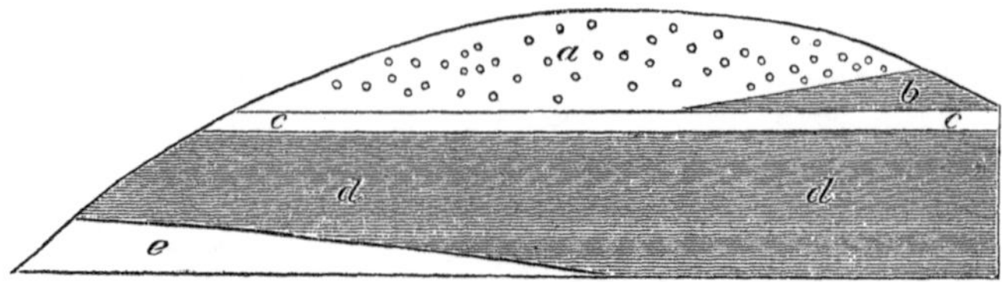

The pebbles in the gravel are of the same kind as those just described. They are very large, the material being rather a coarse shingle, and they are very closely packed. Over this elevation, as in the other, is still seen the layer of broken pieces of sandstone, and the same may be examined in the next cutting, which is a very instructive one.

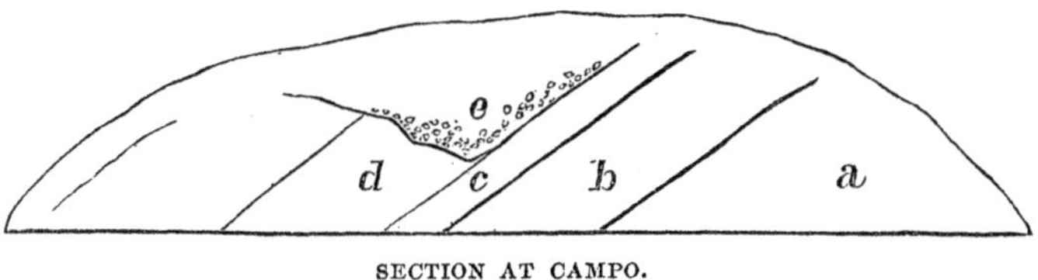

SECTION AT CAMPO.

a. Fine brick-red clayey sand, in some places with scarcely consistence enough to be moulded; in others very clayey.

b. Whitish and yellowish sand.

c. Soft white sands, no consistency under pressure. A thin streak or two in the lower part of this bed is consolidated.

d. White or reddish sands.

e. An irregular mass of boulders, of dark red tertiary sandstone, some of which are three to four feet in diameter.

In this diagram the dip is exaggerated. At the post marking ten leagues from the city of Bahia is a long low cutting, and in a trench by the roadside a stiff sandy clay, mottled with dark red and white, and overlaid by a dark brown sandy soil.

Between this point and Matta station there is much swampy ground bordering the road at intervals, and on this great numbers of piassaba palms grow.

of the railroad I visited some other old washings, from which diamonds had been obtained. I found that they had been dug from the cascalho or gravel sheet on the slope of the hill near the river, which gravel was covered with the common red drift-clay of the vicinity. I take this gravel to be drift. The diamonds appear to me to come from the tertiary beds of the neighboring hills, and this seems to be the opinion of Mr. Nicolay, who shows that the diamonds of the Chapada Diamantina come from a conglomerate and sandstone, which, from his descriptions and specimens, appears to be a tertiary rock of the same kind as that which forms the chapadas of the valley of the Jequitinhonha. Mr. Nicolay says that the cascalho in the vicinity of Bahia differs from that of the chapada, in that it is more siliceous and less metaliferous. There is no Itacolumite in the vicinity of Pitanga. The gravel is made up principally of fragments of quartz and of pebbles of a sandstone like that of the tertiary chapada, though somewhat harder than the kind usually seen along the road. I do not believe that the diamond ever occurs in the true palæozoic Itacolumite in Brazil, but that it is derived from the tertiary sandstones.

At a sugar fazenda beyond Pitanga there is a cutting under a bridge across the railroad, and in a gully made by the rain I found a shale full of cypris, and evidently cretaceous. The beds appear to be horizontal, or with a slight inclination southward. The overlying soil here is red mottled with white.

The next cuttings are not very satisfactory, and appear to be through decomposed sandstones and shales.

I introduce a sketch here to give some idea of the topography, and of the general appearance of the tertiary hills bordering the valley on the west, just below the tunnel.

TERTIARY HILLS NEAR POJUCA TUNNEL.

The sandstone which to some extent, at least, forms the chapadas of the vicinity is a somewhat soft, coarse-grained, reddish and mottled kind, with an argillaceous cement. It has been occasionally used on the railroad for building purposes, but it cannot be very durable.

A ridge crosses the railroad, and a tunnel ("Pojuca tunnel") is made necessary. This ridge runs nearly north-south, and is composed mainly of cretaceous strata, which, according to information received from Mr. Turner, engineer of this section of the road, have a high westward dip. In the cutting at the eastern entrance to the tunnel horizontal beds of a soft brown sandstone are exposed, and this is so soft as to be easily crumbled by the hand. In this occurs a bed, fifteen centm. in thickness, of a yellow clay (decomposed shale?) in which I have found a few Estherians. The stuff thrown out of the tunnel is a bright bluish, very soft shale, containing fish remains, together with a finely laminated blue-gray sandstone, also containing fish remains and fragments of plants. I did not see these strata in place.

In the cutting at the western entrance are thick beds of shale and sandstone, so decomposed that their character cannot well be determined. The strike, as near as I could make it out, is N. 65° E. and dip 73° S. E., but it is to be remembered that Mr. Turner says that in the tunnel the dip is W. The beds of the tunnel and western entrance are cretaceous, I think. Those at the east end appear to overlie the cretaceous beds unconformably, and may be tertiary; but I would not speak positively, on account of the great difficulty of recognizing some of these rocks when they are decomposed. This ridge is covered with red drift.

From Pojuca to Sitio Novo I have been over the road in the train several times, but I was prevented from examining the rest of the road on foot, because of the rain. From Sitio Novo to Alagoinhas I went over the road in a trolly, but in rain and under unfavorable circumstances. Soon after leaving Pojuca the valley becomes narrower. The railroad follows its bottom, so that there are no rocks exposed, and the cuttings are of very little interest. The low lands are very fertile, and are here and there covered by forest, though they are planted to a considerable extent with sugar-cane.

As for the lands of Bahia, those of the gneiss are fat, deep, and exceedingly fertile; the slopes are generally somewhat steep, and in the southern part of the province they are favorable for coffee. The soils of the cretaceous along the railroad are rich; but the rains make fearful havoc with them, carrying them away completely after a few years of cultivation, leaving them barren, as is the case over large tracts along the railroad. Of the low lands, those bordering the rivers are very fertile. The tertiary high lands are sometimes valuable on the slopes, but above, as a general thing, they are dry and barren. The taboleiros are useless, except for grazing. Immense tracts along the railroad are allowed to grow up with the piassaba palm. Considerable difficulty has been experienced by the engineers, in different parts of the road, from the instability of the rocks, and the liability of their sliding on one another, especially when the shale beds become wet. In one case, where the beds dipped with the slope of a hill toward the railway, the upper beds over a considerable area slid down some distance, throwing the track out of shape.

The Pojuca tunnel caved in from the same reason, about two years ago. Some distance south of Alagoinhas the valley opens out and an extensive series of taboleiros is reached, in the middle of which is the station of Alagoinhas. I made a long *détour* over these taboleiros on foot, and found them to be essentially like those at Camassarí. They were great, gently undulating plains of sand, scattered about over which were irregular isolated hills of the horizontal tertiary clays and sandstones. The cutting at the terminus of the railroad is a long one, about five feet deep, through white clayey sands, under which is the ever-appearing pebble line.

In a cutting for a road through a slight elevation near the station, I saw beds of clayey sand slightly consistent and of a white or pinkish color. Here the Saüba ants had formed immense mounds of a clayey sand of a warm pinkish tint, the material having been brought up from a considerable depth by these busy creatures in the excavation of their galleries.

The village of Alagoinhas is on the sand plains, about a

league eastward of the station of the same name. The surface of the plains is not level, and there are differences of elevation amounting to fifty feet more or less, but the slopes are very long, exceedingly smooth and even, and as a general thing their direction is from the hills towards the centre of the plain. The surface is loose white sand, of course a most unfertile soil. The vegetation of the sand plains and taboleiros differs in the most marked way from that of the soil-covered hills. Trees are few, scattered, small, and very often with rough bark, and stiff and contorted branches. One of the most conspicuous of them is the Carahyba, a tree about twenty feet in height, which is scattered all over the plains.

The Murici (*Byrsonima*) is a small scrubby tree, about eight feet high, with bunches of large elliptical leaves covered with hairs like the mullein borne at the ends of the branches. The fruit is about the size of a large cherry, yellow, very fragrant and much esteemed. The Mangaba (*Hancornia speciosa*) is another small tree, with weeping branches and small leaves. Its fruit is of the size of a plum, and very delicious. The Perico is a bush producing a fruit as large as a gooseberry, and very pleasant to the taste. The Bahianos are fond of it, and in the season of fruitage the berries are sold in large quantities in the city.

A small tree called Sambahiba is remarkable for its curled leaves, the upper side of which is so rough and hard as to scratch wood like sandpaper. The Janahuba is a characteristic shrub of the taboleiros, and it is noteworthy on account of bearing at the end of its stem a cluster of large leaves, giving out an abundant milky sap when broken. Among the other common plants of the taboleiros one observes the Alecrim, with its fragrant leaves; the Almescar, furnishing a sweet-scented resin; the Macella, producing a material used in stuffing mattresses; the Purga do Campo and Orelha da Onça, both used in medicine, and the curious Barrigudinha, with its swollen stem. Creeping about over the bushes, one sees here and there tangled skeins of the yellow thread-like stems of the Cipo de Chumbo, a species of *Cuscuta*, or Dodder. A large number of the plants of the taboleiros are medicinal, and very many have aromatic leaves. Small and beautiful flowering plants abound among the tufts of coarse grass with which the plains are covered. These grasses grow in widely separated clumps. My guide indicated the three principal kinds as *Capim agreste*, *Capim de Cheiro*, and *Capim pubo*. I observed one or two species of Melocactus, but I do not remember having seen a single large Cereus, a plant so common on the dry rocky sertões of the Rio de São Francisco. A little palm, Licorí, is very common, and is the only species I saw on these campos.

Ant-hills are common on the taboleiros, looking like scattered boulders. Under one of these I found a large scorpion, an insect which, like the centipede, is common enough in Brazil; but one might travel a year in the country without seeing a specimen of either, unless he made special search for them.

As elsewhere, these campos are burned over from time to time, and the flora has consequently suffered great modification.

To give an idea of the general character of the vegetation, as well as of the topography of the sand plains and the tertiary hills near Alagoinhas, I introduce the sketch on the following page.

THE TABOLEIROS NEAR ALAGOINHAS.

These hills are composed of a soft, reddish, argillaceous sandstone, with some layers, however, of a coarse kind, very hard, of a dark-red color and with a cement of ferric oxide. In some places these rocks form cliffs. Here, as in the taboleiros of Camassarí, the sands bathe the foot of the hills like a sheet of water. The hills are covered from top to bottom with fragments of the coarse red sandstone, with some quartz pebbles. Usually round the base of the hills the sands have a coating of red soil washed down from the hillside. Near Jacaré the soil of some parts of the hilly land becomes better in quality, and is soft and loose. Here we find a liberal forest coating, and some poor settlers cultivate fields of mandioca, tobacco, corn, &c. The soil is sandy, and grayish-brown in color. Farther back comes a belt of forest beyond the taboleiros, and a considerable quantity of sugar is raised in that region.

The height of Alagoinhas station above the sea, according to the survey of Mr. Vivian as laid down on the manuscript map of Mr. Nicolay, is three hundred feet. The tertiary hills of the vicinity I should judge to be one hundred and fifty feet higher, more or less.

As to the extension of the cretaceous beds on both sides of the line, I have very scanty information. The island of Madre de Deus, in the Bay of Bahia, belongs to the same series, as I have seen in Mr. Nicolay's collection several specimens of sandstone and other rocks from this locality precisely like those on the railroad; and the island of Itaparica is also, to a large extent at least, cretaceous. I have not heard of their extension farther south. I believe that the Bahia cretaceous series is confined entirely to the borders and islands of the Bay of Bahia, and that it is for the most part an accumulation within a closed fresh-water basin.*

The sands and gravels of the plains and connecting valleys are certainly newer than the coast tertiary beds, which are denuded in order to form the basins in which these were deposited. They appear to me to be of lacustrine and fluviatile origin, and I believe that the plains were once covered by large lakes, which have been drained by the cutting through of the high lands on the south by the streams flowing from them. From what I have been able to observe, these beds are uniformly overspread by a sheet of clayey sand, mixed with fragments of rock from the neighboring hills, which sheet I consider to be drift, so that the sands and gravels may possibly be of very late tertiary age. It is very probable that they may turn out to belong to the same series as certain similar deposits observed by Mr. St. John in the valley of the Rio de São Francisco, which are overlaid by drift in the same way.

* In Professor Agassiz's collection I find a fragment of greenish shale with cyprids from Parahyba do Norte, which appears to be from a formation similar to that of Bahia.

CHAPTER IX.

THE PROVINCES OF SERGIPE AND ALAGÔAS, AND THE RIVER SÃO FRANCISCO BELOW THE FALLS.

The Province of Sergipe. — Its Division into *Mattas* and *Agrestes*. — The Rio Real. — Estancia; New Red Sandstone, Sugar Plantations, &c. — Sand Dunes. — The Rio Vasabarris. — The Rio Cotinguiba. — Aracajú. — Cretaceous Beds with Inocerami at Sapucahy. — Maroïm. — Cretaceous Limestone with Ammonites. — " Fossil Turtles." — Sugar Plantations. — Messrs. Schramm and Company. — The Bar of the São Francisco. — Sand Dunes of the Pontal. — Character of the River below Penêdo. — Aracaré. — Villa Nova and its Cretaceous Sandstones. — The City of Penêdo and its Geology. — Its Commerce and Fair. — Notes on the Piranha and its Habits. — Propriá. — Morro do Chaves and Cretaceous Fossils. — Traipú. — Iron Ore. — Campos, Vegetation, Cactuses, &c. — Páo de Assucar. — Cattle Fazendas. — Piranhas. — County flat and covered by Boulders. — The River Valley a Narrow Gorge in a Gneiss Plain. — The Falls of Paulo Affonso. — Halfeld's Description. — Liais's Description. — Comparison between Paulo Affonso and Niagara. — Mastodon Remains from near the Falls. — Climate of the São Francisco below the Falls. — Steam Navigation. — Character of the Coast of the Province of Alagôas, South of Maceió. — The Lagôas. — The City of Maceió and the Geology of its Vicinity. — Tertiary Beds. — Harbor and Reefs.

THE Province of Sergipe is a very small one, wedged in on the coast between the province of Bahia and the Rio de São Francisco. Its coast line is only about ninety miles in extent. The eastern part of the province is low and uneven, and there are extensive tracts of sands along the coast. In this region are some lands fit for cultivation. The western half of the province is higher and somewhat mountainous, the principal mountain range being the Serra d'Itabayana, of which I shall have occasion hereafter to speak. The eastern half of the province passes by the name of *Mattas*, because of its forests, while the term *Agrestes* is applied to the western portion, because it is destitute of forest, and is to a large extent barren and dry, being in some parts fit only for pasturage, though during the dry season water fails, the vegetation dries up, and cattle suffer from hunger and thirst.

The principal rivers of the province are the Rio Real, which separates it from the province of Bahia on the south, and empties into the sea at the same mouth with the Rio Piauhy, the Rio Vasabarris, the Cotinguiba, and the Japaratuba.

The Rio Real is a small river navigable for a distance of some nine leagues above its mouth. In its upper course it flows through the dry belt, and it is bordered by cattle fazendas. A little cotton is raised along the river.

The Piauhy, another small river rising in the western part of the province, and to the north of the Real, unites with the latter river just before it empties into the sea. On my return from a voyage on the São Francisco I touched at Estancia, and made a hasty visit to the town. Estancia is built on a rolling country, where the heights of the immediate vicinity are not more than two or three hundred feet. The hills are rounded, and the rocks composing them are coarse red micaceous sandstones,—quite indistinguishable in the hand specimen from the triassic red sandstone of New Jersey. This sandstone covers a large area, and must be very thick. I examined it in several places, but found no signs of fossils. The dip, as a general thing, appears to be but a few degrees to the eastward. This sandstone is covered by a red clayey soil, which bakes very hard, so that the vicinity of Estancia is very arid and of little fertility, and the vegetation is low and sparse. There are, however, very productive lands lying farther to the westward, and probably inside of the sandstone range. Estancia exports every year 7,000 boxes of sugar (48,000 cwt.), together with some cotton, tobacco, etc.

Below Estancia I found a poor exposure of limestone, containing shells, and which was quarried for burning into lime. I could discover no signs of bedding. The *Diccionario Geographico* says that in the neighborhood of the Rio Piauhy a mine of coal was discovered, and that since 1840 no other coal has been used in the forges of the vicinity; but while at Estancia I made careful inquiries about minerals, and heard not a word said about the existence of coal in the neighborhood.

The river is narrow, and at the time of my visit, in August, 1867, it was very shallow. The water below the falls was fresh, but very muddy. As the hot season advances the river falls very low, and the salt water flows up to the port. The lower part of the river is bordered by extensive mangrove swamps; its mouth is obstructed by a bad bar. On shore, on the south side, are some magnificent sand-dunes, forty to fifty feet high, as regular in their outlines and as white in color as snow-drifts.

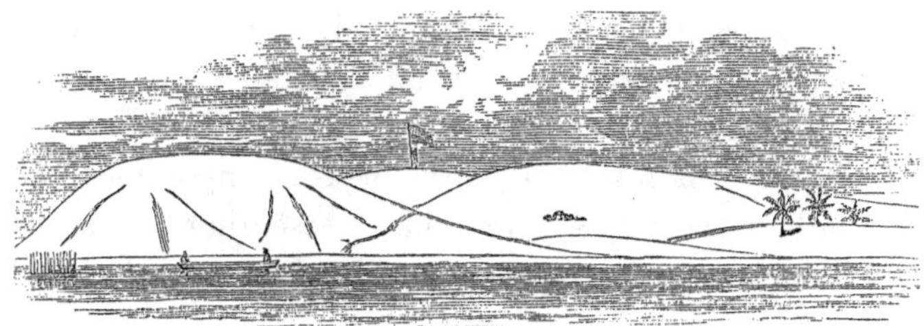

SAND-DUNES AT THE MOUTH OF THE RIO REAL.

North of Estancia low, irregular hills begin to make their appearance, and seem to be tertiary outliers; but away in the interior, twenty or thirty miles from the coast, is seen the blue outline of the Serra de Itabayana. It presents a low, very evenly rounded form, and must be over two thousand feet in height. It is composed of gneiss and mica slate, as I had an opportunity of seeing on the Rio de São Francisco.

The Vasabarris, anciently denominated Irapirang, rises in the province of Bahia, in the Serra Itiúba, according to the *Diccionario Geographico*, and is of very little importance except near the sea, where it is navigable for a distance of about twenty miles. The same authority says that the Rio Sergipe flows into it near its mouth, which is manifestly a mistake. Not far from the mouth is the city of Sergipe d'El-Rei, or São Christovão, on the bank of a minor stream emptying into it from the north. This city was, for many years, the capital of the province, but so very bad is the bar of the Vasabarris that the river was not to be depended upon for navigation. The capital was consequently removed to Aracajú, on the Cotinguiba, a few leagues farther north. The city is now in decay.

The Cotinguiba, or Cotindiba, is a smaller river than the Vasabarris, and takes its rise, as near as I can ascertain, in the Serra d'Itabayana. It is navigable, at high water, for smacks only, as far as Maroïm, a distance of some ten or twelve miles. The river has several branches, on which are a number of towns of more or less importance. At the mouth the river is very wide, presenting a beautiful sheet of water; but it appears to be, after all, only a sort of estuary. Between Maroïm and the sea the banks are largely covered by mangroves, but there are some hills and higher lands.

The bar at the mouth of the river is very dangerous, and the surf beats on it with great fury, sometimes precluding the possibility of entering. Almost across the mouth, from the northern side, stretches a line of sand-banks of a crescent-like shape. Three of these are joined

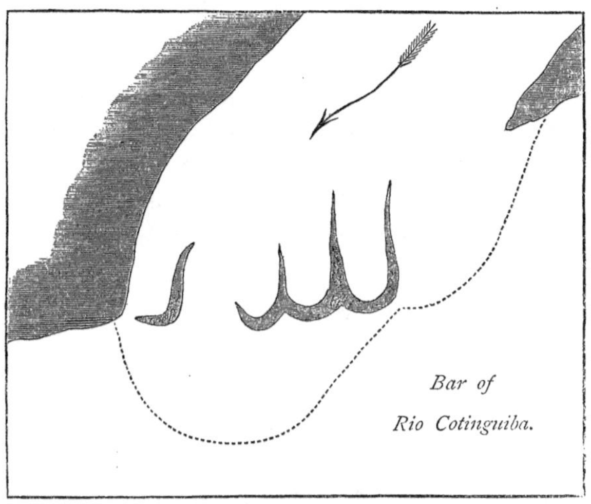

together, their convexities being turned toward the sea, while the extremities are produced up the river as long spits. A similar sand-bank is attached to the northern bank. These hook-shaped spits of sand are made in the struggle between the river and the waves of the sea, in the same manner as the hook at the mouth of New York Bay is made. Like the other rivers on this coast, the Cotinguiba enters the sea very obliquely with a southeast course. The left bank is flat, composed of sand, and continues low for several miles up the river. On the opposite side, at the mouth, there are extensive dunes forty to fifty feet high, flanking a tract of recently elevated sands, stretching along several miles, covered with cocoa-palms as far as the city of Aracajú, a small and uninteresting town, the capital, built on a flat of lower alluvial ground, bordering the river at the base of the sands. In the upper part of the town the dunes are very large and conspicuous. In company with Dr. Brunet, Director of the Agricultural College at Bahia, I walked over the stratified sands for some distance up the river to a hill on which there is a church and little village. I saw no shells in the sands. The hill referred to is tertiary, and we saw in the soil covering it irregular masses of the common dark-red tertiary sandstone. The height of the sand plain above water level was, if I remember rightly, about fifteen feet.

A few miles above the city, and also on the right bank of the river, at a place called Sapucahy, there is quite a large quarry in a little hill composed of a white flaggy and shaly, rather soft and chalky limestone, used quite extensively for building purposes in Aracajú. Of this limestone a thickness of over a hundred feet is exposed. The stratification is remarkably regular, and the quality of the stone is very uniform. As a general thing it is almost pure, white and somewhat soft, but there are bands of a grayish variety, much harder, which have, at first sight, the appearance of a lithographic stone. The thinner shales resemble strongly those of Solenhofen. A large part of the rock is flaggy, and is readily quarried out in large thin slabs. There are some lines of flints in these beds, but these are not rounded as in the English chalk, but tabular and angular. I have submitted some specimens of these rocks to my friend Mr. Arthur M. Edwards, the microscopist, of New York, who has been unsuccessful in discovering in them any microscopic remains. On the surface of some of the layers of limestone I have found great numbers of valves of a pretty

Inoceramus, most probably new, together with a little Ammonite and some teliostian fish-scales. I am told that perfect fishes have been obtained here as well as at Larangeiras, and some specimens were collected a few years ago in this vicinity by his Majesty the Emperor of Brazil, a shrewd geological observer. This white limestone appears to represent the white chalk. The beds at Sapucahy have a moderate dip to the southeast approximately.

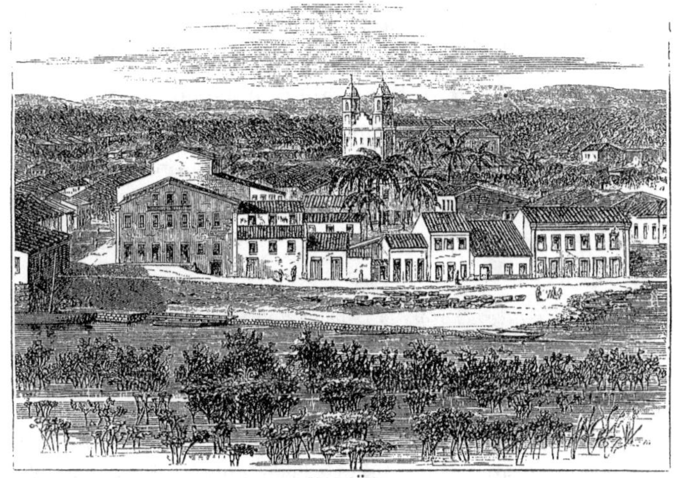

MAROÏM.

Between Sapucahy and Maroïm, a few miles up the river, the grounds are low, and the boatmen informed me that lime abounded in the vicinity. The limestone of Sapucahy is said not to furnish good lime. At Maroïm I was gratified to find the streets paved with large blocks of a coarse yellowish limestone, and to see on them the impressions of some large *Ammonites* and *Ceratites*, some of which I took up and brought away with me.* Besides these fossils I obtained,

* I visited the quarry that supplied the stone, but in the short time at my disposal I found but little.

through the kindness of Mr. Adolph Laué,* several specimens of a large *Natica*. Mr. Nicolay once showed the drawing of a *Cidaris* sent to him from that locality.

At the Salem meeting of the Association for the Advancement of Science, in August, 1869, I exhibited my collection of Maroïm fossils, and expressed the opinion that they were cretaceous. Professor Alpheus Hyatt, in examining the specimens, was struck with the remarkable peculiarities possessed by some of them, and which recalled jurassic forms. Professor Hyatt kindly consented to study the specimens critically, and I am glad to be able to present the following report upon them, which he has prepared for this work. I am especially glad to have the opinion of this naturalist on the Ceratites and Ammonites, because he has made such exceedingly careful studies of these groups.

Report on the Cretaceous Fossils from Maroïm, Province of Sergipe, Brazil, in the Collection of PROFESSOR HARTT. By ALPHEUS HYATT, S.B., *Curator in the Peabody Academy, Salem, Mass.*

In the small collection of fossils from the above locality, placed in my hands for examination, I have been able to make out the following species: —

Natica prælonga Leymerie.

Natica prælonga Leymerie, 1842, Mém. de la Soc. Géol. t. v. pl. 16, fig. 8, p. 13; D'Orb., Voy. dans l'Amér. Mer. t. 3, p. 73, pl. 18; Pal. Fran., Terr. Crét. p. 152, pl. 172, fig. 1.

This species is quite closely allied to *Natica Pierdenalis* of Roemer, collected in Texas, but has a longer and more acute spire. The French specimens were found at Thieffrain and Vandœuvre, in the Lower Néocomien, by M. Leymerie, and the identification

* Since my return home Mr. Laué has been so kind as to send me small lots of these fossils.

was made by a comparison between the Brazilian specimens and the figure given by D'Orbigny in the *Paléontologie Français.* D'Orbigny's figure of the specimen from Columbia, in his *Voyage dans l'Amérique Meridionale,* is that of a young specimen found by Boussingault on the Rio Suba, one of the affluents of the Rio Suarez. This agrees perfectly with the figure of the French specimen. The French, Columbian, and Brazilian specimens are all casts of the interior.

Locality: Cretaceous of Maroïm, C. F. H., 1867, and Mr. Adolph Laué.

Ceratites Harttii n. sp.

This specimen I was at first disposed to consider as a Goniatite in spite of the septa, the characteristics of which are unquestionably ceratitic. It is evidently a fossil cast which has been transported to the beds in which it was found, from some older stratum of precisely the same lithological composition. Serpulæ have incrusted the surface, stretching their long cornucopia-like shell-cases across, or fitting closely into the abraded depressions which mark the former edges of the septal partitions. The presence of these serpulæ show conclusively that the specimens must have been a fossil long before they began to grow upon its surface, and that it had suffered much from the wear and tear of the elements before they could have fitted themselves so accurately into the depressions of its rough and pitted exterior. The umbilicus is filled with the tough calcareous matrix, but its form must necessarily be that of a very deep funnel, the whorls enveloping the greater portion, if not the whole, of the sides of the young. This is certainly the case with regard to the last whorl, which covers nearly the whole breadth of the sides of the immediately preceding volution, leaving only a narrow band exposed. This peculiarity, and the great transverse breadth of the volutions, gives to this species a remarkably close resemblance to *Goniatites crenistria* and other allied forms, and this it was which, at first sight, led me to imagine that the older *Goniatites,* as well as the *Ceratites,* had "colonized" the cretaceous shores of Brazil. The longest diameter of the cast is nearly five ($4\frac{1}{16}$) inches. The latter half of the last volution, occupied by the terminal chamber, is devoid of septal markings, and has suffered too much from abrasion to allow of accurate measurements. Near the last septum, however, the fulness of the original shell is very well preserved, and here the dorso-abdominal breadth of the last whorl was found to be two sevenths shorter than the dorsal breadth, measuring from shoulder to shoulder of the umbilical edge. The abdomen and sides are symmetrically rounded, reaching their greatest width or departure from each other on the umbilical edge. The curvature of the sides is so rapid, however, and the apparent elevation of the abdomen so great, that, without actual measurement, no one would be apt to suspect that the base of the arch was longer than its axis of elevation. The edges of the septa, though worn away to a considerable extent, are sufficiently distinct to allow of the determination of their general characteristics. They are evidently more closely allied to those of *Ceratites (Ammonites) Robinii*, as figured by Von Buch,* than any other species. The abdominal lobe, however, has a siphonal cell of ordinary size in place of the large broad cell occupying the abdomen of that species. The lateral lobes and cells have very nearly the same general outline as in *Ceratites Robinii*, and the superior lateral cell has a similar triple division of the base line, due to the presence of two minor lobes of equal size. The superior lateral lobe is about one third deeper than the abdominal lobe. The inferior lateral cell is very much broader than in *Ceratites Robinii*, and the base, instead of being smooth, is broken by two or more minor lobes, almost obliterated, however, in the cast. In other respects these cells are very like those of *Ceratites Robinii*, having precisely the same hump-like form rising gradually on the abdominal, and bulging out on the dorsal or umbilical sides. The inferior lateral lobes in both species are, on account of the contiguity of the septa, in contact, and

* *Ueber Ceratiten*, Abhand, d. Kong. Akad. Berlin, 1848, p. 176, fig. 4.

set one into the other like a pile of coffee-cups; this lobe, however, in *Ceratites Harttii* is either deeper or of about the same depth as the superior lateral. The first auxiliary cells and lobes are equally imitative, but the basal outlines of the former are divided by two minor lobes similar to those of the superior and inferior lateral cells.

The first auxiliary lobe, also, reaches the umbilical edge, instead of being situated at some distance therefrom, as in *Ceratites Robinii*, and the smaller auxiliary lobes and cells are upon the unexposed, inturned portion of the border, and are not seen upon the sides, as in Von Buch's figure of the latter. Probably no very close comparisons can be made between the form of the shell in these two species, since *Ceratites Robinii* has a shallow umbilicus and narrower whorls. There seems to be no doubt, therefore, that this fossil is undescribed, and as it is the remains of a very remarkable cretaceous animal, its dedication to its discoverer cannot be classed among compliments of a similar sort which are too often commonplace or misapplied.*

Ceratites (*Ammonites*) *Pierdenalis* Von Buch.

No. 4 of the collection is so closely allied to this species in the form of the whorls and the extent to which they envelop each other that I have no doubt of their specific identity. The septa are not apparent, but the hollowing in of the umbilical sides of the volutions and the acuteness of the abdomen are well marked and evidently the same as in *Ammonites Pierdenalis*.

Locality. In cretaceous beds at Maroïm. C. F. H. and Mr. Adolph Laué, collectors.

Ammonites Hallii Meek and Hayden?

No. 1 is probably a fragment of a large whorl of Ammonites Hallii of Meek and Hayden, or a closely allied species. Although

* This Ceratite was obtained from the cretaceous beds of Maroïm. C. F. H. and Mr. Adolph Laué, collectors.

none of the inner whorls are preserved, the umbilicus was evidently deep and comparatively narrow, a characteristic caused by the broad dorsum, small number, and very rapid increase in size of the volutions.

The costæ, or pilæ, as I prefer to call them, cross the abdomen without interruption, and not more than one in four reach the umbilical edge, where they develop large, coarse tubercles and disappear. An outline or section of the whorl would curve like a Roman arch, the abdomen being rounded far down on the sides, the sides flattened only when near to the umbilical edge, which is very abrupt, and in the whorl examined measured about four fifths of an inch from the edge to the side of the preceding whorl. The base of the whorl in its broadest part, from edge to edge, measures three and two fifths inches, and its height, from a line connecting the umbilical edges, three and one fifth inches. The septa are not sufficiently well preserved to afford an accurate description.

No. 2 is a species of the Ligati group. The ligature-like depressions constricting the whorl are plainly visible, and the form reminds one of *Ammonites semistriatus* D'Orb., at least the general aspect of the last whorl somewhat resembles that species.

No. 3 appears to be identical either with *Ammonites Peruvianus* Von Buch, or *Ammonites acutocarinatus* Shumard. The fragment is very much compressed, and the true characteristics of the abdomen obliterated.

No. 3 *a* is a fragment of a young specimen of No. 3. This is not compressed, and shows the prominent keel and broad pilæ of this species much more plainly.

Locality : Maroïm, in cretaceous beds. C. F. H. and Mr. Adolph Laué, collectors.

Ammonites Gibbonianus Lea.

No. 5 is probably the young of No. 6. The pilæ (ribs or costæ) make their appearance on the first quarter of the second whorl.

After this first period the whorls are obscured more or less until the last quarter of the fifth whorl. From this time until the completion of the sixth volution there are large tuberculated pilæ which alternate with others of lesser height and thickness, though in a very irregular manner. The larger pilæ begin to lose their greater proportional height on the latter part of the sixth whorl, and appear about to assume the same form as those of No. 6, described below.

The dorsum of the sixth whorl is much broader than the abdomen; the umbilical edges are rounded and the sides slope evenly to the base of the keel. The pilæ have squarely cut geniculæ, probably tuberculated on the shell, and which bend forward on to the abdomen and terminate close to the keel.

The keel is very prominent, thin, and sharp, and the sides between the pilæ are evenly rounded upon the edge of the abdomen, in those parts not affected by compression.

The septa were too obscure to be observed with any approach to accuracy.

No. 6. The keel is very nearly perfect, and shows to the fullest extent its great breadth and the entirely external position of the siphon. In this respect it resembles Nos. 5 and 7, in both of which the siphon is not present at all in the internal casts of the whorl, but disappears with the removal of the shell.

The latter part of the sixth and the first half of the seventh whorls are exposed, and tolerably well preserved. The umbilical edge is rounded, and the sides slope evenly to the base of the keel. The umbilicus itself rather deep. The abdomen is not quite so broad as the dorsum, measured from edge to edge. The pilæ are depressed on the umbilical border, but the geniculæ are slightly more prominent, but not tuberculated, and bend forward on to the abdomen, terminating near the keel.

They continue to remain straight until near the second quarter of the seventh volution. Here a double curvature begins to be apparent. The lower part bends forward over the umbilical edge with a salient curvature, and is continued by a reëntrant curve, which, also, takes a forward direction over the edge of the abdomen to the base of the keel. The pilæ lose something of their former prominence near the geniculæ, and conform more decidedly to the curvature of the sides of the whorl.

The keel itself on the seventh volution measured nearly one half of an inch, and the whorl nearly two inches.

No. 7. The largest of this lot hardly reaches beyond the sixth whorl. Though differing considerably at first sight, they are really still younger specimens of No. 6 than No. 5, with the casts of the pilæ better preserved. The umbilical edge slopes sharply inwards, the sides incline outward to the edge of the abdomen, and then slope with a reëntrant curve to the base of the keel. The envelopment extends only so far as to cover the abdomen and perhaps the tubercles on the geniculæ. The pilæ conform to the curves of the umbilical edge and then rise gradually to greater prominence and acquire tuberculated geniculæ upon the edge of the abdomen, with extensions which raise folds upon the abdomen reaching nearly straight across to the base of the keel. The keel itself is very prominent, and possesses the same remarkable thinness and prominence observed in No. 6. All the specimens, with one exception, have the pilæ evenly developed and equally prominent, but in this one the same alternation of large and small pilæ may be observed as in No. 5.

Thus there can be little doubt of all four of these varieties belonging to one and the same species. For the specimen last described differs from all other specimens of No. 7 only so far as it agrees with No. 5, namely, in the alternation of the pilæ; and No. 5 differs from it only in those characteristics which it shares in common with No. 6, namely, untuberculated geniculæ much less prominent than in No. 5, and rounded umbilical edges with sides which slope evenly to the base of the keel as in No. 6.

The general aspect of a section of one of No. 5, the amount of envelopment, and the outline of the whorl, closely resemble Marcou's figure of *Ammonites Gibbonianus*, found in Texas, and there is a faint resemblance to Lea's original, but miserably inadequate figure and description of a fragment of the same species found in New Grenada.

It may, perhaps, excite surprise that the Ammonites noticed in this report are not published under different generic appellations from those usually employed, and this indeed calls for some explanation on my part. All the genera described by me in the Bulletin of the Museum of Comparative Zoölogy were collected from Liassic beds, and their characteristics were determined by careful comparisons of the young and adult specimens throughout large series of specimens. This kind of work has led to the conclusion that it will not in most instances promote the knowledge of palæontology to describe isolated genera in other formations. When a series of connected generic groups can be delineated standing in their serial relations to each other, and illustrating natural laws of arrangement, or when the diagnoses of new genera, even though isolated, may indicate important facts of stratigraphical or geographical distribution, the readjustment of the older and more comprehensive names and groups may become imperative. The Brazilian specimens, however, evidently belong to new genera, according to my views of the relations of species among the Ammonites, but for the present any change of their well-known names seems unnecessary.

All of the Brazilian Ammonites are either identical with, or so closely allied to, species already described from the Texas beds by Roemer and others, that they cannot be safely separated.

The presence of such well-characterized species as *Natica prælonga*, *Ammonites Peruvianus*, and perhaps of other species on the western as well as eastern side of the Andes-Rocky Mountain chain in Brazil and Texas, indicates the connection between these slopes, either across the Isthmus or west of Brazil, while a cretaceous ocean still flowed over the whole of the northern portion of South America. These facts, when considered in connection with the discovery of a fossil *Ananchytes* on the Isthmus, as recorded by Mr. Alexander Agassiz, have a direct bearing upon a very important question.

The expeditions of the Coast Survey, as is now well known to all naturalists, have established the fact of a remarkable similarity between the present deep-sea fauna and the species of cretaceous genera; and it has been shown that the surface or littoral animals were more or less represented by identical or closely allied species of the Pacific side of the Isthmus. Thus the question has arisen whether or not the closely allied or identical forms are the descendants of Gulf species, which, having migrated through some ancient channel subsequently closed by the rise of the neck of land forming the Isthmus of Darien. Of course the first step towards the solution of this problem would be to prove the existence of a channel affording a free passage to marine animals at some preceding period. This gives great interest to facts like the above, which appear to confirm the conclusion of Mr. Alexander Agassiz, that during the cretaceous period the Gulf of Mexico and the Pacific Ocean were really continuous seas.

I heard it frequently reported, that large fossil turtles were found at Maroïm. One of them I saw at Bahia, and it was nothing but a huge *Septarium*.

The Maroïm limestone is evidently upper cretaceous, and belongs lower down in the series than the white flaggy limestone of Sapucahy. I saw no trace of red sandstone like that of Estancia, which may perhaps underlie the Maroïm limestone, though it may be wanting, as I have seen it nowhere else, and it seems to be a local formation. The country about Maroïm is hilly, but low. The soil is very rich, and a large trade is carried on in sugar, the greater part of which is in the hands of the wealthy house of Messrs. Schramm & Co.

The shore between the mouths of the Cotinguiba and São Francisco Rivers is low, with a few scattered hills, and is of little interest.

The barra of the Rio de São Francisco is, like that of the Cotinguiba, obstructed by sand-banks, and at times the entrance is dangerous, even for steamers.* A hook-shaped sand-spit is seen extending from the southern side of the entrance.

On entering the river, the southern side is seen to be swampy and for a considerable distance covered by mangrove-trees. The opposite side is sandy, and there are some large areas covered by fine dunes of blown sand. Since Halfeld's survey this has much changed its appearance, the dunes have increased in height, burying some of the cocoa-trees to their crowns, and encroaching on the river.* Strongly in contrast with the smooth sweeping outlines of the growing dunes are the irregular conical hills of sand which occupy a large part of the Pontal. After a dune has grown, some of the coarse trailing plants take root, and shrubs and little trees spring up on its surface. These protect the area over which they grow, while the remainder may suffer removal by the wind, forming little conical hills with tufts of vegetation on their tops. The sand is very fine and of a light brown color. Among the sand-hills are large flats, partially occupied by marshes and ponds, and the resort of great numbers of wading and water birds, cranes, plover, and the long-toed *Parra Jacana*.

Just about the mouth a channel called the Rio Parapuca

* Liais says that near its mouth, after the dry season, the São Francisco delivers 2,800 cubic metres of water per second. (*Bull. de la Soc. de Géog.*, 5me Série 2, p. 390.) Gardner in his Travels, p. 104, draws a very discouraging view of the prospects of rendering the São Francisco navigable, and says that here is seldom more than four feet of water on the bar. Though dangerous, it is regularly crossed by the large coast steamers.

† Gardner describes similar dunes at Peba, five leagues north of the Barra do São Francisco.

leaves the main river and empties into the sea about a league to the south.

From the Pontal to Penêdo, a distance of about twenty-five miles by the river, the shores of the river are very low and flat, and there are many large islands. Ascending some distance the mangrove swamps disappear, and a growth of aninga is common, while the giant ubá grass sometimes covers considerable tracts. These islands and the neighboring flat lands are partially wooded, but the growth is neither so luxuriant nor so thick as that which characterizes the Mucury and Doce in the south, or the Amazonas in the north. The river-banks are quite low, and large tracts are every year covered for a time by the waters of the annual freshet, which, however, deposit a fresh coating of mud over the surface, adding to the fertility of the region. The banks are, in some places, seen to be composed entirely of sand, but usually there is a superficial stratum, more or less thick, of massapé, or yellowish and brownish alluvial clayey soil, which is very fertile. These flat lands are exceedingly well adapted for cultivation, and are especially suitable for sugar-cane, of which, however, little is as yet planted. There are several settlements on these lands, of which Piassabossú is the largest, where there are several *engenhos* for the manufacture of sugar and cachaça.

The higher lands begin a short distance below Penêdo on the right bank of the river at Porteira, and consist, so far as I have been able to see, of cretaceous rocks and outliers of the great coast tertiary sheet.

At Aracaré, a prominent rocky point just below Villa Nova, I found a series of beds much broken up, and about whose stratigraphy, from my time having been occupied in a search for fossils, I do not feel quite sure. The spot is of considerable interest, because rocks are found there that I have not seen elsewhere.

There are beds of a light yellowish or brownish fine-grained, shaly, micaceous sandstone, in which I found a multitude of fossils which are almost, if not quite, undeterminable. Most abundant is a little bivalve which has filled some layers, but which has left only empty moulds of the valves; in addition to these are what appear to be the spines of fishes and fragments of plants. There are some layers of a light-colored shale, in which, however, I found no fossil remains. These beds are considerably inclined, but I omitted to take an observation of dip and strike.

The shore is encumbered by great masses of a considerable variety of rocks, some of which I did not see *in sitû.* Of these is a light, porous, argillaceous, warm red sandstone, which resembles somewhat the sandstone of the tertiary hills near Pitanga on the Bahia Railroad, and with this are associated large masses of a coarse sandstone and conglomerate coated by clay and oxide of iron, in which quartz and agate pebbles are found, which rock also appears to be tertiary, so that I am inclined to think that we have here overlying the fossiliferous sandstone and shale fragments from the now generally denuded tertiary sheet. The fossiliferous beds I believe to be the upper members of the series of sandstones of Villa Nova and Penêdo, about to be described, and which I regard as cretaceous.

I found here numerous fragments of a rock with a sort of oölitic structure which is very interesting; when a fresh, undecomposed species is broken, it is seen to be made up of round or irregularly spherical masses of a granular brown quartz, about the size of coarse duck-shot, filled in with a

VILLA NOVA AND PENÈDO.

cement of a bluish, translucent chalcedony, in which are bedded very much smaller masses. These shot-like grains appear to have been formed by the filling up of globular cavities by quartz, for some of them are minute geodes, still hollow in the centre. Each has a thin concentric coating of milky chalcedony. In decomposing, the cement of the grains becomes white, and sometimes is first removed, leaving the grains projecting. At others the material composing the grains is dissolved out, leaving a honeycombed surface. It is a curious rock, whose formation I do not feel prepared to explain. It looks more like a pseudomorph after oölite than anything else.

At Villa Nova we find the low, rocky bluff point on which the village is built composed of thick beds of a fine, sharp, whitish or slightly yellowish sandstone without fossils, with well-marked oblique and irregular lamination, and a strike N. 50° E., a dip of 15° – 20° S. 40° E.

The city of Penêdo is built on the Alagôas side, at the foot and on the side of a ridge which runs from the left bank of the river towards the northeast. This ridge has a steep slope to the southeast, while for some distance on the northwestern side it is precipitous and about fifty feet high. The rocks exposed on the shore and in the cliffs at Penêdo are of the same general character as those of Villa Nova. A layer of decomposed shale or clay is seen in the cliffs skirting the town, together with some thin bands of a fine shaly micaceous and ferruginous rock. The whole, like the Villa Nova beds, dip to the southeast at a small angle.* I have carefully examined these sandstones for fossils, but have seen only some badly preserved remains of plants.

* Gardner says that the sandstones incline from east to west, which is certainly incorrect.

The hill at Penêdo is covered by red drift-clay, between which and the rock I found a sheet of quartz pebbles, mingled with angular fragments of sandstone.

The Penêdo sandstones are very porous, and of a fine and even sharp grain, which makes them suitable for sharpening tools. The stone is highly esteemed for that purpose, and, owing to the want of sandstones of the same kind elsewhere along the Brazilian coast, it is exported to a small extent, finding its way even as far as Rio.

The same sandstone shows itself at Boassica,* about three miles farther up the river, on the Alagôas side. It has here the same dip and strike and oblique lamination as at Penêdo, and contains occasional layers of pebbles.

On the Sergipe side it is seen again at Coqueiro and Villa Nova, and at one of these places it has been quarried for building purposes at Penêdo.

At Carrapixo, on the Sergipe side, the native civilized Indians manufacture a very good quality of earthenware from the clay of the low grounds.

The city of Penêdo is quite a respectable little town of some 3,000 or 4,000 inhabitants. It carries on a considerable commerce in rice, corn, hides, mandioca-farinha, cotton, &c., &c., which are sent to Pernambuco and Bahia. Every week a fair is held, and a great concourse of people from up and down the river assembles there, bringing hides, coarse sugar, pottery, tobacco, and a host of other articles, which are ex-

* This name, which is applied to a little stream and lake as well as to the settlement, is of Tupí origin, and is derived from *Boya*, a serpent, and *assica*, mutilated, the name of a species of serpent which appears as if it were mutilated. See Tupí Dictionary and *Chrestomathia da Lingua Brazilica.* I do not know what species is meant. The same name is applied to a lake in the province of Rio de Janeiro situated between the Rio Macahé and the Rio das Ostras. *Dic. Geog., sub voce* Boassica.

posed for sale in booths on the broad sand-beach bordering the town. Penêdo is a calling-place for steamers connecting with Bahia, and it has steam communication with the river above as far as Piranhas.

The city is exceedingly well supplied with fish. Among these are Tubaránas, Curimatães,* Piaus, Sarapós (*Carapus*), Piabas of several species, Cachimbaus or Acarís, Piranhas, Pirampebas, &c., &c., of which I made a large collection.†

I extract from my journal some notes on the color, habits, &c., of the Piranha ‡ (*Pygocentrus*), which seem to be of interest.

This species of Piranha, according to the testimony of the natives, is confined entirely to the São Francisco and its tributaries, though other species of the same genus (or Serrasalmo) occur elsewhere in South America; but I cannot vouch for the truth of this statement.

The Piranha of the São Francisco is strictly a fresh-water fish, and it occurs not only below the falls of Paulo Affonso, but also above them. It descends the river quite to the salt water, but never goes into the sea. None of those I saw at

* I have spelled this name as I heard it commonly pronounced. Bates writes it *Curimatá*, and so does Fonseca in his dictionary. The Tupí dictionary gives it *Curymatá*. This fish belongs to the genus Anodus, and several species are very common in the Brazilian rivers.

† All the fishes collected by me on both journeys are in the hands of Professor Agassiz. When the above was written Professor Agassiz intended to contribute to this volume a series of articles and notes on the fresh-water faunæ of the coast I explored, but illness has prevented his preparing them.

‡ In the Tupí dictionary the word "Piranha" is translated *scissors*, and most writers seem to suppose that the name was given to the fish because of its scissors-like jaws. The Tupís knew nothing about this implement before the coming of the Europeans. Piranha (root *Pira*, fish) is an ancient Tupí name, and it was doubtless afterwards applied to the scissors because they bit like the Piranha.

Penêdo were more than twenty inches long, but the fishermen say that they sometimes grow to the length of two feet.

The upper half of the body and head of a specimen just caught were, when seen from above, of a dull and rather dark bluish-gray or lead color. The lower half of the head and body have as a ground color an opaque white, over which is a wash of a clear bright gamboge-yellow, deepening in some spots to a rich orange. All the young Piranhas I saw had the belly of a rich red-orange or blood color. On the sides the yellow is sometimes shaded with light gray, and the yellow or orange tint extends itself upward in irregular lines over the dark gray of the back. The pectorals are light orange-yellow, the tint deepening in the lower and middle part of the fin. The anal fin is on the thickened base grayish, washed with a light clear yellow tint. The border is light purplish-brown. The dorsals and caudal are a dark, dull bluish-gray. The iris is pearly white, with a cloudy patch of black above and below the pupil.* During the freshets the water overflows the low grounds and swamps, and the different kinds of fish leave the river proper and enter the lagoons and quiet overflowed places to spawn.

* Gardner (Travels in Brazil, p. 96) describes the Piranha fish as follows: "It is commonly about a foot in length, but sometimes it is as much as two feet long, being very much compressed laterally and very deep; the back is of a dark brownish color, and the belly yellowish-white, both being thinly marked with reddish spots; the lower jaw projects a little beyond the upper, and both are armed with about fourteen flattish triangular-shaped teeth, upwards of a quarter of an inch in length and very short." This description appears to refer to a different species from the one I describe. Humboldt in his Travels, Vol. II. p. 167, speaks of the Piranhas, or *Caribes* of the Orinoco, as having "the belly, gill-covers, and the pectoral, anal, and ventral fins of a fine orange hue." My specimens from the São Francisco still preserve their orange color. The São Franciscan species appears to be much larger than those of the Amazonas and Orinoco.

The fishermen said that the Piranha also leaves the river and chooses a shallow spot with a sandy bottom to deposite its eggs. Stooping down, the fisherman with whom I was conversing one morning took up a Piranha lying before him, and showed me just how the eggs were laid. The fish having chosen the proper spot, sweeps away the sand with its tail and anal fin, so as to make a little saucer-shaped depression four or five inches wide. The eggs, of about the size of mustard-seed, are then laid in the nest in a ball, two or three inches in diameter. This accomplished, the mother fish takes up her position near the nest, and keeps watch over the eggs until the young are hatched, for the Piranha has enemies in the hungry little Piabas, which swarm about in countless numbers, and from which she must defend her eggs. Looking over a large lot of Piabas in a canoe, I found it difficult to obtain a perfect specimen, on account of the mutilation of the tail and fins. "This is the Piranha's work," said the fisherman, "and the marks of the terrible teeth of the monster." The larger fish of the river also bear similar scars.

The laying, according to the fisherman, takes place principally in October, or soon after the freshets set in. During this time the Piranhas are especially fierce.

One fisherman described in a vivid manner his finding a Piranha watching its nest in a shallow place by the river-side. "I thought to catch it," said he, "and waded softly into the water to put a basket over it, but I was not quick enough. The fish darted at me and took a piece out of my leg. Look there!" and rolling up his pants he showed a pair of crescent-shaped scars left by the fish's jaws. The fishermen are often bitten, and almost every man present had scars to show, either on the arms or legs.

The fishermen united in saying that just as this fish is about to spawn the color of the belly changes from yellow or orange to the same color as the back, but that soon after the eggs are laid the original color returns.

The Piranha frequents rather the deeper parts of the river, and abounds in the eddies among the rocks, but I have seen it caught, as at Penêdo, close in shore, where the water was not very deep and the bottom was sandy.

During my stay at Penêdo a poor little idiot sitting on the pier, having been frightened by a cannon, fell over into the river. The next morning the Piranhas caught in the vicinity were found to contain portions of his body. There are numerous well-authenticated cases where persons have been attacked by the fish while bathing and devoured. Only a short time before my visit to Penêdo a young lady was thus attacked and eaten. A horse fording the river slips and wounds himself on a stone; the Piranhas, attracted by the blood, crowd about in great numbers, each cutting out mouthful after mouthful of flesh, until, in many instances, the voracious creatures have been known to devour the entire animal in a few hours. They sometimes throng about their prey in such numbers that they may be seen leaping one on top of the other out of water, in their eagerness to get at it.*

So far as I could learn, these fish appear to be particularly

* Bates speaks of the great swarms of Piranhas in the Amazonas. Humboldt (Travels, Vol. II. p. 167) says of the Piranhas of the Orinoco: "The Indians dread extremely these caribes; and several of them showed us the scars of deep wounds in the calf of the leg and in the thigh made by these little animals. They swim at the bottom of the rivers; but if a few drops of blood be shed on the water they rise by thousands to the surface, so that if a person be only slightly bitten it is difficult for him to get out of the water without receiving a severer wound."

dangerous only during the spawning season. During my voyage on the São Francisco I saw everywhere bare-legged women standing in the water on the banks washing, and that not only in sandy and quiet localities, but among the rocks, as at Propriá, while I repeatedly saw men wading in the water and boys bathing in the river.

The Piranha is much esteemed for food, and may ordinarily be found for sale in the market at Penêdo.

There is a species of Serrasalmo (?) found at Penêdo called *Pirampeba*, the name being evidently a compound of the two Tupí words *Piranha* and *peba*, or the flat Piranha. This species is smaller than the last, very much flatter laterally, and of a silvery-white below. I did not learn that it possessed any of the voracious propensities of the Piranha.*

Between Penêdo and Propriá,† a distance of about six Brazilian leagues up the river, the Sergipe side is bordered by low hills, some of which are irregular and isolated, and composed of the sandstone above described. The country back from the river is flat, moderately elevated, and appears to be, in part at least, tertiary. On the Alagôas side, for some distance above Boassica, there is a considerable stretch of low meadow-land. At Morro Vermelho, on the Alagôas side, cretaceous sandstones show themselves on the shore, but the dip here is approximately to the N. W. 15°. They are covered by red drift-earth and great quantities of quartz pebbles, and at Prazeres, half a mile farther up, we again see the same sandstones with a dip of about 18°. In this vicinity the massapé lands bordering the river are about fifteen feet in height.

* Gardner speaks of seeing dried sturgeon exposed for sale in the market at Penêdo. There are no sturgeons in South America.

† This is, I believe, the correct orthography, but it is often written Propiá.

Mandioca flourishes well below Propriá, and is planted as well on the alluvial river-banks as on the hillsides; but on the low grounds covered by the enchente it cannot come to maturity. It is usually planted in February or March, and is generally fit for use in a year, when planted on the uplands. It is also cultivated to a considerable extent on the sloping river-banks. Just before the enchente it is pulled up, and is eaten sometimes when not more than six inches long. The river begins to rise in October.

One mile below the town of Propriá, on the Sergipe side, is a small hill called the Morro do Chaves, or, as Halfeld has it, Morro do Eusebio. This hill, which has rocky sides toward the river, I examined in company with my friends Drs. Brunet and Lacerda. The rocks composing it consist of a series of limestones, conglomerates, shales, and sandstones, the whole of which have a strike E. 15° S., and a dip of about 20° to the N. 75° E. The lower strata consist of thick beds of limestone, calcareous sandstone, and conglomerate, some of the layers of which are made up of shells. These beds are well exposed on the side of the hill nearest the town.

This limestone is usually more or less arenaceous, and often contains grains and pebbles of the underlying metamorphic rocks, so as to form a calcareous conglomerate. In some places it is crystalline and metamorphosed. It is used to some extent for burning into lime, but the most of it is very impure.

The shells so exceedingly abundant in some parts are lamellibranchs, about one half to three quarters inch in diameter, and with very thick valves. I fear that even the genus is wholly unrecognizable. Overlying these limestone beds are shales, rather soft, not well laminated, calcareous, micaceous, and of a greenish color, in which I found a large quantity of bones of teleostean fishes, and an impression which had the outline of the tooth of a *Notidanus*. A further examination of these shales may reveal something of more interest.

Going around the hill we come to a break which extends from the top of the hill to the bottom, and in which we find a confused mass of broken pieces of sandstone.

This series of rocks I have referred to the cretaceous. Farther up the river we see no mesozoic rocks.

On Halfeld's map a note says that in the Morro do Eusebio there is a *camada de cal em gneiss granito*, a bed of limestone in gneiss-granite. This is not correct, as is seen from the above description. The limestones, shales, &c., overlie beds of clay slate, which I saw very badly exposed on the river-side. There appears to be a distinct slaty structure to the rock, but I did not see enough of the formation to enable me to make out the bedding. I have seen no rock like this elsewhere on the coast of Brazil.

Propriá itself is built on the river-bank, and has in front of it ledges of flaggy gneiss and mica slate, with a northwest strike* and a dip toward the southwest. It is only a small village of little importance, because the lands of the vicinity are not very productive, yet it exports some cotton, hides, &c. A little lake near the town is said to abound in fish, and to furnish a considerable revenue to the village government. Opposite Propriá the lands are

* This strike is almost at right angles with that of the rocks of the Serra do Mar, but it corresponds to that of the gneiss hills between Cape Corrientes and Tapalquen, and of the gneiss near Montevideo, in which the direction is W. 25° - 30° N. D'Orbigny thinks that the system of upheaval by which these extra-Brazilian rocks were disturbed was very nearly as old as that which disturbed the gneiss of the Serra do Mar.

very low and flat over a very large area. Behind them are seen the tertiary chapadas lying a few miles off. When the enchente comes on the river rises some fifteen to twenty or more feet here, and converts the low grounds just spoken of into a magnificent lake-like expansion.

Between Propriá and São Braz the land is still low, but the hills are of gneiss and rounded. The rocks frequently show themselves along the river-side, the dip being usually up stream, though at Agua Comprida I observed a dip down stream.

The hills in the vicinity of Lagôa Comprida are 300 – 400 feet high, with rounded outlines and steep sides, which are well wooded. Farther up the river-banks grow higher on the Sergipe side. There are very irregular cliffs of gneiss, broken and rough, and almost like those of our northern gneiss regions. The country in the vicinity is, generally speaking, not very fertile, but, on the Sergipe side, two or three miles from the river, the hillsides are highly cultivated, presenting an unusual and pleasant appearance. The same gneissose and mica-schistose rocks, often traversed by large veins, are well seen in the bluffs. Their strike varies little, and the dip is southwest, varying from nearly horizontal to forty-five degrees. The river thus follows the same general direction as the strike of the rocks.

In front of Traipú, on the opposite or Sergipe side, there is a range of high hills, with steep bluff sides towards the river, and a long slope from it. It is composed of gneissose and schistose beds, which, inclined at a rather high angle, dip away from the river. Lying on the southern slope of these hills is a thick bed of a compact rock, which, in the steep rocky sides bordering the river, as well as in the transverse valleys separating the hills, forms a line of bluffs.

Traipú is a little village of no importance on the Alagôas side. At this place a specimen of specular iron ore, said to have come from near Pão d'Assucar, was presented to me. It is equal in richness to the Swedish iron ores, and if in workable quantity would be very valuable.

TRAIPÚ FROM NEAR MARCAÇÃO.

At Marcação flaggy gneiss, very regular in stratification, is exposed on the shore. Strike N. 35° W. Dip 35° S. W. A short distance above Traipú the same rocks with a north-east dip are seen on both sides of the river.

The country above the Serra de Tabanga becomes every mile more and more rocky and barren, while the vegetation becomes more sparse, consisting of small bushes with a great abundance of bromeliaceous plants of several species. Of these last there is a very common one known as macambira, with narrow leaves, bearing along their margin long hooked spines arranged wide apart. This plant furnishes a strong fibre, and during the dry season its roots contribute largely to support the cattle.

Several species of *Cereus*, some of which attain the size of large trees, grow thickly over the rocky hillsides, and form one of the most characteristic features of the vegetation. One of these is the Chique-Chique, so common on barren and arid lands in the interior, and so often described by travellers.

One never sees the Chique-Chique in the zone of forest and fertile lands bordering the coast, but as the forests disappear it makes its appearance, and sometimes grows to the height of forty feet. Melocacti * also abound, together with opuntias.† With these are associated many species of croton and of sapotaceous plants. There is a species of *Bignonia*, *B. Tecoma* Mart., very common on the hills and shores, sometimes growing to a height of forty to fifty feet, with a trunk three or four feet in diameter. Its foliage is very light and the flower is yellow.

The Joazeiro (*Zizyphus Joazeiro*), a very beautiful tree with dense heavy green foliage, is quite frequently seen on the river-banks, together with species of *Azolla*, *Mimosa*, *Geoffroya*, *Peltophorum*, &c.

Along the river here the meadow lands have become very much narrowed down, and occupy only little bays among the hills, though sometimes these are only shut off from the river by beaches, and form lagôas, numbers of which we find all along the river. The elevated country is of use only for pasturing. The thinness of the soil is not the only reason why the vegetation is so sparse and peculiar. The surface is yearly burned over during the dry season, when the cattle

* One of the most interesting of these curious cactuses is a large species described by Gardner under the name of *Melocactus Hookerianus*.

† Gardner speaks of the occurrence of a species of cochineal insect on the leaves of these plants. Dr. Brunet called my attention to the same fact, and assured me that there was no reason why the insect should not be successfully cultivated.

feed on the cactuses and the roots of the bromeliaceous plants.

The meadow grounds are sandy and not very fertile, but during the time when the river is low, mandioca, rice, beans, cotton, mandubi * (*Arachis hypogœa* Linn.) or the pea-nut, are planted, and in front of the numerous *fazendas de gado*, or cattle estates, and villages the green plots of these plantations bordering the river appear in cheerful contrast with the hills behind, which are scattered all over with loose blocks of stone, and bristle with cactuses.

The country in the vicinity of Curral das Pedras is rolling, and not very high.

At Jacobina there is a lagôa where much rice is planted. One man makes more than 1,000$000 per year by renting it for this purpose.

At Intães there is a group of high hills near the river, while others are seen in the distance on both sides, but the country continues with the same general character nearly to Pão d'Assucar. At Lagôa Funda, Alagôas side, strike N. 30° E., dip. N. W. 3° – 40°.† Flaggy gneiss.

Looking ahead from this point there is a very fine view, the country still presenting the same low flat lands, but beyond we see the irregular tops of higher hills, which appear to be of a different structure. At Cajueiro (Alagôas) strike † N. 36° E., dip 30° N. W.

Passing the island of São Pedro the river narrows while

* This curious and well-known plant is a native of Africa. In Brazil it goes by the name *Amendoim, Mandubi,* and *Mandubim.* The second form appears to be the correct one. The name is of African, not Tupí, origin. This plant is largely cultivated in Brazil, and is used for the making of sweetmeats. It produces an oil employed for burning and the manufacture of soap.

† Both these observations were made as we passed close along shore in the steamer, and were carefully taken. Halfeld makes the strike northwest.

the hills become much higher and excessively rocky, their sides being covered with blocks of stone. The rock is still gneiss, traversed by many veins.

A little beyond, the view opens out, and one looks up to the town of Pão d'Assucar, built on the low ground bordering the shore, with a wall of high hills in the background, and over them, seen away in the distance, are the blue tops of the Serra do Pão d'Assucar.

Pão d'Assucar is a considerable village built on the Alagôas side of the river, on a high, narrow strip of alluvial ground. In this vicinity these alluvial lands have a considerable extension, and are well planted.

LOOKING DOWN THE RIVER FROM PÃO D'ASSUCAR.

The rocks at Pão d'Assucar are gneiss, but siliceous and flaggy. At the upper end of the village there is a high, sugar-loafed hill, which gives the name to the locality. This is one of a number of hills which together form a range running southeast, crossing the river here, being continued on the opposite side. The gneiss is vertical, and has a strike of N. 40° W. It is seen in the same position in a prominent bluff opposite the town on the Sergipe side. Between the town and the hills behind are some quite extensive lagôas, on the borders of which rice is planted. I ascended the Pão in company with Mr. Brunet. From the top is one of the finest views I have seen in Brazil.

In the vicinity of Pão d'Assucar are very many cattle fazendas, to which belong great numbers of cattle, and on which cheese (*requeijão*) and hides * are manufactured.

CATTLE FAZENDA AND GARDEN NEAR PÃO D'ASSUCAR.

Immediately above Pão d'Assucar the river narrows much, and the banks become still more steep and craggy, varying in height from two to three or four hundred feet. The hills bordering the river seem to be conical or dome-shaped, but are really in most cases the ends of ridges cut across obliquely by the river.

In proportion as the river narrows, the alluvial deposits on its banks grow higher, and at Entre Montes, an exceed-

* The hides are tanned in stone vats, of which each fazenda possesses one or more. The process is as follows: The hides are cut in two lengthwise and soaked in the river. They are then placed in vats in alternate layers of hides and the ashes of *Barauna* (Melanoxylon) or *Catinga de Porco.* At the end of three days the hair is removed. They are then scraped and placed once more in the vats with the bark of *Angico* stamped with water. This bark is removed three times in the tanning of the best skins, the bark remaining each time a fortnight on the hides. The process is finished by washing the skins and extending them over poles in the sun, when they are pressed into boxes for exportation. Many goat-skins are prepared in the same way. At Penêdo I saw a shoemaker blacking leather by rubbing it over with mud from the bottom of a pond near by. He assured me that he used no other preparation, and that the mud alone gave a rich black color.

ingly picturesque little village built in a notch among hills some seven hundred feet in elevation, these flats are, I should judge, at least fifty feet high. At Allegria and Coleite the rock is red, very homogeneous, and compact. The river

VIEW NEAR ALLEGRIA.

becomes so narrow that in some places it is not more than four hundred feet across, while the precipitous rocky walls, three or four hundred feet high, make the scenery of this part of the river exceedingly fine and beautiful.

Porto das Piranhas is a miserable little village built on a sand and gravel bank on the Alagôas side of the river, at the foot of the hills, which rise with steep sides to a height of about seven hundred feet, if not more. The river here is somewhat tortuous, with rocky shores and occasional rocks and ledges. At the upper end of the village is a sugar-loafed hill which I found to be composed of gneiss, — strike N. 20° W., dip vertical.

I had been told that the country on top of the hills is flat. I climbed in a miserable rain-storm to the summit of the steep slope behind Piranhas, a height of wellnigh seven hundred feet, from which in the intervals between the showers I was able to look over the whole country. Instead of finding a chapada or absolute plain like that of the tertiary regions, as I had been led to expect, I found the general surface country remarkably even, but consisting of

VIEW LOOKING UP TOWARDS PIRANHAS.

a great number of very low ridges whose summits all came to nearly the same level. In the distance on both sides of the river serras or short chains of high, irregular hills were visible. The whole seemed to be formed of gneiss and other metamorphic rocks. The surface of the slope and top was thickly strewn with blocks, usually angular, and of the same material as the surface on which they lay; however, I did observe a number of boulders of a red syenite lying on gneiss; but they could not have come from far, for I saw the same material in place quite close to the edge of the river valley above. Soil there was scarcely any, and the red clay and pebble sheet were absent. Rounded boulders were abundant. It is not possible that these boulders could have been the result of decomposition, for this action has obtained here to only a very slight extent. My stay at Piranhas was necessarily limited to a few hours, which I very much regretted, for I have seen no region on the Brazilian coast where polished and scratched surfaces would be more likely to be preserved than here.

Halfeld gives it as his opinion that gold may be found in the vicinity of Porto das Piranhas, but I do not consider it probable. The same author describes the country from Piranhas to the celebrated falls of Paulo Affonso as being remarkably even, and composed of *granite*, which is, I suppose, compact gneiss, or, in part at least, syenite; the rock over which the falls precipitate themselves being of the same material. My good friend Mr. Franz Wagner of Bahia, who tarried behind our party to visit the falls, said that the rock was sandstone. Burton speaks of sandstone as occurring there, and of a conglomerate at the rapids of Itaparica, a short distance above the falls.

On the road to the falls from Porto das Piranhas is a serra called Serra d'Olho d'Agua, where occurs a sandstone. This rock, according to the reports of some friends who visited the falls shortly after I left Piranhas, is white and much denuded. Halfeld represents it as dipping irregularly from the hills northward and northwestward, and he indicates a locality where sandstone occurs at the *Cachoeira Cancamunhi d'Acima.*

The Falls of Paulo Affonso are situated about 56 leagues, or 168 miles, from the sea.* I have not been able to visit them, so I translate a few selections from Halfeld's description of the cataracts and of the neighboring country: † —

"The first fall, 44 palms 6 inches [about 33 feet] in height, throws itself into a basin garnished by granitic rocks almost perpendicular, and sometimes even overhanging the river; from

* Liais says 300 kilometres.

† I quote from his *Exploração do Rio de São Francisco.*

this basin the river makes a sharp turn to the left at a right angle, and precipitates itself between steep rocks into the bottom of an abyss 66 palms and 1 inch in height, transforming themselves in consequence of this leap apparently into milk foam, casting and

CACHOEIRA DE PAULO AFFONSO.*

blowing up, similar to the explosion of a mine, great masses of water into the air, which are turned into a vapor that rises still higher. . . . . Transferred by this fall into a river of milk, the waters precipitate themselves in great billows and waves, and between towering masses of granite, beating at a right angle against the left bank of the river. This side consists of a native granite rock, which is 365 palms [about 250 feet] high above the surface of the water, and having still 120 palms of depth.†

* I regret not to be able to give a better view of these falls, but the photograph from which it was drawn was not a good one. I am assured, however, it gives a better idea of the falls than any other sketch yet published.

† "Resserré entre deux immenses murailles de pierre, il coule d'abord en torrent et sur un fond dont la déclivité accroît la vitesse, puis tout à coup il se précipite en trois chutes consécutives dont la hauteur réunie et de 84 mètres.

"The impetus with which the waters precipitate themselves against this wall makes them constantly ascend and descend from the point of contact with the rock. On the right they descend in a right angle to the bed of the river below; but to the left, as they have no way of egress, they produce, in consequence of their advancing and retiring movement, a come-and-go like the waves of the sea on the shores, from which has resulted, for thousands of years up to the present time, not only the wearing away of the rock, and the formation of a little bay, but even of a cavern in the rock, which is 444 palms in length, and whose mouth is 80 palms high and 40 broad, divided in the interior into two great halls, the dwelling-place of thousands of bats, and for this reason called Furna dos Morcegos. . . . .

"The rock in which this *furna* is formed, as well as in all the extension of the Cachoeira, is of the hardest granite, of a fine grain, and in truth it is incomprehensible how it should be possible for the waters to form such a cavern in a rock of so great hardness. I am inclined to attribute this fact to the circumstance that the granite in the direction of the cavern, as well as in that of the river, from the mouth of the cavern below to the Riacho da Gangorra, presents many veins of calcareous spar, of flesh-colored feldspar and quartz, which have a thickness of $\frac{1}{4}$ to 5 inches. The granite by the sides of these veins is less hard, and sometimes is decomposed and saturated with muriate of soda [salt], in such abundance that those living near the fall mine this stone on a small scale to extract the salt. . . . . I am inclined to think that the circumstances now indicated with respect to the ready decom-

La dernière de ces chutes, la plus grande des trois n'a pas moins de 10 mètres d'altitude. . . . . La compression de l'air à la surface des eaux après la chute est telle, qu'une pierre lancée avec la plus grande force ne peut résister au vent résultant, de sorte que sa vitesse est anéantie après un parcours de 6 – 7 métres. Cette particularité a répandu, parmi les habitants des environs l'opinion que le lieu de la cascade est enchanté." (Liais, *Bulletin de la Soc. de Géog.*, 5me Série, XI. pp. 390 – 392.)

position of that rock, over the breadth which comprises all the veins mentioned, probably determined the excavation of the cavern below, — a circumstance which would have given place to the formation and present existence of the falls, whose bed is really excavated in the rock, since, for a great distance on both sides of the steep banks of the river, the soil presents a plain without hills or serras, which could have determined the cataract of Paulo Affonso.

"From Paulo Affonso down the river to Porto das Piranhas, the waters of the river are narrowed and run with many falls between steep rocks of 350 to 800 palms in height, which, with the exception of a very few places, as at Porto do Salgado, Monto Escuro, Sitio Novo, &c., are inaccessible to allow of descent to the river. These cliffs are commonly called here *Talhado* [or Talhadão], since the width of the river is generally reduced to a few hundreds of palms and sometimes less, as in the falls of the Garganta, where the river is only 85 palms wide, and runs like a mill-race between perpendicular walls of rock 350 palms in height."

The falls of Paulo Affonso are of the same class as those of the Jequitinhonha at Salto Grande. In both cases a large river just before it reaches the edge of the plateau, and flowing a broad stream with a wide flood plain, dashes first down a slope forming a series of rapids, and then, a few miles farther on, contracting suddenly, plunges in a magnificent series of cascades and rapids into a narrow valley, in which, with a swift and rock-impeded course, it descends to the coast plains, where it spreads out widely and flows calmly, proudly on to the sea. The river is, so to speak, strangled in the *Estreito*, or *Talhadão*. As the rocks of the regions cut through by the falls of Salto Grande and Paulo Affonso are both crystalline and highly inclined, it is manifest that their retrocession must have operated in a very different manner from that of the falls of Niagara, and that the time occupied by the excavation of the gorge below the Brazilian falls cannot be estimated according to the same rule. I should judge that the falls of Paulo Affonso are finer than those of Salto Grande at ordinary times ; but when the Jequitinhonha is swollen the Salto must be a grand sight. Both these falls, as navigation on the coast and on the rivers becomes more prompt and commodious, will erelong become well known to travellers. Burton, who visited Paulo Affonso in 1867, says, that if Niagara be the monarch of falls, Paulo Affonso is the king of rapids. Liais thinks that, seen close at hand, the latter exceed the former in magnificence.*

Halfield, speaking of the Riacho da Vaca, says: † "At its mouth on the *south* side of the river is the Lagôa da Pedra, where I found the fossil bones of a mastodon. The lagoon, which consists of a concavity or basin, is surrounded by great cliffs of the said rock, and is seventy paces in length, ten in width, and ten to twenty-five palms in depth. It was full of earth, sand, and gravel in beds. Of these the lower about twenty years ago contained the bones of a mastodon. The neighbors, residents of the Lagôa da Pedra, had commenced

* "Vue à distance la cascade de Niagara l'emporte donc en magnificence sur celle de Paulo Affonso, mais de près, l'avantage est pour le San-Francisco, dont les eaux furieuses se relèvent avec plus de violence et forment une série d'immenses vagues chargées d'écume. L'éffet de ces grandes vagues, d'où sort, comme de la chute elle-même, une gigantesque colonne de vapeur, ajoute à la splendeur du spectacle, et la force expansive de l'air que les eaux, dans cet étroit canal, entraînent et compriment au pied de la chute, produit une sorte d'ouragan dont la puissance contribue à accroître l'extension de cette immense colonne de poussière aqueuse." (Liais, *Bulletin de la Soc. Géog.*, 5me Série, XI. p. 391.)

† Description of the 328th league.

to dig out this cavity, in order that it might be made to serve as a reservoir for the water of the rains, and as a drinking-place for cattle ; there then appeared at the cutting the bones of a mastodon, which they threw outside the basin, but the intended excavation was never completed, and two thirds of its length remained full of soil in which, at the head of the ancient opening still appeared the points of bones of large dimensions." From this vicinity, a few years ago, an immense collection of bones, teeth, &c., of mastodons were collected and sent to Rio, where they may now be seen in the Public Museum.

"The country for a considerable distance around the hill mentioned presents an extensive plain, over which are found dispersed thousands of enormous loose rocks of granite, and sometimes superimposed one on the other, and set sometimes only upon a point or very small base without falling."

I have already remarked the character of the country above the Serra de Tabanga, the want of a soil, the way in which the surface is strewn by blocks of stone, and also the little amount of disintegration which it has suffered. I cannot conceive how these loose masses could have resulted from decomposition, without showing some evidence of it in a much decomposed surface and an abundant soil. The surface of the country looks precisely like that of our drift-covered regions of the North.

Halfeld states that granite occurs as the foundation rock of the country for many leagues up the river beyond the falls. The term as used by him is rather too comprehensive, and may comprise gneiss and syenite.

The Serra de Itaparica, which forms a long, sharp-backed, narrow ridge 720 palms high, which runs from the river nine leagues above the Cachoeira de Paulo Affonso to the westward for several miles, is, according to the same author, composed of sandstone. "The sandstone in the top of the serra is of a fine, but the lower beds are of coarse grain to the base, which is situated upon granite of extreme hardness, where it forms beds of coarse gravel." * This sandstone is doubtless of the same series as that of the Serra d'Olho d'Agua. Many of the other hills of the vicinity are also composed of this sandstone.

The whole country was evidently once overspread by a sheet of this rock, and has subsequently suffered very extensive denudation.

The climate of the Lower São Francisco presents some interesting features. On the coast rains are frequent and plentiful, and along the shore, as already remarked, there is a belt of forest. Inland, however, the climate becomes more dry; three or four months of the year — June, July, and August — are usually without rain, and everything dries up. The heavy rains, as a general thing, begin in March. The river rises to a great height, and sometimes overflows the high banks on which the towns of Propriá and Piranhas are built, inundating the lower stories of the houses. At both places I saw the muddy line along the fronts of the houses left by the last freshet. Such an immense volume of water pouring tumultuously through the racecourse-like channel between the falls and Piranhas, and, as at Propriá, spreading out over and inundating the extensive low grounds, converting them into a great lake, must be a grand spectacle. The annual overflow takes place in October and lasts until March, during which time it is raining in the highlands.

Nothing strikes one more strongly on the São Francisco

* *Exploração do Rio de São Francisco,* p. 44.

than the regularity with which the winds rise and blow. In the morning I used to look out of my window at Penêdo; all was still, and the river was without a ripple. Canoes and montarias dropped down on the current, and all was repose. About nine o'clock * a breeze from the sea stole over the water, ruffling its surface with ripple-patches; this increased gradually, until at noon a stiff wind was sweeping inland; the montarias spread their picturesque sails and scudded before it up the river; thus steadily the breeze continued until well into the night, when it hushed down, and a calm morning again dawned on the rippleless river. This sea-breeze is perfectly regular. Boats can sail up the river, but they must drift or be rowed down.

The Lower São Francisco below Piranhas admits of navigation by small steamers and sailing craft during the whole year. In August, 1867, the river above Penêdo was formally opened to steam navigation by the Bahia Steam Navigation Company, but Penêdo has been in communication with Pernambuco and Bahia for some time. Through the politeness of my friends, Mr. Hugh Wilson, the able superintendent of the company, and Dr. Anto. de Lacerda, I was enabled to participate in the *fête;* and to them I owe the opportunities I afterwards enjoyed of making the observations recorded above. To Dr. de Lacerda and Dr. Brunet, who accompanied me on the voyage, I am indebted for valuable contributions to my note-book. I take pleasure in expressing here my gratitude to these gentlemen for their kindness.

The country below Propriá is very fertile, and there are large areas of rich lands admirably adapted to the culti-

* Gardner says that the sea breeze reaches Penêdo about noon, which is not correct.

vation of sugar-cane, cotton, mandioca, &c. The region above Propriá is proper for grazing. The whole Lower São Francisco forms a district of much promise.

The coast of the province of Alagôas is formed by a broad belt of tertiary lands of the same general character as those south of Bahia. The country embraced in this belt is a vast elevated plain two or three hundred feet above the sea. The western or interior part of the province is a table-land of gneiss, the continuation northward of the gneiss district of the São Francisco already described. It is dry, barren, and fit only for grazing.

At Jequiá, some miles south of Maceió, the tertiary bluffs come down to the sea line, forming a long range of cliffs of a bright red color, like the Barreiras do Sirí. In the south, I know of only one lake which occupies a basin of denudation in this tertiary sheet, and that is the Lagôa Juparanãa on the Rio Doce; but in the province of Alagôas there are several such lakes, and it is from them that the province takes its name. These lakes in Alagôas are long and narrow, and have usually a northwest trend. The valleys occupied by the Lagôa do Norte at Maceió, and of the Lagôa do Sul a short distance to the south of Maceió, open out broadly to the sea; but a strip of sand more or less wide, extending across their mouths, bars out the sea. The Lagôa do Norte at Maceió is salt. It abounds in fish, and it is said that sea-turtles are found in it. The Lagôa do Norte at Maceió strongly reminds one of the lakes of Central New York, — Cayuga or Seneca, for instance. These lakes of Alagôas, as well as Juparanãa, are very deep, and their basins must have been excavated at a time when the land stood at a greater height than at present.

The city of Maceió is built at the mouth of one of the

largest of these valleys, and at the foot of the Lagôa do Norte. The entrance to the valley is barred by a wide strip of recently elevated sands, covered with a low and sparse vegetation, consisting of clumps of bushes, among which the aroeira (*Schinus terebinthifolius*), as Gardner has remarked, is very abundant, together with species of *Diospyrus*, *Eschweilera*, *Eriocaulon*, *Marcetia*, *Cereus*, *Melocactus*, &c. The lagôa communicates with the sea by a narrow channel across this flat. The tertiary plains are almost perfectly level, and come down close to the coast line. They have steep slopes towards the lakes, and the same towards the sea, where, at their feet, lie sand plains ; but they are precipitous when the sea washes their base. The average height of these tertiary plains along the coast is about one hundred and fifty feet, though they evidently rise gradually toward the interior as they do elsewhere. From the top of the lighthouse at Maceió, built on the edge of this plain, the eye ranges for a long distance both up and down the coast, and into the interior. Save the depression of the lagôa, the country, appears from the sea to be level ; but inland, at the distance of a few leagues, are visible the tops of a few hills, evidently of some earlier formation, but probably gneiss. The slopes of the tertiary lands, as elsewhere, are fertile and heavily wooded, but the higher plain is as usual exceedingly dry, supporting a dense growth of small bushes, with many *Licurí* palms. The character of the vegetation is, as I have already remarked in speaking of the Rio de São Francisco, not wholly due to natural causes ; for this whole country has been repeatedly burned over, and the virgin forest destroyed. During the political disturbances a few years ago large tracts were fired purposely.

A stranger will occasionally observe in the pavement of

the streets palæozoic rocks containing a few fossils, which might mislead one if he did not know that they were brought thither from North America as ballast. Large quantities of the finer qualities of Rio de Janeiro gneiss are carried along the coast, and used both for building and paving; but it is of so peculiar a quality that one learns to recognize it immediately wherever it is found.

The upper beds of the tertiary sheet are well exposed in a cliff and cutting on the side of the spur on which the lighthouse is built, behind the *Matriz* church. The lower bed seen consists of a soft yellowish or reddish argillaceous sandstone, very loose in texture and full of quartz pebbles, which are arranged in layers and lenticular masses. There are some pebbles of a white substance that looks like decomposed feldspar. The fragments of quartz, whether in the form of pebbles or sand-grains, are more or less rounded. Over this lies a thick mass, very indistinctly stratified, of red, pink, and white variegated soft, friable argillaceous sandstone, of the same general character as that found in the southern part of the province of Espirito Santo, and at Pará on the Amazonas. The colors of these beds are very warm, and are distributed unevenly. Over all is a layer of clay and soil with pebbles beneath, like the usual drift coating of the tertiary plains.

The tertiary coast belt extends, to my knowledge, some thirty to forty miles above Maceió, and I have seen the same kind of a coast fifty miles south of Pernambuco.

The flat-topped hills in the vicinity of Cape Sant. Augustinho are portions of this sheet, which is much worn away in the vicinity of Pernambuco. The country back of Pernambuco is quite hilly, and it is probable that cretaceous rocks will be found there.

Northward of Pernambuco I have seen the same kind of coast in the province of Rio Grande, where the lands are precisely like those of Maceió, the valleys having the same steeply sloping sides so characteristic of the tertiary plains.

The city of Maceió is a town of respectable size, built a short distance inland, in part on a slight elevation at the base of the bluffs on the north side of the valley. On the sea-shore is built the town of Jaraguá, which is the port of Maceió. The flat lands in the vicinity of Maceió are covered with cocoa palms, which give to the place a very pretty appearance. Dendê palms are also numerous, and I saw a few date palms which bore fruit. A few specimens of the Assahí (*Euterpe oleracea*) of the Amazonas are found here.

The harbor is formed by a line of coral reefs, which extend off shore at a distance of half a mile or more, and protect shipping from the northeast winds; but the harbor is not well protected during southerly storms, and, what is worse, it is filling up by the drift of sands over the reef. The shore is a sand beach, from which long piers are built out; but owing to the sea, the shipping cannot come alongside, and goods are landed in lighters. The principal trade of the city consists in sugar, cotton, &c., which are principally sent to Pernambuco. As above remarked, the upper plain appears to be very dry; and in the vicinity of Maceió, at least, is not cultivated. The slopes of the tertiary lands are very fertile, and usually very heavily wooded; as also are the alluvial lands bordering the lake, which last are extensively planted.

The water of Maceió is bad; at the time of my visit steps were being taken to supply the city with good water from the river Bebidouro.

The *Companhia Bahiana* was about to put two little steamers on the Lagôa do Norte.

While at Maceió, I heard a *Schisto bituminoso* spoken of as occurring in the low lands near the Lagôa do Sul. I was unable to visit the locality, but, from the information I received, it seems to be an inflammable vegetable deposit of very recent date, underlying the sands of a plain similar to the lower plain of Maceió.

CHAPTER X.

PROVINCE OF PERNAMBUCO.

The Limits, Area, &c., of the Province. — Its Topography, Geology, Climate, Soils, &c. — The Rivers. — Productions of the Province. — The City of Pernambuco or Recife. — Derivation of these Names. — Situation of the City. — The Stone Reef. — The Port formed by it. — Shallowness of Water along this part of the Coast. — Pernambuco a Calling Station for Foreign Shipping. — The Pernambuco and São Francisco Railroad. — Table of Heights along the Line. — Island of Itamaracá. — Fossiliferous Limestones. — Fertility of the Island, Cocoa Palm Groves, &c. — Fernando de Noronha. — Darwin's Description of the Geology of the Island. — Its Dryness and Sterility.

THIS rich and populous province has been so little explored by the physical geographer and geologist, that it is not possible to give more than a very general sketch of its physical features, — a sketch which the writer hopes future explorations may help him to fill out with needed detail.

The province comprises the northern side of the basin of the Rio de São Francisco from the point called Páo d'Arara to the Rio Moxoto, a few miles above the Cachoeira de Paulo Affonso. East of that point the little province of Alagôas is wedged in between it and the São Francisco. The coast line of the province is only about forty-four leagues in length. The superficial area of the province is variously estimated at from 4,467 (Pompéo) to 7,200 square leagues of twenty to a degree.*

The province is separated from that of Piauhy by the narrow plateau known under the name of the Serra dos

* Dr. Almeida, in his Atlas, makes it 5,287.

Dois Irmãos and Serra Vermelha. From Ceará it is separated by the chapadas or plateaus known under the name of Serra Araripé, while the Serra dos Cairirís Velhos separates it from the province of Parahyba, which lies just to the north. Farther on, in treating of the province of Piauhy and Ceará, I shall discuss the facts we possess relative to the structure of the Serras Dous Irmãos and Araripe, which, as already remarked, appear to consist in the main of a narrow strip of horizontally disposed tertiary sandstones lying along the summit of a ridge of metamorphic rock.

We have already seen from Halfeld, Burton, and others, that the part of the province bordering the Rio de São Francisco is composed of much disturbed gneissose and other metamorphic rocks, here and there overlaid by patches of horizontal sandstones and associated rocks, like those of the São Francisco valley farther up. The great mass of the western part of the country appears to be formed of gneiss, mica-schist, &c., and these rocks here and there afford gold, but, so far as I can learn, not to any very large extent. Of the mountain ranges of the province next to nothing is known, and they are very inaccurately represented on the maps. From the description of some who have visited the Serra dos Cairirís Velhos, it would appear that they were in part formed of horizontal sandstones like the Serra Araripe. Some of the mountains of the interior of the province are of considerable elevation, though I cannot learn that any exceed 4,000 feet in height.*

Pompéo † says that the coast for a width of ten to fifteen

* Some of these highlands are visible from the sea, as for instance the Serra Sellada, which one sees lying a few leagues back of Cape St. Augustinho.

† *Geographia*, p. 425.

leagues is low. This portion is covered in part by quite heavy woods. It is very fertile, and is called *mattas*. Beyond this there is another zone of uneven, undulating country covered by *carrasco*, and dry, but it produces large crops of cotton and vegetables. The interior, which goes by the name of *sertão*, is very mountainous, stony, and dry, being fit for nothing but pasturage.

The same author says that "the interior, principally on the borders of the São Francisco and of the province of Piauhy is subject to droughts, like the provinces of the north; still it does not present the sandy deserts and the verdurous *oases* seen by the traveller Koster." *

The coast is low, and, generally speaking, resembles that of the province of Alagôas; consisting along the shore of a more or less wide strip of tertiary beds, which, though sometimes extensively denuded, form high red cliffs presented to the sea, as in the vicinity of Cape St. Augustinho. Underlying these are in some localities cretaceous rocks which have never been carefully examined. The tertiary beds have been swept away over considerable tracts which were occupied by the sea just previous to the last rise of the coast, forming deep indentations in the coast line. On such an indentation, now filled up with sand and alluvial deposits, the city of Pernambuco now stands. I am exceedingly sorry that, though I have three different times visited Pernambuco, I have never been able to examine the high lands in its vicinity. Mr. E. Williamson, in a short paper presented to the Manchester Geological Society, and published in the Proceedings of that society, says, that "at Caxinga, a few miles out from Pernambuco, several fine sections of the sands and marls" of the tertiary "have been exposed by

* *Loc. cit.*

land-slips ; the strata here bear such a resemblance to the new red sandstone of our own districts, that it would be impossible by color and appearance alone to distinguish one from the other."

The lands over large areas near the coast are very fertile, and produce excellent sugar-cane and cotton.

All the drainage of the middle and western part of the province is into the São Francisco. Some of the streams are quite large ; but, as has been remarked by Cazal,* they disappear in the dry season, and the same is the case with the majority of the rivers of the eastern part of the province, which flow into the sea. During the rains the streams, like those of Bahia, swell tremendously, but in dry weather they dry up entirely. The principal rivers emptying into the sea are the following : the Una, whose mouth is a few leagues north of the boundary line of Alagôas ; between this river and Pernambuco are the Serenhaem, Ipojuca, and Pirapama. The Capibaribe empties at Pernambuco. The Una, Ipojuca, and Capibaribe are quite respectable rivers, if we consider their length, the Ipojuca having a course of about one hundred and fifty miles ; but during the dry season they disappear, except in the immediate vicinity of the sea,† so that they are consequently of little or no service for navigation. The climate, of course, varies in different parts of the country. Along the coast it is damp and hot, though

* "Da Villa do Penêdo athé a da Barra do Rio-Grande, em cujo intervallo os viandantes contam acima de duzentas leguas, não sahe para o rio de S. Francisco um só regato no tempo da secca." Corografia Brazilica, Vol. II. p. 158.

† Dr. McGrath of Pernambuco kindly sent me a tracing of a map of the eastern part of the province. In a note he says : " These rivers look very formidable on paper, but, as you are probably well aware, they amount to almost nothing above tide-water during the dry season."

refreshed by sea breezes (Pompéo). In the interior it is very hot and very dry, especially during the rainy season, which lasts from March to June. Von Tschudi says that it is not so oppressingly hot at Pernambuco as at Rio. The population of the province is principally confined to the eastern part near the coast, and to the border of the São Francisco. In the region of the mattas there are numerous and well-conducted sugar estates, which produce a large quantity of sugar, molasses, and rum. Pompéo says, that in 1857 the president of the province reported 1,106 sugar-mills, 18 of which were operated by steam and 346 by water. These produced, in 1856, 18,498:000$000 worth of sugar, and the same year 1,341,354 canadas of aguardente (native rum) were exported, worth 616:000$000. Pompéo, writing in 1864, places the exportation of sugar at over four millions of arrobas. Von Tschudi, in 1866, estimated the yearly export at over one hundred and forty millions of pounds. The cotton of Pernambuco is reckoned very good, and, according to Von Tschudi, it brings the price of good Louisiana cotton. The region of the Garanhuns, lying just north of the middle of the province of Alagôas, is noted for its cotton. Cazal * says that the Serra de Garanhuns is covered by woods, and that streams descend from it, but are soaked up and disappear on reaching the sandy campos of the vicinity. This region produces also maize, mandioca, feijão, and fruits of various kinds.

Among the fruits for which this province is famous are the mango and the delicious giant pine-apple known as the Abacaxi. Cocoa palms are planted in large groves on the coast, and produce a very considerable revenue. In the Sertão a large number of cattle are raised. Pompéo esti-

* *Corografia Brazilica*, Vol. II. p. 159.

mated that in 1864 there ought to be 1,300,000 inhabitants; but there has been no census for many years. Almeida, in his Atlas, estimates the population of the province at 1,220,000 souls, of which the capital has 90,000.

Owing to the flatness of the coast and the small size of the rivers, there are but few ports capable of admitting large vessels. We find in use on this coast, principally for the purpose of fishing, the Jangada, a narrow raft of light logs, carrying a large triangular sail, — a craft which may be launched through the surf on the open coast. Of these, in 1864, the province possessed between seven and eight hundred. The city of Pernambuco,* or Recife,† owes its importance to its consolidated beach or stone reef. Except for this, it would offer no advantages for trade. Its position, in the very easternmost part of the empire, makes the port exceedingly convenient of access; and ships from

* The name Pernambuco is derived from the Tupí. Schalzo says that it means *mare sbucato.* Cazal (*Coro. Braz.* II. 170) makes it a corruption of Paranâbuca, which is said to be the name given to the port by the Cahetés who inhabited the place. The *Roteiro Geral,* cap. 16 (quoted in the *Art de Vérifier les Dates,* Vol. XIII. p. 256), says that the name was given because the reef was broken through by the sea: "Se diz de Pernambuco, por sua pedra junto delle está furada do mar, e quer dizer, Mar furada." D'Orbigny (*L'Homme Americain,* Tom. II. p. 280) makes it a corruption of Paranambu. The French call the place Fernambouc or Pernambouc, and the English used to call it Fernambuco.

† The name Recife was given to the city because of its reef. The Portuguese for *reef* is *recife,* which word is not derived from the Latin *recipere,* as so many authors would have it; as, for instance, Barlæus (*Rerum per Octennium,* &c., p. 66), who says: "Ubi terminatur, pagus fuit, Reciffa dicta, forte ab hoc, quod intra hunc et alium terræ similem tractum oblongum quem Reciffam Lapidosam vocant recepi naves possint et soleant, accipiendis exponendisque oneribus." The authors of the *Art de Vérifier les Dates* (Vol. XIII. p. 33) make the same mistake. The word *recife* is, according to Fonseca, derived from the Arabic *racif* or *razif,* signifying *a pavement.* See note on p. 190.

North America or Europe, bound for South American, East Indian, or African ports, have to go but very little out of their way to stop there.

I have already called attention to the fact that the city is situated on a tract of low ground, occupying a deep indentation in the coast tertiary sheet, and extending from Olinda nearly to Cape St. Augustinho.* It is built at the mouths of the rivers Beberibe and Capibaribe, which unite, forming a sort of delta, composed of a number of irregular islands very difficult to describe, which are enclosed by a net-work of channels. All of these islands are low, and some are marshy. The Beberibe is the northernmost of the two rivers. It takes its rise to the northwest of Pernambuco; at Olinda, about two miles north of Pernambuco, it meets a very narrow strip of sea beach, which extends southward to Pernambuco, a part of the city being built on the end of the spit, which broadens in the manner represented in the sketch-map on the following page. A channel from the Rio Capibaribe joins the Beberibe just above its mouth, and this cuts off from the mainland a large island, on which a second division of the city is built, while the third quarter is situated on the mainland, on the opposite side of the Capibaribe, opposite its mouth. These three quarters are united by bridges, several of which are of excellent construction.

To the southward a bay penetrates deeply into the land, but its waters are very shallow.

That part of the city built on the extremity of the sand beach above described is called Recife, though foreigners almost invariably use the name Pernambuco. It is very closely built up with warehouses, stores, custom-house buildings, &c., and is the centre of the commerce. On the beach

* Kidder compares this tract to the Bahian "Reconcavo."

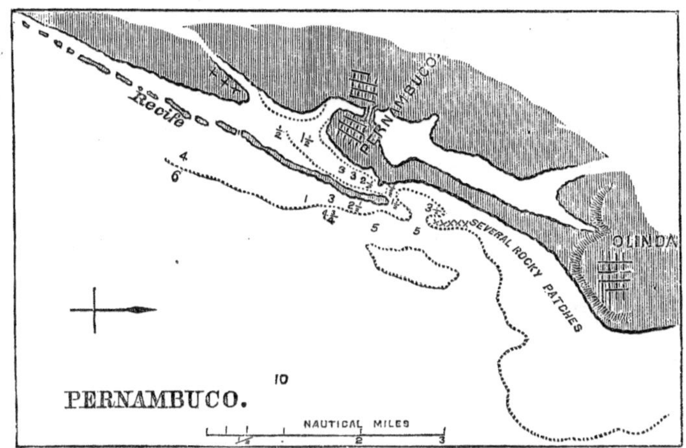

just to the north are the celebrated old fortresses of Brum and Buraco.

Opposite the former, and distant about 250 braças from the shore, the stone reef* begins abruptly, and runs in al-

* The reef has been described over and over again by the old navigators and travellers in Brazil. A very curious drawing of it is to be found in a Dutch work published in 1624, and entitled *Reys-boeck van het rycke Brasilien.* I do not know the author's name. The sketch is, however, so inaccurate as to be of no value as showing the structure of the reef. Barlæus not only speaks of the reef, but he gives a large copper-plate engraving of it, together with an excellent map. There is another large and curious copper-plate of Pernambuco and the reef in the "Historia delle Gverre del regna del Brasile, dal P. F. Gio. Gioseppe di Teresa Carmelitano Schalzo," published 1698. Speaking of the reef, Schalzo says: "Vien reparato da vn come marauigloso molo iui formato dalla natura il quale s'inalza sopra l'acque, distendendosi gran quantità di leghe tagliato dall'istessa natura con tanta egualtà, come si veggono i moli con immensi dispendij fatti dall'arte." Dapper also figures it in his *America.*

This reef is represented in Kidder's "Brazil," and in Kidder and Fletcher's "Brazil and the Brazilians," as if it were two or three miles distant from the city! A comparison between the views of Pernambuco in these works with a good map will prove amusing! What is the use of stating a thing in words, and then giving the lie to it in an illustration?

most a straight line southward, passing within a few hundred feet of the city, extending across the bay, and then skirting the shore for some distance south. I have been able to examine it only in the vicinity of the city, and, since the maps and charts differ so widely among themselves, I can form no definite idea of its real length. Barlæus gives a map and drawing of a similar reef extending across the mouth of a bay to the south of Pernambuco, near Cape Saint Augustine, if I remember rightly.

I have already described with much detail the stone reefs or consolidated beaches of Porto Seguro, Santa Cruz, and Bahia. The Pernambucan reef, so far as I have seen, has precisely the same structure. It is the consolidated core of an ancient beach which has been separated from the mainland by the encroachment of the sea. It is remarkable for its great length and the straightness of its course. It is exposed to a much heavier surf than the Porto Seguro and Santa Cruz reefs, and at high tide and during heavy weather it is usually deluged by the waves. The northern end breaks down abruptly as at Porto Seguro, and forms a wide opening for the entrance of shipping. Near this extremity stands the ancient Dutch fort, known as the Picão, together with a lighthouse. A part of the reef is artificially built up with masonry, to make it a more efficient breakwater. Ships of ordinary draught and small steamers enter the port, but the larger men-of-war and the ocean steamers usually anchor at a distance off the reef.

The water along this coast is very shallow, and the soundings continue for many miles out from the shore. The tides rise at Pernambuco about six feet.

Pernambuco is one of the calling stations of the United States, English, and French mail steamers, and it is the head-quarters of a steamboat company whose boats run to various points on the coast to the north and south.

The Pernambuco and São Francisco Railroad is in about the same condition as the Bahia and São Francisco line. It extends southwestward from the city through the fertile sugar lands, but it is, I believe, completed for a distance of only about eighty or ninety miles, the present terminus being Una on the river of the same name. I owe to the kindness of Dr. McGrath of Pernambuco, and Mr. Mann, the superintendent of the road, the following table of heights, along the line: —

Distances in Kilomètres.		Feet above the Sea.
	Marca 9 no Arsenal de Marinha	10.00
	Turntable de C. Pontas	17.96
$2\frac{7}{10}$	Ponte de Afogados	27.16
$8\frac{7}{10}$	Estãçao de Bôa Viagem	35.11
	Ponte de Jaboatão	26.25
	Ponte de Pirapama	29.12
$31\frac{5}{10}$	Villa do Cabo	56.00
	Centro do tunnel	184.00
	Ponte de Utinga	210.00
45	Estação de Olinda	340.00
$51\frac{8}{10}$	Pedreiras de Timbó-assú	373.00
$57\frac{6}{10}$	Estação de Escada	314.50
70	" " Frexeiras	404.00
	Contendas	413.00
	Ponte de Amaragí	296.18
	Plana	387.30
$113\frac{6}{10}$	Estação d'Agoa Preta	463.00
	Ponte de Formigueiro	418.96
	Excavação no Sitio Gomes	505.96
$124\frac{7}{10}$	Estação Una	437.96

About thirty miles north of Pernambuco is the large and

fertile island of Itamaracá, which is separated from the mainland by a narrow but deep channel like a river, the island being set into the mainland, and not lying off the coast as it is usually represented on maps. It is about ten or twelve miles long from north to south, and very narrow. The land is low, and it is interesting geologically on account of its limestones, which contain fossils, and apparently belong to the cretaceous, if one may judge from a report made to the Brazilian government by the engineer Paulo José de Oliveira, who was sent to examine the island for coal, which had been reported to exist there. Oliveira speaks of a bed of chalky limestone, " containing some fossils of the *turilite* and *carditas* family," and also of other fossils of the " *ananchites* family," found at Porto das Caixas.

Dr. McGrath of Pernambuco has kindly undertaken to have a collection made from these limestones, and I hope that we may soon have their age satisfactorily determined. Lime is manufactured on Itamaracá from these rocks, and also from corals which abound in the vicinity.* Oliveira reports the existence of iron ore on the island. A small quantity of salt is made on the coast from sea-water. The island is very fertile, and is covered with sugar plantations and cocoa-palm groves. The eastern side is thickly planted from one end to the other with these beautiful and useful trees.

Fernando de Noronha belongs to the province of Pernambuco, though it is situated to the north of Cape São Roque. It lies in lat. 3° 55′ S. and long. 32° 40′ W. of Greenwich and is distant from the coast about two hundred miles. It consists of one large island and several smaller ones, the whole being, according to Darwin, nine miles long and three

* See Kidder's Brazil, Vol. II. p. 172.

in breadth. This distinguished observer visited it, and I cannot do better than to quote his own description of the geological structure of the island: —

ISLAND OF FERNANDO DE NORONHA.*

"The whole seems to be of volcanic origin, although there is no appearance of any crater or of any one central eminence. The most remarkable feature is a hill one thousand feet high, of which the upper four hundred feet consist of a precipitous, singularly shaped pinnacle formed of columnar phonolite, containing numerous crystals of glassy feldspar and a few needles of hornblende. From the highest accessible point of this hill I could distinguish in different parts of the group several other conical hills, apparently of the same nature. At St. Helena there are similar great conical protuberant masses of phonolite nearly one thousand feet in height, which have been formed by the injection of fluid feldspathic lava into yielding strata. If this hill has had, as is probable, a similar origin, denudation has been here effected on an enormous scale. Near the base of this hill I observed beds of white tuff, intersected by numerous dikes, some of them amygdaloidal basalt and others of trachyte; and beds of slaty phonolite, with the planes of cleavage directed northwest and southeast. Part of this rock, where the crystals were scanty, closely resembled common clay slates altered by the contact of a trap dike. The laminæ of rocks, which undoubtedly have once been fluid, seems to me to be a subject well deserving of attention. On the beach there were numerous fragments of compact basalt, of which rock a distant façade of columns seems to be formed." †

* This sketch I copy from Ulloa. The water-line has been inadvertently omitted.

† Geological Observations, Part II. pp. 23, 24.

The island is almost deprived of vegetation, resulting from the dryness of the climate, and Ulloa* tells us that sometimes two or three years pass without rain. On this barren rock the Brazilian government has established a penal station. Flocks of sea-birds resort to the island to breed, and sea-turtles in great numbers lay in the sands of the shore during certain months. The harbor is an open roadstead. If the reports of navigators are correct the island is surrounded by coral reefs.

* "La esterilidad de esta Isla no procede de la mala calidad de su Tierra, pues produce todo quanto se siembra en ella proprio de Païses cálidos, sino de la falta de humedad; porque passan dos, y tres años sin llover, ni verse el mas leve aparato de Agua; y su escaséz es causa de que se sequen totalmente todas las plantas, faltando la Agua à los Arroyos, y lo mas pingue de toda la Isla quando las Nubes no la fecundizan con su riego, se vuelve tan árido, y desapacible, como los Peñones, y Rocas; en la casion, que llegamos, se havian passado dos años sin caer Agua alguna." — ULLOA, *Relacion Historica del Viage*, &c., Parte II., Tomo Quarto, p. 416.

CHAPTER XI.

THE PROVINCE OF PARAHYBA DO NORTE.

Limits of the Province. — The Serra or Plateau of the Cairirís Velhos. — The Climate, Productions, &c., of the Province. — Fertile Lands found only along the Coast. — The Rio Parahyba do Norte, its Navigability. — The City of Parahyba. — The Consolidated Beach at the Mouth of the River. — The River and Town of Mamanguape. — The Geology of the Vicinity of Parahyba. — Cretaceous Limestone with Fossils. — Observations of Professor Agassiz and Mr. Williams. — Mr. Williams's Observations on the Geology of the Country between Parahyba and the Gold-Mines of Pianco. — Mode of Occurrence of the Gold. — "The Tasso Brazilian Gold-Mining Company (Limited)."

The Province of Parahyba do Norte lies to the north of that of Pernambuco, from which it is separated by the Serra dos Cairirís Velhos, which seems to be the continuation eastward of the Serra Araripe. As laid down on the maps, the serra is noteworthy for its west-east trend, though it presents a gentle curve to the southward. This is in direct contrast with that of the other serras on the coast, which usually have a more or less northeast direction.

The principal mountain ranges crossing the province, as the Borborema and Teixeira, have a direction considerably to the east of north. These are composed of gneiss. This abnormal trend attributed to the Cairirís Velhos would be sufficient to lead one to suspect that the so-called serra belonged to the same class as the Araripe and the "serras" dividing the São Francisco and Tocantins basins, and the suspicion is confirmed by Pompéo, who says that the Serra

Borborema forms in the south an extensive plateau. So that it is more than probable that the Cairirís Velhos are very erroneously laid down on the maps, and do not form a narrow mountain chain, as represented.

The province forms a regular oblong about one hundred and eighteen * miles in length from east to west, and about ninety miles from north to south. The interior, as in Pernambuco, is uneven, and there are some considerable serras. The climate is very dry. The country is very poorly watered, and is consequently covered with a very scanty vegetation, so that it is fit only for pasturage. Pompéo says that the cattle subsist largely on the *macambira*, a bromeliaceous plant, which not only furnishes food, but is sufficiently juicy to quench their thirst. The coast is low, and much of it is very sandy.† As in Pernambuco, there are some fertile lands along the coast admitting of cultivation, and they are in part covered by forest. These lands produce cotton, sugar, ‡ tobacco, &c.

The climate of this province is hot, but on the coast the heat is modified by the breezes from the sea. It is very dry, and from time to time severe droughts prevail, causing much distress. The province is reputed healthy for Brazilians.

The principal river of the province is the Parahyba do Norte, which takes its rise in the Serra de Borborema, near the southern border of the province, and, skirting the province line, flows east a few degrees north to the sea.

At its mouth the Parahyba forms a sort of estuary, which

* Pompéo says one hundred and ten leagues, which is very far from being correct.

† See Koster's Travels.

‡ Pompéo says that in 1864 there were one hundred and sixty sugar factories in the province.

is quite wide, and opens into the sea from the south, as do many of the other rivers along this part of the coast.

The Parahyba is navigable for large vessels up to the city of the same name, the capital of the province, a distance of three leagues above the mouth (Pompéo).* Small vessels ascend about the same distance above the city, and canoes are said to go as far as Pilar, many leagues farther, but the influence of the tide, according to Pompéo, is felt only six leagues above the river mouth.

The city of Parahyba is situated on the right bank of the river, and is divided into an upper and lower town. It numbers, according to Almeida, 14,000 inhabitants.† Its exports are principally cotton and sugar, which are sent to Pernambuco. Pompéo says that during the year 1862 – 63 there were exported 201,890 arrobas of cotton, worth 3,021:124$800, and 620,270 of cotton, worth 821:120$000.

Barlæus ‡ gives a chart of the mouth of the Parahyba River, of which I give on the opposite page a reduced copy. It is interesting because it shows a stone reef extending across its mouth, leaving an entrance between it and the point on the southern side of the river. The reef begins off the point on which stands the old fort Cabedello, and runs parallel with the shore in a southwest direction for some distance. At its southern extremity it is, according to Barlæus's chart, triple, two short reefs lying inside of it.

I introduce, for comparison with Barlæus, a copy of part of a map of the mouth of the same river by Almeida, which ought to be more accurate, though it does not represent the reef with so much care.

* Judging from Mrs. Agassiz's description of the Professor's visit to Parahyba, the steamer was unable to go up as far as the town.

† The population of the province is about 300,000 (Almeida).

‡ *Rerum per Octennium in Brasilia*, &c., 1647.

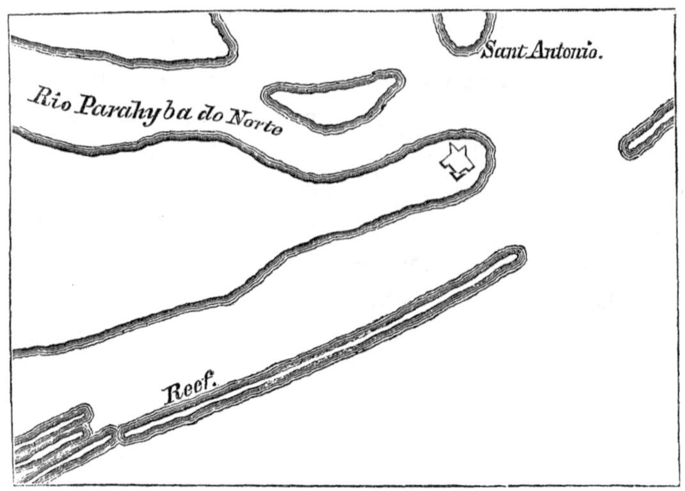

About eighteen miles north of the mouth of the Parahyba is the mouth of the Mamanguape, a much smaller stream than the Parahyba. It is said to admit of navigation for smacks as far as the important town of Mamanguape, which lies about four leagues from the sea. According to the *Diccionario Geographico* there is a stone reef at the mouth of the river, and Almeida represents a line of reefs as running along part of the coast. Cotton is one of the chief exports of the district of Mamanguape.

Of the geology of the province of Parahyba we know very little, except what is to be found in a short paper by Mr. E. Williamson,* from which I condense the following: —

"From Tambalic (Tambahú?) to Parahyba the surface is covered by thick beds of ferruginous conglomerate. . . . . In some

* On the Geology of the Parahyba and Pernambuco Gold Regions, by E. Williamson, Proceedings of Manchester Geological Society. This pamphlet, which I owe to the kindness of Professor Bonamy Price, of Oxford, bears no date.

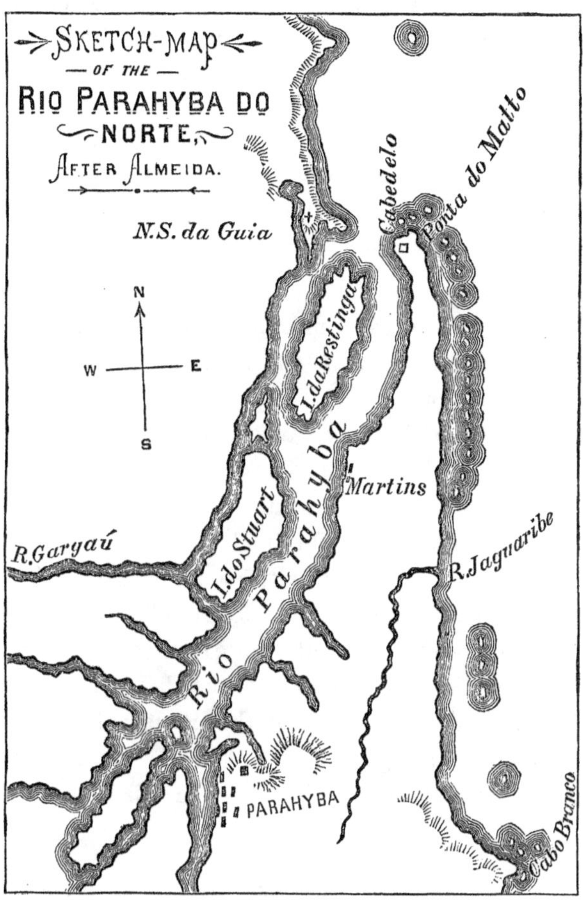

places the conglomerate becomes so very coarse that it is wholly made up of water-worn pebbles, of quartz, gneiss, and the harder schistose rocks, cemented together by peroxide of iron. The size of the pebbles varies from that of a small nut to boulders weighing four or five pounds. This class is well represented below the baths at Tambalic; but as they approach the river towards Parahyba they gradually become mixed with finer and more argillaceous beds, until, at last, at Sta. Rita, a few miles out from Parahyba, they have become divided into regular bands of marls, sands, and conglomerates." These deposits are probably tertiary.

"Immediately underlying the ferruginous conglomerates of Parahyba there occur beds of secondary limestone, having a strike nearly north and south, and dipping gently to the east. For the most part these limestones are siliceous, though at intervals beds of nearly pure limestone and argillaceous bands occur."

Mr. Williamson states that he found a cast of a fish tooth and some estherians in this limestone, and that similar beds abounding in fish remains occur at Minas da Cachoeira. Professor Agassiz touched at Parahyba on his return from the Amazonas and visited exposures of these rocks in the vicinity. He found the limestone of a soft texture, of a white or gray color, and destitute of recognizable fossils. From the green shales he obtained species of estherians which appear to be identical with some occurring in the Bahian beds. This latter observation is of much interest, since it would appear to indicate the existence, on this part of the coast, of fresh-water deposits like those at Bahia.

Mr. Williamson's description of the gneissoid rocks and of the gold deposits I give in full, as it is the only clear account of the way in which gold occurs in this kind of rocks in Brazil: —

"The first clear outcrop of the rocks, from Parahyba, occurs at Batalha on the river Parahyba; it is a hornblendic rock, with numerous small bands of quartz and feldspar much contorted.

"Between the river Parahyba and Pilar a very coarse gneiss occurs, with large crystals of white feldspar and black mica; at Pilar the gneiss is interstratified with mica schists, generally fine in texture; at Mendonça, Mocheira, and Ingá Velho beds of similar character again occur, interstratified with gneiss; at the last-named place the schistose beds become more frequent, until at Ingá the whole of the beds are micaceous and hornblendic schists. A little past Ingá a hard close-grained gneiss rock appears; this flanks the mountains of Lagradoura, which chiefly consist of a white porphyroid gneiss, holding large cleavable crystals of pure orthoclase, interstratified with bands of syenitic and granitoid gneiss, much resembling granite. On the northern flank the hard close-grained gneiss rocks again occur.

"Between Lagradoura and Campinas a well-marked band — granitoid porphyry — occurs, standing out some fifty and one hundred feet higher than the softer rocks which surround them; this porphyry contains large crystals of white orthoclase. At Campinas a series of micaceous beds occur, containing plates of mica. The largest was about two inches in diameter; but I was told that plates a foot square had been found; following this run of micaceous schist is a band of porphyry, in which large cleavable crystals of white orthoclase are embedded in a granular matrix of quartz and feldspar. I could distinguish no true bedding lines in this band, but from its dip and strike, an unbroken outcrop, I am inclined to think that it might be interbedded; the succeeding rocks are mica-schists, and gneiss.

"At Caximba Nova another band of hard granitoid rock occurs; following this there are a long series of mica schists and gneiss; near to Caracol occurs a series of black schists, alternating with bands of granular black rock; the schists are occasionally micaceous. At Caracol a small series of mica schists divide two broad bands of granitoid rock, in places these much resemble the true granites; overlying the upper one is a small band of hornblendic schist, this is followed by a long series of flaggy mica schists. At Carnahúba these are succeeded by bands of hard, close-grained gneiss; this flanks the mountains of Teixeira (rocks similar in character occur at Queimada on the opposite flanks); the rocks of the Teixeira mountains bear such a strong resemblance to those of Lagradoura, that I think they may be but a repetition of the same beds. Between Queimada and the Minas da Caxoeira, another broad series of the same class occurs; the remainder of the rocks on the section are gneiss, alternating with bands of mica schists.

"At various places on the section beds of quartz and quartzite, with plates of mica, were interstratified with the harder rocks; the beds varied from a few feet to two hundred feet in thickness; the smaller bands were often beautifully opalescent, the larger bands granular or amorphous. Hæmatitic and titanic iron ores always accompany them.

"On my journey from Parahyba to the mines I failed to detect any beds of limestone interstratified with the Laurentian rocks, but I was informed that limestone interstratified with the rocks had been observed in other places where the limestones had not been hidden by the covering of ferruginous detritus.

"The rocks at the Minas da Caxoeira, and the position of the gold-bearing veins, will be best understood on reference to the accompanying section, taken along the bed of the Bruscus River, for a length of about six miles.*

"At the southern extremity, divided by a band of softer rock, are two broad and well-marked bands of syenitic gneiss, one of which forms the bed of the beautiful waterfalls of the Bruscus; underlying these is a series of schistose gneiss, and a narrow band of syenite; it is a bluish-gray crystalline rock, and bears a strong resemblance to some of the Welsh upper Cambrian feldspar rocks. Succeeding these are the gold-bearing series, which almost wholly consist of fine-grained micaceous gneiss, passing imperceptibly into mica schists.

"Crossing a bend of the river a little before reaching the Lima lode, a small band of dark brownish-gray feldspar rock occurs; it is subtranslucent, and in places shows chatoyant colors; a little farther on is a band of white crystalline limestone, containing hexagonal crystals of biotite; in the bed of the river it is small,

* This section is omitted.

but about a mile farther east from this point, at a place called Pião, it is said to be a mile broad on the outcrop.

"A little east from where the Descubridora lode crosses the stream there occur a few beds of earthy plumbaginous schists, in which are two lenticular veins of graphite; they seem to be of small extent and of a very indifferent quality.

"At Cacimbinhas, a few miles farther on than the Bôa Esperança lode, another broad and well-marked band of syenitic gneiss occurs, quite as large as that of the waterfalls.

"The auriferous veins which traverse these rocks are very numerous; they appear as irregular lenticular masses, running parallel to the strike, often dipping between, but rarely cutting through, the beds. The matrix of the lodes is a coarse white semi-opaque quartz, containing small quantities of the arsenides and sulphides of iron, sulphides of copper, lead and zinc; most of the galenas contain antimony. The variety of minerals resulting from the decomposition of these ores are very numerous, — carbonate of zinc, carbonate and chloro-phosphate of lead, phosphate, arseniate and carbonate of copper, oxides of antimony and native sulphur are common in some of the lodes; sulphate of copper, sulphate and chromate of lead were more rare; native gold was sparingly scattered throughout nearly all the lodes, and in the Bôa Esperança veins, grains of platinum were found.

"The run of rocks in the valley of the Bruscus are very auriferous, and quartz veins abundant; and though the rocks are greatly contorted, no trace of a true fault can be found anywhere in the whole district; this singularity appears to belong to all the altered rocks which I examined in Parahyba and Pernambuco, for during a ride of 1,000 miles I failed to detect any; it is to this want of true fractures that I ascribe the poorness of the quartz veins, there being nothing favorable to the concentration of the ores; the gold has been equally distributed throughout all the veins. It is well known to miners that no veins are so rich as those in which the faces of dissimilar rocks are brought opposite each other on the walls of the vein.

"On my journey from the mines to Pernambuco I crossed the same run of rocks as those marked on the section, and during my ride was able to trace several anteclinals; this accounts for the vast extent of country covered by rocks of the same age.

"About seventy leagues from Pernambuco I found a band of quartziferous porphyry; a specimèn is exhibited; it has a compact base, composed of an intimate mixture of quartz and feldspar, enclosing crystals of orthoclase and grains of quartz.

"Near to Jerimu there occur, within a few leagues of each other, two bands of crystalline limestone; one a narrow and highly crystalline band, the other a very broad band: in some parts this is micaceous, but none of the beds are so highly crystalline as the small band.

"The country between Jerimu and Pernambuco bears a strong resemblance in character to those marked on the section between Parahyba and Campinas.

"The whole series of these rocks agree in every respect with the characteristic features of the Canadian Laurentian rocks, as given by Sir W. E. Logan, viz.: —

"I. The total absence of anything like argillite or clay slate.

"II. That nothing corresponding to slaty cleavage has ever been remarked.

"III. That the lamination of these masses is apparently in every case coincident with, and dependent upon, the original stratification of the sedimentary layers."

There was formed in 1865 an English company under the name of the "Tasso Brazilian Gold-Mining Company (limited)," to work the mines, not only of Parahyba, but also of Pernambuco. Among the officers of this company appear some very honorable names. I am entirely uninformed as to what this association has done, or what its prospects are.

Sr. José Jacomo Tasso, in an official report made to the

government not long since, said that seven distinct auriferous veins had been discovered, but washing was carried on with great difficulty, because water was scarce, and had to be brought from a distance of some two or three miles.

CHAPTER XII.

THE PROVINCE OF RIO GRANDE DO NORTE.

Limits of the Province, its Position, Mountain and River Systems, &c. — The Rio Piranhas. — Vegetation. — Productions. — The Carnahuba Palm and its Uses. — Cochineal. — Cattle. — Climate. — Natal. — Geology of the Province.

THIS province occupies the extreme northeastern part of Brazil, and forms an irregular quadrilateral about one hundred and eighteen miles in greatest length from east to west, and between eighty and ninety miles in width from south to north. From Ceará it is separated by the so-called serra or plateau of Appodí. Two water-sheds running north-south divide the province into three almost equal parts. The eastern of these water-sheds is formed by a continuation of the Serra Borborema or the Cairirís Novos, east of which a number of little rivers run eastward to the sea; these rivers being of the same character as those of Pernambuco and Bahia, disappear in the dry season. Pompéo says that the Ceará-merim and Trahiry are perennial. None of these rivers are navigable for more than a few miles above their mouths.

The Rio Piranhas, which originates in the province of Parahyba, passes through the middle third of the province of Rio Grande do Norte with a course almost due north, emptying into the sea by several mouths, forming a considerable delta. This river is sufficiently large to allow small vessels to ascend about seven leagues to the town of

Assú, an important place, noted for its cattle, its salines, and commerce in salt. The Piranhas takes its name from the fish of the same name, which abounds in its waters.

The western third of the province is traversed by the Rio Appodí, of which the Upanema forms a branch. These two streams take their source in the serras bounding the province on the south and flow northward, uniting only a few miles from the sea. Cazal says that the lands bordering the Appodí are, for the most part, plains diversified by numerous lagoons that dry up in the rainless years. The river is navigable for canoes only as far as Santa Luzia, six leagues from the ocean.

The greater part of the province is, like the province of Parahyba, dry and largely covered by low, sparse vegetation (*Catingas carrasquentas*), but on the higher lands, where the soil is good, there are considerable areas covered by forest and adapted for cultivation. Forests are also found in the eastern part on the low, swampy grounds, and along the banks of some rivers.

Cotton and sugar-cane are the principal productions of the country, and are cultivated on the serras and river margins, particularly of the rivers Ceará-merim, Carimataú, Potengí, Trahiry, &c.

Pompéo says that there were, in 1862, one hundred and eighty-five sugar factories in the province, producing 375,000 arrobas. The *carnahuba* palm (*Copernicia cerifera*) is cultivated largely on the low grounds, and is one of the most important vegetable productions of the country. This beautiful palm, which is met with all over the northeastern provinces of Brazil, is of so much interest that I will give a short description of it and its many uses.

It grows to a height of thirty or thirty-five feet. The

leaf-stalks remain persistent to a height of six feet, more or less, from the ground, the rest of the stem being smooth. The fronds are fan-shaped, furnished with thorns, and disposed in a close ball-like head, so that the tree presents a very different appearance from that of any other species of palm.

The uses to which the different part of the plant can be put are exceedingly numerous.

The roots are used as sarsaparilla. The stem, when about six feet high, is furnished with a pith which, treated with water, gives a sort of meal used for food in famine times. The adult stem is an excellent timber employed in building. The midribs of the fronds are used for making fences, &c., the leaves for thatching, the fibre for cordage. The fruit, properly cooked, tastes like boiled Indian corn, and is used for food. The gum is edible.

Soon after the young leaves have opened, they are cut and dried in the shade. Scales of a waxy substance are then easily dislodged from their surface, and may be melted over a fire into cakes. This is the carnahuba wax. The same substance is said to be furnished also by the berries. It is very brittle and brown in color, but it may be bleached. Mixed with common wax or tallow, it is made into candles of a fair quality. The proportion is usually three parts carnahuba to one part wax, or one eighth to one tenth of tallow. (Burton.) This palm is so very abundant that its wax is likely to become a very important article of export from the northern provinces.

The carnahuba may be occasionally seen in the province of Bahia. Burton met with it in the valley of the São Francisco, just above the Barra do Rio Grande, but it is most abundant growing on the low lands bordering the streams in the provinces of Pernambuco, Parahyba, Ceará, and Piauhy. The wax is collected to a considerable extent, and one may buy carnahuba candles or wax all along the coast.*

In addition to sugar and cotton and the carnahuba, rice, mandioca, beans, tobacco, &c. are cultivated. A species of cochineal insect is found in Brazil, living on the leaves of cacti, and it has been supposed that it might be cultivated successfully in this province, and made an article of commerce. Pompéo says that it is found in the Sertão of Seridó, but he does not say whether it is made use of or not.†

In the interior large herds of cattle are raised, and in 1862 there were 2,013 cattle fazendas,‡ with about 59,630 head of cattle.

The climate of the province is very hot and dry, and droughts of great severity sometimes prevail.

The most important place in Rio Grande do Norte is Natal, situated at the mouth of the Potengy. It is a small town, but of some little importance. The entrance to the river is very difficult, owing to shoals and rocks. On the southern side is a stone reef represented in the following sketch-map from Almeida.

* See Von Martius, Palms, p. 49; Dr. Manoel Arruda da Camara in Koster, Travels in Brazil, Vol. II. p. 311; Brande, An Account of a Vegetable Wax from Brazil, Phil. Trans., 1811, p. 261; Boussingault, *Ann. de Chimie*, Vol. XXIX. p. 330; Sir M. A. de Machedo, *Notice sur le palmier Carnauba*, Paris, 1867; Agassiz, Journey in Brazil, p. 453.

† Spix and Martius (Travels in Brazil, Vol. II. p. 19) speak of the occurrence of the cochineal insect in the province of São Paulo.

‡ Accioli, *Corographia do Brazil*, p. 61, says that cheese and butter are manufactured and exported to a considerable extent.

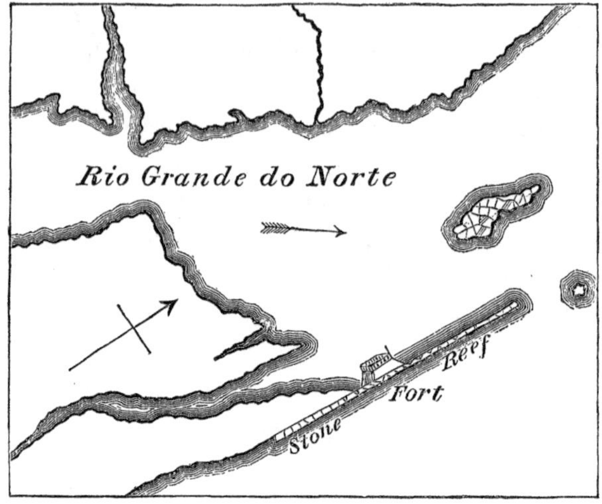

The coast is bordered by reefs, but they have never been examined by a competent observer.

So far as the geology of the province is concerned, the southern part appears to be largely composed of gneissose and other metamorphic rocks, which form in the interior a high and more or less mountainous country like that of the provinces immediately south. Bordering this is a considerable band of tertiary rocks. The immediate vicinity of the sea is flat, sandy, and often, as in the vicinity of Natal, covered by dunes of blown sand. I know of no mines or mineral deposits in the province, though gold is said to occur there. In the *Exposition Universelle* of 1867, a specimen of native sulphur, said to have come from this province, was exhibited. I know of no localities in Brazil which afford native sulphur, though it has been affirmed to exist on the Itatiaiossú. One of the principal products of the province is sea-salt, which is largely manufactured at the Salinas d'Assú and Mossoró.

CHAPTER XIII.

THE PROVINCE OF CEARÁ.

Geographical Position and Limits of the Province. — The Serra da Ybiapaba. — Its Topography and Geological Structure. — Serra de Araripe. — River Basins. — General Sketch of the Geology of the Province. — Climate. — Character of the Soil. — Productions. — City of Fortaleza. — Population of the Province. — Gardner's Sketch of the Geology of Ceará. — Character of Country in the Vicinity of Aracaty. — Description of Country between Aracaty and Icó. — Serra de Pereira. — Villa do Icó and Vicinity. — Country between Icó and Crato. — Gold Washings. — Crato. — Serra de Araripe. — Villa da Barra do Jardim. — Description of Fossil Fish Locality. — The Fishes noticed by Spix and Martius and Others, and described by Professor Agassiz. — Glacial Phenomena of Vicinity of Fortaleza spoken of by Professor Agassiz. — Mammalian Remains. — Minerals. — Meteorolites.

THE province of Ceará is bounded on the north by the Atlantic, on the east by the provinces of Rio Grande do Norte and Parahyba, on the south by the province of Pernambuco, and on the west by Piauhy, and, according to Dr. Pompéo, has an area of 4,681 square miles. Its western boundary line is formed by a narrow range of high lands called collectively the Serra da Ybiapaba.* Pompéo says that this serra begins near the coast not far from the eastern mouth of the Parahyba, and runs, under the names of Serra Grande, Serra da Ybiapaba or Crathéus, in an almost southerly direction, to the confines of the province of Pernambuco, where it ties in with the great serra running

* This name is spelled in a variety of ways, as Ybiapaba, Hibiappaba, Ipiapaba, &c. Some would make it mean *land of precipices.* According to the Dict. da Lingoa Tupí, *Ibý apába* means *terra talhada,* or *gashed land.*

southwest, forming the water-shed between the basins of the Parnahyba and São Francisco, and called the Serra dos Dous Irmãos or Borborema.

In all this extension the serra presents steep slopes and often precipitous sides towards the east, but its top is perfectly flat, forming a chapada, according to Feijó, 2,000 to 2,400 feet high. This chapada is in some places from thirty-two to fifty-six miles in width. On the western side the slope is not so marked, and the country appears to grow gradually lower all the way eastward to the river Parnahyba. At Cratheús the serra is abruptly broken through by the valley of the Poty. As to the exact structure of any part of this serra in detail our information is very scanty, but from the observations of Gardner, Capanema, Feijó, Pompéo, and others, there can be no doubt that it is composed of comparatively modern stratified rocks lying horizontally on disturbed gneissose and other metamorphic strata.

The horizontal strata appear to be principally sandstones. Gardner supposed that the whole range was cretaceous. On this head I shall have more to say farther on. The serra separating the Province of Ceará from Pernambuco is called Araripe. This range is of the same general character as the serra just described. At the end of the *Termo do Jardim* it grows very low, and Dr. Pompéo describes the water-shed between the Riacho dos Porcos, which flows into the Rio Salgado in Ceará from the Riacho do Mundo Novo, which flows towards the São Francisco, as of very little elevation, and it has been thought practicable to connect over it the waters of the Rio de São Francisco with those of the Salgado in Ceará. Beyond this point the serra, after suffering continual interruptions, stretches northward along the eastern border of the province, until near the sea it meets with the Plateau of Appodí.

The province is divided into two portions by a line of serras which runs from the sea-shore near the capital southwest or south-southwest to the Serra da Ybiapaba. The southeastern half of the province forms a single river basin, and is watered by the river Jaguaribe and its tributaries. The western is watered by a host of little rivers, all flowing directly into the sea.

The coast forms a broad belt of sands but slightly elevated above the sea. This belt varies in width from four to six leagues. (Pompéo.) The sands are light and piled up more or less in dunes, while in some places they drift inland to a considerable distance. The low coast lands, which have a coating of alluvial matter, are very fertile, and are planted with cane, mandioca, &c.

From the coast the land rises gradually towards the serras, and is uneven though not properly mountainous, being diversified by hills and plains. According to the descriptions of Dr. Pompéo, the sertão, or the greater part of the area of the lower country, is composed of mica-slate, gneiss, and other metamorphic rock.* The country appears to have been once covered by a sheet of sandstone similar to the great sandstone formation of Piauhy, to be described in the next chapter. This has been denuded to a great extent, the sandstone in many cases being left capping the hills.

The climate in the interior is very hot and dry, but on the sea-coast it is moist and more tempered by the refresh-

* Pompéo, *Ensaio Estatistico da Provincia do Ceará*, p. 145, says: "In the sertão limestone and marble occur in many places, as in the Serrote de Cantagallo (*Caminho de Baturité*), where the rock is a primitive marble, Giboia, four or five leagues from the capital, &c., &c. Capanema reports graphite in connection with limestone.

ing breezes that blow in from the ocean. The greatest heat experienced on the sertão is 35° Cent. (95° Fahr.), and the temperature does not fall below 18° Cent. (64.4 Fahr.)

The rainy season begins in January or March and lasts until June. The rest of the year is without rain, the rivers and streams dry up, and occasionally the want of water over certain tracts is such that the inhabitants are obliged to leave for more favored districts. Accioli * says that in 1792 a drought prevailed for four months, and caused the inhabitants of seven frequezias to abandon them entirely.† Other noted *seccas* prevailed during the years 1825 and 1845.

Pompéo divides the soil into *beira-mar*, which is fitted for agricultural purposes; *montuoso*, productive and covered with forest; and *sertão*, dry, cut up by taboleiros, hills, &c., and with brooks dry during the summer.

The sertão is valuable for pasturage, and Ceará is noted for its cattle, of which, according to official papers, there were, in 1854, about 310,000 head, with nearly 40,000 horses. Cheese is manufactured in considerable quantity, not only for home consumption, but for export.

Among the natural vegetable productions may be mentioned the following: The Quina, ipecacuanha, tatajuba, jacarandá, cedro, páo d'arco, and a host of other species of woods valuable for building and dying purposes, and the balsamo, jatobá, almescar, maneçoba, carnahuba, cajú, mangaba, &c., &c.

Cotton, coffee, sugar-cane, and the other vegetable productions of Northern Brazil, are here cultivated. ‡

* *Corografia do Brasil*, p. 49.

† A very graphic account of the droughts of the northeastern coast of Brazil is to be found in Koster's Travels in Brazil.

‡ Pompéo gives the following amounts of cotton, sugar, and coffee exported between July, 1857, and April, 1858: Cotton, 52,552; sugar, 121,697; Coffee, 304,009 arrobas.

The capital city is Fortaleza, or Ceará, a city of some 16,000 inhabitants, situated on the coast about two leagues distant from the mouth of the river Ceará. Its principal importance consists in its export of coffee. Besides the capital there are seven other small cities in the province, namely, Aracaty, eight miles above the mouth of the Jaguaribe on the east bank, a place next to the capital in importance, and noted, amongst other things, for its manufacture of carnahuba candles, sole-leather, &c., &c; Icó, a fearfully hot place, about thirty leagues from Aracaty, on the Rio Salgado; Crato, situated in a fertile district eighty leagues from the sea; Granja, at the mouth of the river Camocim, a good port and stopping-place for steamers from Pernambuco; Quixeramobim, a little city in a cattle district in the sertão; Baturité, sixteen leagues from the capital, and noted for its activity in the cultivation of coffee, sugar-cane, &c.; and Maranguape, also noted for its coffee and sugar plantations.

The whole population of the province is probably about 540,000.

In the year 1841 Gardner published, in the "Edinburgh New Philosophical Journal,"* a short paper on the geology of Ceará. This article gives so clear and minute a description of the country, and contains so many important geological facts, that I think it worthy of being reproduced almost in full, particularly since I shall have to examine critically some of his general conclusions arrived at, not only in this paper, but also in his since published "Travels in Brazil."

* Geological Notes made during a Journey from the Coast into the Interior of the Province of Ceará, in the North of Brazil, embracing an Account of a Deposit of Fossil Fishes, by George Gardner, Esq., Edinburgh New Philosophical Journal, April, 1841, p. 75.

Mr. Gardner says: —

"I landed at the northeast corner of the province, at the town of Aracaty, which is situated on the east bank of the Rio Jaguaribe, at three leagues from the coast. The first thing that struck me on my arrival was the flatness of the country around it, reminding one of the descriptions which are given of the pampas of Buenos Ayres. With the exception of a few low sand-hills towards the sea, and a round, isolated one about eight hundred feet high, situated two and a half leagues to the southwest of the town, called the Serra de Areré, there is nothing to interrupt the uniform level. The soil for many leagues around is of a sandy nature, and the characteristic vegetation is a beautiful species of palm called carnahuba by the Brazilians. It is the *Corypha cerifera* of Martius, and is so abundant that, on my journey south to the Villa do Icó, I rode for about two days through a forest of almost nothing else. Two and a half leagues to the south of Aracaty I first met with rocks. This was on crossing the river at a place called the Passagem das Pedras. I found them to consist of thin strata of gneiss, almost in a vertical position. The little inclination which they had was towards the northwest, in the direction of the above-mentioned Serra de Areré. From this place to the Villa do São Bernardo, a distance of a little less than eight leagues, the country continues perfectly flat, but the ground among the carnahuba palms, and in several large open spaces almost destitute of vegetation, called vargens, is covered with abundance of gravel; and this, which extends over large tracts, gives it the appearance of the dried-up bed of an immense river. Intermingled with this gravel there are numerous boulders of various sizes, the largest I saw not being more than four feet high. They are all more or less rounded, and consist of granite, gneiss, and quartz. For the next ten leagues the country continues nearly of the same character, with the exception of a low range of gravelly hills running from east to west, and wooded with shrubs and small trees, the most common of which is a species of mimosa. During the next ten leagues a slight but perceptible rise of the country takes place. The soil is generally a yellow-colored clay, in many places thickly covered with gravel and boulders, while in others gneiss rocks are seen cropping out and forming often long slightly elevated ridges covered with a species of cactus and a large bromelia. Their strata, like those farther down, are almost vertical. In this tract carnahuba palms become less numerous, and small dicotyledonous trees and shrubs more abundant, but all of them in the dry season, at which time I passed, destitute of leaves. These deciduous woods, which often cover large tracts of country, are called catingas by the Brazilians. These tracts are still farther characterized by three large species of cacti, belonging to the genus Cereus. During the next twenty leagues, which brought me to the Villa do Icó, the nature of the country differs in again becoming more level, consisting of large open campos or vargens, the vegetation of which, during the dry season, is quite burnt up, but they are said to yield abundance of grass during the rains; and the catingas or deciduous woods are much larger than they are farther down. The rocks are gneiss and quartz, and in several places large tracts are covered with fragments of the latter, more or less rounded. At about ten leagues below Icó, the monotonous level of the country is varied by a mountain range, which makes its appearance to the eastward. This is the Serra de Pereira. It runs from the southwest to the northeast. It is sixteen leagues in length, but its greatest height is not more than one thousand feet above the level of the plains in which it is situated. The structure of its southwest extremity at least is entirely primitive, but near its base I observed a coarse red conglomerate, containing rounded fragments of both primitive and secondary rocks.

"The Villa do Icó, which is one of the finest in the interior of the north of Brazil, is situated on the east bank of the Rio Jaguaribe, in the middle of one of the large open campos which I have already described, and during the dry season is one of the most miserable places imaginable to live in. The country around it is then so much dried up that not a green leaf is to be seen; and the river, which during the rains is of considerable size, becomes quite dry. The houses are all built of brick, which are made from a very good kind of clay found in the neighborhood, and are all whitewashed on the outside with a white limestone, which is found about ten leagues to the west of the villa.

"From Icó I went to the Villa do Crato, which is about thirty-four leagues to the southwest of the former place. Between these two places the country is of a more hilly, undulating character, more abundantly wooded, the trees larger, and many of them evergreen. Owing to these circumstances but few of the large campos which exist below Icó are met with. The carriage of goods between Aracaty and Icó is effected in large wagons, generally drawn by twelve oxen; but the hilly nature of the country between Icó and Crato does not admit of this mode of conveyance, the backs of horses and even of oxen being made use of instead. Shortly after leaving Icó I passed over the southwest end of the Serra de Pereira at a place where it has but a slight elevation and consists entirely of gneiss. From this place to the Villa das Lavras da Mangabeira, a distance of about ten leagues, the country is of a gently undulating nature, and in many places well wooded. This villa, which is situated close to the Rio Jaguaribe, takes its name from a number of small gold-workings (*lavras*) which, from time to time, for many years past, have been wrought in its neighborhood. Nothing, however, was done to any extent till about two years ago, when two English miners were sent for by a company in the city of Ceará, the capital of the province. They continued their labors till about two months before I passed through the place, having been recalled by their employers. I could not learn what amount of gold they had obtained, but the persons of whom I made inquiries remarked, with apparently much truth, that they did not believe it was sufficient to repay the expense, or the work would not have been abandoned. The gold is here found in small particles, in a dark-colored diluvial soil, at a considerable depth; but the place being shut up, I had not an opportunity of examining it.

"At about eighteen leagues below Crato I lost sight of the gneiss rocks, and for the next four found them replaced by a gray-colored primitive clay-slate. At the termination of this, the secondary stratified series begins, the few rocks which I met with from thence to Crato consisting of a white coarse-grained sandstone.

"The small Villa do Crato stands in the middle of a large undulating valley, which is bounded to the south, to the west, and to the north by mountains which, in their highest parts, do not rise more than from 1,200 to 1,500 feet above the level of the town. The country around is very fertile, producing abundance of cane, from which an impure sugar, in the form of small square cakes, is made, mandioca, Indian corn, rice, cotton, and tobacco, besides all the varieties of fruit which are to be met with on the coast. The great cause of this fertility is the numerous springs which exist along the foot of the mountains. The small streams which proceed from these are divaricated in a thousand directions, for the purpose of irrigating the plantations. The mountains are branches of the long range which separates the provinces of the coast from that of Piauhy to the west, which here receives the name of Serra de Araripe. Their tops are perfectly level, and extend so for many leagues to the westward and southward, forming what the Brazilians call Taboleiras. I have ascended this range in all directions, and have universally found it to consist of a generally white-colored sandstone, but in many places it is of a reddish tinge. In the bed of one of the largest streams which proceed from it, where a section of the rocks to a considerable depth is formed, I found a stratum of limestone about three feet thick, immediately below the sandstone, and below it another of an impure coal, two feet thick, resting on another stratum of limestone. Nothing seems to have disturbed the strata, as they all lie in a perfectly horizontal position, and the level nature of the serra proves that this is general. In the limestone I could meet with no fossil remains. The temperature of two of the springs, which rise at the base of the serra I found on examination to be 75° Fahrenheit.

"That part of the serra which lies to the south of Crato is a branch which runs about ten leagues to the eastward. On the south side of it there is another small villa called Barra do Jardim, distant from Crato about fourteen leagues. I went to this place partly for the purpose of botanizing and partly to make a collection of fossil fishes, which I was informed were found in great plenty in its neighborhood. The road skirts along the base of the serra in a southeast direction for about five leagues, at the termination of which it is necessary to ascend it for the purpose of crossing to the other side. The ascent is far from being good, it being left entirely in the hands of nature. The only rock I observed was sandstone, similar to that which exists at Crato. The breadth of the serra here is nearly eight leagues, and during the whole of this distance the road is as level as a bowling-green; and, as no water is to be found on it, travellers are obliged to supply themselves with it before ascending. For small parties it is carried in calabashes, but when many pass together a horse is provided to carry two large leather bagfuls. These Taboleiras are generally thinly wooded, with small trees, the principal of which are a species of *Caryocar* called Piké, a small tree belonging to the natural order Apocynaceæ, which produces a delicious fruit called Mangaba, a fine species of *Brysonema*, the Cashew [cajú] (*Anacardium occidentale*), a purple-flowered *Qualea*, and several small leguminous trees belonging to the division Rectembriæ.

"The Villa da Barra do Jardim stands in a small valley, upwards of a league in length, and in its broadest part about half a league in breadth. It is bounded to the north and east by the branch of the serra which I crossed over, and to the west by another, but neither so broad nor so long. Having made inquiries for the place where the fossil fishes were to be found, I was directed to a rising ground which extends along the foot of the serra. On my arrival at an open place of this gently sloping ridge to the north of the villa, I found the ground covered with great abundance of stones of various sizes, and I was informed that almost every one of them on being broken presented some part or other of a fish. These fragments I soon found to consist of compact fawn-colored limestone. They are of all sizes, from pieces not larger than an egg to blocks of several feet in circumference, and are almost all rounded and smoothly polished, having apparently been for a long time under the influence of a current of water. They, in general, split very readily, and almost all of them present parts of a fish in a more or less perfect state. But by far the greater number of them are so much broken that it is with considerable difficulty tolerably perfect specimens can be obtained. The spot which these stones occupy is not above an hundred yards square, and almost no other stone is mixed with them; but on every side of this deposit the ground is covered with little rounded sandstones, similar to the rock of which the serra is composed. Besides this I afterwards visited other deposits, one half a league to the south of it; one at a place called Maccapé, five leagues to the east of Jardim; and another at Mundo Novo, three leagues to the west: all perfectly similar to the one I have described, being all situated on the declivity of the low hills which stand between the valley and the serra, and all occupying places which are almost altogether free from other kinds of stone. From these places I have obtained a suite of specimens, embracing upwards of a dozen species of fossil fish.* They vary in size from those of a few inches in length to others which must have been several feet; and all of them, so far as my limited knowledge of the subject allows me to judge, except two species, belong to the order *Cycloideæ* of M. Agassiz. The most abundant species is one of those which do not belong to

* Mr. J. E. Bowman in a note says: "Agassiz makes them but seven species, and refers three of them to the Ctenoid group."

this order. Of it I possess a nearly perfect specimen about a foot and a half long, but, judging from other fragments of the same species, it must have attained a much larger size.* It has the head very much elongated, and the scales of the back and abdomen are angular, while those of the sides consist of but one row of long, narrow ones, arranged vertically. Of the other species I only possess the tail and a very small part of the body. It differs from the last in appearing to be entirely covered with small angular scales. Both of them I have no doubt belong to the order *Ganoideæ* of M. Agassiz.†

"On breaking these stones, some of them exhibit abundance of a minute bivalve shell; and at Mundo Novo I met with a very perfect specimen of what I believe will prove to be a species of Turrilites, about an inch and a half long, and a single valve of a Venus about half an inch in length and in very excellent preservation. Both of them were found in the same fragment of limestone. I was informed by a person in Jardim that a few years ago he found a small serpent coiled up in a stone which had been split, but this, no doubt, was a species of Ammonites. In the several hundred stones, however, which I broke in search of fish, I met with nothing of this description. During my excursions in the neighborhood of Barra do Jardim I nowhere met with limestone *in situ*."

When we come to discuss, in the next chapter, the geology of the neighboring province of Piauhy, we shall have an opportunity of seeing how correct Dr. Gardner's conclusions are with reference to the age of the sandstone beds.

* "The fish here described is the *Aspidorhynchus Comptoni* Agass. — J. E. B."

† Specimens of these fish find their way all along the coast, and it may be well for future observers to remember this. I had a specimen of *Aspidorhynchus* given me at Penêdo, and specimens of another genus were sent me recently by Mr. Laué from Maroïm.

The above paper was followed in the same journal by a communication by Professor Agassiz, in which the species of fossil fish were named and described, the opinion being expressed that they were of cretaceous age.

The first notice I have seen of the fossil fishes of Piauhy and Ceará is made by Spix and Martius in their Travels, one of the species being figured in the atlas accompanying their work.

In 1838 Mr. Nicolet placed in the hands of Professor Agassiz a few specimens he had received from Pernambuco.

Gardner's specimens, in the collections of Mr. Bowman, the Marquis of Southampton, Lord Enniskillen, and Sir Philip Egerton, were examined by Agassiz in 1840. Two years later M. F. Chabrillac sent to M. Elie de Beaumont from Pernambuco a few more specimens he had obtained from Ceará. These were placed in the hands of Professor Agassiz, who made a long report on them in a letter addressed to M. Elie de Beaumont, published in the *Comptes Rendus*, Vol. XVIII. p. 1007. In this letter Agassiz enumerates seven species as known to him from the province of Ceará, viz.: —

Aspidorhynchus Comptus Ag.
Lepidotus temnurus Ag.
Rhacolepis buccalis Ag.
Rhacolepis Olfersii Ag.
Rhacolepis latus Ag.
Cladocyclus Gardneri Ag.
Calamopleurus cylindricus Ag.

Professor Agassiz in the above-mentioned letter restates his belief in the cretaceous age of the fishes; and recent examinations of considerable collections in Rio confirm him in this opinion.

From Ceará, the capital of the province, Professor Agassiz made a journey to the serras, and studied the glacial phenomena there exhibited. Between the sea and the Serra de Aratanha he reports that he found everywhere on the higher lands a morainic soil with boulders.

"On this very serra of Aratanha," says Mrs. Agassiz, "at the foot of which we happen to have taken up our quarters, the glacial phenomena are as legible as in any of the valleys of Maine, or in those of the mountains of Cumberland in England. It had evidently a local glacier formed by the meeting of two arms, which descended from two depressions spreading right and left on the upper part of the serra and joining below in the main valley. A large part of the medial moraine formed by the meeting of these two arms can still be traced in the central valley. One of the lateral moraines is perfectly preserved, the village road cutting through it; while the village itself is built just within the terminal moraine which is thrown up in a long ridge in front of it." *

At the close of the chapter from which I make the above extract Professor Agassiz himself says: —

"I spent the rest of the day in a special examination of the right lateral moraine and part of the front moraine of the glacier of Pacatuba; my object was especially to ascertain whether what appeared a moraine at first might not, after all, be a spur of the serra decomposed in place. I ascended the ridge to its very origin, and there crossed into an adjoining depression, immediately below the sitio of Captain Henriques, where I found another glacial bottom of smaller dimensions, the ice of which probably never reached the plain. Everywhere in the ridges encircling these depressions the loose materials and large boulders were so accumulated and imbedded in clay or sand that their morainic character is unmistakable. Occasionally, where a ledge of the underlying rock

* A Journey in Brazil, p. 456. The Serra de Aratanha is composed of gneiss.

crops out, in places where the drift has been removed by denudation, the difference between the moraine and the rock decomposed in place is recognized at once. It is equally easy to distinguish the boulders which here and there have rolled down from the mountain and stopped against the moraine. The three things are side by side, and might at first be easily confounded, but a little familiarity makes it easy to distinguish them. When the lateral moraine turns toward the foot of the ancient glacier, near the point at which the brook of Pacatuba cuts through the former, and a little to the west of the brook, there are colossal boulders leaning against the moraine, from the summit of which they have probably rolled down. Near the cemetery the front moraine consists almost entirely of small quartz pebbles; there are, however, a few large blocks among them. The medial moraine extends nearly through the centre of the village, while the left-hand lateral moraine lies outside of the village, at its eastern end, and is traversed by the road leading to Ceará. It is not impossible that eastward, a third tributary of the serra may have reached the main glacier of the Pacatuba. I may say that in the whole valley of Hasli there are no accumulations of morainic materials more characteristic than those I have found here, not even about the Kirchet; neither are there any remains of the kind more striking about the valleys of Mount Desert in Maine, where the glacial phenomena are so remarkable; nor in the valleys of Loch Fine, Loch Augh, and Loch Long in Scotland, where the traces of ancient glaciers are so distinct."

From Dr. Felice, a land surveyor familiar with the Serra Grande, Professor Agassiz learned that "there is a wall of loose materials, boulders, stones, &c., running from east to west for a distance of some sixty leagues from the Rio Aracaty-assú to Bom Jesus in the Serra Grande"; and this wall Professor Agassiz believes to be a part of the lateral moraine left by a great Amazonian glacier.

Bones of huge Mammals, Mastodons and Megatheria, are abundant in various parts of the province, and perfect skeletons have been found. Some of the localities are the following: Santa Catharina, (lagôa,) Sitio Cronzó, at the foot of the Serra Ybiapaba at Inhamuns; between Cratheús and Quixeramobim; Timbauba; in the place called Sucatinga a skeleton was found in an excavation, and part was sent to Rio; at Sta. Cruz, in a lagôa, another skeleton was found.

The following notes on the mineral productions of the province are taken from Dr. Pompéo's *Ensaio:* —

Amethysts. Serra do Tauá. Some of the crystals from this province are very large and of a beautiful color.

Gypsum. Araripe (*Fibrous*) Cairirí.

Saltpetre. Found all over the interior, but more particularly at the following places: Tatajuba, where it was extracted by the government; Pindoba (government works); Tagycioca em Curú; Carnahubal em São Pedro de Villa Viçosa; Boassú na Granja; Conceição, Curú; Pirangi, Choró; Uruburetama; Ipú, &c.

Salt. In various parts of the Jardim, of the Sertão, and especially in the Aracaty-assú, the waters are impregnated with salt. As elsewhere there are, in argillaceous soils, salt licks much frequented by cattle. Salt is largely manufactured on the coast from sea-water.

Alum. Feijó says that a considerable surface in the Inhamuns is covered with alum. Capanema reports it from Araripe.

Magnesia. Cafundó, Inhamuns.

Carbonate of Potassium. Ipú, Serra Grande, Crato, S. Gonçalo.

Amianthus. Cairiri, in veins. Quixeramobim near Lavras.

Lignite. Quixeramobim.

*Gold.** Granja, Baturité, Crato, Termo de Milagres, Ipú, Rio Sal-

* Feijó, quoted by Dr. Pompéo, *Ensaio*, &c., 152, says: "De ouro encontram-se mais ou menos vestigios por todos os riachos, corregos e vertentes das montanhas, que formão as costaneiras da serra grande, desde a Timonha até Cari-

gado from Missão-Velha to Lavras. In all these places the gold occurs in grains or powder in sands, gravels, or clays; its source is not known.

Copper. Said to occur in the Serra Grande and elsewhere.

Zinc. São Pedro, near the Serra da Mãosinha, Termo de Milagres. "Dr. Theberge says that he encountered near Milagres a large quantity of *blende* (sulphide of zinc), so abundant that in certain localities it was only necessary to burn a clearing to reduce the metal, which ran into the hollows, where pounds in weight were collected." (Pompéo.)

Galena. Ipú, Quixeramobim.

Molybdate of Lead. (Capanema.) Near Villa Nova.

Sulphide of Antimony. (Capanema.) Near Villa Nova.

Graphite. Baturité, Quixeramobim, &c.

Specular Iron. Cangatí. Iron ore occurs in many localities in the province.

In the *Comptes Rendus*, Tome 5me, p. 211, I find a statement that on the 11th of December, 1836, a large meteor passed over Ceará and exploded over the village of Macão, at the entrance of the river Assú, showering down over a large tract of country fragments of stone, many of which penetrated houses and destroyed cattle. One of these fragments was sent by M. F. Berthou at Paris for analysis.

ris, com particularidade nas vertentes do Salgado, Acaracú e Jaguaribe, no Inhamuns, Banabuihú, Quixeramobim, e cabeceiras de Juré. Em todas essas vertentes e terrenos visinhos basta lavar a terra que se acha debaixo do cascalho para pintar o ouro."

CHAPTER XIV.

PROVINCE OF PIAUHY.

Geographical Position, Limits, &c., of the Province. — The Rio Parnahyba and its Tributaries. — Description of its Basin. — General Geological Structure and Topography of the Province. — Table-topped Hills of Sandstone. — The Serra dos Dous Irmãos and its Structure. — Discussion of Gardner's Observations on the Geology of Piauhy and Ceará. — Gardner mistaken in referring the great Sandstone Sheet to the Cretaceous. — Sandstones probably Tertiary. — Their great Extension over Brazil. — Distribution of the Cretaceous Beds in Brazil. — Climate of Piauhy. — The Campos Mimosos and the Campos Agrestes. — Peculiarities of their Vegetation. — Productions of the Province, Population, &c.

The province of Piauhy forms a rather long and irregular strip lying west of the province of Ceará, and to the northwest of Pernambuco and Bahia. Its area is about equal to that of Bahia, and it embraces all the country watered by the tributaries of the Parnahyba on the eastern side.

This river takes its rise in the Serra da Tabatinga in Goyaz, nearly on the same parallel as that on which the mouth of the Rio de São Francisco is situated, and in the angle formed by the union of the two hydrographical basins of the Tocantins and São Francisco. Its course is approximately north-northeast; and Pompéo gives its length as 330 leagues, which appears to me to be altogether too high an estimate. The same author says that it is navigable for a distance of 260 leagues. It has, according to Pompéo, six mouths, but I cannot learn whether it has a regular delta.

On the west side, in its very upper course, a few small

streams, namely, the Balsas, Balsinhas, Penitente, &c., unite, and with one or two other little rivers enter the main stream, but for the remainder of the distance to the sea the Parnahyba flows so close to the western rim of the hydrographical basin that it does not receive another affluent of importance from that side. On the east, however, it receives a host of little rivers which take their rise in the Serra dos Dous Irmãos and its continuation the Serra da Ybiapaba. Most important among these is the Gurgueia, which drains the extensive lake of Paranaguá,* the Canindé, and the Poty.

The Rio Parnahyba is a white-water stream, flowing for the most of its course through level, more or less swampy lands, grown up with thick bushes and groves of carnahuba and piassaba palms. It is without obstructions, and navigable for a great distance.

The basin of the Parnahyba is a one-sided one, the drainage being towards the east.

As the little coast streams east of the Parnahyba are comprised within the limits of the province of Ceará, Piauhy has a coast line of only about twenty miles, extending from the mouth of the Parnahyba to the Barra do Iguarassú. The base of the country consists in the south and southeast, at least, of gneiss and other metamorphic rocks, much inclined, but the greater part of the country is overspread by a thick sheet of sandstone in horizontal strata, extending southward to the edge of the basin.

This sheet of sandstone has been very extensively worn away by the rivers, and between them are isolated table-topped hills or extensive chapadas. Such is the character

* The same name is applied to the bay on which the principal seaport of the province of Paraná is built.

of the country in the vicinity of Oeiras, where the sandstone is of a reddish tint, and is sometimes exposed in perpendicular cliffs. The Arraial de São Gonçalo is situated at the foot of one of these sandstone hills, which is four hundred feet high.* The Serra da Topa is another sandstone mountain, the rocks being of a white or pale red color and disposed in terraces, the top being perfectly flat.

These sandstones lie on a basis of metamorphic strata, which has a gentle slope northward or northeastward, and just south of the Lake Paranaguá they lap up over these older rocks, abutting against the Serra dos Dous Irmãos, reaching, according to Mr. St. John, a level of about 1,500 feet. Westward they rise toward the so-called Serra de Ybiapaba. The question now is, whether they tie in with the sandstones of the Ybiapaba, and of the Chapadão da Mangabeira, or whether they are newer than the sandstones of Jacobina and the São Francisco-Tocantins divide. From all that I have been able to learn I think that they will be found to form part of the great sandstone sheet of Minas, Goyaz, &c., and that the table-topped hills of Piauhy will be found to be the exact equivalents of those of Santarem, Monte Alegre, &c. The clays and sandstones of the lower grounds bordering the coast would seem to belong to the same series as those of the vicinity of Pará.

The water-shed between the hydrographical basins of the São Francisco and the Parnahyba is, according to the testimony of Spix and Martius, Gardner, and other travellers, a low, very gentle swelling, composed of gneiss, mica-schist, and other similar rocks, all more or less disturbed and denuded down very evenly, as is the case with the similar strata forming the divide between the São Francisco val-

* Alcide d'Orbigny, *Voyage Pittoresque*, 149.

ley and that of the Paraguassú, and with a gentle slope both to the east and west. This gneiss ridge is about 1,250 feet in height where the road from Oeiras to Joazeiro crosses it.

Gardner, in describing the journey from Paranaguá, across the water-shed, southward into the province of Pernambuco, says that, shortly after having passed the boundary of the province of Piauhy, an elevated table-land is reached called the Serra da Batalha, which he describes as being covered on its slopes with huge blocks of sandstone, of which he supposed the serra to be composed. Two other serras of the same character lie to the south, and Gardner supposes them all to form part of one great range.

Padre Cazal, in his *Corographia*, says that in some parts the serra dividing Pernambuco from Piauhy has two or three leagues of chapada on top.

From Mr. St. John's observations it would appear that in some places the sandstones are completely removed, as is the case in Bahia also.

That the Serra dos Dous Irmãos should have its capping of sandstone is not wonderful, when we find the Serra da Ybiapaba so capped to a height of over two thousand feet, and when on the highest land on the São Francisco-Paraguassú divide we find patches of the same rock, forming a series of chapadas traceable southward, and tying in with the chapadas of the Pardo and Jequitinhonha valleys.

The age of the sandstones of the Serra da Ybiapaba and the Serra de Araripe, and also of the great sandstone sheet covering so large an area in the province of Piauhy, remains to be determined. Gardner has called them all cretaceous. Let us examine upon what grounds he has based his conclusion. The whole matter turns upon the relation the fish-bed bears to the strata composing the serra. He says in his "Travels" (p. 202) that "the place where these [fish] were found was on the slope of a low hill about a mile from the serra," and that the specimens all come from loose masses, rounded, as he erroneously supposed, by the action of water, and scattered over a very limited surface. He consequently did not see them in place. Owing to his finding in the immediate vicinity pieces of sandstone * like that which form the serra, he arrived at the conclusion that the fishes come from the sandstone, and that because the fishes were cretaceous, therefore the serra, and not only the serra, but all the great extension of sandstones covering the northeastern shoulder of Brazil, must be cretaceous also, and he even went so far as to divide these rocks into a series of groups, referring them to European horizons.

The fishes do not occur in rolled masses, as Gardner says, but in concretions, as Mr. Bowman has remarked in a note to Gardner's paper, and as I can also testify after an examination of specimens in my own collections. The occurrence of these concretions on a surface unassociated with other rock would lead one to suspect that they had weathered out of some softer rock, and this is confirmed by a statement made by Pompéo, on the authority of Dr. Theberge, that the fossils are found in "an extremely sticky clay." Theberge says that they either occur in this way or at the bottoms of deep valleys in brooks whose bed is a schistose limestone, so that we have no evidence whatever that these fossils occur in the sandstones of the serra; on the contrary, they occur in a band of rocks lying well below the sandstones, and bared in the valleys by the denudation of the sandstones, so that we may safely conclude that they are

* In his paper in the Phil. Trans., Gardner says that these were little rounded sandstones. In his Travels he speaks of them as *rounded blocks*.

older than the sandstones.* Gardner was misled, in the first place, into the belief that the sandstones were cretaceous from finding rocks having some resemblance to flint, and by mistaking for chalk a white tabatinga clay on the top of the serra, dug by the inhabitants for use in white-washing. That the cretaceous rocks form an extensive series underlying the sandstones is very probable, because Gardner speaks of a number of localities, all of which he describes as being situated on the declivities of low hills skirting the base of the serra. On the west side of the Serra da Ybiapaba he found a similar deposit of fossil fishes, which would lead one to suppose that this serra was also underlaid by the same deposit.

Gardner, in his "Travels," says that the series of cretaceous rocks in Ceará and Piauhy forming the serras, &c., consists in descending order of—

1. White chalk with flints exposed in pits, and partially overlaid by red diluvial clay.

2. Sandstone with ichthyolites, equivalent to the English upper green sand.

3. A series of marls, soft and compact limestones, and lignite, equivalent to the English gault.

4. A ferruginous sandstone deposit, equivalent to the lower green sand or Shanklin sands.

Now, of this series, the white "chalk" is a Tabatinga clay. There is no evidence that it contains flints, and he himself says that he found none in any of the "chalk-pits" he examined. In the second place, there is no evidence that the fossil fishes occur in the sandstones, and if

* In a paper read at the Salem meeting of the American Association, I ventured to state my strong suspicion that the beds affording the fishes would turn out to be disturbed and inclined, like the other cretacean beds in Eastern Brazil, when Professor Agassiz remarked that this had been reported by Dr. Coutinho.

they did they would not prove the deposit to be upper green sand, since Professor Agassiz has called attention to the resemblance borne by *Aspidorhynchus Comptoni* and *Lepidotus temnurus* to allied species of the chalk of Kent. So far as the third and fourth series are concerned, Gardner found no fossils in them, and their cretaceous age is unproved. Pompéo, with Capanema most probably as authority, gives the following as the succession of rocks in the serra of Araripe: —

The uppermost beds, A, consist of beds of *Psamenito*, sandstone of a reddish color, with bluish, sometimes black nodules. Below these comes a bed, B, of an exceedingly foliated limestone, under which lies a bed, C, of black clay (?), (Tauá), with layers, a palm in thickness, of a bluish and very hard sandstone, containing veins of pyrites and galena (?), or with a very bituminous schist containing the same sulphides and spherical nodules. This is in turn underlaid by a series of sandstones, D, less argillaceous in character. In a sandstone similar to this occurring at St. Pedro, Dr. Gonçalves Dias found fossil wood.

Of this series, A corresponds to No. 2 * of Gardner's series, B and C to No. 3, while the lowest sandstones of his series may or may not correspond to D.

Both of the above sections appear to have been made in the Serra de Araripe, near Crato. Gardner says he saw no limestone *in sitû* at Jardim.

That these horizontal strata, and especially the sandstones, have anciently had an immense extension over the surface of Ceará there cannot be the least doubt, as Dr. Capanema has remarked, for the sandstones are often

* Gardner numbers the beds from below upwards; accordingly this would correspond to his No. 3.

found capping isolated hills, at a long distance from the serra. Indeed, there seems every reason to believe that the provinces of Ceará and Piauhy were covered with it even beyond the coast line. It has been swept away very largely from the coast and the basin of the Jaguaribe almost to Crato.

The upper part of the Serra da Ybiapaba for apparently its whole length is composed of the sandstone, and the plateau of Apodí seems to be formed in part of it. It is true that where Spix and Martius crossed the Serra dos Dous Irmãos they found, up to the highest point of the pass, 1,250 feet, only metamorphic rocks; but they found overlying these rocks thick beds of laminated clay (*schieferthon*), and they report the Serra da Topa as composed of *Quadersandstein*. Mr. St. John did not meet with the sandstones, but Gardner did in crossing the Serra da Batalha, and he describes the top as a flat chapada. The Chapada de Sta. Maria, lying between the São Francisco and the eastern branch of the Tocantins, is another dead-level table-land composed of sandstone, south of which comes the great Chapadão de Urucuia, evidently only the prolongation of the Chapadão de Sta. Maria. On the opposite or southern side of the valley of the Urucuia we see the chapadas once more.

In the southern part of Minas, as already remarked, there are the chapadas between Piumhy and Passos on the Rio Grande, east of which is the immense Chapadão de Tabatinga, some two hundred miles in length, between the two branches of the Paraná, the Paranahyba, and Rio Grande. Then we have the elevated plains of the provinces of São Paulo and Paraná, composed in great part of argillaceous sandstones. The valley of the São Francisco to the Cachoeira de Paulo Affonso is filled with horizontal deposits, calcareous in the upper part of the valley, silicious and calcareous in the lower part. East of the valley of the São Francisco we find the hills of the water-shed between the São Francisco basin and the streams flowing eastward into the Atlantic capped here and there with horizontal deposits, worn away on every side; the basins of the Rios Pardo and Jequitinhonha are filled with these beds, and everywhere they rise to a very uniform height over the country, — a height of 2,000 to 3,000 feet above the sea. Westward we know that an immense tract of country in the province of Matto Grosso and the Amazonas is covered in like manner with similar deposits. All these facts speak of a very uniform submergence of the whole country to a depth of at least 2,000 or 3,000 feet below the present sea level, during which the valleys were filled up with beds of clays, sandstones more or less argillaceous, limestones, &c., to a greater or less height. These deposits appear in great part to have been rapidly formed in the bottom of a muddy sea, the material being derived from the decomposed rock crust covering the country.

The fossil fishes, according to Agassiz, resemble those of the European *sénonien,* so that if the sandstones are cretaceous they must belong to the very uppermost division of that formation. We have seen how this same great formation extends over almost the whole Empire, but nowhere that I have heard of affording a single fossil. But on the coast outside the edge of the plateau, lying on the extension of the gneiss basis, we have, beginning at the Abrolhos and extending northward through Bahia, Sergipe, Pernambuco, and Parahyba, a great series of cretaceous rocks belonging to different epochs in that period. Wherever I have seen them they are always disturbed, while they moreover form border deposits abruptly abutting the edge of the plateau, showing that the plateau was out of water when they were deposited. These cretaceous beds are compact sandstones, shales, limestones, conglomerates, &c.

The beds of the great sandstone formation approach the coast in very many places, as on the Rio de São Francisco, where they make their appearance near the Cachoeira de Paulo Affonso. But they are everywhere horizontal, and are nowhere disturbed by the foldings which tilted the beds of the cretaceous a few leagues to the east. But the cretaceous beds of the coast are overlaid by a series of clays, sandstones, &c., which, though strictly a coast formation, bears in lithological character a very close resemblance to the great sandstone sheet of the interior. This latter, which is overlaid by the drift I have referred, as already stated elsewhere, to the tertiary; but this coast formation is, I think, without doubt, younger than the similar formation of the interior, so that the latter must be either upper cretaceous or tertiary, and I must give it as my firm conviction that it will be found to be the latter.

As to the occurrence of drift in the province of Piauhy, I have no information whatever.

I know of no workable mines, though gold is said to occur at Olho d'Agua, near Oeiras.

The climate of Piauhy is hot and, according to Pompéo, rather damp. It is apt to be very prejudicial to foreigners, especially in the low grounds along the banks of the Parnahyba, Poty, and other streams, where intermittent fevers are prevalent.

The inhabitants make a distinction between the vegetation of the eastern part of the province and the central and western part. That of the former region is called *mimosa*, and is characterized by catinga forests, while its plants are furnished with an abundance of hairs and prickles, stiff leaves, small flowers, a very tender fibre, and very often a milky juice. Gardner says that the grasses of the mimoso pastures are annuals, their color is a brighter green, and they have more pliant leaves than those of the agrestes. Spix and Martius give a long list of grasses which are characteristic of the campos mimosos.

The campos agrestes of Lower Piauhy consist in part of woods, in part of quite open plains. The trees are, according to Gardner, almost all deciduous, and many are gnarled and stunted. Swamps are not infrequent and support clumps of Buritî palms. The grasses of the open plains are coarse and perennial.

The rains begin in October and last until April, heavy thunder-storms prevailing during that season.

The principal industry of the province consists in the raising of cattle, and agriculture is pursued only to a small extent.

The population amounts to about 250,000, of whom about 30,000 are slaves. The capital is Theresina, which has about 6,000 inhabitants. The other cities are Oeiras and Parnahyba. The latter is particularly unhealthy.

CHAPTER XV.

THE PROVINCES OF MARANHÃO, PARÁ, AND AMAZONAS.*

Sandstones of the Coast of Maranhão. — The Interior composed of Metamorphic Rocks. — Gold-Mines of Turí and Maracassumé. — Climate of the Province. — Rains. — Cities of Maranhão, Caxias, &c. — Pororoca or Bore at the Mouth of the River Mearim. — Professor Agassiz's Sketch of the Geology, of the Amazonian Valley. — His Theory of the Mode of Deposition of the Amazonian Beds. — Discussion of this Question. — Cretaceous Rocks in the Amazonian Valley.

THE coast of the Province of Maranhão is low and flat, and consists of a tertiary ferruginous sandstone passing into conglomerate, and overlaid, as in the vicinity of the port of Maranhão, by a series of sandstones and clays. Gardner, apparently on the principle that all sandstones must be cretaceous, refers this series to that formation, while Spix and Martius, as usual, call the rock *Quader-sandstein.* The sandstones and associated rocks form a line of high red cliffs along the shore of the island on which Maranhão is built, just north of the city, east of point São Marcos. On the mainland west of the channel a similar line of cliffs stretches from the village of Alcantara to the curious landmark, Mount Itacolumí. The same rocks extend far up the valley of the Itapicurú, on the banks of

* That this name was given in commemoration of the supposed tribe of female warriors described by Orellana there cannot be the slightest doubt. The attempt to derive it from *amassona,* a word not to be found in the Portuguese dictionary, falls into the same category as the derivation of *Maranhão* from *mar ou não?* or *Alexander the Great* from *all eggs under the grate!* It is not the Rio Amazonas, but the Rio das Amazonas, the river of the Amazons.

which at Mangue Alto they were observed by Spix and Martius to lie on granite containing pistacite, and at Cachoeira this rock passed into a syenitic form.

Mr. St. John, who descended the Itapicurú and made a considerable stay at Maranhão, will doubtless, in his report on the geology of the country, furnish us with valuable information concerning the character and extent of the sandstone deposits.

In the south and west the country is higher, more unequal, and very largely composed of ancient metamorphic rocks.

Gold occurs in the province, and is or has been worked by a mining company; but I have been unable to obtain facts bearing upon the nature of the deposits or their yield. The two principal mines are those of Turí and Maracassumé.

The climate of Maranhão, — situated as the country is on the edge of the great Amazonian valley, of which we are told by Professor Agassiz it anciently formed a part, — has the same general character as that which prevails on the Amazonas, being hot and damp; the greatest heat is about 31° Cent., 97.8° Fahr., and the lowest 21° Cent., 69.8° Fahr. (Pompéo.) "The rains begin with great regularity at the end of December, although from October on showers occur, commonly called the *chuvas de cajú* [the cashew-rains]. The rains are very abundant, and accompanied by much thunder and lightning, becoming more frequent and heavy in May, the end of the winter season." * From June to December the winds called *ventos geraes* blow steadily from the northeast or east-northeast during the day, and during the night-time from the east.

* Pompéo, *Geographia*, p. 391.

The climate, except in the vicinity of the Paranahyba River, is said to be quite healthy.

A great part of the country is heavily wooded with the virgin forest, but in the interior there are some extensive campos and alluvial flats often inundated during the rainy season. The principal products of the country consist of rice and cotton, although a little coffee is planted. A large number of cattle are raised.

The population of the province amounts to about 390,000 inhabitants.

The capital is Maranhão, a beautiful city of 35,000 inhabitants, built on an island lying off the mouth of the Itapicurú River. Alcantara is another considerable town situated on the mainland opposite the capital, and in the midst of a region noted for producing a most excellent quality of cotton.

The city of Caxias, on the navigable river Itapicurú, about three hundred miles from Maranhão, is a large town, the centre of an important trade with the interior. Carolina, on the Tocantins, is a town of but little importance.

About twelve leagues west of the capital of the province is the mouth of the river Mearim or Meary. This river has so strong a current, and its channel is so shaped, that it causes the tide to enter with a bore. Cazal says that the river suspends for a long time the rise of the tide, then it comes in with great fury, rising in a quarter of an hour the distance it had taken nearly nine hours to fall, and then running for three hours with the rapidity of a mill-race.†

* The river is navigated by steam.

† Speaking of this river, Cazal says : " Seu alvo he profundo, e largo ; e sua corrente tão rapida, que suspende a enchente da maré por largo tempo ; resultando desta opposição ondas encapelladas, chamadas *pórórócas,* que depois de

It will be remembered that a similar phenomenon is witnessed at the mouth of the Amazonas, and in the vicinity of Pará, where, as in Maranhão, the bore is called pororoca.

Professor Agassiz has treated so largely of the physical geography of the provinces of Pará and Amazonas, or the Amazonian valley, in the "Journey in Brazil," that it is not necessary for me to repeat here any of his conclusions; besides, the limits of this volume forbid that I should at this time enter upon the discussion of so fruitful a subject. I shall therefore confine myself, so far as the Amazonas is concerned, to a very condensed statement of Professor Agassiz's views with reference to the origin and stratigraphy of the various formations which occupy the Amazonian valley, and with a few remarks thereupon; this seems necessary in order to complete my sketch of the Geology of Brazil. In the *Bulletin de la Société Géologique de France* (2me Série, T. 25, p. 685) is a short article on the Geology of the Amazonian valley, by Professor Agassiz and Dr. Coutinho, presented by Professor Jules Marcou, which gives Professor Agassiz's views with great conciseness and clearness; and as it is not accessible to general readers, I have reproduced the most important part of it here, together with the section accompanying it.

Professor Marcou says: —

"Mr. Agassiz thinks that the whole valley of the Amazonas was formed at the end of the cretaceous period, which has left traces of deposits in the province of Ceará and on the Upper Purús. Here

vencidas, tudo quanto vazou em quazi nove horas, enche em menos d'hum quarto; ficando a maré caminhando para cima tres horas completas com uma rapidez semelhante á calha d'hum moinho. Este fenomeno occupa o espaco de cinco leguas com grande roído. Ha sitios, denominados *espéras*, onde as canoas esperão a decizão do combate, e continuão a viagem sem perigo." — *Cor. Braz.*, Tom. II. p. 260.

and there, whether by denudations or by anterior dislocations, one sees more ancient rocks. Thus Major Coutinho has found palæozoic brachiopods in a rock which forms the first cascade of the Tapajos; carboniferous fossils have been collected on the banks of the rivers Guaporé and Mamoré, in Matto Grosso; and finally, at Manáos, Coutinho has recognized slates or *phyllades* in a very inclined position, and beneath the formations of red sandstone of the Amazonian valley."

Professor Agassiz supposed that during the tertiary the Amazonian region was above water, and that the sandstones and clays that now fill it are drift.

The following is a copy of the ideal section of these later deposits by Professor Agassiz, forming a *résumé* of the observations of M. Coutinho and himself: —

"I. Coarse sands (*Sable grossier*) forming the base of the drift throughout where the level of the water has uncovered the lower beds of plastic clays.

"II. The streaked plastic clay (*Argile plastique bigarrée*) shows itself on a large scale along the sea-coast at Pará, at the Island of Marajó, Maranhão, and here and there in the hollows along the course of the Amazonas.

"III. Foliated clay in very thin beds, with frequent indications of cleavage. This deposit appears to be more considerable in the banks along the course of the Rio Solimões than in the lower part of the Amazonas. It is in these beds at Tonantins, on the Rio Solimões, that M. Agassiz has found leaves of dicotyledonous plants, which appear to be identical with species at present living in the valley of the Amazons.*

* These leaves occur in a fine, soft gray clay, resembling very closely the recent alluvial clays of the Brazilian rivers. They are excellently preserved. The leaf is partly carbonized, but it curls up from the surface on drying, and may be detached, leaving a beautiful impression of the venation, &c.

"IV. A crust of sandy clay, very hard, moulded in the inequalities of the foliated clay.

"V., VI., VII., VIII., and IX. *Sandstone formation*, sometimes regularly stratified and compact, especially in the lower beds (V.), such as one sees on the borders of the *igarapés* of Manáos; sometimes cavernous and intermixed with irregular masses of clay (VI.), especially well developed at Villa Bella and at Manáos; at others all the characters of a torrential stratification (VII., VIII., and IX.). The deposits of this last nature are only seen in the elevated hills of Almeirim, Ereré, and Cupatí, and in the most elevated cliffs of the borders of the river, as at Tonantins, Tabatinga, São Paulo, and on the borders of the Rio Negro.

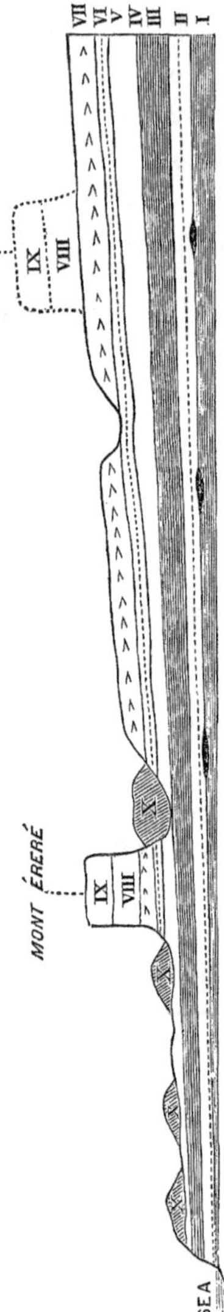

"X. The argilo-arenaceous, unstratified drift, occupying all the inequalities of the soil resulting from the denudation of the sandstone with torrential stratification. It is in this drift that MM. Agassiz and Coutinho have found true erratic blocks of diorite, of a metre in diameter, at Ereré. This formation is never met with on the cliffs elevated several hundreds of feet in height. There is not a trace of it on the summit of the hills of Ereré.

"The fact that the coarse sand, No. I., appears throughout at the level of low water, that it follows the general slope of the valley, shows incontestably that the deposition of this formation does not reach back to an epoch anterior to the excavation of the valley itself. The total thickness of the Amazonian drift does not exceed three hundred metres; it covers the whole basin of the Amazonas, from the Andes of Peru and Bolivia to Cape São Roque; or, in other words, it is the most colossal drift formation known."

Professor Agassiz believes that the beds I., II., III., IV., or the coarse sands and clays, were deposited in a lake or sheet of fresh water occupying the valley of the Amazonas, and sustaining on its surface a glacier, descending eastward from the Andes, and furnished with a gigantic moraine in front stretching across the mouth of the valley and converting it into an inland fresh-water lake. After the ice had broken up and become more or less disintegrated, and the waters of the lake had swollen, the sandstone formation V., VI., VII., VIII., IX. was laid down, then the barrier was burst; the waters of the lake, suddenly released, furrowed and wore down the sandstone beds, sweeping them entirely away over an immense area, leaving only isolated hills, like those of Éreré, Obydos, Cupatí, Almeyrim, &c., standing as remnants of the once universal sandstone sheet. After this period of turbulence and denudation came on an epoch of quiet, and in the bottom of the diminished lake the clays, No. X., were deposited, while ice-rafts floating on its surface dropped here and there boulders, to be buried in the accumulating material. Then the moraine was destroyed; the drainage of the waters furrowed deeply these clays, and even cut through them into the sandstone below, in which the various channels of the system of the Amazonas are excavated. Professor Agassiz believes that the great barrier stretched across the Amazonian valley far eastward of its present extremity, and he has called attention to the similarity between the formations found spread over the coast of Maranhão and Piauhy and the Amazonian formations here described, showing conclusively that these deposits were once continuous. It is his belief that the Amazonian formation formerly extended a hundred leagues out to sea beyond the present mouth of the Amazonas. There can be no doubt that there is a rapid waste of land going on along the sea-shores of the mouth of the Amazonas and of the coast eastward for a long distance,— a waste amounting to even so much as two hundred yards in ten years in the bay of Braganza, or a mile in twenty, as on the coast near Vigia, where an island a mile wide disappeared in that time.*

I have three times visited Pará, and have had an opportunity of seeing something of the Amazonian formation. The rock underlying the town, and exposed at ordinary low-water level at the base of the bluff under the fort, is a coarse dark-red sandstone with an abundant cement of iron oxide, and precisely like the red sandstone I have so often described as occupying a similar level and underlying the tertiary clays of the coast of Rio, Espirito Santo, &c. Over this sandstone is a considerable bed of red, white, and particolored felspathic clays, with a greater or less admixture of sand, which clays present exactly the same appearance and level as the tertiary clays of the provinces just referred to.

Before I knew anything of the conclusions of Professor Agassiz as to the age of the Amazonian deposits, I had satisfied myself that the clays and sandstones extending along the whole eastern coast of Brazil, from the Bay of Rio to

* Since the tertiary, at least, and I believe for the greater part since the drift, the whole eastern Brazilian coast has suffered denudation by the sea to an immense amount, and a very wide strip of tertiary rocks has been removed. I believe that these deposits once extended beyond the Abrolhos, and that south of Cape Roque the sea has cut them away for a mean width of fifty miles or more.

the Amazonas, were older than the drift clays which overspread them; and as they are stratified deposits on an open coast, there can be no doubt of their marine origin. At first, misled by what I had read of the geology of Brazil, as well as by the strong resemblance the sandstones bore to the new red sandstones of the Basin of Minas in Nova Scotia, with which I have been familiar since my boyhood, I was disposed to regard the Brazilian formation in question as triassic; but I soon found that it was underlaid unconformably by cretaceous rocks in Bahia, and I came to the only conclusion possible, — that it was older than the drift and newer than the cretaceous. I can see no reason, therefore, for considering the coast beds as anything but tertiary, though they may be, and probably are, very late tertiary. I have attempted no comparison between these beds and the tertiary beds of the pampas, because, in the absence of fossils, and having never seen the pampian tertiaries and post-tertiaries, I have nothing to aid me in instituting such a comparison. It has seemed to me that the fact of the occurrence on an open sea-coast of clays and sandstones precisely similar to those occupying the lower plains of the Amazonas, as at Pará, and in fact tying in with them, relieves one of the necessity of looking to a fresh-water origin for the Amazonian beds.

There cannot be the slightest doubt that the beds forming the mountains of Ererê, Almeyrim, once covered the whole valley, and have been enormously denuded. I should never have doubted whether the red sandstones at Pará really belonged to the series of beds forming the Monte Alegre-Ererê hills, if I had not found along the coast of Espirito Santo the same sandstones with precisely similar overlying clays, with no table-topped hills piercing them as at Ererê.

Only once have I seen what I thought to be a table-topped hill standing on the coast tertiary plain. That was on the coast south of Rio de Contas, the observation was a doubtful one, and I have felt more like instituting a comparison between the chapadas on the Amazonas and those of Ybiapaba and Minas Novas; or, in other words, I am disposed to regard the chapadas of Ereré as the outliers of the great tertiary sheet that once covered the great Brazilian plateau, and now lies unbroken over such an immense extent in the province of Matto Grosso. According to the observations of Dr. R. P. Stevens and others, the plateau of Guyana is covered by an extension of the same great sheet, while the valley of the Orinoco is occupied with clays precisely similar to those of the valley of the Amazonas.*

It is with much hesitation that I express an opinion at variance with so distinguished an authority as Professor Agassiz; but the facts have seemed to need a different interpretation from that which he has given them. My conclusions, after all, do not affect his theory of the former existence of glaciers under the tropics, down to the present level of the sea, — a theory which I hold as firmly as he.†

* Professor Orton found west of Tabatinga tertiary shells in beds which he considers to be a part of the Amazonian formation. These fossils were described by Professor Gabb in the *Journal of Conchology*. The species are *Neritina pupa, Turbonella minuscula, Messalia Ortoni, Tellina Amazonensis, Pachydon obliqua,* and *P. tenua.* It is much to be regretted that Professor Orton has not given a description of the locality where these fossils are found, and of their mode of occurrence.

† I have confined myself in this chapter to a short discussion of the question of the age of the Amazonian sandstones, and I have attempted to give no description of the great river and its wonders. I would refer the reader desirous of knowing more of the Amazonas to the "Journey in Brazil."

It is indeed surprising that after several hundred volumes, classic in science,

Professor Agassiz has called attention to the fossils from cretaceous beds discovered by Mr. Chandless on the river Aquiry, one of the affluents of the Purús. These beds consist of hardened clay and "pseudo-conglomerate," * — the latter being a sort of clay rock full of concretions, which give to the formation the appearance of a conglomerate. Associated with it is a sandstone. The fossils are said by Mr. Chandless to be very abundant, and, according, I believe, to the determination of Professor Agassiz, they consist of the bones of Mosasauri † and turtles, together with fossil wood. These remains appear to be principally confined to the clays and "pseudo-conglomerate."

Professor Agassiz regards these fossils as indicating a horizon like that of the Maestricht beds in Europe (*Maestrichtien*, — upper chalk). Judging from the description of

have been written on Brazil, and some scores of works have been published on the Amazonas, by such writers as De la Condamine, Humboldt, Spix and Martius, Prince Adelbert, Bates, Wallace, Agassiz, and a host of others of greater or less note, the idea should be so generally prevalent that the country is unexplored, a perfect *terra incognita*, and that every year or two some traveller never before heard of should astonish himself, if not the world, by rediscovering the river. Perhaps after spending a month on its waters, the greater part of which is consumed on board his canoe or the steamer, he writes a book, or at least a magazine article or two! Few countries have suffered more in America at the hands of superficial travellers and writers than Brazil. I would especially recommend to the attention of my readers the excellent little book by Mr. Bates, "The Naturalist on the Amazonas." Its author is a good naturalist, and his eleven years of residence in the country have enabled him to write with great accuracy.

* Chandless's papers on the Rio Aquiry. (Jour. Roy. Geog. Soc.)

† Professor O. C. Marsh, in one of his papers read before the Salem meeting of the American Association, called attention to the rarity of the Mosasauroid forms in the European cretaceous and their great abundance in the cretaceous of North America. It is interesting to observe the occurrence of this same type in South America.

Mr. Theberge, the Aquiry beds must resemble those of Ceará, in which the fossil fishes occur.

On the Purús Mr. Chandless * found the same beds at about lat. 7° 15′ S., long. 66° W., with bones and an abundance of fossil wood.

* Ascent of Purús, Jour. Roy. Geog. Soc., Vol. XXXVI.

CHAPTER XVI.

PROVINCES OF GOYAZ AND MATTO GROSSO.

The Geographical Position of the Province of Goyaz. — The Chapada da Mangabeira. — Geology of the Vicinity of Natividade. — Gold-Washings of the Serra da Natividade, the Arraial da Chapada, and the Arraial da Conceição. — Structure of the Serra at the Town of Arrayas. — The Serra Geral. — Subterranean Streams. — Western and Southern Goyaz composed of Metamorphic Rocks. — Distribution of Gneissose and Granitic Rocks in Western Brazil. — The Montes Pyreneos and their Height. — The Rio Araguaya and its Navigation. — Dr. Couto de Magalhães. — Ilha do Bananal. — Note on Piranhas. — Gold, Diamonds, Iron, and Chrome Ores. — Climate, Forests, Population, &c. — The Western Part of the Plateau of Brazil composed of undisturbed Beds of Sandstone, &c. — The Amazonas-Paraguay Water-shed a Plain without Serras.

THE materials for writing a sketch of the Geology of the Province of Goyaz are very meagre, since it has never been explored by any competent modern geologist. St. Hilaire, Pohl,* Burchell, and several other naturalists visited the province in the early part of the century. Gardner made a journey through the eastern part in the year 1840, making a few geological observations; but Castelnau, in 1844, travelled very extensively through the length and breadth of the province, furnishing us with connected and valuable geological sections, though these sections and the accompanying text rarely ever do more than indicate the litho-

* Pohl's (G. E.) *Reise im Innern von Brasilien in den Jahren* 1817 - 1821, 2 Bde., *mit Atlas* (Wein, 1831 - 37), and his *Beiträge zur Gebirgskunde Braziliens* (München, 1832) I have never seen.

logical character of the formations. The stratigraphy is vaguely given, and no attempt is made to show the age of the different deposits.

Goyaz lies west of the provinces of Piauhy, Bahia, and Minas Geraes, and is very long from north to south, and narrow from east to west. It comprises the basin of the Tocantins above its junction with the Araguaya, the part of the basin of the Araguaya east of that river, and the right side of the basin of the Paranahyba, from the Rio Jacaré to the Rio Apuré.

The Chapada da Mangabeira is, as already remarked, the southward continuation of the table-land separating the province of Piauhy from the valley of the São Francisco. It is in some places quite forty miles in width. Its top forms a plain, and it consists of horizontal beds of sandstone lying on metamorphic rocks. Between the chapada and Natividade the country is composed of these latter strata, while the serra at Natividade has, according to Gardner, the centre of granite, overlaid by schistose rocks. The western side of the serra is bounded by beds of a very compact grayish-colored limestone, which extends northwards for several leagues, forming a range of low hills. The surface deposits on this serra, which are largely composed of a ferruginous gravel, doubtless like the drift *cascalho* of Minas Novas, contain gold, anciently mined to some extent. Gardner says that the view east and north from the serra is bounded by several low ridges, but that to the west and south the country appeared to be one vast plain. The same traveller tells us that the whole country about the Arraial da Chapada, a few leagues west, has been turned over in search of gold. The Arraial stands on a low chapada, but Gardner does not describe its structure. Gold also occurs in the vicinity of the Arraial da Conceição. Gardner's account of the structure of the serra on which the town of Arrayas stands is interesting. The rocks are all metamorphic and almost vertical, the inclination tending toward the east. Gardner says: "The most westerly of these rocks have an arenaceous, schistose structure, and these overlie a very compact, grayish-colored, stratified rock very much resembling gneiss, in which are imbedded innumerable rounded pebbles of granite and quartz, of all sizes, from one to three or four inches, and which is probably equivalent to the graywacke rocks of the Old World." Limestones, which occur both to the north and south, were not observed here. Whether the limestones mentioned belong in the same series with the slates does not appear from Gardner's statement. The Serra Geral, eastward of Arrayas is described as being of no great elevation, and presenting a level top as far as the eye can reach, being evidently the continuation of the Chapada da Mangabeira, and, like it, composed of horizontal sandstone beds.* In speaking of the road from Bonita to Arraial de São Domingos, we are told that "the top of the serra still continued to be level, with a precipitous face, the rock being of a reddish yellow," and that "shortly after leaving Bonita an elevated pyramidal peak of the same elevation as the serra is descried to the southeast, presenting a remarkable resemblance to some enormous work of art," so that there can be no doubt about the general structure of the serra.

Castelnau represents on his map a little stream just north of São Domingos as flowing in a subterranean channel.† Gardner describes a river near the Fazenda de São João,

* This is confirmed by Mr. Thomas Ward, who has travelled over it.

† Castelnau, Vme Partie, *Geographie*, *Atlas*, Planche 4.

that disappears in an opening in limestone strata, and runs underground for several miles, when it reappears.*

On his journey to São Romão Gardner followed the Serra Geral from near São Domingos to the head-waters of the Urucuia, and he describes it as one great elevated plain or chapadão. Of the geological structure of the southern part he gives but few hints, but as he occasionally mentions the occurrence of limestone it is very probable that it may be, in part at least, composed of strata of that rock, the continuation of the horizontal limestone deposits of the Rio das Velhas. Most maps represent a narrow mountain-chain separating the basins of the São Francisco and the Tocantins, and Gerber, in his map of Minas Geraes, though he represents correctly the Chapadões de Santa Maria and do Urucuia, with their great level tops, draws along the water-shed a mountain-chain on top of the chapada, calling it the Serra das Araras and the Serra do Paranan. Along this whole region we have no evidence that I have seen of the existence of any extensive elevations breaking through the great table-land. Castelnau's map of the southern part of the province shows the chapada as extending southward nearly to Catalão. Gerber represents a chapada on the opposite or south side of the valley of the Urucuia. When we come, however, to the southern part of the province, we find, just over the limits, in the Province of Minas Geraes, the immense Chapada da Tabatinga lying between the Rios Grande and Paranahyba, and composed of horizontal beds of sandstones, &c.

According to Castelnau, Saint Hilaire, and others, the

* There are other examples of subterranean streams in Brazil. Gerber represents the Rio Pardo, an affluent of the São Francisco on the left side, as flowing under a ridge.

foundation rock of western and southern Goyaz is everywhere gneiss, mica-schists, clay-slates, and limestones, evidently belonging to the same metamorphic series we find in the eastern part of the Brazilian plateau. These rocks are much folded, rising in mountains comparing in elevation with those of Minas Geraes. The ridge dividing the basin of the Tocantins from that of the Paranahyba branch of the Paraná is of the same character. Clay slates and other metamorphic rocks are seen at Cuiabá and Diamantino in Matto Grosso, and Chandless * speaks of granite seen in the bed of the river Tapajos, ten miles above the Rio de Peixes, just below the Rio das Tropas, at the shallow of Mangabal Grande, and at various points below on the same river. All these observations go to show that the great Brazilian plateau, like that of Guyana, was originally wholly composed of gneissose and schistose metamorphic rocks, much disturbed throughout. It has been supposed by some that, going westward from the Serra do Espinhaço, the signs of metamorphism disappeared, and that rocks which in eastern Minas Geraes might have been highly metamorphosed spread out flatly westward, as in the eastern part of the United States the palæozoic rocks which were folded along the Alleghanian region spread horizontally over the west. This is not the case, so far as I can learn, in Brazil. The metamorphic part of the Brazilian plateau, so high on the east, in Minas slopes off to the north-northwest, and southwestward from the vicinity of Ouro Preto, and dips under the great sheet of tertiary rocks, showing itself only where these are denuded, or where an occasional and rare prominence pierces these strata, but a ridge of these rocks stretches off in a series of high lands from Ouro Preto and Barbacena into Goyaz.

* Journal Royal Geographical Society, Vol. XXXII.

Mr. Thomas Ward, in a note addressed to the author, very aptly describes the Province of Goyaz as a metamorphic island in a sea of sandstone, and such seems indeed to be the case. The sandstones have been swept away from the greater part of the Araguaya-Tocantins basin, leaving the irregular surface of the metamorphic rocks exposed. The highest points in Goyaz are the Montes Pyreneos, near the city of Goyaz, which are said to be over 9,500 feet.*

The highlands in Southern Central Goyaz, collectively known as part of the Serra dos Vertentes of Baron von Eschwege, form the water-shed between the Tocantins-Araguaya basin on the north and the Paraná basin on the south. The Araguaya and Tocantins above their junction are both large rivers, but the Araguaya is very much longer, and should rank as the main river. It flows for the greater part of its length at a much lower level than the Tocantins, and it offers much greater facilities for navigation. The Araguaya has been several times explored. Castelnau † in 1844 descended the river from the mouth of the Crixas to its union with the Tocantins, and then ascended the Tocantins, making plans of both rivers. He found the Araguaya navigable and with but few obstructions. In 1856 the President of the Province sent Sr. Vallée to explore the same river, and he reported that it might easily be made navigable.‡

* I find in the Bahia *Interesse Publico* for the 21st of November, 1868, a letter from Sr. H. R. dos Genettes, describing an ascent of the Pyreneos. This gentleman says that he ascertained the height of the most elevated point to be 2,932 metres, or about 9,619 feet, which is much greater than had been supposed.

† Castelnau, *Expéd. dans l'Amér. du Sud, Hist. du Voyage,* Tomes I. et II.; also Atlas.

‡ E. J. C. Valleé, *Exploração do Rio Araguaya.* Published in the Annexo P, to one of the Government Reports, published, I think, in 1865. My copy wants the title-page.

Dr. Couto de Magalhães, formerly President of Goyaz, espoused the cause of the Araguaya, and navigated it in a little steamer from Jurupencem, fourteen leagues from the capital on the Rio Vermelho, a branch of the Araguaya, to Pará. The president published not long since an excellent memorial on the advantages to be gained by the navigation of the river. Mr. Ward tells me that a steamer now makes regular trips from Pará to Goyaz. I have told the story of Araguaya to show that Brazil is not wholly without the spirit of enterprise. She is exploring her great rivers and establishing, slowly it is true, steam navigation upon them, and in a few years the interior of Brazil, so long shut out from the world, will be accessible to commerce.

The lands bordering the Araguaya are in great part flat, low, and composed of sands, clays, and other very recent deposits. An interesting feature in the river is the Ilha do Bananal, formed by an arm leaving the main river on the east in latitude about 12° 30′ (approx.), and entering it again in about 9° 30′. Castelnau determined the length of the island to be seventy-five leagues. Almeida, in his map of Goyaz, does not represent it as quite so long.

The Araguaya is very rich in fish,* and a species of dolphin occurs in it.

Gold is found in many localities in the province. The

* Castelnau says that piranhas — he calls them *pirangas* — are very numerous and voracious. According to him, "leur voracité est telle, que presque tous les oiseaux aquatiques que nous procurions avaient les pattes en partie dévorées par eux. . . . . Un de nos compagnons de voyage poussé par l'excès de la chaleur, se mit imprudemment à l'eau, et fut presque aussi tôt attaqué par des légions de ces animaux ; immédiatement les eaux furent teintes de son sang et il fut heureux pour lui qu'il se trouvât très près du rivage, vers lequel il se précipita avec rapidité, échappant ainsi à une mort certaine et affreuse." — *Hist. du Voyage*, Tome I. p. 404.

country in the vicinity of the capital is largely auriferous. Castelnau speaks of the occurrence of the precious metal at the following localities, Rio Vermelho, Rio Bagagem, Serra Dourado, Districto de Ouro Fino, Morro do Calisto, Districto da Anta, Thesouras, Rio Claro, Julgado de Crixas, Natividade, Trahiras, &c.

Diamonds have been found on the Rio Claro; iron occurs at Ouro Fino, Anta, Aldêa de São José, and chrome at Ouro Fino, where it is said to have been found by Pohl.

The Province of Goyaz is, generally speaking, dry, consisting of campos and catingas. Forests have but a small extension. There is a large tract covered by the virgin forest between the capital and Meia Ponte. The province is especially adapted for grazing. The climate is dry, but varies much, according to latitude. The population is, according to Almeida, 250,000 souls.*

The western part of the high lands of Brazil, forming the Amazonas-Paraguay water-shed, and comprised in the province of Matto Grosso, is completely covered with undisturbed tertiary beds, and forms a low swelling plateau, on which the rivers take their source.

This is well shown in the maps and geological sections accompanying the great work of Count Castelnau. The rivers Xingú, Tapajos, and Paraguay all take their rise in this plain † within a few miles of one another, near Diaman-

* Castelnau devotes Chapter XVII. of the second volume of his *Hist. du Voyage* to a description of the Province of Goyaz.

† Chandless says that the water-shed between the Amazonas and the Paraguay, "though commonly called a serra, has nothing of a mountainous character. It is simply a high range of country, varying but little in its general elevation, though deeply grooved by the valleys of the rivers. Around them one finds more or less virgin wood; the rest is campo, that is, pastures sprinkled more or less thickly with stunted trees, in parts including the *quina* tree. This

tino, and the water-shed is so low that wooden canoes ascend the Tapajos from Santarem, cross over, and embark on the Paraguay, descending to Villa Maria.

In descending the Tapajos, on his way from the Diamantino to the Amazonas, Chandless found the river bordered by sandstones, which, as at Creporé, he describes as soft in character.

On the road from Goyaz to Cuiabá one passes over an immense plain of sandstone in horizontal beds. The valley of the Paraguay at Cuiabá and Diamantino is excavated in this sandstone sheet down to the metamorphic rocks lying beneath. In the valley of the Paraguay, near Cuiabá and Diamantino, diamonds and gold occur in considerable abundance.

So little is definitely known about the geology and physical geography of the Province of Matto Grosso, that I content myself with these few general remarks. Castelnau has written more than any one else on the physical features of the province, and the reader is referred to his *Histoire du Voyage* for more details.

range seems to consist mainly of sand-rock and clay. In general it drops steeply and often precipitously to the lower country, the plain below appearing as a sea with deep bays and inlets."

CHAPTER XVII.

PROVINCES OF SÃO PAULO, PARANÁ, SANTA CATHARINA, AND RIO GRANDE DO SUL.

The Serra do Mar of São Paulo a Plateau. — Its Character. — Drainage in the Provinces of São Paulo and Paraná to the Westward. — Eastward-flowing Streams of no Importance. — The São Paulo Railroad. — Description of the Country, along the Railway between Santos and São Paulo by Major O. C. James. — Geology of Vicinity of São Paulo. — Mawe's Description of the Gold-Mines of Jaraguá, and the Method of extracting the Gold. — Country westward to Campinas. — Iron-Mines at Ypanema — Serra Arassoiava or Guaraçoiava. — Climate, Products, &c., of the Province of São Paulo. — General Topographical Features of the Province of Paraná, its Climate, Productions, &c. — Matte or Paraguayan Tea. — Tea-Planting in Brazil. — Rivers. — Colonies. — Paranaguá. — Coal Basin on the Rio Tuberão in the Province of Santa Catharina. — General Description of the Physical Features of the Province and that of Rio Grande do Sul. — History of the Coal-Mines of Brazil. —Observations of Perigot, Bouleich, Avé-Lallemant, Plant, &c. — Description of the Coal-Fields of the River Jaguarão. — Engineer McGinty's Report on the Candiota Coal. — Coal Basin on Rio São Sepé. — Basin near São Jeronymo.

THE so-called Serra do Mar, seen in sailing along the coast of the Provinces of São Paulo and Paraná, is the edge of the great Brazilian plateau, which along the coast of São Paulo has a height of 2,500 – 3,000 feet.* Towards the sea it presents a very steep declivity, but on the opposite side there is no corresponding slope. Climbing the serra at Santos, one finds himself on an immense table-land of gneiss, roughened by a line of considerable hills a few miles from its edge, but

* Mawe estimated the height of the plateau on the Santos and São Paulo road at 6,000 feet.

soon growing gradually lower in going westward, until at Campinas broad plains are reached, that stretch off with more or less interruption toward the Paraná, tying in with the great plains of Paraguay and the Argentine Republic. The united provinces of São Paulo and Paraná lie, like Ohio in North America, on the western slope of the border of the great interior continental basin of South America. As the bluff edge of the plateau so nearly coincides with the coast line, the drainage in these two provinces is principally westward into the Paraná, while the rivers flowing eastward are of very little importance. The province of Santa Catharina, just south of Paraná, lies partly on the seaward side of the serra and partly behind it. The streams flowing eastward are of no importance, while those flowing westward form the head-waters of the Uruguay. South of the province of Paraná the water-shed bends inland some two hundred and fifty miles, and then runs southward through the province of Rio Grande do Sul, ending as the high lands break down and disappear on approaching the Rio de la Plata.

The city of São Paulo, the capital of the province of the same name, is situated on the top of the plateau, at a distance of forty-five miles from the sea. The principal port of the province is Santos, a considerable town of some 7,000 inhabitants, and noted for its export of coffee, which reaches 160,000 sacks yearly. From this place to São Paulo runs a railway, which is continued westward beyond Campinas. This railway was constructed by American engineers, and my friend, Major O. C. James,* was one of the corps. On my last voyage, while at Rio, I

* I wish I could fully express my indebtedness to Major James: I owe him a thousand acknowledgments for his kindness.

obtained so many facts of interest from him with reference to the topography and geology of São Paulo, that, as he was about to return to São Paulo, I asked him to make observations on the surface deposits along the line of railway, which he did, furnishing me with a report, from which nearly all the facts relating to the route of the railway have been taken. Major James says that Santos is "within a league or two of the foot of the great back-bone of mountains, — a league or two of soft, oozy mud, a few feet above the sea-level, the base of a sort of estuary, of which the spur and main range form the bounds. This interlying marsh or lagoon is overgrown with very shabby palms, large moss-covered trees, looking like spectres in the landscape, and with a very heavy, tangled undergrowth, — lazy, sinuous lines of water, having their beginnings and endings in the sea, traverse it, as if it were just for a stroll inland beneath the umbrageous foliage and back again to the ocean refreshed." * Just before the railroad reaches the base of the serra the mud is left, and a broad band of gravel is crossed. This lies at a higher level than the mud, and appears to slope away from the serra. The materials are very coarse, and there were no sands seen. Major James says: "We dug into it ten feet or so, to get gravel for ballast. There are no shells in it. The height above sea-level is not great, — say ten or twelve feet, not more than twenty." This deposit evidently corresponds to the raised beaches of Rio and northward.

A few small streams flow into the estuary, and the railroad follows one of these until a deep gorge in the mountain is reached, — a gorge formed by two spurs jutting out at right angles to the line of the serra, placed like abutments

* It is hardly necessary to add that the locality is very unhealthy.

to support the mountain. Taking the southern slope of the northern spur, the railway creeps up at an angle of one in ten until it reaches the summit, five miles from the plain, having attained an elevation of twenty-six hundred feet.

CUTTING ON THE SÃO PAULO RAILWAY, SHOWING DRIFT LYING ON DECOMPOSED ROCK.*

Beginning near the foot of the serra, an examination of some of the cuttings through the massive corrugations of the hillside reveals a yellow unstratified clay,† not very compact, interspersed with pebbles, stones, and rocks, nearly all well

* The decomposed rock is seen gullied away underneath the drift.

† Major James says that this is generally the color of the drift-clay even to its whole depth. He compares the drift-paste on the São Paulo Railroad on the slope of the serra to that exposed in the deep cutting near Rodeio on the Dom Pedro Segundo Railroad. At São Paulo, however, it is more red, as we shall presently see.

rounded, overlying the rock *in sitû*, a decomposed gneiss. Here and there in a few cuttings a thin sheet of pebbles may be traced between the yellow clay and the rock, losing itself now and then. This description will apply generally to all the cuttings except a few where the excavation is made through partially decomposed rock.

Major James informs me that the clay is thicker on the top of the ribs or corrugations of the hills cut through by the railroad than on the sides, which is owing to the denudation of the slopes. This clay sheet may be fifty or more feet in thickness. On reaching the top of the serra one finds himself on an elevated country, a table-land diversified by hill and valley, the hills being generally low and rounded, the valleys wide and with a luxuriant vegetation flourishing on marshy bottoms. On the old mule-road near the edge of the plateau the country is gently rolling, and for some eight or nine miles the soil, when not covered by a sparse vegetation, is of a gray color, and presents a dreary aspect. On the surface is a layer of white sand, only two or three inches in thickness generally, but when drifted of course more. Under the sand is an almost white clay, of about the same consistence as potter's clay. Adobe houses built with it are nearly white. Under this is the drift, which is exposed in the banks of the Rio das Pedras, that flows through this tract and runs toward the coast. The highest point in the serra is to the west of this region.

Near Tamanduatahy the land spreads out between the hills level as a lake, and about two miles wide, covered with deep layers of black soil, which Major James describes as "fibrous and woody like peat." He informs me that the railroad was built over the surface of this bog, and he did not know what kind of a soil underlay it, but he felt satisfied that it occupied a shallow valley in the drift, which he believed extended underneath. Beyond this are some cuttings through higher ground, in which the reddish drift-paste, nearly one hundred feet in thickness, is cut through. In this clay are "boulders like those of the serra, of a very hard, bluish-gray rock scattered through the loose material, and in one or two instances exposed on the surface."

The city of São Paulo lies at a distance of forty-five miles from Santos. It is a large and important city of 20,000 inhabitants built on high ground, nearly surrounded by a low plain, through which flows, on the west, the river Tieté, one of the affluents of the Paraná. According to Mawe, the hill on which São Paulo is built consists of the following deposits in descending order. First of all, a coating of red soil more or less thick, impregnated with iron oxide; under this is sand, together with other materials associated with pebbles, the whole being from three to six feet thick. Under this comes a layer of purplish or variegated clay, with thin layers of sand; then follows a layer of stratified materials, the whole resting on decomposed gneiss-granite.* At Itú, a short distance from São Paulo, the Rio Tieté is represented by Major James as cutting through horizontal deposits of red sandstone and conglomerate, and this is the material which is used to pave the streets of the capital; as Fletcher and Mawe have remarked, the rock contains gold.†

* M. Pissis gives a similar description of the deposit, and illustrates it by a section. He calls the horizontal deposits of São Paulo and Itú fresh-water tertiary, and says that a similar lacustrine deposit occurs in the upper part of the valley of the Parahyba do Sul. (*Mém. de l'Institute de France*, Vol. X.)

† Mawe (American edition, p. 73), speaking of the streets of São Paulo, says: "The material with which they are paved is lamillary grit stone, cemented with oxide of iron, and containing large pebbles of rounded quartz, approxi-

Westward of São Paulo are some high hills; the most conspicuous among which is Jaraguá, in whose vicinity gold-mines were anciently worked.

Mr. Mawe, during his travels in Brazil, visited these gold-mines, and described them in his "Travels in the Interior of Brazil." I am unable to refer to the English edition, but I translate a few paragraphs from a French abstract,* since it gives very neatly the mode of occurrence of the gold and the ancient method of extracting it.

After speaking of the discoveries made by the Paulistas, he goes on to say: —

"The gold-mines of Jaraguá, being situated at a distance of four leagues from Sao Paulo, were the first that were discovered. This part of the country is unequal and mountainous. The rock forming the principal base of the soil rarely ever shows itself. It appears to be a granite † passing into gneiss.

"This primitive rock is immediately covered over in many points by a bed of a not very solid agglomeration, formed principally of pebbles of quartz and gravel. This itself is covered only by the vegetable earth.

"It is this conglomerate that is intermingled with grains of gold. They give it the name of *cascalho*.‡

"The mining takes place in open cuttings, and the extraction of the gold is carried on by washing; negroes are employed in this work.

mating to the conglomerate. This pavement is an alluvial formation, containing gold, many particles of which metal are found in the chinks and hollows, after heavy rains, and at such seasons are diligently sought for by the poorer sort of people."

See also Spix and Martius, "Travels in Brazil," English Translation, London, p. 21.

* *Annales des Mines*, 1817, Vol. II. p. 202.

† In the German edition Mawe describes the gneiss as containing some hornblende and an abundance of mica.

‡ Mawe, in the original English edition says, *cascalhao*.

"When a current of water can be found whose level is sufficiently elevated, steps are cut in the earth, each twenty to thirty feet long, two or three wide, and a foot in height. At the base a trench two or three feet deep is dug.

"On each step are placed six or eight negroes, who, as the water slowly descends from above, stir up continually the earth with shovels, until it is all converted into a liquid mud and carried lower down.

"The particles of gold contained in the soil descend into the lower cutting at the bottom, to which they soon settle because of their greater specific gravity. The workmen are continually employed in removing the stones from the ditch, and in cleaning the surface, an operation which is facilitated by the current of water which falls there. After five days of washing they remove the sediment from the bottom of the cutting. It is of a deep carbonaceous tint, and composed of iron oxide, pyrites, ferruginous quartz, and scales of gold.

"This sediment is transported to another current of water, there to undergo another washing operation. For this purpose funnel-shaped wooden bowls or *gamellas,* two feet large at the mouth and five or six inches deep, are used. Each workman, standing upright in the brook, takes about five or six pounds of the auriferous sediment in his gamella. He then causes a certain quantity of water to enter, and agitates it with dexterity, in such a way that the scales of gold soon fall to the bottom and on the sides of the vessel, uniting themselves together and separating from the other lighter substances, which the water holds in suspension and carries away little by little with it. He then rinses the gamella in another of larger size, full of water. The gold is deposited there, and he recommences a similar operation. The washing of a gamella takes eight to nine minutes.

"The gold taken out varies in the number and the size of the scales; some are so small that they float, while others are as large as peas and even larger.

"This operation, of which the result is of the greatest importance, is watched by inspectors. The gold-dust is carried to a mint, where the impost of the fifth is taken out and the rest melted. . . . .

"The mines of Jaraguá have been famous for two centuries for their great yield. This district was regarded as the Peru of Brazil; * but its riches are to-day infinitely less."

The occurrence of gold in the gneiss regions of the Serra do Mar, at São Paulo and Cantagallo and in the same eozoic belt on the Mucury, as reported by the engineer Schieber, is interesting, as these rocks are rarely ever rich in the precious metal, and the whole gneiss belt of Brazil is remarkably barren of metalliferous deposits of all kinds.

Gold is said to occur in the Villa de Guarapuava, to the west of the river Tibaji and elsewhere in the province.

Major James reports that the cuttings beyond the Tieté show the line of pebbles and stones overlying the rock.

The high hill called Cabellos brancos is bare of soil and vegetation on top, whence the name.†

At Jundiahy, according to Major James, the rounded hills are nearly bodily composed of the unstratified paste, but behind the station the line of pebbles may be distinctly traced, like a thin sheet or veil over the rock. A few leagues farther on, near Campinas, the country becomes less rugged, — one might say it is heavily undulating, — and here from a slight eminence the gradual descent into the great basin of the interior is plainly perceptible. The hills are

* Pompéo says that the gold-mines of São Paulo produced, up to the beginning of the present century, 4,650 arrobas of gold. (*Geographia*, p. 480.)

† Von Tschudi, Vol. III. p. 231: "Sie führt ihren Namen 'Das gebirge des weissen Haares,' weil der Kamm während der kalten Jahreszeit in den Frühstunde oft mit Reif bedeckt ist."

now composed of a very dark-red ferruginous earth resembling in grain coarse brown sugar; a mass squeezed by the hand retains the impress of the palm. Major James says that "travellers in dry weather get to look like Indians!" This earth goes by the name of *terra roxa*.* The terra roxa of Campinas Paulo is, according to Major James, the continuation of the drift-paste of the higher lands and seaward slope of the serra. It varies much in thickness, and lies usually on the tops of ridges between the rivers, not descending into the valleys, which, as in the basin of the Jequitinhonha, are very deep, steep-sided, and narrow. This red earth forms a most fertile soil, and the country covered by it is clothed by an exceedingly luxuriant vegetation. Bamboos are very numerous, but there are very few cacti. No soil is better suited to the coffee-tree, and in this part of the country it is extensively cultivated on the upper lands, but never on the slopes or on the intervales. The terra roxa rarely ever contains pebbles. It lies on horizontal beds, which, according to my friend, consist of "a soft friable rock, generally of a light-gray color, often hard and laminated like that used for flagging our sidewalks; sometimes it is a red sandstone, very soft; this latter is what Fletcher says São Paulo is paved with." This same material forms the great plain extending west to São João do Rio Claro, and, as I have already remarked, tying in with the formation of the plains of the Paraná and the south. This formation I believe to be the same as that occupying the valley of the Jequitinhonha, and which I have referred to the tertiary.†

* Literally, deep-red earth; the ordinary red soil of Rio goes by the name of *barro vermelho*.

† Since writing the above I have received from Major James a specimen of this rock, which is precisely like the clayey sandstone of the chapadas of the Jequitinhonha.

At São João de Ipanêma, near Sorocaba and about twenty leagues southwest of São Paulo, there are beds of sandstone and limestone associated with diorites and porphyry, with heavy deposits of magnetic iron ore.* This ore is mined to a considerable extent, and is smelted almost on the spot.

Von Eschwege says that the sandstones are modern-secondary.† I would suggest a comparison between them and the sandstone and iron deposits of the São Francisco River, described by Burton.

These iron deposits were discovered in 1578 by one Affonso Sardinha, who is said to have found at the same time ‡ "a vein of silver (?), of whose extraction the government took charge; but as the expense was great all was soon abandoned, and these sites remained unpeopled until 1803, an epoch in which some naturalists, exploring the serras of the district of Sorocaba, came to recognize the true importance of the iron-mines of Guaraçoiava. After seven years the prince-regent brought over from Sweden, at no small expense, a company of miners, under the direction of an individual of the same nation, named Hedberg, who set up four forges, which, owing to their bad arrangement, were of no service. . . . . In 1815 new forges were constructed by order of the same prince, together with a *fabrica* on a larger scale than the first, and the Count of Palma . . . . was charged with the direction of the work of the engineers and the superintendence of everything. This governor ordered two enormous furnaces to be built beside those which

* A collection of specimens from this locality made by Olfer existed in 1828 in the Königl. Mineralogische Museum at Berlin.

† *Ann. des Mines*, 8me Vol. 1823, p. 405.

‡ *Dic. Geog.*, art. *São João d'Hipanêma.*

still existed." * "The iron manufactured is excellent, and the ore gives fifty to eighty-five pounds of metal to the quintal. At present there are two high furnaces measuring eight mètres in height, producing regularly 3,000 kilogrammes of cast-iron in twenty-four hours with uninterrupted work."

The region in which these iron-works are situated is covered with forest, and wood is the fuel used. The flux employed is limestone and diorite. In the immediate vicinity of the furnaces is found an excellent quality of sandstone, of which a refractory kind is employed in the lining of the furnaces.

The serra *Araassojava* (Guaraçoiava), according to Spix and Martius,† is an isolated ridge, about one thousand feet high above the level of the river Ipanêma, and a league in length from north to south. The ore occurs in a large deposit, and our authors speak of seeing a mass of it forty feet perpendicular. It is associated with a yellow sandstone with a scanty argillaceous cement, and a clay slate or shale of a dirty lavender color. Its strike is east-west. In the same hill a porous quartz rock of a light-brown color was observed, containing light-blue chalcedony. On the Rio Tieté at Araraitaguaba Spix and Martius report a sandstone similar to that of Ipanêma.

The soils of the Province of São Paulo are exceedingly fertile, and the climate is favorable to the growth especially of coffee, sugar, and tobacco. Coffee flourishes exceedingly well on the campinas west of São Paulo, and it is probable that there is no more valuable coffee region in Brazil. The climate is so mild in the vicinity of the city of São Paulo

* *Catalogo da Segunda Exposicão Nacional de* 1866, p. 69.

† *Reise*, Vol. I. pp. 253, 254.

that many European plants may be successfully raised, such as flax, wheat, the vine, peaches, &c., &c. The climate on the higher lands is very healthy and agreeable, and well suited to Europeans. The province has now nearly 800,000 inhabitants, and counts a considerable number of flourishing towns. There are some German colonists in the province, and I understand that a considerable number of Americans from the Southern States have settled there.

The Province of Paraná, which lies just south of that of São Paulo, has very nearly the same topographical features, being low along the coast, rising more or less steeply to the summit of the plateau, and then sloping off towards the Paraná in extensive campinas. The capital is Curitiba, situated, like São Paulo, at some distance west of the edge of the plateau. The greater part of the country is covered by heavy forests, though in the northwest there are extensive plains. An important business of the province consists in the raising of cattle, of which large numbers are exported. Coffee, cotton, potatoes, sugar-cane, Indian corn, wheat, vegetables of different kinds, &c., are cultivated to a considerable extent. The "*herva mate*," or *matte* of the Brazilians, — *Ilex paraguayensis*, or Paraguayan tea, largely grown in the Paraguayan republic, and used in lieu of Chinese tea, — is grown in large quantities and forms an article of export. Bousquet* says that the Chinese tea grows well, but that it is not as yet much cultivated. Tea will also grow at Rio, São Paulo, and elsewhere in the south. Bousquet speaks of the vanilla as growing spontaneously in the vicinity of Paranaguá. The province has never been explored by a competent modern geologist. It is just possible, as admitted

* *Note sur la Province de Paraná*, par M. Bousquet, *Bull. de la Société de Géographie*, 5me Série, T. 9, p. 528.

by Mr. Plant, that coal may be found on the low lands between the Serra do Mar, or Cubatão, as it is commonly called in Paraná, and the sea. Bousquet says native mercury occurs near Paranaguá. Gold and diamonds are found on the borders of the Rio Tibagy.*

This province, like São Paulo, is well watered by large streams. Between the two provinces runs the Paranapanema, on the west is the Paraná and to the south the Uruguay, while the Ivahy and Tibagy, both rivers navigable for canoes,† run through the province, one emptying into the Paranapanema and the other into the Paraná. The Paraná is navigable for about ninety leagues from the falls of Urubupunga in the Province of Goyaz to nearly opposite the Tieté in São Paulo, where navigation ends at the island of Sete Quedas. The Salto Grande on the Paraná is described as being comparable to the Caxoeira de Paulo Affonso.

So far as water highways are concerned, the province is well furnished.

There is a little Brazilian-French colony, called Santa Theresa, established on the Ivahy, and another colony of some five hundred inhabitants, called Superaguy, near Paranaguá, of which the inhabitants support themselves by agriculture and fishing.

The seaport of the province is Paranaguá, a considerable town, situated on a large and beautiful bay which forms an excellent and spacious harbor.

The province of Santa Catharina lies just south of Paraná. It is one of the most fertile of the provinces of Brazil, and is

* Oliveira, *Exploração de Mineraes*, p. 11.

† Vereker "*On Brazilian Province of Paraná*," Jour. Roy. Geog. Soc., Vol. XXXII. p. 137.

blessed with a delightful and temperate climate; but except on the seacoast it is not well settled, and the population is probably not more than 150,000. The capital is Desterro, a large town, delightfully situated on the western side of the island of Santa Catharina. There are several flourishing colonies in the province.

Between the edge of the plateau and the sea there exists in Santa Catharina a coal-basin, in which along the banks of the Rio Tuberão beds of bituminous coal of fair quality are exposed. The Visconde de Barbacena is interested in the development of this region, and we may hope erelong to see some results of his labors. I am not aware that any report on this field has yet been published. I understand that the coal-beds lie quite flat, as elsewhere to the south.

The Province of Rio Grande do Sul is the most southern of Brazil, lying just north of Uruguay, and situated between 27° 50′ and 33° 45′ south latitude. Much of the country is hilly, particularly in the eastern and northern portions, but towards the west and south it consists of plains covered with herbs and forming the pasture-grounds of herds of cattle. The northern and eastern portion is more or less heavily wooded. The interior has never, so far as I can learn, been scientifically explored, and I am unable to find any reliable description of its physical features.

Of the geology of the province we have little information except with reference to the coal-mines.

In an article in the Quarterly Journal of Science, No. 11, 1864, Mr. Edward Hull says that the first notice of the coal deposits of the Province of Rio Grande do Sul was taken by one Guilherme Bouleich in the year 1859. This is not quite correct.* In May, 1858, Dr. Avé-Lallemant visited coal-

* The real discoverer of the Brazilian coal-fields was Dr. Perigot, of Flush-

mines on the Arroio dos Ratos, which were at that time worked on a small scale. He refers to an examination of the locality made some time before by a Bacharel Vasconcellos.* Dr. Lallemant describes two horizontal seams of coal, — an upper, worked, about four feet thick; a few feet below this seam a second was found of the same thickness. I have not had the opportunity of examining specimens of this coal; but I learn that it is a fair bituminous variety, with a more abundant ash than the English coals sent to Rio. It has been in use some ten years on the steamers of the Companhia Jacuhy, and has been found more economical to employ it than the English coal. That of the Arroio dos Ratos sells for 13$000 – 17$000 ($6.50 – $8.50 American currency) per ton, which, as we shall presently see from Mr. Plant's report on the Jaguarão coal-field, is a much lower price than that of the English coals. There are three separate coal-fields in the Province of Rio Grande do Sul,† all of which Mr. Nathaniel Plant has studied with care. This gentleman informs us that the basins are separated from one another by rolling hills of syenite, mica-schist, and granite, together with trachytic and basaltic rocks.

The largest of these basins is situated in the Jaguarão and Candiota valleys, between lat. 31° and 32° S. and long. 53° and 54° E.

ing, Long Island, who was engaged in 1841 by the Brazilian government to make geological observations in the Empire. This gentleman reported the existence of a coal-field in Santa Catharina, some three hundred miles long from north to south and twenty to thirty miles wide.

* Avé-Lallemant, *Reise durch Süd-Brasilien im Jahre* 1858, Theil I. p. 478.

† The Brazilian Coal-Fields, by Nathaniel Plant, F. R. G. S., &c., Geological Magazine, Vol. VI. No. 4, April, 1869. I owe a copy of this paper to the kindness of Mr. Plant's brother, Mr. John Plant, Curator of the Royal Museum of Salford. Much credit is due Mr. Plant for his long-continued exertions in working out the structure and limits of these coal-fields, and for his persistence in endeavoring to bring the coal into market.

Mr. Plant presented Professor Agassiz in 1865 with a short description of this coal-field, together with specimens of rocks, fossils, and coals, which I have examined. The fossils are of true and characteristic carboniferous genera, and no doubt can exist as to the equivalency of the deposits.* I give Mr. Plant's paper in full.† One of the photographs mentioned I have had engraved as a frontispiece to this volume.

I may add that Mr. Plant has been for several years exploring the coal-basins of the province with a view to having them worked by a mining company. It is to be hoped that his efforts may be abundantly successful.

Coal-Fields of the River Jaguarão, and its Tributaries, the Rivers Candiota and Jaguarão-chico in the Province of Rio Grande do Sul, Brazil.

The coal-basin of the river Jaguarão is situated in the southern part of the province of Rio Grande do Sul, between lat. 31° and 32° S., and long. 324° and 325° (French meridian) in the valley of the Jaguarão and its tributaries, the rivers Candiota and Jaguarão-chico. It covers an area of about fifty miles by thirty, its greatest diameter being from north to south. The coal strata which the geological section illustrates, and whence the accompanying specimens have been obtained, and the thickness of the beds determined, are exposed in an elevated escarpment on the banks of the river Candiota, at a place called "Serra Partida,"‡ where they appear in the following order of superposition: —

* Strange to say, that, after all the explorations made of these coal-fields, their carboniferous age was long in being recognized. Affonso Mabilde, in a report made to the president of Rio Grande do Sul, says that the coal is a lignite of tertiary age.

† This paper also appears in Fletcher's work. I presume Mr. Plant furnished him with a duplicate copy of the MS.

‡ See frontispiece, which is engraved from a photograph presented to Professor Agassiz by Mr. Plant.

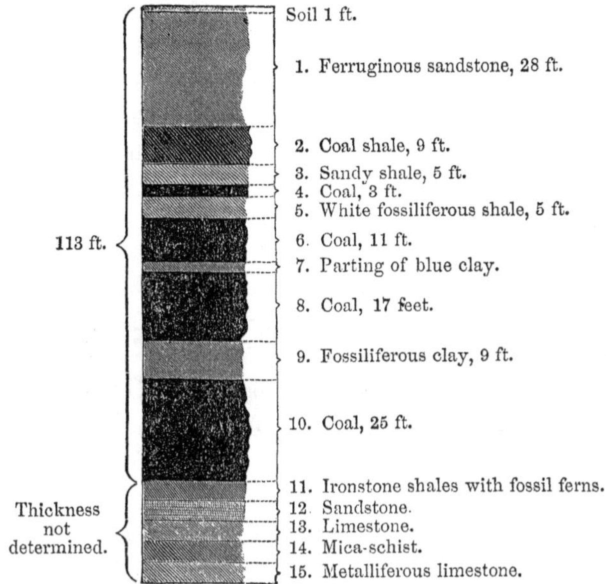

No. 1. The uppermost bed is composed of sandstone of a highly ferruginous nature, resembling in its appearance the "Grés Bigarré" of Europe; it contains nodules of a silicious peroxide of iron, yielding from twenty-five to thirty-five per cent of metal. It varies considerably in its thickness, in some places being completely worn away, and in others attaining a depth of upwards of two hundred feet. Immediately below this occurs a bed,

No. 2, of coal-shale, very argillaceous, and perhaps unfit for fuel; it possesses a thickness of nine feet, and can be seen cropping out wherever the superincumbent bed has been denuded; it rests upon a bed,

No. 3, of sandy-ochreous shale containing septaria of an ochreous oxide of iron, which, together with the ironstone found in the sandstone, will, in all probability, be turned to profitable account when the coal-beds are worked; underneath this is,

No. 4, a bed of bituminous coal, three feet thick. The mineral, although it leaves a high percentage of ash, will be found useful in smelting the iron ores from the interstratifying beds; and there is every reason for supposing that it will be found of a better quality when the bed is fairly worked. The samples tested were taken from very near the surface, which may in some measure account for its apparent impurity; it rests on,

No. 5, a bed of white clay or schist, containing innumerable impressions of fossil plants (perhaps aquatic), the general appearance of which would lead one to conclude that these carboniferous deposits belong to a later period than that assigned to the coal-measures of England and the United States, were such a conclusion not confuted by the fossil ferns found in the other interstratifying shales; it has a thickness of five feet, and overlies a,

No. 6, bed of good coal eleven feet thick. This coal resembles very much in its appearance the Newcastle, and may be traced for many miles along the banks of the river Candiota, sometimes forming the bed of that river and of the small streams falling into it; it is separated from another seam by a thin parting,

No. 7, of blue clay.

No. 8. The coal of the lower bed appears to be even of a better quality than the one above; it has a clean, shining fracture, and in some places thin seams of pure cannel-coal may be traced along the bed. It is highly inflammable, boiling up like oil during combustion. This coal has been used as fuel in various ways with marked success. It has been tried on the steamers navigating the Lagos dos Patos in the province of Rio Grande, and although it left a greater portion of ash than the Cardiff coal, it was found to be a good caking coal, and served every purpose of a steam fuel. Below this is another bed,

No. 9, of blue clay, containing vestiges of fossil plants; in everything else it is similar to the upper bed of the same mineral. It reposes on,

No. 10, the thickest seam of coal exposed in the escarpment at the Serra Partida. This is the lowest bed of coal exposed in any part of the coal-fields of Candiota, but, in all probability, other beds will be found nearer to the centre of the basin, or this, as well as the incumbent beds, may become thicker, judging from the fact that all the beds appear to thicken as they approach the middle of the valley of the river Jaguarão. The great thickness (twenty-five feet), the good and homogeneous character of the seam, are important features in this coal-field. The mineral (although taken from near the decomposed face of the cliff on the river Candiota) was found to leave even less ash than that from the seam above. It has frequently been used on steamers with the same success as that obtained from Newcastle coal. The coke obtained from this coal by Mr. W. G. Ginty of the Rio Gas Works (*vide* Mr. Ginty's report) was even better than that derived from Newcastle coal. It overlies a bed,

No. 11, of ironstone shale, which, in a scientific point of view, is the most important deposit of the coal-measures of the Jaguarão, from the fact of its containing impressions of organic remains, by which the geological age of the coal-beds can be determined; the fossil plants found embedded in this shale all belong to the same genera as those which characterize the coal-fields of Britain and the United States, the most abundant belonging to the genera Lepidodendron * and

* Three species of plants have been described by Mr. W. Carruthers, viz.: —

Flemingites Pedroanus CARRUTHERS. Geol. Mag., Vol. VI. No. 4, p. 5, Pl. P. V.

"Stem lepidodendroid, scars small, obovate, without any markings; base of the petiole permanently attached to the stem; leaf slender, linear. Fruit a cone (?) the scales of which support numerous sporangia."

The stems of this plant at first sight resemble those of some species of *Lepidodendron;* but they may be readily distinguished from Lepidodendron by the character of the scars, which never show any impression from articulating surfaces. Associated with the stems and leaves of this species are found in great abundance minute flattened bodies which Mr. Carruthers considers to be the sporangia of the plant under consideration.

Odontopteris Plantiana CARRUTHERS. *Loc. cit.*, p. 9, Pl. V. Figs. 2 & 3.

"Pinnules broad at the base, irregularly lobed, obtuse at the apex; basal

Glossopteris; others have been recognized as being similar to the ferns found in the very oldest secondary rocks, thus leaving no uncertainty as to the true carboniferous character of the coal-measures of the river Candiota. This seam is very prolific of fossils, and there can be no doubt that when these immense beds of mineral treasure are worked many new and interesting forms of vegetable life will be brought to light to enrich our knowledge of the coal-fields of the Southern hemisphere. The ironstone shale is very rich in metal, and will doubtless be worked as an iron ore when the mines are opened. Below this there occurs another bed,

No. 12, of sandstone, similar in all respects to the uppermost bed. After which is a bed,

No. 13, of fine crystalline limestone, containing small fragments of graphite, disseminated throughout the mass. It is traversed also by veins of a very pure carbonate of lime in the form of double-refracting spar, which, in some places, attain a considerable thickness. This limestone will not only be of immense value for manufacturing into lime, but also as a flux in smelting the iron ores. The three things essential for the erection of smelting-works are thus found in the same district interstratified with each other: the ore, the fuel, and the flux, all of the very first quality,—a combination of mineral riches (only waiting for the hand of man to realize them) scarcely to be found together in one spot in any other part of the globe. Evidently,

No. 14, the two lowest beds of these coal-measures are mica-schist,

No. 15, and another limestone rock of a very dark and compact

pinnules large, much and irregularly lobed; nerves arcuately parallel, dichotomous."

Nœggerathia obovata CARRUTHERS. *Loc. cit.*, p. 9, Pl. VI. Fig. 1.

"Frond sessile, flat, entire, elongate obovate, attenuated towards the base; nerves dividing dichotomously, parallel."

I have seen *Calamites* and a *Sphenopheris* (n. sp.) in specimens of coal shale from one of south Brazilian coal-fields; but the labels having been lost, I have been unable to determine the exact locality.—C. F. H.

nature. It is scarcely possible to determine which is the lowermost, as in some places the mica-schist * is seen lying on the syenite which surrounds the coal-basin, and in others the limestone; the name of "metalliferous limestone" has been given it, owing to the innumerable crystals and thin veins of sulphuret of iron which appear in it. In all probability other metalliferous veins will be found in this limestone.

Nearly the whole of the coal-basin of the valley of the Jaguarão is enclosed by syenitic hills of from two hundred to three hundred feet high; the sides towards the coal-field slope gently downwards till they disappear under the sandstone overlying the coal; on the other side the syenite, after presenting an uneven and undulating aspect for some three or four leagues, gradually subsides into an even country, which continues on almost perfectly plain † till the seaport city of Rio Grande do Sul is reached, so that the company (already formed for making the survey for a railway to carry the mineral riches of the valley of the Jaguarão down to a seaport, where the coal can be shipped to the different ports along the coast of Brazil, and the River Plate) will find no difficulty in discovering a route along which a cheap line of rails can be laid down.

The photographic views of the different escarpments in which the coal-beds are shown along the river Candiota will show the great facilities afforded for working the coal in almost any part of the basin by open cuttings. Tramways can be laid down branching off in different directions from the main trunk line, along which the coal-wagons can be run right into the seams of coal, thereby rendering the sinking of expensive shafts quite unnecessary.

The general dip of the beds is from 5° to 10° S. W., and in no place are there signs of subsequent upheavals or dislocations of

* The mica-schist is without doubt much older than the carboniferous.— C. F. H.

† Mr. Hull speaks of a gently sloping plane of basalt stretching from the coal-mines toward the port on the Rio Gonçalo. (Quarterly Journal of Science, No. 11, April, 1864.)

strata visible, so that very little obstruction will be met with in carrying the tramways along the seams as the working of them goes forwards.

It is almost unnecessary to dwell upon the immense value of these coal deposits as a commercial enterprise, when it has been already ascertained, by a "running survey" of the country between the seaport of Rio Grande do Sul and the coal-mines of Candiota, that in all probability the coal will be delivered on board vessels lying in the port of Rio Grande at, perhaps, less than 7$000 per ton, where it is at the present moment being sold at 24$000; and as soon as a bill is passed allowing vessels of all nations to trade between the Brazilian ports, there will be no lack of enterprising shipowners to carry the Rio Grande coal to Rio de Janeiro—in which port alone the enormous amount of 180,000 tons of coal are annually imported—for a price which will enable the coal-mining company to sell the Candiota coal in the market of the capital of the Brazilian Empire for about 15$000 per ton,—a price which will annihilate any competition from foreign markets, seeing that the foreign coal is seldom sold for less than 22$000 per ton.

The consumption of coal in the river Plate is perhaps as great as that of Rio de Janeiro, and the facilities for supplying the markets of Buenos Ayres and Montevideo from the coal-mines of the river Candiota are still greater than those for supplying Rio. The coal can be sent from the mines, put on board colliers, and delivered in Montevideo in three or four days, at about half the cost of delivering the same article in Rio, which is a market where coal is never less than fifteen dollars per ton (or 30$000).

The consumption of coal along the Brazilian coast and in the River Plate increases yearly, and in all probability it will be found, after the coal-mines of Candiota have been opened for a few years, that a single line of railway will not be found sufficient to carry the supply of coal to meet the increasing demands.

NATHANIEL PLANT.

RIO DE JANEIRO, 20th July, 1865.

Report on the Candiota Coal by W. G. Ginty, Engineer-in-Chief of the Rio de Janeiro Gas Works.

MR. NATHANIEL PLANT: —

DEAR SIR, — I have received and examined your samples of Brazilian coal from Candiota with great interest, and I am glad to be able to congratulate you on its really good quality.

The samples you sent me were too small for complete and satisfactory analysis in the apparatus at my disposal. I found also that the samples varied a good deal in appearance and quality. This has arisen, no doubt, from their having been obtained from various positions on the nearly perpendicular face of the immense stratum, and from variable periods of exposure, as, owing to the crumbling away or disintegration of pieces under the incessant action of the weather, these samples may have been exposed for periods varying from each other as seconds do from centuries.

The Candiota coal resembles the Newcastle steam-coal (which comes to this market at least) very much in structure, cleavage, and general appearance, nor does it differ very much from Newcastle coal in its useful properties, except that it contains more than double the quantity of ash, which is detrimental to its heating powers; but this objection is likely enough to disappear altogether in samples from the deeper parts of the mine.

The coke from the Candiota coal is, however, very different in appearance from that of the Newcastle coal, and resembles the coke of (what is sold here as) Cardiff coal in its silvery-colored laminations.

Some of this Candiota coal, however, especially that of the lower seam is very friable, and is evidently what is called caking coal; that is, it boils or becomes molten during the process of carbonization. However, all the qualities of the coke from the Candiota coal are very good.

As you say the dip or inclination of the seams or strata of this Candiota coal is 5° from the plane of the horizon, I think it most reasonable to presume that a much finer, more compact, and equable quality of coal may be calculated upon at lower depths. 5° is a gradient of 1 in 11.4 or 8.77 per cent, or 462 feet per mile. Thus in such an immense field as you have described to me there is ample margin for obtaining other than surface coal, which for obvious reasons in Brazil, as elsewhere, cannot be as pure, as compact, or as uniform in quality as that obtained at great depths. I shall watch the prosecution of your explorations in this direction with great interest.

The following are the results of my examinations (as far as they went) on the Candiota coal, the samples of Newcastle, Cardiff, and Wigan cannel, with which it is compared below, having been tried at the same time in the same apparatus : —

	Specific Gravity. Water 1,000.	Per Cent of Coke.	Cubic feet of Gas per Ton	Illuminating Power in Standard Candles.
Candiota coal (mean of three quantities)	1,240	63	6,900	5.00
" " Lower seam	1,230	60	8,198	5.80
Newcastle	1,250	62		
Cardiff	1,275	80		
Gas, or Cannel Coal (Case and Morris)	1,240	62	9,600	20.50

From the appearance of the lower seam I do not despair of your finding a good gas coal for us in the Candiota district, and thus freeing the Brazilian gas companies from the fearful tax they have to pay in the shape of freights from England, amounting to from 200 to 300 per cent on the value of the *materia prima.* I send you labelled samples of the different qualities of coke above referred to. I remain your obedient servant,

(Signed) W. G. GINTY, Mem. Inst. C. E.,
Engineer Rio de Janeiro Gas Company.

Mr. Plant describes a second coal-basin exposed in the valley of one of the tributaries of the Rio Jacuahy, called the São Sepé, in lat. 30° 20′, long. 53° 30′.

There are two beds of coal exposed in this basin, one fourteen, the other seven feet in thickness. The sandstone beds overlying these veins are disturbed and overflowed by trachytic eruptions. The area of this basin, so far as is known, is about fifteen miles.

Near the town of São Jeronymo, on the banks of the Jacuahy, is a third coal-basin, situated in lat. 30°, long. 51° 30′. At a depth of fifty-seven feet below the surface there is at this locality a bed of highly bituminous coal six feet in thickness. This is underlaid by a bed of ironstone shale with fossils like those so common at Candiota. Several other coal-beds of from two to six feet in thickness, and interstratified with clay and ironstone, have been passed through by the shafts of the mines, which have reached a depth of one hundred and eighty-three feet.

Since the foregoing was sent to press I have had the opportunity of examining the valuable paper of Weiss on the collections made by Sellow in Southern Brazil.*

I regret that want of space prevents my giving a short abstract of this paper. I confine myself, therefore, to a quotation from the introduction.†

" On the 30th degree of south latitude," says Weiss, " there extends from the sea in a direction west a few degrees south (in der 7ten Stunde ‡) obliquely into the country for more than 5° of latitude a range of hills of 'basalt,' that is of amygdaloid and black porphyry or melaphyr, § but nowhere is the rock true basalt.

* *Über das südliche Ende des Gebirgszuges von Brasilien in der Provinz S. Pedro do Sul und der Banda Oriental oder dem Staate von Monte Video; nach den Sammlungen des Herrn Fr. Sellow,* Abhandlungen der Königlichen Akademie der Wissenschaften, zu Berlin, aus dem Jahre 1827.

† *Loc. cit.*, p. 222.

‡ The German miner's compass, according to Gehler's *Physikalisches Wörterbuch* (see article on Compass) was divided from north to south and south to north into twelve hours (*Stunden*) instead of degrees.

§ Melaphyr is the name given by Von Buch to a species of porphyry found

"On the southern side of its round and roof-shaped hills lies the valley of the Guaiba or Jacuy, with Porto Alegre at the junction of four rivers with the Lake Viamão. Nearer the Uruguay it slopes to the west, but even there it is united with a range of the same character extending southward (in der 12ten Stunden) separating the head-waters of the Ibicuy from those of the Caavera, forming between the Caavera, Ibirapuitam Grande and Ibirapuitam Chico considerable ridges, and when it turns to the west toward Salto Grande it sends streams southward to the Daiman and Rio Negro, northward into the Arapey, Quaraim and Ibirapuitam. This chain of amygdaloid is the source of the great quantities of chalcedonies, agates, carnelians, rock-crystals, and amethysts which cover the banks of Uruguay downward below the Rio Negro. These uniting chains divide the country in a natural manner into a northern and southern half, while farther south there is no mountain-chain, as formerly supposed, to form the boundary between the Portuguese and Spanish possessions. The amygdaloid formation is probably continued on the Uruguay above, since it forms reefs and cliffs on the river at Salto Grande and Salto Chico, as well as fourteen leagues farther up at the Capella de Belem, and among others the map of Nuñez gives just above the Saltos Grande and Chico a 'Monte Grande del Monteil.' On the slope of the amygdaloid formation there is spread an extensive clayey sandstone formation extending into the country, and it ascends to the foot of the granite coast mountains. It is certainly very young, and much younger than was supposed by one traveller (Sellow), who referred it principally to the Permian (Rothliegende). From its character as well as its situation, it is extremely probable, not to say certain, that it is tertiary, and may be provisionally referred by us to the Molasse or Braunkohlen sandstone."

in the Alps. The ground mass is black or blackish gray with embedded crystals of Labradorite (?) and Augite, with occasionally mica, hornblende, and pyrites. Sometimes it becomes amygdaloidal. See *Melaphyr* in *Handwörterbuch der reinen und angewandten Chemie.*

CHAPTER XVIII.

THE GOLD-MINES OF BRAZIL.*

Geological Distribution of Gold in Brazil. — Gold in Gneiss, at Jaraguá, Cantagallo, Pianco, and elsewhere. — The richest Deposits found in Veins traversing Clay Slates. — Character of Auriferous Quartz. — Granular Quartz, or Cacó. — Gold when associated with Sulphides rarely visible. — The auriferous Iron Ore, Jacutinga. — Gold-Mines of São João d'El-Rei. — The Morro Velho Mine, Mode of Occurrence of the Gold, Method of Extraction, Yield, &c. — The Gongo Socco Mine. — The Rossa Grande Gold-Mining Company. — Mines at Morro de Santa Anna, Congonhas do Campo, São Vicente, Cata Branca. — The Gold-Mines of Brazil not yet fairly developed.

THE gold of Brazil † occurs in the ancient metamorphic rocks, and in drift gravels and clays, and alluvial sands and gravels derived from the wear of these rocks.

The eozoic gneiss of the coast-belt furnishes gold at numerous localities along its whole extent. ‡ The mines of

* In this chapter I have not tried to treat exhaustively of this subject. I have endeavored rather to present such facts as will enable the reader to obtain a fair idea of the mode of occurrence of gold in Brazil, and of the character of the mines. A great amount has been written on the gold-fields of Brazil, but it is for the most part lacking in scientific accuracy, and much of it has rather tended to obscure than throw light upon their real structure.

† For a short and interesting sketch of the early history of gold-mining in Minas Geraes, see Burmeister's *Reise in Brasilien*, p. 590, "Zur Geschichte der Goldminen und ihrer ersten Entdeckung." Consult also Von Eschwege's *Pluto Brasiliensis*, and Mawe's "*Travels in the Interior of the Brazils.*" Spix and Martius, St. Hilaire, Castelnau, and almost all writers on the interior of Brazil, have had more or less to say about the gold-mines.

‡ I think that the auriferous deposits occur in the upper part of the gneiss

Jaraguá in São Paulo, of Cantagallo in Rio de Janeiro, and of one of the tributaries of the Itapémerim, are among the most important in this region. Gold also occurs in the gneiss of the Mucury Basin and in the north, as at Piancó. Over this whole region the gold is found rather sparingly, and appears to be derived from the quartz veins traversing the gneiss; but the only instance I know of where gold has been extracted from a quartz vein in the gneiss-belt is the mines on the Rio Bruscus in Parahyba. With this exception mining has been confined entirely to the washing of the cascalho * underlying the drift-clays, and of the gravels and sands of the rivers. These washings have all been abandoned. Between the coast gneiss-belt and the sea I know of no auriferous deposits, but in very numerous localities in the interior of the country the newer metamorphic rocks are rich in gold.

The formations affording most gold are clay slates traversed by auriferous quartz lodes, the itacolumite rock which is also veined with auriferous quartz, and certain iron ores variously known under the names of Itabirite and Jacutinga.† All these formations are, I believe, of lower silurian age.

In the clay slates the quartz veins sometimes show free

series. According to Mr. Wall, " Geology of a Part of Venezuela," Quart. Jour. Geol. Soc., Vol. XVI., p. 460, gold is found disseminated in gneiss near Valencia, Venezuela. Gold occurs in the gneisses of the Itacama Mountains. (Tate.)

* The auriferous cascalho of Brazil is quite a different thing from the *cascajo* of Venezuela. The Spanish word has the same derivation as the Portuguese, but according to Mr. Le Neve Foster, it is applied in the Caratal gold-field to a decomposed schist, on which the pay-dirt rests.

† This term is derived from two Tupí words, — *Jacú*, a kind of bird (*Penelope*), and *tinga*, white. The name was given to the rock because of its resemblance in color to the feathers of the above bird. Sometimes foreigners spell it jacotinga, which is not correct.

gold with very little pyrites associated. The auriferous quartz varies much in character. Sometimes it is compact and milky, at others, as at the Cata Branca mines, it is very granular and sugary. A specimen of vein-quartz with gold, from the mine of São Vicente in Minas Geraes in the Museum of Comparative Zoölogy, is composed for the most part of clear, colorless quartz in rather coarse granules, giving to the rock the appearance of a pure quartz sandstone, or of white lump sugar; * but in the same specimen the quartz passes into a more compact rock, which has a bluish look.

In the quartz veins, as elsewhere the case, the rock is not all auriferous, but the gold runs in streaks. The São Vicente specimen above described shows well-marked streaks, rich in free gold, which appear to have run parallel with the side of the lode.

Where the vein rock is rich in sulphides the gold is, as a rule, not visible, but intimately mixed with the rock. This is the case at Morro Velho. The sulphides consist of magnetic iron pyrites, which is the most abundant and yields a little gold; common iron pyrites is less abundant, and gives more gold; and the mispickel or arsenical iron pyrites, which is the principal gold-bearer.

Of the ferruginous auriferous deposits none is more interesting than the so-called jacutinga formation. Heusser and Claraz † say that the jacutinga is a pulverulent variety of

* This sugary quartz goes in Brazil by the name of *Cacó.*

† Heusser and Claraz, *Ann. des Mines,* Tome XVII. p. 290 : —

"L'itabirite est simplement une variété de fer oligiste schisteux qui est accompagnée de quartz et de mica. Elle présente quelquefois des couches puissantes et très-étendues qui peuvent être exploitées comme minerai de fer. Quand elle est pulvérulente, on la désigne sous le nom de *jacotinga.*"

Itabirite, a name given by Von Eschwege * to a rock composed of micaceous specular iron ore, compact specular iron, rarely laminated, a little oxide of iron and quartz disseminated. It is the rock of which the Peak of Itabira and the Serra da Piedade are composed.

Burton describes the jacutinga as follows: † "This substance, of iron black with metallic lustre, sparkles in the sun with silvery mica; the large pieces often appear of a dark reddish brown, but they crumble to a powder almost black. The constituents are micaceous iron-schist, and friable quartz mixed with specular iron, oxide of manganese,‡ and fragments of talc. . . . . The floor rock at Cocaes is fine micaceous peroxide of iron (specular iron), thin and tabular. . . . . Much of the jacutinga is foliated. . . . . It shows great differences of consistency; some of it is hard and compact as hæmatite, and this must be stamped like quartz. In parts it feels soapy and greasy, not harder than fuller's earth; it is easily wetted and pulverized, but it is hard to dry."

Gardner describes the jacutinga as a soft friable greenish-colored, micaceous iron-schist. §

The gold of the jacutinga is free. Castelnau ‖ says, that at Gongo Soco it is always confined to a little vein which winds about in the rock. This is never more than five to

* "Itabirit, — Eisenglimmer, Eisenglanz, meist dichter, auch blättriger, hin und wieder magnetischer Eisenstein und wenig Quarz erscheinen entweder als festes, dichtes Gestein, oder haben ein Körnigschiefriges Gefüge." — Von Eschwege, *Geognostisches Gemälde von Brasilien*, 28ste Seite.

† Burton, Vol. I. p. 301.

‡ Castelnau speaks of the occurrence of manganese in the jacutinga of Gongo Soco.

§ Travels in Brazil, p. 373.

‖ *Op. cit.*, p. 246.

seven millimetres in width, and sometimes it is as thin as a hair; it contains much manganese.

In the gold region is found an auriferous superficial deposit of broken fragments of ferruginous rocks cemented together, and called *Tapanhoacanga* or *canga*.*

I have already described in previous chapters the mode of occurrence of gold in the drift gravels and clays, and in the sands and gravels of the river bottoms.

With these introductory words, let us now examine some of the more noteworthy gold-mines of Brazil.

At São João d'El-Rei and São José, situated a few miles west of Barbacena, on the Rio das Mortes, are auriferous deposits formerly worked for many years with great profit. In 1830 an English company called the "São João d'El-Rei Mining Company" was formed, and these mines were leased; but in 1834 it was found that they were unprofitable, and they were abandoned. Captain Burton says the gold was principally obtained from a lode, which however he does not describe. The jacutinga formation is said to occur here.

The country over a very large area in the vicinity of Ouro Preto is very auriferous, and here are situated the richest gold-mines of the empire.

The gold occurs primarily in quartz veins traversing various metamorphic rocks, such as clay-slate, mica-slate, iron-schists, &c., and also disseminated through the rock in some places; and secondarily it is found widely distributed in drift and alluvial sands and gravels.

The celebrated Morro Velho mine is situated on the western side of the valley of the Rio das Velhas, not far from Sabará. It was long worked by native miners, but on the

* For a detailed description of this formation see page 559, next chapter.

failure of the mines at São João d'El-Rei, it was purchased by the company of that name, and has been worked with remarkable success ever since. The gold is extracted from a lode of quartz enclosed in clay-slate. The following account of the vein and the mine, with the accompanying statistics, I extract from the work of Mr. Arthur Phillips: * —

"The formation affording the gold is a strong, well-defined lode, though irregular in direction, dip, and dimensions; its inclination or underlie has also been found to vary at different depths and in different parts of its extent; the vein-rock is mostly composed of quartz, with iron pyrites disseminated more or less regularly throughout its mass, and the lode is not unfrequently traversed by clay-slate and barren white quartz. When pyrites is absent in these rocks gold is seldom present.† In some places the vein is cavernous and less close in its texture than in others; but where drusy cavities are frequent the yield of gold diminishes; the most productive matrix for gold is a compact mixture of quartz and pyrites, with varying quantities of slate. The great metalliferous deposit called the Cachoeira, Bahú, and Quebra Panella, is one continuous, very irregular vein varying in width from seven to seventy feet, and at one point reaching one hundred feet.‡ The

* J. Arthur Phillips, The Mining and Metallurgy of Gold and Silver, p. 82, which see for additional information concerning these mines.

† "Arsenical, magnetic, and ordinary iron pyrites predominate at different points and in varying quantities: carbonate of lime, dolomite brown spar, and, very rarely, copper pyrites are also present in the vein."

‡ Burton, Vol. II. p. 234, says: "The breadth of the lode varies from four to sixty feet. The general direction when worked is west to east with northerly shiftings. The dip is 45°, rising to a maximum of 46° 30′ or 47°. The strike is from S. 82° E. - S. 58° E. The cleavage planes of the killas are in some places transverse to, in others parallel with, the lode. In certain sections of the mine walls they bear N. 36° E., but the average is more easterly. The direction is S. 46° E., and their dip is at angles varying from 43° - 70°. . . . . The underlay or underlie dip, or inclination of the mineral vein, is 6° in the

average thickness, 176 fathoms perpendicular on the Cachoeira and 165 fathoms on the Bahú, is nineteen feet. The stoping space extends over 807 square fathoms, ... the enclosing rock is a clay-slate of tolerably uniform texture. . . . . The mineral brought to the surface is first freed from slate and other unproductive stone on the spalling-floors, and the ore, after being broken to a uniform size, is stamped fine. . . . . The stamping-mills, as is also the pumping and other machinery, are moved by water-power. The pulverized ore, issuing from stamp coffers through finely perforated copper grates, passes over bullock-skins in the first instance, and, lower down the inclined tables, over woollen cloth. The bullock-skins are taken up and washed in vats every hour, and the woollen cloths at longer intervals. The subjoined table shows the quantities of rock raised and stamped, the amounts of gold produced, and annual net profits made since 1868."*

The following account of the Morro Velho mines I translate from Gerber.†

"This company, after having worked for some time the gold-mines in the vicinity of São João d'El-Rei, acquired the auriferous lands of Morro Velho, which since then they have explored with great profit to its stockholders, owing this happy result not only to the richness of the formations, but especially to the perfection of the method of extracting the gold, and to the great skill with which the establishment is regulated. The auriferous lodes of Morro Velho are in general pyrites, contained in argillaceous schist, and inclined to the southeast, about 45°. The principal mines explored are those of Cachoeira and Bahú. The first has in horizontal section a length of 1,120 palmos, a width which varies from

Bahú and 8° in the middle Cachoeira. Its dip varies from S. 82° E. – S. 58° E., and the inclination from 42° – 47°, but everywhere parallel with the striæ."

* For a minute account of the process of separating the gold at Morro Velho, see Burton, Vol. I. Chap. XXVI.

† *Noções Geographicas*, &c., *de Minas Geraes*, 1863, p. 33.

	1849.	1850.	1851.	1852.	1853.	1854.
Stone raised, tons, ...	69,336	67,106	79,810	82,642	85,698	86,048
Stone and Ore stamped, tons,	69,004	64,313	81,629	81,236	86,866	86,433
Gold produced, pounds Troy,	2,583	2,517	3,057	3,323	3,623	3,464
Net profit, ...	£38,136	£35,880	£51,586	£55,391	£49,273	£44,740

	1855.	1856.	1857.	1858.	1859.	1860.
Stone raised, tons, ...	87,297	89,877	86,407	88,901	88,968	91,361
Stone and Ore stamped, tons,	86,848	87,424	86,335	87,270	82,880	74,528
Gold produced, pounds Troy,	3,325	2,992	2,539	2,733	3,294	3,974
Net profit, ...	£34,466	£32,233	£787	£8,545	£38,058	£60,460

	1861.	1862.	1863.	1864.	1865.	1866.
Stone raised, tons, ...	96,612	90,896	84,758	65,435	78,883	In the year, 522,119 oitavas of gold. (Burton.)
Stone and Ore stamped, tons,	71,902	67,508	65,697	62,147	59,607	
Gold produced, pounds Troy,	5,051	5,182	4,713	2,852	4,153	
Net profit, ...	£96,769	£87,531	£63,285	. . .	£80,438	
Loss, ...	. . .	. . .	. . .	£14,629	. . .	

13 to 85 palmos, and it had in March, 1861, a perpendicular depth of 1,190 palmos, and in February, 1862, of 1,480 palmos. The mine of Bahú averages the same dimensions. Both possess six inclined planes for the transport of the ore. In the year 1859 there were occupied in the service of the mine 274 free natives and strangers and 407 slaves; in all, 681 men; among them 242 broqueiros [blasters], who extracted in 311 working-days 89,000 tons of ore, of which 6,119 were rejected as poor, and the remaining 82,881 tons were stamped in 6 mills with 134 hands. With the breaking of the stone, in the service of the mills and the amalgamating establishment, were occupied in this year 9 Europeans, 21 native men, and 24 native women, 79 men-slaves,

and 254 women-slaves; in all, 387 persons. The mean yield per ton was 3.9 oitavas.* The amalgamation has been made at the rate of 70 pounds of mercury to the ton of stamped ore, and the loss of mercury was 0.58 ounces for each cubic foot amalgamated, which makes about 6%. The total product of the establishment was in the year referred to 342,885 oitavas of gold, with whose extraction was incurred an expense of 115:808 $067 Reis [about $ 58.000 more or less], i. e. 357 Reis [less than 20 cents] per oitava. The state of health of the establishment is flattering; the mortality in the whole of its population amounted during 1859 to only 2.76%, and, excluding accidental deaths, only to 2.14%, a circumstance very noteworthy for a population composed for the most part of miners, which proves the solicitude the superintendency feels for the welfare of its employees. The capital of the company subscribed since 1830 is £ 128,400 sterling, which has during thirty years, up to 1860, produced a net income of £466,874 6 *s.* 1 *d.*, the possessions of the establishment being worth, beside this, £ 100,000. The last dividend of the company was £ 2 for share of £ 15."

The Morro Velho mine is an example of successful vein-mining to a great depth. There is no appearance of a diminution in richness of the ore in descending.

Phillips states, on the authority of Mr. Hockin, the managing director of the mines, that

"The rock treated at the Morro Velho mine is principally a mixture of magnetic, arsenical, and common iron pyrites, finely disseminated in, and intimately mixed with, a quartzose gangue. The composition of what is called pure ore may be taken at about 43 per cent of silica, and 57 per cent of pyritous matter. Of these minerals, arsenical pyrites is usually the most auriferous, though it does not occur in large quantities. Pure specimens of

* An oitava is 7.343 grains Troy.

this substance afford by assay from four to six ounces of gold per ton, and wherever crystals of this mineral make their appearance the yield of the precious metal is large. Cubical pyrites is of more frequent occurrence, but is far less rich in gold; solid specimens of this substance, but slightly mixed with quartz, yield about an ounce and a half of gold per ton by assay.

"Magnetic pyrites constitutes the largest proportion of the sulphides found, but this is very slightly auriferous since pure specimens generally yield rather less than four pennyweights per ton. Branches of clay-slate are often found in the principal veins, and this rock, by assay, affords from five to seven and a half pennyweights of gold per ton. Quartz without any admixture of sulphides has never been found to be auriferous, and it is a remarkable fact that the smallest speck of gold is rarely seen previous to concentration in any of the ores of the mine."*

The gold-mines of Gongo Soco lie about twenty miles east of Morro Velho, on the opposite side of the valley of the Rio das Velhas. They were once very productive, and became famous; Weddell says that the old miners once took out one hundred pounds in three hours! but through bad management the company that worked them failed. The income of the company during the thirty or more years of its existence was £ 1,388,416, of which £ 375,163 was profit.†

Gardner says, on the authority of Helmreichen, that at the Gongo Soco mines there is the following succession of rocks: a bed of itacolumite, underneath which is a bed of auriferous jacutinga fifty fathoms thick; ‡ then a thick layer

* For the details of the method of treatment of these ores at Morro Velho see Phillips, *op. cit.*, to which there is a frontispiece representing the Morro Velho establishment.

† Lieutenant Moraes quoted by Burton.

‡ This section does not agree very well with Castelnau's description of the

of ferruginous itacolumite, with a dip of 45°, lying on clay-slate, containing great masses of ironstone. Underlying the slates is granite. Castelnau says that these beds dip to the south. The gold occurred free in the jacutinga, and was separated by washing. M. Weddell * describes the Gongo Soco jacutinga as black and friable as coal; it was said to be very soft, so as to admit of being worked with a pick; blasting was unnecessary. When the gold was not visible, the ore was stamped and washed without using mercury. When the precious metal was visible it was treated first in a mortar, and then washed in a *bateia* or wooden washing-bowl. The Gongo Soco gold is said to contain palladium,† and is deep yellow in color. Burton says that he has seen specimens " of a bright brassy tint, and sometimes dingy red, like worked unpolished copper."

The Rossa Grande Gold-mining Company owns a tract of land twenty-one square miles in area, not far from the mines of Morro Velho. The gold occurs in a mixture of quartz sometimes associated with iron-ore, at others with arsenical pyrites, or ferric oxide. Some of these ores are said to be very rich, producing even as much as fifty oitavas to the ton. Besides the vein-rock, gold occurs in jacutinga, and in alluvial washings. Burton visited the mine, and reports it as looking very much like a failure.

The Morro de Santa Anna, where the Dom Pedro North

mines: Castelnau says that the jacutinga is ordinarily only sixteen centimetres in thickness.

* Castelnau, *Hist. du Voyage*, Tom. I. p. 243 (*bis*).

† The gold of Brazil is always alloyed with silver, and occasionally with platina; sometimes it contains a considerable percentage of iron, when it is very dark in color. Iridium and irid-osmium occur in the gold-washings of Minas.

d'El-Rei Company was established, is a mountain about four thousand feet in height above the sea, and some two thousand feet above the neighboring valley.

I give Phillips's description of the geological structure of this mountain, but I must confess that it is somewhat enigmatical: * —

"The face of the mountain is covered with *canga*, an iron conglomerate, about four feet thick; this is auriferous, and will probably pay for stamping. Beneath the *canga* is the first jacutinga formation, about sixty feet in thickness, containing veins rich in the precious metal; the jacutinga partakes more of the character of mica-slate than of iron-sand, and the auriferous veins in it are more like quartz than ironstone. This rests on a stratum of hard ironstone about three feet in thickness, which is the second jacutinga formation, but quartz is the predominating constituent, and rock is, according to Captain Treloar, a more correct name for it than jacutinga. This lode averages about four feet wide; it opens and contracts, and where it expands it is generally found most productive. Subjacent to the layer is a layer of hard clay and mica-slate, of about five feet in thickness, and then comes the rock formation which has yielded the chief returns of gold. In the present workings it is about ten feet wide; but in its longitudinal course it so expands and contracts as to become in some places extinct. . . . . The general direction of the lodes at Morro de Santa Anna may be said to be easterly and westerly, and their underlie northerly; but both vary, owing to the lodes hugging the contour of the mountain."

Near Santa Anna is another mine called Maquiné, worked by the company owning the Morro de Santa Anna, which has been abandoned. Out of all the gold-mines of Brazil only two have paid, — Morro Velho and Maquiné.

* Phillips, *op. cit.*, p. 85.

The rest have failed, some of them after a more or less prosperous career, and notoriously in most cases from bad management and an imprudent outlay of funds.

At Congonhas do Campo, Burton describes the gold as occurring "in the pores and cavities of friable or rotten quartz injected into greenstone," * and states that "Mr. Luccock detected dust-gold 'among schist-clays and the other component parts of the ground,' and the latter contained the ore 'with equal certainty, and nearly equal quantity, whether of the prevailing red hue, or any of the shades of brown or yellow.'"

The gold-mine of São Vicente belongs to the East d'El-Rei Company, which is, we understand, about to be reconstructed. The gold is found in a quartz vein that strikes east-west with a dip of 28°, and whose rock character we have already described. Burton says that "failure is its actual state," but that "the little lode may pay if worked safely, that is to say, scientifically and economically." A specimen of the São Vicente gold was examined for me by Mr. Clarke of the Cornell Chemical Laboratory for palladium. There was not even a trace.

The Cata Branca mines belong to the Morro Velho Company, and are situated two miles east of the village of Corrego Secco. The following notes on the mines are from Burton †: —

"The serra of Cata Branca trends where mined from east of north to west of south. The containing rock proved to be micaceous granular quartz with visible gold, as in California. The strike was N. 15° W., and the dip 80° to 85°; in some places the stratification was nearly vertical, in others it was bent to the slope of the

* Captain Burton, Highlands of the Brazil, Vol. I. p. 174.

† Vol. I. p. 182.

mountain, and generally it was irregular. The lode, narrow at the surface, widened below from six to eighteen feet, and the greatest depth attained was thirty-two fathoms."

He describes the vein-rock as varying granular to compact, and states that canga and jacutinga occur here.

"The lode, which could not be called a 'constant productive,' abounds in *vughs*, or vein cavities, tubes, pipes, and branches, called by the Brazilian miner 'Olhos,'—eyes, surrounded by a soft material, mainly running vertically, and richer in free gold than the average. Near these pockets, but not disseminated through the vein, was a small quantity of auriferous pyrites, iron and arsenical. A little fine yellow dust, oxide of bismuth, ran down the middle of the lodė and gave granular gold. The best specimen averaged from 21.75 to 22 carats, our standard gold.

"The Santa Antonio lode lay parallel with and east of the Cata Branca. The Arédes mine, eight miles to the southwest, was beyond the peak; here the serra is covered with boulders of hard quartz, very numerous at the base of the great vein. They rest on the common, soft, variously colored clays of the country, and are intersected with lines of sugary quartz, which gave a very little very fine gold. This formation extends far to south and west of Itabira; openings were made in it, and one, the 'Sumidouro,' was successful. Arédes showed also a small formation of jacotinga containing red gold, sometimes alloyed with palladium and accompanied with oxide of manganese."

In 1843 M. d'Osery, geologist of the expedition of Count Castelnau, visited and examined the Cata Branca mine. He reported* that at the locality the rocks consist of itacolumite and clay slates alternating, and in strata almost perpendicular. He described the auriferous vein as running nearly

* Castelnau, *Expédition dans l'Amérique de Sud, Hist. du Voyage,* Tome I. p. 244. By a blunder in pagination, pp. 241 – 256 are repeated in this volume.

north-south and traversed by fissures or faults in which the gold occurred. It was also found in the fissures for a distance of two or three palms on each side of the line of the faults, together with bismuth. Sometimes gold was found where the vein quartz came in contact with the enclosing rock; the interior of the vein was very barren. D'Osery thought that originally the vein was composed of pure quartz, which was afterwards disturbed and the gold and bismuth introduced by sublimation.

The mine proved a failure because of bad working and a want of economy.

Gold-washings occur, as already described, in almost every province in the empire.

In Maranhão is a mine belonging to a company called the Montes Aureos Gold-Mining Company (limited). I know nothing of it further than that it is said so far to have been worked with but little success.

The generally received opinion that the gold-mines of Brazil are exhausted is a very great mistake. There are still surface deposits of great extent which, with modern appliances, could be successfully worked. The underground wealth of the country is almost untouched,* and if the mining public of America knew Brazil better, I am persuaded that the gold-fields of that country would not be neglected by American capitalists.

* In this belief I am supported by Burton, and Liais, in treating of the head of the basin of the São Francisco, says: "Quant aux filons pyriteux qui abondent dans les régions montagneuses circonscrivant le bassin du San Francisco, et où ses divers affluents prennent leur source, ils sont été à peine attaqués. C'est là cependant que réside la grande richesse aurifère de la province de Minas-Geraës. Car c'est de la surface décomposée de ces filons pyriteux qu'était provenu l'or qui fut jadis retiré des dépôts meubles" — *Le San-Francisco au Brésil*, par M. Liais, *Bull. de la Société Géographie*, 5me Série, E. 2, p. 309.

CHAPTER XIX.

RESUMÉ OF THE GEOLOGY OF BRAZIL.

Eozoic Rocks and their Distribution in Brazil. — Absence of Limestones. — The Silurian Age in Brazil. — The Auriferous Clay-Slates of Minas probably Lower Silurian. — Note on the Silurian of the Andes. — The Distribution of Marine Animals in the Palæozoic. — The Devonian Age in Brazil and South America. — The Carboniferous of Brazil and Bolivia. — The New Red Sandstone. — The Cretaceous, its Distribution in Brazil and South America. — Several distinct Periods represented. — Tertiary Rocks. — Drift. — The Glacial Phenomena of Patagonia. — Tapanhoacanga. — The Drift of Rio and of the Region of Decomposition. — The Drift of the Dry Region of Bahia, Sergipe, and Alagôas. — Examination into the Merits of the various Theories proposed to account for the Formation of the Brazilian Drift. — The Theory of Subaerial Decomposition. — Wave Action during a Subsidence. — Wave Action during Elevation. — All these Theories unsatisfactory. — The Glacial Hypothesis the most reasonable.

Eozoic. — The gneiss of the province of Rio de Janeiro is an orthoclase variety, varying from schistose to coarse-grained and porphyritic, or homogeneous and granitic; so far as I have been able to observe, it is everywhere stratified, and consists of metamorphic sedimentary deposits. Though much of the rock would be described as granite if seen in the hand specimen or in a single quarry, I have never failed to find the large masses stratified, so that in this work I have included all the varieties under the general head of gneiss. These rocks in the province of Rio are of great thickness, and the Serra do Mar and the Serra da Mantiqueira are wholly composed of them. According to Pissis they are divided into two groups, an upper and lower, subdivided as follows: —

Upper.	Gneiss, with numerous heavy beds of quartzite, and an abundance of mineral veins which are not found in the lower division.
Lower.	1. Fine-grained gneiss, without garnets and with some subordinate beds of quartzite in coarse grains with a little mica. 2. Gneiss fine grained, very rich in mica, and with an abundance of garnets. 3. Porphyritic gneiss.

On crossing the Mucury district I found the gneiss, which was at first coarse and porphyritic, becoming finer on going westward, and finally giving way to heavy beds of mica-slate or mica-schistose gneiss with bands of quartz. The same succession seems to obtain elsewhere in Brazil, the gneisses proper being overlaid by mica-slates, the older rocks along the coast generally lying to the eastward. This corresponds very well with what D'Orbigny says of the succession of similar strata in Bolivia and the Andes, where the gneisses are immediately overlaid by mica-slates. Elie de Beaumont and other geologists of note have long since signalized the gneisses of the Serra do Mar as among the very oldest stratified rocks of the globe. The system of upheaval of the gneiss of the Serra do Mar D'Orbigny calls the Brazilian; and Elie de Beaumont, in his report on the "*Considérations générales sur la Géologie de l'Amérique méridionale*," by M. D'Orbigny,* says that this system is one of the oldest known, and that perhaps it preceded the *soulèvement* of the most ancient system of mountains hitherto described in Europe. It is certainly the oldest of the rock formations of the Brazilian plateau. When we come to compare the Brazilian gneisses with the Laurentian rocks of Canada and

* *Comptes Rendus*, 28 May, 1843.

Europe we find such strong resemblance in lithological character, and in the system of the upheaval, I can see no reason why we should not refer them to the eozoic. The axis of upheaval is the same as that of the Laurentides.

In North America heavy beds of limestone are interstratified with the gneiss of the Laurentian. In the Serra do Mar beds of limestone are very rare, and the thin bed I examined at Pirahy is the only one I have seen in the Serra do Mar. This contained only faint streaks of serpentine. Limestones appear to occur interbedded with the gneiss at Cantagallo. Some of the limestones of the interior of Bahia may belong to the same series.

The absence of clay-slates among the gneisses in Brazil recalls also the Laurentian of North America. There can be little doubt that the great mass of the gneisses of the coast provinces north and south of Rio are eozoic, but these rocks have in the northern provinces been so slightly examined that it is impossible to describe them with any detail, and some of the mica-schists associated with them may be Lower Silurian or Cambrian. Along the coast of the Province of Bahia there are dioritic gneisses in the series, and on the São Francisco and elsewhere we find syenites. The study of these old rocks in the southern provinces is attended with immense difficulty, owing to the forests, the decomposition of the surface, and the thickness of the drift. But in the drier northern provinces, where the rocks are more exposed and less affected by decomposition, they may be well examined. In the preceding chapters I have shown that gneiss is found in every province of the empire. Not only does it form the great coast belt extending from Maranhão to the mouth of the Rio de la Plata, but it sends off a band from southern Minas Geraes into Goyaz, and the Montes Pyreneos and a considerable part of the mountainous region of Central Goyaz are composed of it. The same rock shows itself in the cataracts of the Tocantins, the Xingú, the Tapajos, the Arinos, and the Madeira, showing that the tableland of Brazil is everywhere underlaid by it.

In the present state of our knowledge of the stratigraphy of these rocks it is quite impossible to do more than guess at their thickness, for as in Canada and elsewhere there are numerous reversed folds, and one may travel for miles over the surface of the Brazilian gneisses, finding them always highly inclined, and all dipping in the same way. The Serra do Mar, where crossed by the Dom Pedro II. Railroad, is a monoclinal ridge, but it must be composed of several reversed folds, else the thickness would be enormous.

The highlands of Venezuela and Guiana are largely composed of gneiss similar to that of Brazil, and disturbed by the same system of upheaval as has been remarked by Humboldt, D'Orbigny, Agassiz, and others, and this gneiss area, bounding the Amazonian valley on the north, was doubtless an island at the opening of the palæozoic time. The highlands of Brazil formed another island, while the Chiquitos gneiss region to the southwestward was probably another.

Since the foregoing was written and sent to the printer, I have been honored by a visit from Dr. T. Sterry Hunt, who has examined with care the large suite of metamorphic rocks I brought home from Brazil. Dr. Hunt has kindly furnished me with the following note for publication: —

"The gneissic rocks of Rio de Janeiro and the Serra do Mar present the characteristic types of the Laurentian of North America, including as they do coarse granitoid and porphyritic varieties with red orthoclase and fine-grained gray and white banded gneisses, often hornblendic. The white crystalline limestone with pale green serpentine which occurs with these Brazilian gneisses is not distinguishable from that of the North American Laurentian. The fine-grained, tender micaceous and hornblendic schists, which in Brazil succeed the gneisses, are very like the similar rocks which in some parts of New England and Acadia appear to follow the Laurentian, and are associated with staurotide, cyanite, and chiastolite slates; while the auriferous argillites and quartzites which follow these schists in Brazil strikingly resemble those which in Nova Scotia occupy a similar stratigraphical position. This triple parallelism in lithological and mineralogical character in the rocks of regions so widely separated is in itself a strong argument in favor of their geological parallelism."

Silurian. — Notwithstanding all that has been published by the various geologists who have studied the gold region of Minas Geraes, the exact succession of the different members of the metamorphic series lying just inside of the gneiss belt has never been satisfactorily worked out. The clay and talcose schists, the itacolumite,* itabirite, and other associated metamorphic rocks of this region appear to be lower palæozoic in age. I have called attention to the striking resemblance borne by the clay-slates and associated quartzites to the gold-bearing rocks of Nova Scotia, and I have suggested that they may be the equivalents of the Quebec group of North America. The gold-bearing rocks in Minas Geraes resemble the similar auriferous series of the southern Atlantic States in which itacolumite occurs.

* On page 149 I have spoken of the occurrence on the Rio Gavatá of a schistose quartz rock resembling a sandstone. I had not a good opportunity of examining the locality, and I doubted whether it was a metamorphic rock. A further examination of a specimen of the rock in company with Dr. T. S. Hunt has proved it to be a true itacolumite.

Clay-slates with auriferous veins occur in other parts of Brazil besides Minas, as, for instance, in Goyaz, and in the vicinity of Cuiabá in Matto Grosso.

These rocks are everywhere so metamorphosed, that all trace of fossils has been completely obliterated.*

* The Silurian rocks of the Andes of Bolivia and Peru have been examined by the English geologist, Mr. Forebs (Quart. Jour. Geol. Soc. Vol. XVII. p. 53), who thus describes their distribution: "The rocks which I have grouped together as pertaining to the Silurian epoch show themselves continuously, or very nearly so, over an area from northwest to southeast of more than seven hundred miles; and the area occupied by them cannot be estimated at less than 80,000 to 100,000 square miles. They form the mountain-chain of the high Andes, rising to an absolute height of 25,000 feet above the sea, and, in the part of South America more particularly the subject of this memoir, continuous through Peru from the north of Cusco over the snowy ranges of Carabaya and Apollobamba, across the provinces of Munecas, Larecaja, La Paz, Yungas, Sica-Sica, Inquisivi, Ayopaya, Cochabamba, Cliza, Misque, Chayanta, Yamparez, Porco, Tomini, and Cinti throwing off spurs along the eastern side of the main chain, right through the province of Caupolican down to the river Beni in Mojos, into Yuracores, Valle Grande, Santa Cruz and Chuquisaca, and to the east into the provinces of Oruro, Potosi, and Chichas." The rocks consist of clay-slate, shales, and graywackes, and, — according to Mr. Forbes, they probably represent the whole Silurian from top to bottom. D'Orbigny had already described ten species of Silurian fossils from the Central Andes; to this list Mr. Forbes has added nineteen new species, described by Mr. Salter in his paper following that of Mr. Forbes. The genera represented in the Andean Silurian are Cruzeana, Lingula, Orthis, Graptolithus, Phacops, Asaphus, Boliviana, Patella, Bellerophon, Arca?, Ctenodonta, Cucullella, Strophomena, Tentaculites, Beyrichia, Homalonotus. Two species of Phacops described by D'Orbigny are doubtful, being probably Devonian. If we subtract these we have left only twenty-seven species of fossils known for the Andean Silurian. In the course of his paper Mr. Salter makes an interesting remark that species in the Silurian had a very limited range, those of India, Australia, and Europe being entirely different. On the other hand, the species of the Devonian, especially of the upper part, had a very wide range, while the carboniferous types are almost cosmopolitan. I must confess that, after a careful study of the carboniferous Brachiopoda of Nova Scotia, I am hardly prepared to go quite so far as to admit that the Producti from Bolivia, Nova Scotia, Ireland, and Belgium, thrown together under the name of *Cora,* are all the same species, and I may

This series offers a large number of crystallized minerals, among which are the topaz, enclaze, &c. The topazes of Minas appear to be found in the cascalho formed from the *débris* of these rocks.

Devonian. — Some of the metamorphic rocks of Minas Geraes or Bahia may be devonian, but I have seen no rocks referable to that age on the coast, unless it be that the slate conglomerates, sandstone, and shale, with fossil plants found on the Rio Pardo, may belong to it.*

Carboniferous. — There can be no uncertainty about the existence of true carboniferous strata in Brazil, for besides the coal we have an abundance of fossil plants of carboniferous genera. The coal-basins lie just south of the tropics, but within the range of the palm, and they are a coast formation, corresponding in this respect to the coal-basins of Acadia, Massachusetts, and Rhode Island. I know of no carboniferous strata north of Rio on the coast. It would seem as though the depression of the coast which allowed the accumulation of the coal-beds of the southern provinces had not extended to the north. The very slight disturbance of the coal-beds is noteworthy, as is also their bituminous character.†

say the same of the other Nova Scotian Brachiopoda referred by the distinguished Mr. Davidson to European forms. But whether the species were or were not absolutely cosmopolitan during the carboniferous, the resemblance of the marine animals was much greater during the carboniferous than before that time.

* Messrs. D'Orbigny, Salter, and Forbes refer to the devonian certain fossiliferous rocks of the eastern plateau of Bolivia. See Forbes, *op. cit.*, p. 51; and Salter, *op. cit.*, p. 63. The Falkland Islands, described by Darwin, are composed of rocks probably belonging to the Lower devonian. (Quarterly Journal Geological Society, London, Vol. II. p. 267, 1846.)

† Carboniferous rocks are found in the Rio Guaporé, one of the branches of the Madeira. The carboniferous rocks of Bolivia have been studied by D'Or-

*Triassic.** — I have referred to the triassic a thick series of red sandstones, lithologically identical with the Connecticut River and New Jersey new red sandstone, apparently barren of fossil remains, and which occupy a large area in the Province of Sergipe, underlying the cretaceous. These rocks are more or less inclined. I know of no trap associated with them.

Jurassic.† — I have seen no rocks on the Brazilian coast referable to this age. I can explain their absence only by

bigny and Forbes. I extract a short account of these last by D'Archiac, *Géologie et Paléontologie*, p. 499 : —

"Les roches carbonifères de la Bolivie, situées à l'ouest des Andes, se rencontrent, par places, comme de petits bassins allongés généralement du S. O. au N. E. situés au milieu de la grande plaine quaternaire qui entoure le lac de Titicaca, se montrant aussi au nord du lac, et plus au sud dans les provinces d'Arque et d'Oruro. Le point le plus bas où on les observe est à 3,800 mètres d'altitude et on peut les suivre jusqu'a 4,000 et 4,500 mètres. À l'ouest du lac, entre Tiquina et la Guardia, M. Forbes donne une idée complète de la série des assises disposées en bassin renversé et présentant successivement des plus anciennes aux plus récentes qui occupent le milieu du plissement, des grès blancs, des conglomérats et des grès rouges, des argiles blanches panachées, des calcaires en bancs épais, bleus et jaunes, des argiles schisteuses panachées, un calcaire bleu puissant, enfin des grès jaunes et blancs." And in a note he says: "Les fossiles de ces assises, étudiés par M. Salter, sont; *Productus Semireticulatus* (*P. Inca d'Orb.*) *P. longispinus* (*Capacii d'Orb.*) *Spirifer Condor*, *S. Boliviensis*, *Athyris subtilita*, *Orthis resupinata*, *O. Andii*, *Rhynchonella* nov. sp. *Euomphalus* (*Phanerotinus?*), *Bellerophon* voisin de *B. Urii*, des polypiers et des crinoides indéterminés. Des provinces d'Arque et d'Oruro ont été obtenus les *Spirifer Condor* et *lineatus*, les *Productus Cora*, *semireticulatus*, *Boliviensis* et *l'Orthis Andii*."

* Mr. M. D. Forbes refers to the Triassic or Permian a series of red and yellow sandstones, saliferous and gypseous marls, clays, gypsums, cupriferous sandstones, and red conglomerate found in the Andes, but which contain no determinable organic remains. (Quart. Jour. Geol. Soc., Vol. XVII. p. 36.)

† Jurassic rocks containing Ammonites, Terebratulæ, Spirifers, and other fossil characteristics of that epoch have been found in the Andes by MM. Crosnier, Mayen, D'Orbigny, Darwin, Domeyko, Forbes, &c. They extend from Chili to Peru.

supposing that during the Jurassic the coast stood higher than at present. In this respect the Brazilian coast would resemble that of Eastern North America.

Cretaceous. — The cretaceous rocks of Brazil are unknown on the coast south of the Abrolhos, which islands I believe to be outliers of this formation. Properly speaking the cretaceous deposits begin a few miles south of the Bay of Bahia, and occur at intervals along the coast northward, occupying, at least in several instances, separate basins, some of which are fresh-water. We find cretaceous rocks in Bahia, Sergipe, Alagôas, Pernambuco, Parahyba do Norte, Ceará, and Piauhy. It is difficult to estimate their exact extent, because they are largely covered up by tertiary beds. It is very probable that marine cretaceous beds underlie the tertiary deposits throughout the whole valley of the Amazonas, but the only place where they show themselves, so far as I can ascertain, is on the Aquiry, an affluent of the Rio Purus, where they have been examined by M. Chandler, as is stated by Professor Agassiz.* I am not aware that they are exposed anywhere to the eastward on either side of the valley.

Among the cretaceous rocks of Brazil several periods are represented.

The fossil mollusks of the fresh-water beds of the Bahia Basin have a very strong wealden look, but they are associated with teleostian fishes and other remains, which are certainly cretaceous. They evidently belong low down in the series, and they may represent the Néocomien.

The compact limestones at Maroim affording Ammonites, Ceratites, Natica, &c., are probably middle cretaceous. Over these are the flaggy white and grayish limestones with Ino-

* Journey in Brazil.

ceramus, Ammonites, fish, &c., apparently representing the white chalk, *Sénonien.* For the fresh-water beds at Bahia I would propose the name Bahian group; for the Maroïm limestones, that of Sergipian group; for the flaggy limestone beds near Aracajú, the Cotinguiban group, and for the Aquiry beds the Amazonian group. The cretaceous of Brazil would then be divided as follows: —

Amazonian group (Aquiry) with Mosasaurus, *Mæstrichtien ?*
Cotinguiban group with Inoceramus, Ammonites, &c., *Sénonien ?*
Sergipian group with Ammonites and Ceratites, *middle cretaceous ?*
Bahian group, Crocodilus, Pisodus, species of Melania and other fresh-water shells, cyprids, &c., *Néocomien ?*

The sandstones, shales, and limestones of the Abrolhos and the lower São Francisco I believe to be cretaceous, but I have no fossils by which to determine their exact age. They may correspond in part to the Sergipian and Cotinguiban groups. Those of the São Francisco are mud and sand deposits instead of limestone, which accumulated along the coast elsewhere in clearer water. It will be remembered that I have described the limestones near Propriá as sandy and even pebbly.

The cretaceous rocks nowhere form very high hills. They appear to have been deposited in a shallow sea, which was not deep enough to penetrate into the São Franciscan valley above the falls. The sandstones above the falls, described by Burton as cretaceous, I am persuaded will be found to be tertiary. The cretaceous rocks have suffered slight disturbance, and at the Abrolhos it is worthy of note that they are associated with volcanic deposits.

At the time of the deposition of the cretaceous, the northern part of South America was depressed more than at present, while the coast of the southern provinces of Brazil seems to have been higher than now.

In speaking of the cretaceous of South America, M. D'Archiac, *Géologie et Paléontologie*, p. 624, says: —

"Nous avons fait voir que, d'après les recherches de M. H. Karsten, confirmées depuis par celles de M. Wall, on pouvait présumer que l'étage inférieur de la craie tuffeau, le gault et une partie du groupe néocomien étaient représentés dans le Venezuela, particulièrement dans les cordillères de Mérida et Truxillo. Tous les calcaires crétacés des chaînes de ce pays, comme ceux du même âge, que l'on suit jusqu'au Chili, sont d'ailleurs entièrement noirs, bitumineux et semblables à ceux des grandes montagnes de l'Europe."

Tertiary. — The clays and ferruginous sandstone forming the coast plains outside the cordilheira are undisturbed, and overlie the cretaceous unconformably. They are overlaid by the drift-clays, which descend from the cordilheira and cover their glaciated surfaces, so that, though I have nowhere found fossils in them, I have felt justified in referring them to the tertiary. The horizontal beds of clays, sandstone, &c., of the Jequitinhonha and São Francisco valley are everywhere undisturbed, even where they closely approach the coast where the cretaceous rocks have suffered upheaval. They resemble the coast beds, except that they are thicker, stand at a very much higher level, and in some cases form beds of pure sandstone and conglomerate with limestone and iron ore. They, too, are covered by the drift-clays. I suppose that they are also tertiary, but older than the coast clays. To the same group evidently belong the horizontal deposits of the plateau of São Paulo, similar strata occupying the upper part of the valley of the Parahyba do Sul, and the clays and sandstones of the elevated plains of the north. These beds must have been deposited when the continent stood at a level full 3,000 feet lower than at present. The material was evidently derived from the wearing away of the decomposed gneissose rocks, and it appears to have been deposited rapidly in a muddy sea, not favorable for the existence of life. After these beds were deposited the coast rose very uniformly, and they suffered very extensive denudation. Along the coast outside the cordilheira there were deposited, probably in a large part made up of the results of the older beds, the coast sandstones and clays. According to my own observation the upper level of the coast clays south of Bahia is always much below the level of the lowest beds of the older beds. I have never seen them tie in with one another, but I strongly suspect that in the vicinity of Monte Pascoal an outlier of the older beds lies surrounded by the newer.

The stratified and loose sands and clays of the Taboleiros at Alagoinhas appear to be older than the drift. They are certainly newer than the Coast Tertiary group. They need much more study, and I must confess that there are some puzzling points in connection with them.*

Drift. — In South America from Tierra del Fuego northward, to at least 41° S., glacial phenomena have been observed and reported by Darwin and others, and these phenomena appear to be identical with those so well studied in the northern hemisphere. Drift occurs in the Falkland Islands (Darwin), Australia, and New Zealand. The Antarctic Continent is buried in ice and snow. No doubt can exist that a drift period prevailed over the southern part of

* Tertiary rocks are found over large areas both north and south of Brazil.

the southern hemisphere. D'Archiac* has already called attention to the fact that no mention of either striæ, furrows, or polished surfaces has been made by those who have studied the drift of South America, which seems very remarkable. He suggests that it may be perhaps owing to a want of attention on the part of the travellers.

It is not to be wondered at that, when Professor Agassiz claimed in 1865 to have found glacial drift in the vicinity of Rio, scientific men were astonished and doubted the correctness of the Professor's deductions ; and when from under the equator he reported the discovery of glacial moraines the statement seemed past belief.

In the preceding pages, in connection with a careful description of the Brazilian coast, I have noted with much detail the occurrence of certain surface deposits northward to Pernambuco, at least, which deposits I have claimed to be glacial drift. I propose now in this chapter to bring together as concisely as possible all the facts bearing upon this subject, and then to discuss them for the purpose of showing that no other hypothesis than that of the glaciation of the coast is sufficient to account for them.

Von Eschwege describes a formation which is known in Minas Geraes as *Tapanoacanga*.† It consists of angular

* *Géologie et Paléontologie*, p. 719.

† Von Eschwege, in his *Geognostisches Gemälde von Brasilien*, p. 30, gives so interesting an account of the Tapanhoacanga that I translate it almost entire. He says: "This rock is composed of sharp-cornered, angular, rarely slightly rounded fragments of micaceous iron (*eisenglimmer*), specular iron, and magnetic oxide of iron, held together by a red, yellow, or brown ochreous cement. These fragments are from several lines to eight inches in diameter. It is often very auriferous, and contains sometimes scales of talc, chlorite, and here and there fragments of itacolumite. The cement becomes in some places

or rounded fragments of micaceous iron and other rocks cemented together by an ochreous paste, which sometimes exists without the gravel. This formation, with a thickness of from six to nine feet, more or less, he states, wraps the highest mountains round about like a mantle. The same material is found elsewhere. Mawe, in his description of the mines of Jaraguá in São Paulo, speaks of the gold as

so abundant that the embedded pieces are not visible; this then forms distinct deposits of red ironstone in thin layers, containing ordinarily many little flakes of mica. This rock is not only found in the valleys and on the slopes of the mountains, but it covers their most elevated ridges and flanks like a sort of mantle from half a toise to a toise and a half in thickness [a toise equals 6,395 feet]; it is in general superimposed upon the ferruginous schist and clay-slate. The most important foreign mineral deposits found in it are brown hæmatite and wavellite, which occur in considerable masses near Villa Rica. . . . . The Serra do Tapanhoacanga near Congonhas do Campo (Province of Minas), whose summit rises to a height of 4,800 feet, is completely covered, over an area of several miles; all the flank of the mountain where Villa Rica is situated is incrusted with it; the surface is overturned by the mining works. The Campo de Saramenha, *vis-à-vis*, is as if paved with it. It is abundant along the route from Villa Rica to Serro do Frio; it is probably also found in the Province of Goyaz. *Tapanhoacanga* signifies in an African idiom * negro's head; the miners have given this name to the rock in question because of its uneven, knotty surface, which appears concretionary like an hæmatite. It is difficult to explain the origin of this conglomerate. The angular fragments, the irregularity with which they are piled up one on the other, the manner in which this rock covers, like a coat or glazing, the top and flanks of the mountains, tend to make one believe that it is not the result of the rapid degradation of the ferruginous mountains which formed only the most elevated points in the country, and of which the peak of Itabira, the Serra da Piedade, and others are the remains, but that it is due to the extremely prompt drying up of the liquid, which sojourned formerly on the mountains, and has brought there the fragments which to-day cover them. These could not follow the liquid to the bottom of the valleys, and were arrested, like solidified lavas, on the midst of the slopes. The disorder with which they are piled up proves sufficiently that the deposit was not gradually made." Castelnau says that the canga is certainly of plutonic origin!

* This is not African, but Tupi, *Tapanhuna* meaning *negro*, and *acanga* head.

occurring in a layer of cascalho,—he writes it incorrectly *cascalhao*,—or gravel of rounded pebbles, principally quartz, which wraps the hills round about, and is covered by a sheet of soil. This surface deposit rests on gneiss. At Minas Novas the gold, as we have already described, has been mined from a similar gravel composed of rounded quartz pebbles, &c., with a ferruginous cement, and overlaid by a similar bed of clay, the whole resting on clay-slate decomposed in place.

At Rio the rounded surface of the decomposed gneiss is covered by the same sheet of quartz pebbles and overlying clays, and all the province, except the flat alluvial plains, such as border the coast, and whose elevation is usually less than twenty feet, is covered with the same deposit to the tops of the highest hills I have examined. The pebble sheet, it is true, varies in thickness, and in some localities is absent, especially over areas in which quartz veins are not abundant. In some places the pebbles are coarse, in others fine, and occasionally we find intermingled with them fragments of gneiss, trap or tertiary sandstone. We find these surface deposits everywhere lying immediately over a rounded surface of gneiss, albeit the rock may be decomposed to a great depth. The pebbles and rock fragments are not confined to the pebble sheet alone, but, sometimes rounded, sometimes angular, they are frequently found in the overlying clay. This last may vary greatly in thickness and color, but the general composition is very uniformly the same. The whole deposit is everywhere without structure, presenting no trace of stratification. The same layers extend over the provinces south of Rio. To the northward they are found all over Minas. I have seen them covering uniformly the hills of Espirito Santo, and the coast of

Bahia, Sergipe, Alagôas, and Pernambuco. Northward I have myself not seen these deposits, but Professor Agassiz reports their existence in various localities on the coast north of Pernambuco, and in the valley of the Amazonas westward to the confines of Peru.

At Rio, as described by Professor Agassiz in the "Journey in Brazil," and by myself in the chapter on Rio de Janeiro, there are, in the valley of Tijuca, near Rio, and elsewhere, deposits of immense boulders of trap, gneiss, &c., which are evidently morainic and the work of local glaciers; and the Professor has described similar moraines as existing in the Province of Ceará. The pebble and clay sheet covers a large portion of the province of Minas, and is found not only on the hills, but on the campos. In the provinces of Bahia, Sergipe, Alagôas, and of the north there is, as I have described, a zone of dry country, lying just behind the coast forest belt, and largely composed of gneiss, mica-slate, and the like, over which the surface deposits consist of boulders of rock of all sizes, rounded and angular, scattered over the surface, and sometimes piled up in confusion, with very little soil, the rock frequently being bare; with these occur rounded quartz pebbles. The surface of the gneiss country in this dry zone in Bahia, as observed by Messrs. Allen and Nicolay, and in Sergipe and Alagôas by myself, is remarkably even, and over large tracts forms a plain. The topography is remarkable for shallow depressions without outlets, forming ponds during the rainy season. Mr. Allen describes having seen immense pot-holes worn in the rocks by the action of falling water on some of the highest swellings and elevations of these plains, now far away from any obstacle over which water could be precipitated.

Such, in brief, are the characteristic features of the surface deposits of the Brazilian coast, to which I have applied the name of drift. I have many times called attention to the rounded surface on which the drift rests, though I have nowhere seen either polished or striated rocks, which is not at all wonderful, as the surface of the rock, wherever I have examined it, even where the decomposition was least, as on the Rio de São Francisco below the Falls of Paulo Affonso, was always more or less decomposed.

Nowhere over the whole region covered by the drifts do we find other water deposits than those clearly referable to the action of rivers or lakes. Above the old sea level of Rio, Victoria, Bahia, &c. there are neither raised beaches nor any other testimony of the action of the sea.

All this immense sheet of structureless clays, gravels, and boulder deposits stretching along the whole coast, and covering alike the coast tertiary plains, the elevated campos, and the serras from bottom to top, belongs to the same formation, and is referable to the work of the same geological agent. We have claimed with Professor Agassiz, to whom belongs the honor of the first announcement of the occurrence of drift in Brazil, that that agent was glacier ice. This hypothesis has been much disputed, and many other ways of accounting for the formation of the sheet of detritus have been proposed. Among them the most important are the following, the respective merits of each of which we propose to examine in detail: —

I. Sub-aerial decomposition.

II. Wave action acting over the surface of the country during a slow subsidence of the coast.

III. Wave action extending over the surface during a slow rise of the land.

I. *Decomposition.* — We have seen how decomposition

may, as at Rio in the case of trap-dikes, at Victoria, and on the islands of the Abrolhos, produce, with the aid of rains, not only a soil, but boulders of decomposition, which may be rounded or angular, and resemble drift boulders so closely as to make it exceedingly difficult to distinguish them from erratics. One can easily conceive how, in the gradual decomposition of a bare surface of rock, — gneiss, for example, — as the rock wasted, the resulting clay and sand may be washed away, and spread over the surface of the soil on the lower grounds. Such is indeed the case, and one sees at the foot of the gneiss precipices, not only at Rio, but elsewhere, a soil of this kind. It closely resembles the drift, but is more washed, the sandy portion remaining near the foot of the precipice, while the muddy part is carried farther off.

Where the surface of a hill is very uneven, unfurnished with soil, and strewn with blocks of rock, as in the dry zone of Bahia, and the São Francisco, one may readily see how by decomposition a structureless soil might be formed covering the surface; but there are a few facts which make this whole hypothesis of the formation of the drift of no value. In bluffs, natural or otherwise, in the vicinity of Rio, as well as in the cuttings on the Dom Pedro Segundo and Cantagallo railways, and on the União e Industria road, and in the Minas Novas region, one may see a great thickness of the surface deposit lying on rock decomposed *in sitû*, and lying undisturbed on the solid rock. Though in old excavations it is very difficult, in any fresh cutting it is the easiest thing in the world to point out the line of separation between the surface detritus and the decomposed rock,* which, by the by, may have only the very thinnest coating of clay, or may

* This line is shown in the engraving on page 508.

be bare. The distinction between this drift and the decomposed rock is of the sharpest kind. The surface deposit is without structure, and has the same appearance that the decomposed rock below would have if it were ground up, and intimately mixed together without washing, while in a mass of the decomposed rock one sees the relative arrangement of the materials preserved undisturbed, with the quartz veins, &c., in place. The veins invariably terminate abruptly at the line separating the decomposed rock from the overlying deposit.

One never sees a quartz vein traceable through the clays, as it certainly would be if these had resulted from decomposition alone. The clay is usually remarkably free from quartz pebbles or boulders, and one rarely sees even a pebble on the surface of the ground in the gneiss regions near Rio, which would certainly not be the case if it were simply a product of decomposition. The greatest objection to the theory under discussion is presented by the sheet of rounded and angular quartz pebbles, for that could never have originated through decomposition. It is evidently the result of mechanical action of some kind, and I am convinced that we must refer the overlying clay-sheet to the same cause. It is evident that the agent, whatever it was, that rounded the pebbles and ground up the clays must have had some part to play in the moulding of the country, though it is to erosion and decomposition that I should attribute the broader topographical features of the coast, and I would refer to the agent that formed the drift the moulding of the actual surface on which the superficial deposits now rest.

Let us now discuss the merits of the hypothesis that the surface detritus has been the result of water action, and examine the two theories of wave action brought to bear over the country during a gradual rise of the land from the sea, and of similar action exerted in like manner over the country during a subsidence.

II. *Wave Action during a Rise of the Land.*—It has been suggested that, in a rise of the land, wave action brought to bear over the surface might leave a coating of loose material similar to the drift strewn over the whole country, but I must confess that, even for a single isolated hill, I cannot see how this theory would have the slightest weight, for it could never produce an arrangement of the materials such as actually exists. Suppose, for instance, that we have

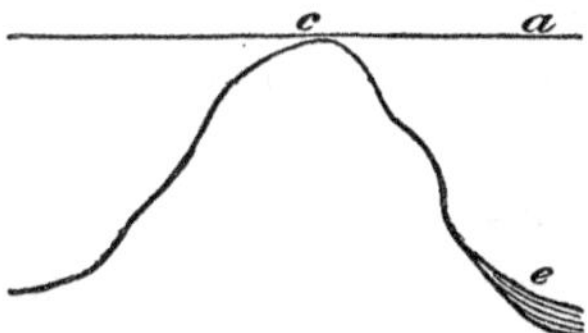

a hill a few hundred feet in height, and which is rising. At a certain time the sea level is found at *a*, so that the waves will wash the hill-top. No one who has been on the sea-shore will doubt that the effect will be to remove the finer materials, and carry them off to be deposited somewhere in quiet water; and this is the natural effect of the sea where it breaks against a slope or over a ridge. It may be that on the sides of the mountain these materials, if in sufficient abundance, may be deposited, and we will suppose that they are so deposited at *e*, but on steep slopes we should not expect to find them. Gravels and sands would be washed by the waves into deeper water, but could rest only on gentle slopes. If the slope from *c* to *e* were twenty or more degrees, and the land should rise gradually so as to bring wave action to bear successively over that slope, we should expect to find the rock surface swept quite clean of loose materials, only sand and gravel being thrown beyond the reach of the waves into hollows or flat places on the rocks; and it seems to me that the result of such a rise would be, even if there were already a coating of decomposed rock on the surface, only to wash it over and gully it out on the hill tops and sides, if it were deep, and cover it with sand and gravel, or to remove it entirely, leaving the hill tops and sides bare, the loose materials being transported into the valleys, and there left as stratified deposits; and not only in the valleys, but upon flat places on the hill-sides, should we expect to find such deposits. Now we find nothing of the sort. Associated with the drift are neither sands nor stratified deposits of any kind. Besides, this hypothesis does not touch the question of the formation of the great angular and rounded boulders unassociated with sands strewn over the dry zone, nor does it explain the transport of boulders over an irregular surface. The tendency would have been to sweep the hill-tops and the steeper and seaward slopes bare, which is not the case.

III. *Wave Action during a Subsidence.* — The hypothesis of the action of waves over the country during a slow subsidence of the land is equally untenable. On a gentle slope sands and gravels would be formed, and perhaps deposited immediately upon the rock, and this sheet one might suppose drawn up like a curtain over the hills as the land sank. The lighter materials settling in deeper water might, at the same time, form a sheet drawn up over the first, so that we might have the surface covered by a sheet of sand and pebbles and over this a sheet of clay.

But we have the pebble sheet unaccompanied by washed sands, which is something incomprehensible under this hypothesis, and, what is of more importance, we find them lying on slopes so steep that it would be impossible for them to remain unless supported by the weight of the overlying clay. It is incomprehensible how water could have formed these deposits without at the same time laying down stratified beds of sand, gravel, &c. in the valleys, or of leaving sand deposits on the plains. It is impossible that the sea should have been without currents and without tides, and the inevitable effect of their action would be to sweep away the lighter material and deposit it along the shore. In the tertiary beds, not only of the coast, but of the interior, we have an example of a sandy deposit rapidly thrown down in a muddy sea, and which, besides sand, contains a very large percentage of clay, passing even into pure white clay; but nowhere is it difficult to distinguish these tertiary deposits from the drift. The tertiary clays were the products of the destruction of gneiss and other metamorphic rocks, and so was the drift, but in the former the material has been washed, though rarely ever arranged. The mica has been completely destroyed, and one sees nothing of it, while in the drift-earth it is constantly to be found. One would expect that, if the Brazilian drift were a sedimentary deposit, the clays would resemble those of the tertiary, which is nowhere the case. They are everywhere just such a material as would result from the mechanical trituration of the rocks, and are wholly without stratification or signs of having been deposited by the action of water. There has been within recent times a slow elevation of the Brazilian coast. In sheltered as well as exposed situations the deposits which it has brought above the surface consist of sand; gravels and clays are exceedingly rare along the exposed coast. Nowhere in river, lake, or sea deposits have I ever seen on the Brazilian coast anything resembling the drift-clays. I have studied with care the effect of the action of the sea on the solid rocks of all kinds along the coast, with the view to ascertain, if possible, whether the peculiar evenly moulded surface covered by the drift could in any way be due to water action joined to the effects of decomposition. Where the rock is gneiss and very homogeneous in structure, and not well bedded, as is the case in the range of hills skirting the seashore on both sides of the entrance to the bay of Rio, the rocks swept by the waves may have a very smooth and regular outline, but where, as on the shores near Bôa Viagem at Rio, Os Busos, Ilhéos, or Bahia, the rock is well bedded and the strata are very highly inclined, the softer beds give way first and leave the harder projecting, and the rocks within reach of the waves are worn in the most irregular manner. Now I do not see how we can resist the conclusion that, if the surface clays and gravels were the products of wave action, we ought to find the surface of the rock on which they rest showing some signs of that action in the wearing away of the softer beds, leaving the harder standing up; but this is never the case. There can be no transition more abrupt than that from a wave-washed, rock-bound shore, and the smooth, even outlines of the hills above the line of wave action. Take, for instance, the coast between the lighthouse at Bahia and the Morro do Conselho, which is to a large extent rocky, and examine the moulding of the rocks washed by the Atlantic surf, and then compare it with the moulding of the rock on the seaward side of any of the exposed hills, where, if washed by the sea, the rock would have been subjected to the pounding of the same surf, and you will be convinced that the moulding of the drift-covered rock-surface was due to an agent that did not respect so thoroughly the difference in hardness between the beds as water does. It is useless to suppose that the sea might have washed against a decomposed surface which might have been rounded down by the surf. The wash of the waves would have removed it entirely. The rounded wave-washed rocks on the shores of Rio are bare, and it is inconceivable how, in the face of the tremendous Atlantic surf, they could ever be covered by detritus by wave action, as all the hills lying along the coast of Bahia and Rio invariably are. Along the coast where decomposition prevails, from the wetness of the climate, the clays are very abundant, and it is a rare thing to find boulders of any other rock than quartz; but in measure, as one goes inland and approaches the dry zone, as is beautifully seen on the São Francisco, the clays grow less abundant while the pebble-layer gradually passes into a sheet of boulders of rock scattered over the surface with little admixture of earth. The rock being of the same general character over large areas, it is usually a difficult matter to decide whether a boulder is travelled or not; * but I have seen at Piranhas syenite boulders lying on gneiss, though evidently coming from not far away, and I have seen gneiss and quartz boulders lying in the clays on the tertiary plain on the Mucury. I have already called attention to the intermixture of greenstone and gneiss boulders of immense size in the valley of Tijuca, occupying situations into which water could not have brought them, and into which they could not have fallen.

* This is an important point to bear in mind. The geology of Brazil is so very simple that we find the same kind of rock over immense areas.

No one seeing the boulder-scattered surface of Bahia and the São Francisco, where the decomposition is exceedingly slight, would ever, I am persuaded, seek for an explanation of the distribution of these masses over the surface in running water or wave action, which last would have been powerless over so uniformly level a surface. I must insist upon the fact that the unarranged materials are precisely like our unmodified drift in the north, and that the surface of the rock on which they lie has the moulding of the surfaces on which our northern drift lies, and that if we refer the northern drift to the action of glacier ice, we must do the same thing for the Brazilian surface detritus, contrary as it is to all our preconceived opinions of the distribution of drift. The fact that neither Professor Agassiz nor myself, nor any one else of our expedition, has been able to discover glacial striæ in Brazil is of very secondary importance. The drift itself exists all over the country, and it cannot be explained away. I have looked carefully for striæ, but there has been everywhere enough decomposition of the surface of the rock as well as of the boulders scattered over it to have destroyed all trace of them. Once I thought I had found striæ. On the Dom Pedro Segundo Railroad, near Mendes, while engaged in making an examination of the cuttings, I found one in which the drift-clay had been removed from over the decomposed gneiss, exposing the glaciated surface. This appeared to be quite fresh, and to my surprise was deeply furrowed with parallel striæ. I took pains to inquire of the engineers of the road, and learned that the drift had slid off from the upper part of the cutting, which was a sufficient explanation. I speak of this only in order to put other observers on their guard against being deceived by any similarly striated surfaces. I can offer but one explanation of the formation of the pot-holes observed by Mr. Allen, and that is that they were formed by glacial cascades in the same way as the pot-holes seen so often on the surface of ridges in the north have been formed during the drift; for, according to the testimony of Mr. Allen, the pot-holes of the Province of Bahia occur on the gneiss plains, far away from any present obstacle over which the water may have flowed. Mr. Allen describes them as being exceedingly well preserved, and having smooth sides.

The drift is, as above stated, removed everywhere down to the limit of wave action before the later elevation of the coast; but the occurrence of the drift on some outlying hillocks of the tertiary clays on the line of the extension of the Cantagallo Railroad, between Porto Novo and Porto das Caixas, which are now surrounded by recent sands, made me suspect that the clay was once continuous below the present sea level between them and the mainland. The fact, too, that it extended uniformly down to the same level everywhere was almost sufficient proof that it formerly extended to a much lower level. At Bahia, as already described, recent sands blown or washed over the drift have been cemented and have protected it from the action of the sea, so that it may be seen extending beneath them down nearly to low tide. This fact seems to prove satisfactorily that formerly the land stood at a higher level even than now.* Drift occurs on some of the islands off the coast.

* From the observations of Darwin and others we know that this recent uprise has been much greater in the south than in the north, and it seems to increase in going south from Rio to the Straits of Magellan. It would seem that the great movements just antecedent and posterior to the drift period in

I believe that during the time of the drift the country stood at a much higher level than at present, and that it was covered by a general glacier. Over the coast region, where decomposition of the rocks had largely obtained, and where the surface of the rock, rendered even by this agent, had been covered by a thick layer of loose material, the glacier reworked this loose material, and when it disappeared left it as a paste, in which the harder materials, such as fragments from quartz veins, &c., more or less rounded, were embedded. The layer of quartz pebbles underlying the paste appears to have consisted of coarser fragments borne along by the bottom of the glacier, while the paste seems to have been more or less distributed through the body of the glacier. A glacier moving over the gneiss regions of Rio or Espirito Santo to-day would find few loose rocks to transport, for the precipices are smooth and unbroken, and little falls from them, so that one could not expect to see moraines of coarse materials formed by the glaciers of that region, and if the ancient glaciers moved over a country whose surface was decomposed, it is not wonderful that the drift consists of paste with but few boulders. On the contrary, over the dry zone the cliffs are ragged and broken, and the rock surface is apt to be broken up, and we should expect to find over such a region drift of a different character from that which obtains over the moist coast region, and resembling more closely the drift of North America.

In the drift-paste I have never seen the slightest trace of organic remains of any kind.

Post-Tertiary. — To this epoch belong the cavern depos-

South America have corresponded with those of North America during the same period. In North America the oscillation of level was greater in the north than in the south; in South America it was just the reverse.

its in Minas Geraes, affording the remains of Mastodon, Megatherium, &c., and the lagôa deposits on the borders of the Rio de São Francisco already described.

Recent. — To recent times belong the sands containing recent shells, &c., exposed by the late uprise of the coast, the solidified beaches, rock reefs of Pernambuco and elsewhere, the coral reefs, the peat deposits, and the alluvial beds of the rivers and lakes.

APPENDIX.

ON THE BOTOCUDOS.

Origin of the Name Botocudo. — Stature. — Physical Form and Characteristics. — Manner of Wearing Hair. — Lip and Ear Ornaments. — Professor Wyman's Description of Skull of Botocudo from São Matheos. — Comparison with other described Botocudo Skulls. — Color of Botocudo. — Manner of Painting the Body. — Dislike to being clothed. — Bows and Arrows described. — Gerber's Enumeration of the Tribes. — Von Tschudi's Description of the Tribes. — Ranchos and Huts. — Food. — Mode of procuring Fire. — Manufactures. — Marriage Customs. — The Botocudos cruel Husbands. — Facility with which Wounds heal. — Treatment of Children. — Religious Ideas. — Belief in the Bad Spirit, Janchon. — No Belief in a Supreme God. — Burial Customs. — War Customs. — Cannibalism. — Dance. — The Botocudos fast disappearing. — Botocudo Character. — Geographical Distribution of the Botocudos. — Peculiarities of their Language, Pronunciation, Grammatical Structure, &c. — Botocudo Vocabularies.

No Indian tribe of Brazil save the Tupís has been more celebrated than that known as the Aimorés, Aimborés, or Botocudos, the latter being the name by which it is known in Brazil, as well as in most recent works on the country. They call themselves *Engeräckmung*,* a word which I cannot translate. *Mung*, in Botocudo, means *to go*, and the termination, which is more likely to be a separate word in the proper name, probably has the same meaning, but I have not been able to find a definition for the remainder of

* I give Prince Neuwied's orthography. Gerber, also a German, in his *Noçoes Geograficos*, &c., p. 24, spells the word *Endgerekmung*, and he says that Guido Marlière gives it *Crackmun*. The name Botocudo is spelled by different authors in all possible ways, as, for instance, Botokoudy, Botokude, Bootoocudy, &c.

the word. The name Botocudo was without doubt applied to the tribe by the Portuguese, because of the custom of piercing the under lip and the ears, and inserting therein round, flat pieces of wood, like barrel-corks, or *botoques*, as they are called in Portuguese. The termination *udo* in Portuguese has the signification of *furnished with*, as in the words *cabelludo, velludo.* It is true that *Bodoque* means a pellet of clay, such as is thrown from a sort of bow in use among the Indians of Brazil, and that the same word means also a kind of stone or earth, employed by the Indians to ornament the body. Some have thought that the name of the tribe was derived from this word, but the derivation I have above given is without doubt the correct one. The *Corografia Brasilica* * gives the same origin for the word, and so do Neuwied and Von Tschudi.

In Espirito Santo and in the Mucury region they are commonly called *Bugres*, a name which Von Tschudi derives from the French. In São Matheos and on the Doce I heard them called Tapuyos, a Tupí word applied to savages generally.

Judging from the Botocudos I have seen, I should, with Von Tschudi and M. Serres, describe the race as of middling height. I have seen many who were five feet ten inches in height, and I remember especially one powerful fellow, who could not have measured less than five feet eleven inches. D'Orbigny makes the

* "As outras nações convizinhas, ao menos algumas chamam-lhes Aymborés, e os conquistadores por corrupção Aymorés; mas de muitos tempos por cá quasi não tem outro nome entre os Christãos senão o de Botocúdos pelo extravagante e ridiculo costume de furarem as orelhas e os beiços e dilatarem-nos notavelmente com rodellas de páu, parecendo-lhes que ficam assim mais gentis e airozos." — *Corografia Brasilica,* Tome II. p. 72.

The *Diccionario Geographico* derives the name from *Boto* and *codea,* "because the Indians of this nation were *rolhos* [short and thick], and went with the body covered with a coating of gum-copal, with which they were accustomed to paint themselves, to preserve them from the stings of mosquitoes and other insects." This is a custom which at present, at least, does not seem to be in use among the Indians.

mean height of the male Botocudo 1.620 metres, and his extreme height only 1.000 metre, but this must be a typographical error.* According to M. Porte, the height of the male Botocudo varies from 1.85 metres to 1.18 metres, and that of the women from 1.35 metres to 1.16 metre.†

The limbs and body of the Botocudo, though exceedingly strong, look soft and effeminate, and the muscles have not the same prominence and knottiness seen in the muscularly developed white or negro.

They are generally broad-shouldered and large-bodied, but their arms, and especially their legs, are apt to be thin, though very muscular, and the latter strike one as being disproportionately small, when compared with those of the negro and white man, the calf being but slightly developed. Von Tschudi calls attention to this, and Agassiz speaks of the small size of the legs of the Indian in comparison with his square, heavily built trunk, but I have seen Botocudos as well proportioned as the whites.‡ In all the males

* *L'Homme Américain*, Tome I. p. 102.

† *Comptes Rendus*, Tome XXI. p. 5.

‡ Specimens of Naknenuks were carried to France by M. Porte and were examined by M. Serres, who published in the *Comptes Rendus* (Tome XXI. p. 7) a description of them. I am not aware that any other scientist has made a more detailed study of the Botocudo than he, and I quote a paragraph or two from him relating more especially to the configuration of the trunk. M. Serres says: —

"La poitrine était bien conformée chez l'homme; un peu aplatie sur le devant elle paraissait d'une seule venue et ne présentait pas l'espèce de voussure que l'on remarque au niveau du grand pectoral chez les hommes de la race caucasique developpés au même degré; voussure qu'offraient d'une manière marquée les Américains Ioways, comme on le remarque chez les hommes les plus forts de la race caucasique. En revanche elle paraissait plus allongée chez le Botocudo et plus large que l'ordinaire à la région inférieure. La poitrine de la femme était, en arrière plus arquée, que celle de l'homme; en avant, elle s'inclinait en bas d'une manière si marquée, qu'il m'a fallu la mésurer plusieurs fois pour m'assurer qu'il n'y avait rien d'exagéré dans le portrait qu'en a fait notre peintre si distingué du Museum M. Verner. De cette inclinaison de la poitrine résultait l'abaissement du sein, abaissement qui rappelait celui des femmes éthio-

the pelvis seemed extraordinarily narrow, and the hinder parts very small. The hands are well formed but small. Those of the women are particularly so, as M. Serres has remarked. The feet are smaller than in the Caucasian race.

The physiognomy of the Botocudos varies so extraordinarily that it is exceedingly difficult to describe its peculiarities. Of a dozen or more Botocudos in the fazenda of Capitão Grande no two looked alike.* There were two young men who were partially civilized, and spoke Portuguese, and I should never have taken them for anything else than very light-colored mulattoes. They all have low foreheads, as Von Tschudi, Neuwied, and M. Serres have remarked. Their eyes are black, usually small and full of life. Neuwied says that blue eyes sometimes occur. M. Serres says that those of the women he saw were more open than those of the men. The exterior angle of the eye is sometimes a little oblique. As for the nose, it is usually rather short: and in the four profiles of Botocudos at São Matheos in my note-book it is represented as having a concave outline, the extremity being large, while the alæ are rather wide; but I have seen examples of narrow and arched noses. I give two of these profiles in the following wood-cut.†

piques, et qui pourrait devenir un caractère de grande importance s'il n'y avait rien d'individuel dans cette disposition.

"Comme celui de l'homme, le thorax de la femme était très-élargi inférieurement; cet élargissement ne paraît avoir sa cause dans l'abaissement de la volume du foie, qui je reconnus par la percussion dans les limites inférieures que n'atteint jamais cet organe dans son état naturel chez la femme caucasique.

"Avec cet abaissement du foie coïncidait un abaissement de l'ombilic, et à celui-ci répondait un abaissement du pubis, que je reconnus avec peine, à cause de la saillie graisseuse du mont de Vénus. L'abaissement de l'ombilic faisait saillir l'abdomen en bas et sur les côtés et celui du pubis inclinait en bas et en arrière le bassin; de là résultait ampleur de la région fessière, déjà moins dévelopée que chez la femme caucasique."

* Nothing can be more false than the oft-quoted and sweeping assertion of Ulloa: "Visto un Indio de qualquer region, se puede decir que se han visto todos en quanto el color y Contestura." — *Noticias Americanus*, p. 252.

† I do not offer these sketches as accurate portraits. They were drawn from

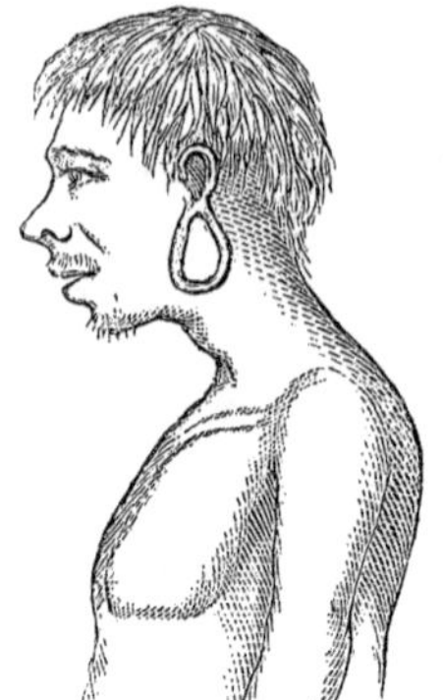
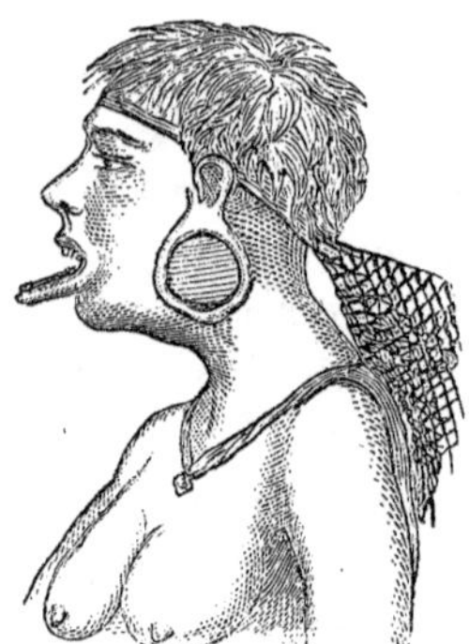

BOTOCUDO MAN AND WOMAN.

M. Serres describes the nose of the men as straight, and those of the women as slightly arched at the base. In both, according to the same author, the alæ are large, but more so in the men than the women.

Neuwied's plates of the Botocudos are well drawn, but they give one absolutely no idea of the race. The chief Krengnatmuck, barring his abominable head, has the figure of a Caucasian, while his wife might have posed for a Venus. Neuwied's figures were evidently drawn from Caucasian models. The Botocudos as a race are very ugly, but some of the young girls might, by a very liberal construction of the word, be called pretty. As a general rule, the women have the abdomen very large, the breasts flabby and pendent, and not unfrequently they are bow-legged. The children, like all Brazilian children, are apt to fall into the habit of dirt-eating, and are very often stunted, swollen, and sickly.*

nature, and the outlines are not far from correct. I have introduced them also for the purpose of showing the appearance of the pierced ear and lip. The woman carries a bag on her back.

* I saw in Brazil a large number of children and adults who were addicted to this habit. In most cases the clay is not eaten because of want of nourishment, but from a morbid appetite. Among some nations, however, as the

The cheek-bones are generally quite high, though not so much so as among the Tupí descendants. They appear especially prominent in the emaciated old women, who are wretchedly ugly in appearance. The face is somewhat flattened. The mouth is always very large, and the lips are quite thick.

The hair is black, coarse, and straight. The beard is of the same character, and very sparse.* They generally pull it out; but I have frequently seen men with a very sparse beard,† as, for instance, the one whose profile I have given. They sometimes cut off or pull out the eyebrows, and the women at least suffer hair to grow on no part of the body except the head. The hair is always worn short, and falls over the forehead. Sometimes it is shaved away for two or three finger-breadths all round, with a razor made from bamboo; but this custom is not general, and none of those I saw at São Matheos were shaven in this way. The women usually wear in their perforated lips and ears round disks of wood (*botoques*), like the cork to a large, wide-mouthed bottle. Of the many Botocudos I have seen on the Rio Doce, at São Matheos, at Colonia Leopoldina, Urucú, and Philadelphia, only the adult women had both ears and lips pierced. The old men invariably had the ears perforated, but I do not remember ever having seen a male with a hole in his lip, and I never saw a child with either ear or lip perforated, which leads me to suppose that the custom is going out of use. The piercing of the lip and ear is performed, according to Neuwied, when the child is seven or eight years of age. Neuwied says that it is done with a sharp piece of wood; other writers

Ottomacs on the Orinoco, large quantities of clay are eaten in times of great scarcity of food. Humboldt has investigated this subject, in his usual exhaustive manner. See his Travels, Bohn's Edition, Vol. II. p. 495.

* M. Serres says (*loc. cit.*): "Leurs cheveux étaient noirs, épais, courts, lisses et limités en demi-cercle sur le front. Ceux de l'homme étaient plus rudes que ceux de la femme."

† Dr. Karl August Tölsner describes the Botocudos of the Colonie Leopoldina as wearing a sparse beard. (*Die Colonie Leopoldina in Brasilien*, Göttingen, 1860, p. 65.)

say that the sharp spine of the Airí palm is used. The openings once made, small pieces of wood are inserted to distend them, afterwards larger and larger ones being used until the opening of the ear may be, according to Neuwied, even four inches in diameter! I have never seen a lip-plug in use more than two inches in diameter. The ear-plug is much larger. The lip or ear orna ment consists of a thin section of the stem of a Barrigudo tree (*Chorisia*), which furnishes a wood quite as light, if not lighter than cork, and of a white color.* The lip-plug is usually about three fourths of an inch thick. The lip surrounds it like a thick red cord of flesh. It is usually worn the most of the time, but may be, and is from time to time, removed. The lip then hangs of course against the chin, a hideous loop of flesh, comparable more to a great worm than anything else, displaying the teeth with a horrible grin. The pressure of the plug against the lower incisors in front pushes them out of place, and even causes them to fall out, so that an old woman with the lip ornament always wants the lower front teeth, and not infrequently the upper. Neuwied describes and figures the jaw of a Botocudo in which the alveolæ of the front incisors had completely disappeared, leaving the bone as sharp as a knife.

Neuwied, in the Atlas to his Travels, on Plate 17, represents four heads,—three profiles and one full face. These figures are really of very little value, as they have evidently not been drawn from nature. In Figure I. the ear-plugs are represented twice as thick as they ought to be, and the under lip is represented as touching the upper, which is absolutely impossible. The position of the plug is better represented in Figure IV. It is usually carried, in the repose of the features, nearly horizontally. In a smile it is inclined upwards, and often touches the nose. In eating it may be taken out, but none of the Botocudo women I have seen eating removed it. A more comical sight than an old woman sucking a stick of sugar-

* Fletcher says that these plugs are made from the wood of the aloe, which is incorrect. Ewbank speaks of the plugs as made of *pito* wood, doubtless meaning the same thing.

cane can scarcely be imagined. In quarrels the perforated ears and lip are apt to suffer, and it is no uncommon thing to see them broken. In this case the ornament is not necessarily discarded. The two ends are then tied together with a bit of bark, or something of the sort, and the plug is replaced. At Urucú I saw a rather young woman whose lip had been torn and tied up. Usually the ear-plug is not worn, and the loop of flesh is left dangling, sometimes reaching to the shoulder. When the plug is removed the opening generally appears very irregular, as in the man whose profile I have given. In travelling through the forest this loop would be likely to be caught against limbs of trees and be torn, so it is very often turned up and laid over the ear, which shortens the organ in the first place, and produces a horrible deformity. I observed that two old women at São Matheos wore the ear-flap in this way even in camp.

Neuwied* cites a number of examples of nations that pierce the ear and lip. The Aguitequedichagas, Lengoas, and Charruas of Paraguay wore large blocks of wood in their ears and lips, but the lip-plug was smaller than that of the Botocudo. The Gamellas of Maranhão used immense wooden lip-plugs, and Major O. C. James informs me that the Bugres of São Paulo have the same custom, though it is, however, now going out of use as the Indians become civilized. Major James says that the civilized Indians close up the opening in the lip with wax. The Murás, on the Amazonas, used to pierce the lip, but the custom is now abandoned. The Tupinambas wore ornaments of nephrite stone in the lip. Mr. George Gibbs has called my attention to the fact, that the Koloshians of Alaska pierce the lip and wear a plug. It is very interesting to know that this custom obtains among savages so widely separated. Wood describes a nation in Africa that pierces and distends the upper lip by inserting a ring, a custom more hideous than that of the Botocudo.

I give figures of the skull of a male Botocudo I obtained for

* *Reise nach Brasilien,* Band II. seite 7.

the Museum of Comparative Zoölogy at São Matheos. The man's name was Kūpārā́ck, or the Onça. He had died of disease, and had been buried in the vicinity of the fazenda, but the rains had uncovered the body, which his relatives had left to rot in a swamp. A half-civilized Indian led me to the spot, and himself procured for me the skull, which we carried to the house and placed on a table. When the Indians came in to supper the Botocudos gathered around and made sport of it, thrusting their fingers into the eyeless sockets and laughing at it, though at the same time they knew that it belonged to one of their near relations. This skull I placed in the hands of Professor Jeffries Wyman, of Cambridge, Massachusetts, who has kindly furnished me with the following interesting and valuable notes upon it. "From the references in the *Thesaurus Craniorum* of Dr. J. Barnard Davis, page 235, it appears that only a few crania of Botocudos have been described, — not more than five in all; and of these but one has been measured, and this very imperfectly by Dr. Davis, as he had only a cast, the original being in Stockholm. The specimen from São Matheos is, therefore, a valuable addition to the previous collections. It is that of a man somewhat advanced in life, the teeth gone and the alveoli largely absorbed; the sagittal and

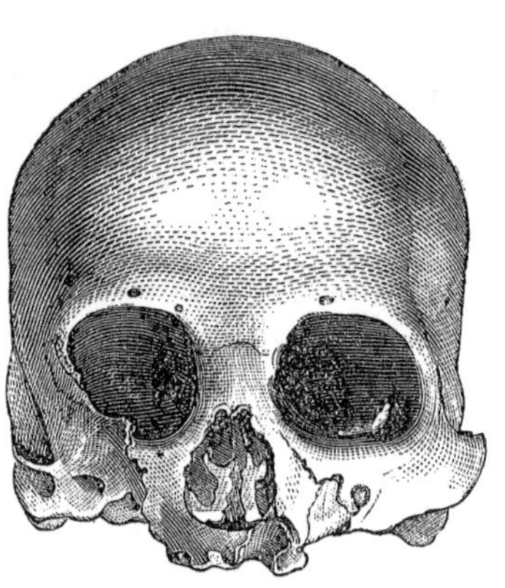

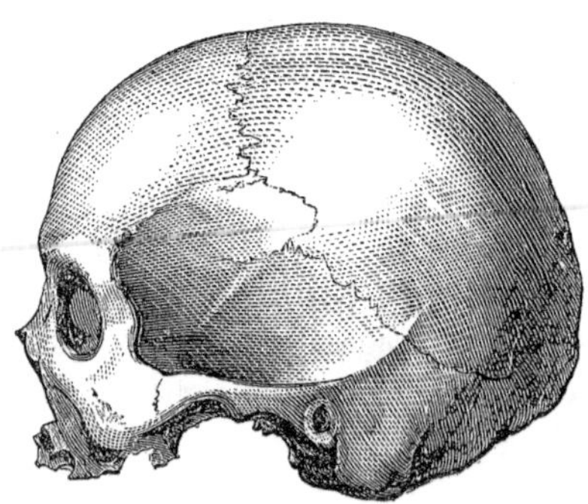

lambdoidal sutures are closed in those portions where the two join. The side walls of the head are vertical and the top somewhat roof-shaped. The *foramen magnum* has about the same position as in the American aborigines generally, its index being 40.6, while in these it is 40.9. The breadth across the malar bones, together with the roof-shaped top, give to the whole, when viewed in front, a somewhat pyramidal form compared with that of the other barbarous tribes generally. The size of the cranium is large, its length being 510 millimetres, and its capacity 1,435 centimetres, or 88 cubic inches; while theirs is only 1,376 centimetres, or 84 cubic inches. The length of the skull being taken as 100, its breadth is 72.8, and it is, therefore, decidedly elongated or dolichocephalic. The whole is massive and heavy, and, at the hinder part especially, quite thick.

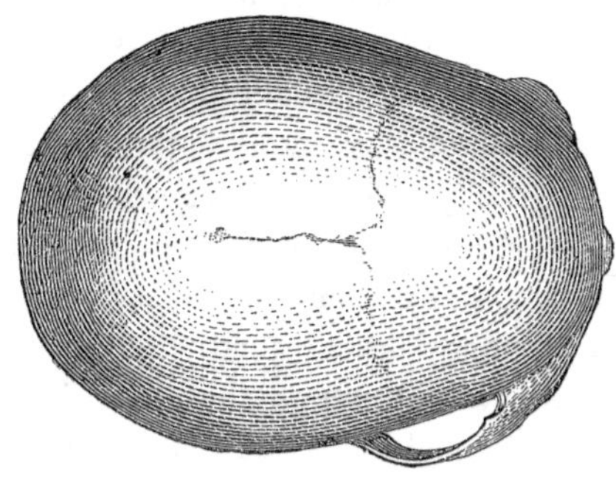

"Though somewhat smaller, this skull, as will be seen by the following table, agrees very nearly in its proportions with that described by Dr. Davis in his *Thesaurus*.

	Circumference.	Length.	Breadth.	Height.	Length.
	m.m.	m.m.	m.m.	m.m.	m.m.
São Matheos	510	187	136	138	380
Davis's collection . . .	525	190.5	139.5	144.5	401
Blumenbach's collection	. .	165	. . .	139.5	. .

"When compared with Blumenbach's specimen there is a wide difference. The one figured by him in his *Decades Craniorum*,

Plate LVIII., was brought from Brazil by the Prince Wied Neuwied, and is also figured in Morton's *Crania Americana*, Plate XV. In the first case the view is oblique, and in the second exactly in profile, this view being taken from a drawing furnished by the Prince.

"In its thickness and massiveness it agrees with that of São Matheos. As far as one can judge from the figures of Blumenbach and Morton, the one described by them is short and high, and, according to Blumenbach's description, remarkable for its brutality, or, to use his own words, 'If you disregard for a moment the under jaw and interval of the orbits, the projecting nasal spine, and the other particulars peculiar to man, the general aspect approaches nearer to that of the orang-outang than that of any other skull from a barbarous nation to be seen in my collection.' His figure seems to justify his words, and represents the jaws more projecting and simious than does that of Dr. Morton.

"Of the original shape of the jaw of the cranium of São Matheos it is now hardly possible to form a correct idea, since it is somewhat broken. The teeth are gone, and the alveoli partly absorbed. There is nothing, however, in what remains to indicate great size or forward projection. The whole cranium will compare favorably with the crania of other barbarous tribes of America. Certainly there is nothing indicative of extreme degradation."

In a letter accompanying the above notes Professor Wyman says: "It is quite curious to see what an entirely different looking thing the skull of São Matheos is when compared with the famous one described and figured by Blumenbach, and which has hitherto given the key-note to all that has been written about the skulls of Botocudos. If there were only your São Matheos skull and his, and they had fallen into different hands for description, one would have given us the connecting link of man with the apes, while the other would have given us a highly respectable American savage."

The skull described by Blumenbach* is figured as a vignette to the second volume of Prince Max. zu Neuwied's *Reise nach Brasilien.* It seems to me to be most extraordinarily short and small compared with the size of the jaw. Among all the Botocudos I saw in Brazil there was not one with so ape-like a head and such projecting jaws.

Von Tschudi† figures the skull of a Botocudo chief, named Porokum, from the Mucury. He gives a front but somewhat oblique view of it, which, however, shows that it agrees with my specimen in the perpendicularity of the sides and in the roof-shaped top.

The color of the Botocudo is a light yellowish-brown, like that of a very white mulatto, or perhaps more correctly speaking that of a white man somewhat tanned, not burned, by exposure to the sun. Neuwied says that they are of a reddish-brown color, Von Tschudi a dirty, nearly bronzed brown.‡ The color varies much. I should say that the bodies of the Botocudos I have seen were, on the average, much lighter in color than those of the white canoe-men of the Jequitinhonha, accustomed to work almost naked in the sun, and that as a race they were much whiter than the Tupí descendants along the coast, who differ most markedly in their whole physiognomy, stature, &c., from their uncivilized neighbors.

D'Orbigny has called attention to the yellowish skin-tint of the Brasilio-Guaraní races. It is interesting to observe that the Boto-

* Prince Neuwied quotes at the end of the chapter on Botocudos, Vol. II. p. 70, a few words from Blumenbach relative to this skull. Blumenbach says: "Der Botocude womit Ew. meine ethnologische Sammlung bereichert haben, und der eben so sehr zu den merkwürdigsten als zu den seltensten Stücken derselben gehört, ähnelt in seiner Totalform (doch ohne den Unterkiefer) dem vom Ourangutang mehr, als einem der acht Negerschädel die ich besitze, wenn gleich bey manchen von diesen die Oberkiefer stärker als an dem brasilianischen Cannibalen prominiren."

† *Reisen durch Süd-Amerika*, Zweiter Band, 328te Seite.

‡ M. Serres describes the color of the Botocudos he examined as "brun rougeâtre, un peu plus rosée que celle des Ioways."

cudos, a nation fitted for a life in damp shady forests, and unable to bear the sun in the open campos of the interior, are of a very pale color, and very much less dark than those races which live on the plains or on open grounds. Southey* thought that their pale color was the result of their life in the shady forests, and D'Orbigny† held the same belief, stating in confirmation of it that, while the Guaranís of the plains of Corrientes and the Gran Chaco are dark colored, the Guarayos and Sirionos, though belonging to the same race of the Guaranís, who for about four hundred years have lived in the damp and shady forests, are very light colored.

Gumilla says, that the people living in the forests of the Orinoco are almost white, while those of the plains are dark.‡ I am told that the Botocudos are capable of blushing. I never detected anything of the sort, and I doubt whether it is possible in any except those considerably civilized.

Among the Indians I saw on the coast only one young woman had her cheeks painted, though the custom seems, at least formerly, to have been quite common with the Indians in the forest not only to paint the face but the body. This young woman had a bright red spot on each cheek, painted with a tint prepared from the seeds of the Urucú (*Bixa Orellana* Linn.), a fruit common in the Brazilian forests, and from which anatto, or cheese-coloring, is prepared. This same color is also used by other tribes in Brazil to paint their bodies. A deep blue-black of greater durability is extracted from the fruit of the Genipapo (*Genipa*), and is also used for ornamenting the body.§ The style of ornamentation appears not to be fixed, but to vary according to the caprice of the individual. Neuwied describes three principal styles. In one the face from the mouth upwards is painted red with urucú. In

* History of Brazil, Chap. XIII.

† *L'Homme Américain*, Tome I. p. 79.

‡ *Hist. de l'Orénoque*, Trad., Avignon, 1752. Tome I. p. 108. Cited by D'Orbigny. See Humboldt, Travels, Bohn's edition, Vol. II. p. 463.

§ Henderson, in his usually inaccurate style, says that the Botocudos paint themselves green or yellow.

another the body, upper arms, and thighs to just below the knees are stained black, the colored portion being separated from the unpainted by a red stripe. Sometimes one half the body was painted black, the rest being left uncolored. Occasionally a black line like a mustache was drawn across the upper lip, and extended to the ears, the rest of the face being red, and Neuwied says that sometimes the sides of the body are blackened from the shoulders to the feet, the middle of the body being without color. The colors are usually prepared in the upper shell of a turtle, and are carried in a joint of bamboo.

As Von Tschudi has well remarked, a naked Botocudo warrior, with his black painted body, red face, and lip and ear ornaments, must present a most demoniacal appearance. In the forest the Botocudos go naked. Not a single Indian I saw wore any ornament on the head or body, unless it were a simple string of beads. When at work on the fazendas they go as nearly naked as possible; the men often tie a shirt by the arms around the waist, letting the body hang down in front; the women wear only a tattered skirt. These garments are immediately discarded as soon as they return to the forest, and one may see little bands entirely naked in the vicinity of Philadelphia and other settlements in the Botocudo region. Like other savage nations, the Botocudo shows no symptom of shame on exposing the person in the presence of those who are dressed.

The ornaments of the wild Indian woman consist of a band sometimes worn about the head, from which on one side depend a number of long strings to the ends of which are attached hoofs of capibaras; of collars made of hard berry-like fruits strung on threads, together with the teeth of monkeys, &c., or of strings of the hoofs of wild pigs; and of armlets of beads and teeth. The chiefs sometimes ornament themselves with feathers, but this is rare.

I never saw a savage Indian woman wearing a flower, though the civilized Indians are very fond of them, as Mrs. Agassiz tells us in the "Journey in Brazil."

The arms of the Botocudo consist of the bow and arrow; clubs are rarely used. The bow in ordinary use is about five feet in length, and is made from the wood of the Airí-palm. This wood is exceedingly hard, heavy and strong, and of a very dark reddish-brown color. The bow is thickest in the middle, where it is round, and it tapers regularly to each end. So difficult are these bows to bend, that no one but an Indian can use them. Mr. Copeland brought one from the Guandú with him, which not even our strong canoe-men could bend.* These bows vary somewhat in size, some being even seven feet in length.

The arrows are usually about six feet in length, and are made of the *Ubá*, *Cannachuba*, &c., which furnish light and strong reed-like stems.

The arrow used in war is tipped with a javelin-like head, five or six inches long, and sometimes two inches broad, which is made of a piece from the side of a joint of Bamboo, and is consequently convex on one side and concave on the other. This is cut into a sort of elliptical shape, and sharpened to a long acute point. It is then hardened in fire, and the arrow-head is prepared with an edge as sharp as a knife. The head is bound into the arrow-shaft with bark. This arrow, which is employed not only in war, but in the chase of the Tapir, is, like the other kinds in use among these Indians, tipped with the feathers of some large bird, a single feather being bound on each side. It makes a terrible wound, and one particularly dangerous, because of the concave shape of the arrow, which facilitates bleeding.

Another arrow in use sometimes in war, but usually in the chase, is furnished with a thin, narrow head, about a foot long, with backward projecting points cut on one side, — a terrible weapon.

For birds and small game an arrow is used whose tip is made

* Speaking of the strength of the Indians of the vicinity of Rio, and of the stiffness of their bows, Lery says, Cap. XIII.: "Si longitudine et crassitudine nostros adeo superant, ut eos nec lentare nec adducere ullus nostrum possit; quin potius immo totis viribus puerorum decem annorum arcubus curvandis opus esse."

from a stem cut at the node where several little branches have their origin in a circle; the stem is fashioned into a blunt point just above the node and the branches are cut off short. An arrow of this kind, of course, does not tear but only bruises. I once encountered several Indians near Urucú returning to their camp from hunting small lizards with these blunt arrows. Guns are not much in use among these savages, though they soon learn to use them very expertly.*

The Botocudos use a sort of speaking-trumpet made of the skin of the tail of the great armadillo (*Dasypus gigas*) to call one another in the forest.

Before the discovery of America the Indians of Brazil, both Tupí and Botocudo, used cutting instruments of stone of various shapes, and at Linhares on the Rio Doce the subdelegado presented me with a stone instrument, probably of Tupí origin, of the shape of a saddler's cutting-knife. It was made from a very hard gray stone. I unfortunately lost the specimen, so that I am unable to figure it. I have never seen any of the stone implements of the Botocudos. Their principal cutting instrument, besides their bamboo razors, consists of a common stout knife, like a butcher-knife, though they make knives from hoop-iron, or anything else that will serve the purpose. This knife they always carry slung over the back by a cord around the neck.

The nation of the Botocudos is divided into a number of little *tribus*, as they are called, or collections of a larger or smaller number of families, each tribe inhabiting a certain region in which they have their *Aldeamento* or head-quarters. Each one of these tribes is governed by a chief usually selected for his strength and bravery, and the tribe frequently takes its name from the leader. Thus a tribe in the Mucury region, headed by a chief of Herculean strength named *Pojichá*,† goes by his name. Henrique

* Von Tschudi says that the Indians not only shoot more successfully, but to a greater distance with their arrows, than the European can with his gun.

† A son of Pojichá was a servant in the house of Signor Gazinelli, in Santae Clara.

Gerber* says that "they divide themselves into several tribes, of which some are domesticated and gathered together into villages; others, still wild, wander through the forests of the valleys of the Mucury, Doce, Tambacury, Urupucá, &c. All of them, however, distinguish themselves disadvantageously from the Machalalis and Malalis by the inferior degree of their intellectual faculties. The principal domesticated tribes are: —

a. The Naknenuks (dwellers in the Serra), a confederation of various tribes, who occupy the valleys of the upper Todos os Santos, Poté, and Mucury, in the *Aldeamentos* of the Capitão Felippe, in the forest of São João, of the Captain Poté, on the margin of the brook Poté, of the Captain Timothéo, on the head-waters of the Todos os Santos, &c.

b. The tribes of Pojichá, encamped three leagues below Philadelphia.

c. The tribes of Giporok, on the margins of the Urucú and Lower Mucury.

d. The Bakués, on the left margin of the Mucury.

e. The Aranaús on the margins of the Surubim and Sassuhy.

Von Tschudi † has given the distribution of the tribes with so much detail and apparent precision that I quote what he has to say on the subject: —

"The nation of the Botocudos is broken up into a multitude of tribes, of which the most divide themselves again into independent hordes. On the head-waters of the Mucury and Todos os Santos live the *Naknenuks.* From my researches into the meaning of this word two entirely opposite explanations have presented themselves to me. According to one version, the name should mean 'Lords of the land,' according to the other, 'Not from this land.' I am not in the position to determine which translation is the more correct. ‡ To the Naknenuks must be reckoned as belonging the

* *Noções Geograficas,* &c., pp. 24, 25.

† *Reisen durch Süd-Amerika,* Vol. II. p. 264.

‡ See Gerber, quoted above. Von Martius translates the name "*homines terræ.*"

Americanos d'Agoa Branca on the Rio Preto, belonging to the basin of the Jequitinhonha, where they possess a considerable *aldea.*

"The Naknenuks of the Mucury consist of the following known hordes, which are called after their chiefs: The horde of *Poté,* probably the strongest of all, only two leagues distant from Philadelphia, that of *Cracatan, Braz, Poton, Timothéo, Inhome, Felipe, Ninkate,* and *Nortete.* The last, which formerly numbered over a hundred 'bows' fit for war, has lately melted away to only a few families. . . . .

"South of the Serra Mapmap-crak, which separates the basin, live the *Aranaús,* the bitterest enemies of the Naknenuks, on the Rio Aranaú. From the northern tributary of the Mucury, the Rio Pánpan [Pampam or Pampão] to Santa Clara stretch the *Bakués,* and west from these to near the shore the tribe of *Urufú.* On the source of the Rio Preto are the bands of *Joao Ima, Casimiro, Maciel,* and other subordinate chiefs, *Jumerai, Caporá, Ampaquejú.* In the southern basin of the Mucury we meet at the Ribeirão de Saudade with the tribe of the redoutable Captain *Poschischá,** somewhat farther east of the Riberao das Lages, the tribus of *Mekmek, Shiporok,* and *Potik,* and still eastward toward the coast, in the basin of the São Matheos, the hordes of *Pokorun, Batata,* and others. On the Rio Urucú, or the largest southern affluent of the Mucury, the *Shiporoks* and the chiefs *Juquirana* and *Maron.*

"On the subject of the name '*Shiporok*' I have no more light than on that of Naknenuk. According to some, Shiporok means *enemy,* and with this name the Indians commonly designate their adversaries. The tribe, however, does not name itself so. By what name it designates itself I could not learn. An Indian soldier, and one excellently acquainted with the language of this race, and of whom I made numerous inquiries concerning it,

* Von Tschudi, like a German, has mistaken a sound very near the French *j* for an *sh.* Shiporok is pronounced *zhĭpŏrŏ'k,* the sound *zh* representing a sound somewhat intermediate between the French *j* and the Spanish *ch.*

assured me that Shiporok meant '*from this side of the mountain,*' or '*from behind the mountain,*' and that this was the only name of the race. In the language of the Botocudos, Shiporak means *brother*, and Shiporok, *arm.* In the year 1816 Prince Max. zu Neuwied met with a brave Botocudo chief, called Jeparak, at the Quartel dos Arcos on the river Belmonte. It is not unlikely that his band went later southward, and settled in the basin of the Mucury. About thirty years later the Shiporoks were at the Lagôa d'Arara, on the north bank of the Mucury, and about eight years after we find them on the Rio Urucú, a southern tributary of this stream. By the indistinct, often suppressed sounds of the vowels in many words of the language of the Botocudos, I believe that the mode of writing the word on the part of Prince Maximilian is no hindrance to the opinion that the Indian race of the Jeparaks which he met with at Belmonte are identical with the Shiporoks at Urucú.

"All the Indian bands on the basin of the Mucury, with the exception of the Malalis, Machacalis, and perhaps the Aranaús, belong to the race of the Botocudos or *Engeräkmung*, as they call themselves. Some bands have settled down in permanent dwelling-places, Aldeamentos or Aldeas, and we have especially to notice the Aldeamento do Poton, the Aldo. do Poté, Aldo. do Cracatan, Aldo. de Curiença (das Cursiumas), Aldo. do Nortete, Aldo. de São João, Aldo. d'Agoa Boa, Aldo. dos Aranaús. In the year 1817, out of these eight Aldeamentos one hundred and four individuals were converted to Christianity (of course only in name); among them the three chiefs, Poté, Poton, and Cracatan. By the inhabitants of some of these villages some cultivation of the ground is carried on, but it is confined chiefly to the cultivation of maize and mandioca. The number of individuals in these bands varies much. Some count several hundreds, others scarcely eighty to a hundred, with only about twenty efficient fighting men. From the number of the bow-bearing warriors one can with surety estimate the number of souls in a band, because this last, on the average, amounts to four times the number of warriors. I believe that I am not far from the truth when I estimate the total number of Indians in the basin of the Mucury at from twenty-eight hundred to three thousand souls.

The Botocudos, when travelling in the forest, build for themselves shelters of palm-leaves, which they stick in the ground in a half-circle, the tips of the fronds arching together forming a sort of roof. In travelling through the forest between the Mucury and Peruhype, I saw great numbers of deserted ranchos of this kind in the forest.

Where they encamp long in one place they make their ranch more substantial with a better roof, and often of sufficient size to hold several families. The whole furnishing of the cabin of a Botocudo is of the simplest possible kind. The fire is made in the middle. Rarely ever are earthen pots used for cooking. They make use of gourds and the cup-like receptacle of the Sapucaia (*Lecythis*) for drinking purposes and for the preparation of their food. Water they carry in the joints of the Taquara-assú, in which they also keep their painting materials. Beds are made of *estopa* or bast-fibre.

The food of the Botocudos consists of sapucaia nuts, palmito buds, and the fruits of the Ingá, Jaboticaba, Araça or Goyaba, Maracujá (passion-flower), &c., with the roots of Cipós and other plants. They are fond of Indian corn, bananas, and mandioca, which they steal whenever they can from the plantations.

They hunt game of all kinds, but they are particularly fond of monkeys, whose flesh, as I can myself testify, is exceedingly savory. They even eat the onça and other carnivores, the ant-eater, alligators and lizards, and the boa-constrictor.

Among birds they are particularly fond of the Mutum, the Jacupemba, &c., and they also eat their eggs. Fish are usually shot with small bows, which are used with great dexterity. Sometimes they employ a poisonous root, which, put into the water of a pool, kills the fish.

They are very fond of the great, fat larvæ of certain insects which burrow in decaying wood. Among these, according to Neuwied, is the larva of the *Prionus cervicornis*, which, with other species, live in the trunks of the *Bombax*, or Barrigudo. Numbers of these disgusting grubs are impaled on a sharp stick and toasted at the fire.

Usually all animal food is cooked in this way. Bananas, potatoes, &c., they sometimes bake in the hot ashes. Ants also are eaten.

They are fond of honey, and formerly they used to cut down hollow trees with stone axes to obtain it. To-day steel axes and hatchets are occasionally to be seen among the Indians.

Fire is to the Botocudo an object of much care, because if it is lost it is only to be rekindled with great difficulty. In order to obtain it the Indian procures a stick of some light, dry wood, and makes a small hollow in it. This stick he places on the ground and holds securely with his foot. He then takes a long dry stick, one end of which is somewhat blunted, and places it in the hollow above mentioned. The other end is taken between the two palms, the stick is held vertically, and by a rapid motion of the hands it is caused to twirl until the friction of the lower end in the hollow of the other stick has caused it to take fire, when estopa or bast is ignited, and a fire is speedily made. This method of procuring fire is also employed by some of the aborigines of North America. We find the same custom in Africa among the Bushmen and Caffres,* and in the Aleutian and Caroline Islands.

The only things manufactured by the Botocudos consist of bows and arrows, a few little ornaments, and bags made of the bast-fibre of different plants. These last they barter with the whites for food, &c. They bring in to the fazendas the wax of wild bees, ipecacuanha, skins, &c., but this barter is conducted only on the very smallest scale.

* Alberti, *Descrip. Phys. et Hist. des Caffres*, p. 36. Campbell, *Reise in Süd-Afrika*, p. 37.

The Botocudos take usually but one wife. Von Tschudi says that when a man has chosen a woman for his wife, he agrees with the father as to a certain tribute which he shall pay in game or something else, when the woman is handed over to him and with no further ceremony is thereafter his wife.

Neuwied says that a man may have as many wives as he can take care of. Adultery is rare, and is visited with heavy punishment on the woman. The husbands are very cruel and unkind to their wives. The husband, when angry with his spouse, beats her unmercifully, and cuts her with his knife. I never saw a married woman who was not covered with scars, on her face, back, breast, and arms; it is the commonest thing to see them six inches or more in length, and one woman may bear the marks of many terrible wounds, which it seems marvellous she should have survived.

The good health of savages, and the facility with which they recover from injuries which would have proved fatal to an ordinary civilized man, have been often commented upon by many authors. Numerous instances are on record of Negroes, Malays, Pacific Islanders, and American Indians who have survived horrible wounds and mutilations, showing that Nature's power of healing is greater among savage than among civilized nations.*

The woman is really the slave of the husband, and all hard work falls to her lot. On the march she carries the family goods, or the larger share, packed in a bag, which is slung on the back by a band which passes over the forehead. The mother carries her child on her back sitting in a loop of bark which passes over her forehead, the child clasping her neck.

The children are kindly treated, at least when young, but the tie between parent and child is not strong.† At São Matheos

* Waitz, Introduction to Anthropology, p. 126.

† The women carry their children on their backs sitting in a loop of bark which passes over the mother's forehead, the child clasping the mother about the neck.

there was on the fazenda a young woman who had two children; one, a boy several years old, was sick from dirt-eating, and was stunted in growth, yellow, and swollen; the other was a babe at the breast. The mother was anxious to sell the elder, and I could have bought him for a trifle. One day the babe suddenly died. The mother immediately dug a grave for it in the floor of the rancho, and went pleasantly about her work as usual, the only effect on her being to make her determine not to sell the boy. Children are frequently bartered away to the fazendeiros, who in reality hold them as slaves.

Of the religious ideas of the Botocudos we have not so much information as we could wish. Most writers agree with Neuwied that the Botocudos believe in and fear a bad spirit called *Janchon*, and Neuwied says that they recognize many of them, which they distinguish as great and small. The great devil comes in the guise of a black man visiting the camps; sometimes he sleeps awhile by the fire and then goes away, but all who see him die. This same devil is accused of beating the dogs to death with sticks, and of killing children.

The Tupí race seems to have some idea of a God, and they called Him *Tupa* or *Tupan*, which name is derived from the word *tuba*, father. It is a difficult matter to-day to arrive at any clear knowledge of the primitive religious belief of the Botocudos, because they have derived so many ideas through intercourse with the Portuguese. I was unable to learn that the Botocudos had any idea of a God. The moon, which they call *Taurŭ́*, is an object of fear, the Indians believing that occasionally it falls upon the earth, destroying men, and that it sends storm and famine. No worship is offered to any of these things, and they have neither priests nor medicine-men like the Tupís.

The dead are buried in the immediate vicinity of the camp, or even in the wigwam, and this last appears to be usually the case. The child that died at São Matheos was buried in the earth forming the floor of the ranch in which the Indians were quartered. When a death occurs they usually desert the camping-ground for another, but I feel sure that I saw the Indians occupying the same ranch in which the dead child was buried. The relatives gather together and howl for one day after the death, the women taking a specially prominent part in the ceremony; but the next day they all go about their work as if nothing had happened. In some places the hands of the defunct are bound together before burial. The corpse is buried in a horizontal position, and a fire is lit to keep away the devil, for they believe that if this evil spirit should find no fire at a grave he would dig out the body. Deserted wigwams in the vicinity of graves are a common sight in the forest. Sometimes a shelter of palm-leaves is built over the burial-place, as Neuwied has remarked. Nothing is put in the grave with the dead body.*

The different tribes or Aldeamentos are frequently at war with one another. At the time of my visit to the Mucury there was a skirmish between two bands near Cannas Brabas on the Philadelphia, the particulars of which I was unable to learn. I had sent my baggage on ahead of me to Philadelphia in care of Signor Battista, who was conducting a train of ox-carts with salt, &c. Near Cannas Brabas he was attacked by the Botocudos and shot at, but escaped by plunging into the forest, from which he emerged nearly naked. On arriving at Cannas Brabas we found the place nearly deserted, but we passed the place of danger in the night without seeing an Indian.

Their mode of warfare is the attack by night or from ambush. A victory is celebrated by song and dance. I found the belief everywhere current that they sometimes ate the bodies of the slain, and from all the information I have received, I think there can be no doubt that cannibalism is one of the customs

* The feitor at Capitão Grande told me that when a little child died it was the custom to put by the grave a bottle of milk drawn from the mother's breast, together with the bones of some wild animals.

of the Botocudos. At present the Indians are very peaceful, but from time to time they have committed outrages on the whites, murdering, and burning houses on the Mucury, at São Matheos, and on the Doce.

When at the Fazenda do Capitão Grande I witnessed one evening the dance of the Botocudos. A bottle of cachaça, or native rum, had been given them, over which they were very merry. Four of them took their position naked in the yard in front of the house, and formed a square, facing one another, then all four placed their left feet together, the right leg remaining stretched out. All bowed their heads together and placed their arms on each other's shoulders. Then they began a monotonous song,* *Călănī́-ā-ā́*, *Călănī́-ā-hā́*, to which they all kept time by hitching a step forward with the right foot, keeping the left quiet. This dance they kept up for some time. They are apparently very fond of it, and the overseer of the fazenda informed me that they sometimes kept it up until, exhausted by fatigue, they fell to the ground.

These Indians used to be very numerous at São Matheos, living in the forests even quite close to the sea; but they have been so killed off that at present none are found in the vicinity except on the head-waters of the river. They were hunted down by the Portuguese settlers like wild beasts, and one gentleman told me at ——, that during his life he had, either with his own hand or at his command, been the means of putting to death by knife and gun and poison over a thousand of these poor creatures! The injuries committed by the Botocudos on the whites are as nothing compared with the wrongs inflicted upon them by those who have dispossessed them of their home, and have almost destroyed the race.

The Indians have learned the use of rum and tobacco, of both of which they are very fond, and which are rapidly working their ruin. In the Mucury, Colonia Leopoldina, São Matheos, and on the Doce,

* St. Hilaire compares the song of the Botocudo to that of the Chinese.

the Indians come into the settlements to beg, and they not unfrequently are employed to work on the fazendas, their service being voluntary. At São Matheos they are paid in victuals and rum; but on the Mucury they have learned the use of money, which they call *pataca*. At São Matheos I saw both men and women at work with the negroes. They are not much to be depended upon, usually remaining but a few days on the fazenda, and then returning to their wild life in the forest. They are very lazy, and half a dozen are scarcely worth an able-bodied negro. They seemed very docile and good-natured; indeed, I was particularly struck with this last feature in their character. At their work they laughed and played jokes on one another, and in the house at their meals were quite as merry as the negroes. They have nothing of the gravity, stolidity, and want of curiosity of our Northern Indians. They have no idea of *meum et tuum*, and they are particularly addicted to stealing bananas, corn, or anything else they happen to take a fancy to. They often come almost stark naked into Philadelphia and Urucú to beg.

One may occasionally find a civilized Botocudo on a fazenda, but the children sold by the parents and employed as servants on the plantations scarcely ever grow up.

At present the Botocudos are confined to the virgin forest between the Rio Doce and the Rio Pardo. They are very rarely seen near the coast, and never frequent the campos of the interior. The race is fast diminishing, and in a few years will pass out of existence. The Indians of the coast tribes of the Tupí race, — thanks to the labor of the Jesuits, — have become civilized and converted to Christianity, and now form an integral part of the Brazilian population, but the Botocudos resist civilization and the influence of Christianity, and are sunk in the lowest barbarism.

The language of the Botocudos is entirely different from the Tupí, and from that of the other coast tribes, as the Patachos, Machalalis, &c. Although spoken by all the Botocudos, there are dialectic differences observable in each band, different words being sometimes used by different bands to distinguish the same object. The language of the Naknenuks, Jiporoks, &c. is one and the same. Latham's division of the Botocudo into the Botocudo proper and Naknenuk is incorrect, and the differences observable between his vocabularies are traceable to the collectors of these vocabularies, who have spelled the words incorrectly, as they caught the sound from the lip of the native.

At present we know very little of the grammatical structure of the language, the vocabulary being very incomplete.

It is a very simple language, with very few or no inflexions. The first words the stranger is likely to hear uttered by a native — namely, *gling-gling* — bear one of the most distinguishing features of the language. It is extremely rich in reduplicated words. Thus we have Tŏn̄-tŏn̄ = bad; Kiacu-käck-käck = a butterfly; Ong-ong = to sing; Naak-naak = a gull; Encarang-cuong-cuong-gipakiŭ = the great boa, &c., &c.

This reduplication appears rarely ever to extend itself to words of two syllables. Instead of doubling a word of two syllables, only the last syllable is reduplicated, as in the sentence, *min-yan̄-yan̄-rī-mā'-hă-ū'm* = I am thirsty. These constantly recurring doubled syllables give to the language a stuttering character. The Tupí-Guarani is also rich in reduplicated words, but not so much so as the Botocudo.

The principal points in the structure of the language thus far noted by Neuwied and others are these: —

There is but one gender, namely, *neuter*. There are two numbers, a singular and a plural, and perhaps a dual. The plural is formed by adding *uruhŭ* or *ruhŭ* (many) to the singular, thus: *Kjiem* = house; *Kjiem-uruhŭ* = houses, also village; *Tyŏn* = tree; *Tyŏn-uruhŭ* = trees or forest. In writing my vocabulary at São Matheos, my interpreter, a native Botocudo who spoke Portuguese quite well, gave me what appeared to be a dual form for several words, and these I subjoin exactly as I noted them from his lips.

Man	cuáh-hāh
Two men	'nī't-chŏ-vŏ
Woman	pŏ-chī'k
Two women	'nīt-chŏ-vŏ-ŏ'n̄
Eye	kĭ-tŏ'm
Two eyes	nĭk-ĭ-tŏ'm-chŏ-vŏ'
Ear	ñŏn̄-hŏ'n
Two ears	ñŏn̄-hŏn-chŏ-vŏ'
Arm	yĭ-mŭn
Two arms	yĭ-mŭn-chŏ-vŏ
Hand	ïp-ä'
Two hands	ïp-ä'-chŏ-vŏ
Leg	ïp-māk
Two legs	ïp-māk-chŏ-vŏ

The word for man in this vocabulary corresponds with that given by Neuwied, but in no vocabulary can I find the word *pŏ-chī'k*, meaning woman. The dual forms of both are strange, yet my Botocudo insisted that *'nī't-chŏ-vŏ* meant *dous homens*, and *'nīt-chŏ-vŏ-ŏ'n̄, duas mulheres.* There may possibly be some mistake here, but I have given the words in the hope that some one else may be able to explain them. With the other words the dual was formed by adding the termination *-chŏ-vŏ*, but in the case of *Kĭtŏ'm* a prefix *nĭ* was added.

One thing is certain, the Botocudos cannot count. Their only numeral adjectives being *mokenam*, which means, rather, *single*, and *uruhú, many.** The Botocudos at Capitão Grande kept an account of their days' work on their fingers and toes, and I was assured that the largest number they could reckon was twenty. At the expiration of ten days' work, for instance, an Indian who wished to settle with the feitor would go to him and count off *tĕmplān*, day, ten times on the fingers. In the examples above given there is

* The Tupís count only to three; for higher numbers they use the Portuguese.

a termination which signifies two, and which answers to a dual. The subject is one of much interest, and I regret much that during my stay in the Mucury I was unable to give it more study. Von Tschudi gives numerals up to ten, which he obtained from a Nak-nenuk through the aid of an Indian soldier; but he himself expresses a doubt as to the veracity of his interpreter.

There are in the Botocudo language two cases, a nominative and an objective. The latter is marked by the syllable *te* (ti or de) between two substantives coming together and governing one another (Gottling quoted by Neuwied); thus, Tarú means *moon,* but also sun, sky, and time; Tarú-ti-pó, literally, *sky-runner.*

The adjective always follows the substantive, as *cuiáñ-cudgí,* the small ant-eater. Neuwied says that the comparative is formed by adding the termination *uruhú,* and the superlative by the addition of the adverb *gicaram,* as *cuang-mah,* the stomach is empty, *cuang-mah-gicaram.** The only pronouns we know are *Kjick* = *I,* and *Hä* = *he, she,* or *it.* We have *Kjuck* for *my,* which, by the way, may be used before the noun it qualifies, as *Kjick-kjuck magnán-joóp,* but Neuwied says that his Botocudo *Quäck* used *Kjick* as well as *Kjuck.* So far as the verb is concerned it seems to be very simple in its construction, and to have only two forms; namely, infinitive and participle. The third person singular present is formed by prefixing to the verb *he, het,* or simply *a,* which appear to be only different forms of the third person singular present of the verb *to be.* We at present know too little of the language to speak at all positively about other points in its construction.

Prince Max. zu Neuwied gives in the second volume of his *Reise* a German-Botocudo vocabulary of several hundred words. Latham (*Elements of Comparative Philology,* p. 509) gives four short vocabularies, and Von Tschudi (*Reisen durch Brasilien,* Vol. II. p. 288) has another short vocabulary. Other vocabularies have

* I should hardly call this a superlative.

been collated by Guido Marlière, Von Eschwege, Jomard, Renault, Von Martius, St. Hilaire, D'Orbigny, and others. Neuwied's is prepared with great care, and he had the advantage of long intercourse with a lad, whom he retained in his employ. Von Tschudi's was hastily written with the aid of an Indian interpreter, and the sounds are very imperfectly represented, a soft *ch* or *zh* sound being represented by the German *sch*, &c. I do not know where Latham's vocabularies were collected. They are very inaccurate. While at São Matheos I spent a long time with a young Botocudo, who spoke Portuguese, and collected quite a vocabulary, using a phonetic alphabet of my own, by which, with an ear accustomed to the pronunciation of many languages, I think I have been able to express very nearly the true pronunciation of the words. The pronunciation is exceedingly indistinct, and the words are very hard to catch. The stupidity of my *pundit* was discouraging, and the labor of getting together the vocabulary was very great.

This vocabulary is too voluminous to be inserted in this volume. I hope to publish it elsewhere.

The language is usually spoken on a high key, but in a weak tone and rather rapidly. It is particularly rich in nasals, but has neither gutturals nor sibilants.

The sounds of *s* and *z* do not occur in it, nor those of *f* and *x*. I have observed the sound of *v* only in the dual termination -chŏ-vŏ. In many cases it is impossible to distinguish between *l* and *r*.

INDEX.

A.

Abacaxi pine-apple, 431.
Abrolhos Islands, derivation of name; danger in passing, 174; situation of, 175; geology of, 175-179; depth of water in vicinity of, 192; coral reefs of, 192-199; navigable channel inside of, 201.
Acanthastræa Braziliensis Verr., 62; described, 195, 205.
Agaricia Agaricites Edw. and Haime, 62, 195.
Agassiz, Prof. Louis, on glacier of Pacatuba, 469; on the structure of the Organ Mountains, 15; on the Amazonian sandstones, 487.
Agassiz, Madame, 469.
Agassiz, Mr. A. E., on the former connection between the Atlantic and Pacific across the Isthmus of Darien, 393.
Agate, 531.
Agoas Marinhas, 151.
Agrestes, name given to Western Sergipe, 379.
Aimorés or Aimborés, 577.
Airí palm (*Astrocaryum*), 94, 216, 298.
Alabaster, 298.
Alagôas, general description of province of, 422.
Alcobaça, 224.
Aldêa Velha, 85.
Allen, Mr. J. A., 275, 295, 301; notes on the geological character of the country between Chique-Chique and Bahia, 309-318; on salt deposit of São Francisco Valley, 331.
Alluvial lands on Rio Pardo, fertility of, 243; not proper for coffee, 245.
Alto dos Bois, excursion to, 142.
Alum, 471.
Amazonas, derivation of name, 484.
Amazonian formation described by Professor Agassiz and Major Coutinho, 488; Prof. Agassiz's theory of the, 490; probably marine and tertiary, 491; of Pará, 491.
Amazonian group, cretaceous, 550.
Amethysts, 471, 530.
Amianthus, 471.
Ammonites, 384.
Ammonites acutocarinatus Shumard, 389; *A. Gibbonianus* Lea, 389, *A. Hallii* Meek and Hayden, 388; *A. Peruvianus* Von Buch, 389; *A. semistriatus* D'Orb., 389.
Ampullaria, eggs of, 47.
Amygdaloid, 530.
Anacardium occidentale, 465.
Ananchytes, occurrence of, on Isthmus of Panama, 392.
Anchieta, letter on Manati, 75.
Angico, bark of, used in tanning, 411.
Aninga, 395.
Anta (*Tapirus Americanus*), 94.
Antedon Dubenii or *A. Braziliensis*, 62.
Ants a pest on the Ilha do Governador, 5.
Ant-hills, 82, 375.
Antimony, 448.
Aplysia Argo D'Orb, 204, 214, 230.
Aplysia Braziliensis Sander Rang, 204.
Aquitequedichagas, custom of piercing the lip and ears among, 584.
Aracajú (city), 381, 383.
Aracaré, fossiliferous shales, &c. of, 396.
Aracaty, 460; country in vicinity of, 461.
Aranaús, tribe of, 593.
Aranha caranguejeira (*Mygale*) of Abrolhos, 180.
Arara (*Psittacus macoa*), 95.
Araraitaguaba (*Arara* macaw, *ita* stone, *guaba* eat, locality so called from a stone which the parrots are said to eat), 516.
Aratú land crab, 239.
Arédes gold-mine, 545.
Aricurí palm (*Cocos schizophylla*), 237.
Armação, 40.
Armação, locality, 40; a trying-house, 182, 183, 184.
Arms of Botocudos, 590.
Aroeira (*Schinus terebinthifolius*), 423.
Arraial da Chapada, gold-washings of, 497; their present condition, 161; da Conceição, 498; da Conquista, 255; d'Itinga (*Yg* or *hy* water, and *tinga* white), 165; das Queimadas, want of rain, 323; de São Gonçalo, 475.
Arroio dos Ratos, coal-mines of, 520.
Arrows of Botocudos, 591.
Arsenical pyrites, 540.
Article, definite, use of before names of places in Brazil, ix; use of before name Brazil, viii.
Artocarpus Braziliensis Gom., 245.
As Azeites, 115.
Aspidorhynchus Comptoni, 467, 468.
Assahí (*Euterpe oleracea*), 425.
Assú (town), 452.
Astrocaryum Airi Mart., 216.
As Trovoadas, 144.
Ateles, 286.
Ateles hypoxanthus, 95.
Attalea palm, 144.
Augite, 297.
Ave-Lallemant, Dr., on coal-mines of Rio Grande do Sul, 520.
Avicennia, social plant, 256.

B.

Bahia, description of point on which city is built, 333; description of city, 334; climate of, 337; population, 335; bay and harbor of, 335; exports of, 335, 336; steamship connections, 337; Bahia de Todos os Santos, 267; compared with Bay of Rio, 272.

Bahia Steam Navigation Co., 337, 421.
Bahia and São Francisco Railroad, 338; geology of, 353.
Bahia de Camamú described, 262.
Bahia de Espirito Santo described, 65; depth of, 72; littoral fauna of, 73.
Bahia de Nova Almeida, 85.
Bahia de Santa Cruz, 85.
Bahian group, 556.
Bahú, isolated hills or gneiss plain at, 303.
Bakues, tribe of, 593.
Bamboo, 84, 141, 216.
Barauna (Melanoxylon), ashes used in tanning, 411.
Barbacena, 4.
Barbacena, Visconde de, 519.
Barlæus, 432; on mouth of Rio Parahyba do Norte, 442.
Barnacles, 73.
Barra do Jardim, 465.
Barra do Commandatuba, colony at, 246.
Barra do Rio de Contas, 260.
Barra do Patipe, 241.
Barra do Poxim, 241.
Barra do Rio Grande, width of São Francisco valley below, 332.
Barra do Rio de São Francisco, 393.
Barra da Vareda, country in vicinity of, 252.
Barra Secca, 106; lands at, 107; consolidated beach, 107.
Barreiras do Siri, tertiary beds of, 56.
Barrel quartz at Minas Novas, 157.
Barrigudo trees, 91, 166, 251.
Barro vermelho, 514.
Basin of Rio Parnahyba one sided, 474.
Bateia, wooden pan for washing gold and diamonds, 160.
Bates's "The Naturalist on the Amazonas," 494.
Beach ridges, 113, 114, 115, 123; formation of, due sometimes to storms, 220.
Beach, raised, at Santos, 507.
Beaches, solidification of, at Bahia, 342; at Rio Vermelho, 344; at the Abrolhos, 179.
Beaches solidified, 189.
Baturite, 460.
Bats, bones of, in caves at Lagôa Santa, 285.
Bauhinia, 250.
Bay of Rio, description of, 6; depth, 7; shell deposits of, 7; corals of, 7; tide of, 7, 8; saltness of, 8; entrance to, 8; once filled with tertiary beds, 22.
Beans, 120, 454.
Beira-mar, 459.
Belmonte, 173; sand plain near, how formed, 235.
Bemdêgo meteorolite, 325.
Benevente, harbor of, &c., 60.
Bignonia Braziliensis, 93; *B. Tecoma*, 408; species of, social plant, 256.
Biotite, 447
Birds, bones of, in caves at Lagôa Santa, 285.
Bismuth, occurrence of in auriferous vein at Cata Branca, 545.
Black water streams, 122, 217, 227.
Blumenbach on skull of Botocudo, 587.
Boassica, cretaceous sandstone at, derivation of name, 398.
Bôa Vista, height of pass of, 12, 26.
Bolerio, 141.
Bom Fim, cretaceous strata of, 346.
Bone of whale valuable for manure, 184.
Bones, fossil, in caves, mode of occurrence of, 284; immense number of recent species, 285.
Bone caverns, extent of, how formed, stalactites and clay deposits in, 283; number of, 284; of Rio das Velhas, 280.
Bodoque, 578.
Bore or Pororoca, 486.
Botocudos, 216, 577; origin of name, 578; stature of, 578; physical characteristics of, 579; number of, 595; liveliness of, 612; skulls of, 585; color of, 588; dislike of clothing among, 590; houses of, 596; food of, 596; method of making fire, 597; cruelty of, 598; treatment of wives and children, 598; religious ideas of, 599; funeral customs, 599; want of numerals, 604; warfare of, 600; language of, 602; Botoque, or lip and ear ornament, 578, 583.
Boulders, scarcity of in drift, 24; occurrence of, at Tijuca, 28-30; paucity of, owing to decomposition, 573; boulders of decomposition, 594; at Victoria, how formed, 69.
Bows of Botocudo, 591.
Braço do Norte of River São Matheos, 117.
Braço do Sul of Rio São Matheos, 117
Brachiopods, palæozoic, at first cachoeira of the Rio Tapajos, 488.
Brackish streams, 255.
Bradypus tridactylus, 94.
Bradypus torquatus, 94.
Braganza, waste of land on bay of, 491.
Brazil, derivation of name, x; use of article before the word, viii.
Bromeliaceous plants, 249, 250.
Brunet, Dr., 272, 421.
Bugres, 578.
Buriti palm (*Mauritia vinifera*), 277.
Burning of forest, injurious effects of, 78; change of flora caused by, 423.
Burton, 276; on use of definite article before word Brazil, viii; on Jacutinga, 535; on occurrence of cocoa palm at Brejo do Salgado, 119.
Byrsonema, 465.

C.

Cabbage palm, 94.
Cabral, discovery of Brazil, 226.
Cacáo (*Theobroma Cacáo*), 120, 236, 244, 259, 260, 261.
Cacaoeiros, cacáo plantations, 244.
Cachoeira (Town), 272
Cachoeira (Bahia), county in vicinity of, 322.
Cachoeiras, 21.
Cacó, sugary quartz, 534.
Cacti, 237; arborescent, 152.
Caçeló whale, 182.
Caiauhé (*Elæis Guineensis* L.), 270.
Caititú, 94.
Cajú or Cashew, 459, 465.
Cajueira (*Anacardium occidentale*), 116.
Caladium, 249.
Calamites, 525.
Calamopleurus cylindricus, 468.
Caldeiroes, 314; of probable glacial origin, 315; mastodon remains at, 325; occurrence of diamonds in, 307.
Calháo, 152
Callithrix, 95, 286.
Camamú, geology of, 262.
Camassari, geology of vicinity of, 361.

Campinas, plains of, 506.
Campo, section at, 365.
Campos, city of, 46; vicinity of, 47.
Campos, character of their floræ largely due to fires, 320.
Campos of Alto dos Bois, 146.
Campos, Liais's picture of, 290.
Canal uniting Macahé with Campos, 43; between Cannavieiras and Commandatuba, 246; between Rio Doce and Victoria impracticable, 80; connecting Rio Jecú with Victoria, 64; projected along coast south of Caravellas, 223.
Candles of Carnahuba wax, 453.
Candona candida Müll., 348.
Canga, gold in, 536, 543.
Cannibalism among Botocudos, 600.
Cansanção (*Jatropha urens*), 250.
Cantagallo, old gold-mines at, 50.
Cañon of the Jequitinhonha, 163.
Cape Frio, 39; height of, 39.
Capibara (*Hydrochærus Capabara*), 94.
Capibara, fossil (*Hydrochærus sulcidens*), 287.
Capoes, 146, 147.
Carapato (*Ixodes ricinus*), 153, 155, 257.
Carapina, 82.
Caravellas, 223.
Carboniferous in Brazil, 553; fossils, occurrence of, on Guaporé and Mamoré, 488.
Cariocar, 291, 465.
Carnadeiras, fever of Upper São Francisco, 292.
Carnahuba palm (*Copernicia cerifera*), description of, 452, 446, 459, 460, 461; wax of, 453.
Carnivora more abundant in Post Tertiary than at present, 288.
Carnelian, 531.
Carrasco, 147, 253, 429.
Cascalho, meaning of word, 51; auriferous, of Minas Nova region, 159; auriferous, at Jaraguá, 512.
Cascate Grande, 27.
Cassis Madagascariensis (*C. Cameo*), at Os Busios, 39, 198, 203.
Castelnau, 496, 501, 504; on chapada diamantina, 307; on whale fishery at Bahia, 185.
Cata Branca gold-mines, 544; auriferous vein at, 545.
Catinga de Porco (tree), ashes used in tanning, 411.
Catingas, 144, 151, 250, 251, 296, 322, 452.
Cattle and cattle fazendas in Province of Rio Grande do Norte, 454.
Cavia (*Cœlogenys rupestris*), 257; bones of, in caves, 285.
Caximba Nova, 446.
Caxoeira do Angelim, 168; do Campo, bone caves at, 281, 285; de Dorma, 167; da Farinha, 168; do Inferno, 88, 169.
Caxoeirinha do Jequitinhonha, 171.
Cachoeira de Paulo Affonso, distance from the sea, Halfeld's description of, 414; Liais on, 415, 418; compared with those of Salto Grande on the Jequitinhonha, 417; Burton on, 418; compared with Niagara, 418.
Caxoeirinha do Rio Pardo, 242.
Caxoeira de Santa Anna, 169.
Cazal, 226, 430, 486; on mastodon remains, 261; on fossil bones, 280.
Cebus, 95.
Cecropia, 94; social plant, 256.
Cedro cedrela, 94, 261, 459.
Cemeterio of the Abrolhos, 180.
Centipede, 375.
Ceratites Harttii, 386; *C.* (*Amm.*) *Pierdenalis* Von Buch, 388.
Cereus, 250, 296, 407, 423.
Cerithium, 229.
Cervus, 95.
Cervus rufus, fossil with megatherium, &c., 286.
Cervus simplicicornis found with megatherium, &c., 286.
Cesalpina echinata, 94.
Chalcedony, 531.
Chama, 229.
Chandless, 500, 503; discovery of Mosasaurian and turtle remains on the Purús, 494
Chapada, meaning of term, 132; at Agua da Nova, 149, 150; magnificent view from edge of, 150.
Chapada Diamantina, geology of, 295; height of, 301; da Mangabeira, 497; at Santa Rita, 144; do Sincorá, 306; between Sucuriú and Agua Suja, 153.
Chapadas of Piauhy, 474; of Rio Jequitinhonha, 165, 168; of Rio de São Francisco, 277, 291; between the basins of the Tocantins and São Francisco, 499.
Chapadão de Santa Maria, 499, do Urucuia, 499.
Chapeiroes, 211, 212, 213; dangerous to navigation, 201; distribution of, 211; mode of growth of, 199; coalescence to form reefs, 200; appearance of, 200
Charruas, custom of piercing lip and ears, 584
Chelonia Mydas, 112, 113.
Chiefs, Botocudo, 592.
Chique-Chique, cactus, 408; town, origin of name, 309; occurrence of diamonds near, 307.
Chlamydotherium, 286.
Chlorite, 328.
Chlorite slate, 299.
Chrome localities in Goyaz, 503.
Cidaris, 385.
Cladocyclus Gardneri, 468.
Claussen, Mr., 281.
Clays of Amazonian formation, 488.
Clays of Tertiary, 57, 124, 225.
Clay-eating among Botocudos and Brazilians, 581.
Clay-slate, 151, 298, 328, 464, 500; auriferous, distribution of, 552.
Clay-slates, auriferous, of Brazil, resemblance borne to the gold-bearing rocks of Nova Scotia, 551; probably Lower Silurian, 551; auriferous veins in, 533; absence of, in eozoic of Brazil, 549.
Climate of campos, 253; of Ceará, 458; moistness of, dependent on forests, 321; of Colonia Leopoldina, unhealthiness of, 217; of Fernando de Noronha, 439; of Maranhão, 485; of Rio Grande do Norte, 454; of São Francisco below falls, 420; of Pernambuco, 429; of Rio Doce, 98; of Rio Grande do Norte, 441; of Lower São Francisco, 420.
Coal, Candiota, report on, by Mr. Ginty, 528; facilities for mining in Rio Grande do Sul, 526; cost of, 527; said to occur on Rio Piauhy, 380.
Coal Basin on Rio Jacuahy, 530; on Rio São Sepé, 529; of Rio Tuberão, 519.

Coal-mines of Rio Grande do Sul, discovery of, 519.
Cochineal, 408, 454.
Cocoa palm (*Cocos nucifera*), distribution of, &c., 118, 238, 271, 431, 437; oil of, 238.
Cocos coronata Mart., 298; *C. flexuosa*, 146; *C. nucifera*, distribution of, &c., 118; *C. schizophylla*, 237, 298.
Coffee, 79, 101, 123, 244, 261, 265, 459; export of, from Bahia for year 1864-65, 336; plantation of Campinas, 514; of the Colonia Leopoldina, excellence of, 217.
Colonia do Guandú, 90; Leopoldina, 217; do Mucury, history and present state of, 131; de Santa Leopoldina, 78; poverty of soil, 79; de Santa Isabel, 65; de Santa Maria, 76; de Santa Theresa, 518; do Urucú, productions, climate, scenery, &c., 133.
Colony, American, on Jequitinhonha, 172; on Rio Doce, 105.
Color of skin lighter in forest-dwelling tribes, 588.
Comatula, 62.
Commercio, 306.
Compass, miner's, 530.
Conglomerate, 242, 301, 462; cretaceous, 354; tertiary at Agoa da Nova, 150.
Congonhas do Campo, gold at, 544.
Conocarpus, social plant, 256.
Consolidated beach at As Pedras, 113; at Barra Secca, 107; at mouth of Rio Parahyba do Norte, 442; at Guarapary, 62; of Pernambuco, 434; at the islands of the Abrolhos, 179.
Copaifera officinalis, 93.
Copeland, Mr., 142, 242.
Copocabana, gneiss of, 10.
Copper, 472; boulder, found near Cachoeira, 300; localities for, in Brazil, 300; sulphate and other salts of, 488.
Corals at Itamaracá, 437; dead, near Rio Sant' Antonio, 234; distribution of on submerged border of Recife do Lixo, 210; used for manufacturing into lime, 213.
Coral banks near Ilha Itaparica, 269; in Bay of Camamú, 265; dead, in Lagôa near Ilhéos, 259.
Coral islands, 211, 212.
Coral reefs, height of edge of, 209; in vicinity of Pernambuco, 214; near Camamú, 213; around Quieppe Island, 213; of Abrolhos, 187; of Pernambuco, 188; of Maceió, 188, 213, 425; near Cape São Roque, 188; of Roccas, 214; off Periperi, 213; of Santa Barbara dos Abrolhos, 192; species that contribute most to build up, 214; raised border to, composed of barnacles, serpulæ, &c., 205; submerged border of, 205.
Corcovado, description of, 9; magnificence of view from, 11.
Corn, 152.
Corôa Vermelha, 211.
Coronula, 112.
Corrego de Santo Ignacio, diamonds in, 307.
Cotigipe, cretaceous rocks at, 359.
Cotinguiban group, 556.
Cotton, 225, 336, 441, 442, 443, 459, 486; of Calhão, excellent quality of, 152; of Pernambuco, quality of, 431; of Rio Grande do Norte, 452; value of, exported from Bahia during year 1864-65, 336; factory, 266.
Crabs, burrows made by, in shales, 353.
Cratheus, 457.
Crato, 460, 463.
Crax, 95.
Cretaceous, of Abrolhos, 175; of Monserrate and vicinity of Bahia, 346; beds at Bahia of fresh-water origin, 358; soils of Bahia, 372; near Aracajú, 383; at Maroïm, 384; fossils described, 385; at and near Penêdo, 396; limestones at Propriá, 404; of Itamaracá, 437; Parahyba do Norte, 445; fossil fishes of Ceará, 466, 476; of Purus, 494; Résumé, 555.
Crocodilus Harttii Marsh, 355.
Crustaceans, abundance of, on reefs; list of species occurring on the Abrolhos reefs, 203; list of species occurring elsewhere in Brazil, 203.
Crysoberyls, 151, 164.
Cuiabá, 500; diamonds at, 504.
Curimatães (fish), 399.
Curitiba, 517.
Cuscuta, 375.
Cutia (*Dasyprocta*), 94.
Cypræa exanthema, 40.
Cypris (?) *Allportiana* Jones, 348; *C. conculcata* Jones, 348; *C.* (?) *Monserratensis* Jones, 348.

D.

Dance of Botocudos, 601.
Darwin on phosphate incrustation of Abrolhos, 178; on corals of Brazil, 187; on Island of Fernando de Noronha, 437.
Date palm, 425.
Davis, Dr. J. Barnard, on skull of Botocudo, 585.
D'Archiac, carboniferous rocks in Bolivia, 554.
De Beaumont, Elie, on the gneiss of the Serra do Mar, 548.
Decomposition of gneiss, 24; Darwin, Agassiz, Pissis, Heusser, and Claraz on, 24; cause of, 25, 26.
Decomposition of rocks in India and North America, 26.
Decomposition, drift not referable to, 563.
Decomposed rock separate between drift and easily distinguished, 564.
Deer of Campos, 147.
Dendê (*Elæis Guineensis*), 270, 425.
Denudation, subaerial, effects of, 33.
D'Orbigny on gneisses of Bolivia, 548.
D'Osery, on Cata Branca mine, 545.
Desterro, 519.
Devonian rocks in Brazil, 553.
Diamantiferous rock from chapada, 304.
Diamantino, 500, 504.
Diamantina, diamonds at, 156.
Diamonds, discovery of, at Diamantina, 156; mode of occurrence of, at Chapada Diamantina, 302-309; yield of, at, 308; of Sincora, 306; of Pétanga, Bahia, mode of occurrence of and method of washing, 369; probable source of, 370; localities for in Goyaz, 503; occurrence of in Province of Paraná, 518.
Didelphys, 94.
Dinosaurian remains, cretaceous, Bahia, 356
Diorite, 296.
Diploria, absence of, in Brazil, 214.
Diplothemium maritimum, 237.
Disthene, 325.

Dolphin, species of, found in the Araguaya, 502.
Dom Pedro II., the Emperor, geological observations of, 35.
Dom Pedro II. Railroad, geological observations on, 14.
Dom Pedro, North d'El-Rei Company, 542.
Drawings, Indian, 297, 326.
Drift on Cantagallo Railroad, 20; of Rio, 23, 561; at Tijuca, 27; of Victoria, 69; on Philadelphia road, 132; at Alto dos Bois, 145; near Fazenda da Lagôa, 148; near Calhão, 151; of Minas Nova and vicinity, mode of occurrence of gold in, 160; at Bahia, 341; once continuous down to the surface of the sea, 342; on Bahia Railroad, 366; on Lower São Francisco, 419; of Ceará, 469; on São Paulo Railroad, 508; of Patagonia, 558; Résumé of arguments to prove existence of in Brazil, 558; distribution of in Brazil, 561; clays of, totally unlike those of the tertiary, 568; former extension down to level of sea, 572.
Drought in Ceará, 459.
Dry season, abundance of life during, 155.
Dual form in Botocudo language, 604.
Dunes, 124, 332, 345, 380, 382, 394; in Province of Rio Grande do Norte, 455; Ceará, 458.

E.

Ears, perforation of, by Botocudos, how performed, 582.
Ear-plug, 583.
Echinaster crassispina, 62.
Echinometra Michelini Desor, 62, 203, 214; nests of, in rock at island of Maricás, 36.
Edwards, Prof. A. M., on turba of Camamú, 263.
Eggs of loggerhead turtle, 110.
Elephants in Post Tertiary in Brazil, 288.
Elevation of coast within recent times, evidence of, at Rio, 35.
Embaüba (*Cecropia*), 94.
Ema (*Rhea Americana*), 146, 254.
Encope emarginatus, occurrence of, at Rio Sant' Antonio, 235.
Engeräckmung, 577.
Entomostraca, cretaceous, 347, 348.
Entre Montes (village), 411.
Eozoic rocks, occurrence of gold in, 532; distribution of, in Brazil, 547; of Venezuela and Guiana, 550.
Eretmochelys imbricata, notes on, 112.
Eriocaulon, 423.
Erosion, topographical features produced by, in moist and wooded region, 318.
Erratics, 489.
Escadinhas, Cachoeira das, 89.
Estreito on Jequitinhonha, 166.
Estancia, Geology of vicinity of, 379.
Estherians, 445.
Eugenia, 249.
Eunicea humilis Edw. and Haime, 62, 74, 196, 229.
Exports of city of Bahia during year 1864-65, 336, 337.

F.

Falls of Paulo Affonso, 414; of Jequitinhonha, 170.
Farinha, 120; exportation of, from Rio de Contas, 261.
Farrancho, 167.
Faults, apparent absence of, 448.
Fauna, marine, of Brazil, resemblance to that of West Indies, 198.
Faunæ, comparison of Brazilian radiate fauna with the West Indian, 214.
Favia conferta Verr., 194; *F. gravida* Verr., 194; *F. leptophylla*, 207.
Feira da Conceição, country in vicinity of, 322.
Felis concolor, 95, 286; *F. macroura*, 95; *F. onça*, 95; *F. pardalis*, 95; *F. mitis*, 286; *F. protopanther*, 287.
Ferns, fossil, of Candiota coal-mines, 522.
Fazenda do Capitão Grande, 117; do Tenente Honorio Ottoni, drift in vicinity of, 148; at Escravania, immense number of bones in cave at, 285.
Feijão (beans), 79.
Feijó on structure of Serra da Ybiapaba, 457; on gold of Ceará, 472.
Fevers of Rio Doce, 98; of São Matheos, 118; on the Jequitinhonha, cause of, 169.
Fiddle crabs, 230.
Fires, effect of, on vegetation, 83, 320, 375.
Fish, excellence of Brazilian, 186; fossil, 347, 353, 354, 355, 359, 371, 384, 445, 466, 477.
Fisheries of Espirito Santo, richness of, 81.
Fishery of Garoupa in Abrolhos region, 185.
Flemingites Pedroanus Carruthers, 524.
Fletcher, Rev. J. C., 514.
Folds, reversed, in Eozoic, 550.
Forbes on Siluria of Bolivia and Peru, 552.
Forest, luxuriance of, on Rio Doce, 93; on Rio Pardo, 245.
Forests, belt of, coincidence with belt of rains, 319; losing their hold, giving way to campos, 319; effect of fires, 320; decomposition dependent on forests, 320.
Fortaleza (city), location and population of, 460.
Fossils (foreign), carried in ballast, necessity of caution, 269.
Fossils of Siluria in Bolivia and Peru, 552.
Fossil bones, 280, 311; plants of Abrolhos, 176; plants of carboniferous, 523; shells of recent species in reef rock at Porto Seguro, 229; shells, reptilian remains, &c., in cretaceous of Monserrate, 345; of cretaceous of Maroïm, 385; shells near Propriá, 404.
Frade de Macahé, height of, 42; de Itapémerim, height of, 58; de São Leopardo, 68.
Francylvania, location, history of, &c., 95.
Freshet (annual) of São Francisco, 420.
Frigate-bird (*Tachypetes aquilina*), cemetery of the, 180.
Frogs, bones of, in caves, 275.
Fruita pão (*Artocarpus incisa*), 245.
Funeral customs of Botocudos, 599.

G.

Galena, 448, 472.
Gamella (*bateia*), 512.
Gamellas, Indian tribe, custom of piercing lip and ear, 584.
Gardner on mythical reef, 189; note on Piranha, 400; Paper on the Interior of the Province of Ceará, 460; probably wrong in referring the sandstone of Piauhy and Ceará to the cretaceous, 476.

Garnets, 13, 21, 49, 50, 151.
Garrafão, near Rio Itabapuana, height of, 53.
Gavia, height of, 10.
Gazzinelli, Signora, 129.
Gelasimus Maracoani Latreille, 229; *G. palustris* Edwards, 230.
Genipapo, black color, furnished by, 589.
Geoffroya, 408.
Gerber, 3.
German colonists at Santa Clara, 132.
Gibbs, Hon. George, 584.
Glaciated surfaces, character of, 24.
Glacier of Pacatuba, 469; of the Amazonas, 490.
Glacial phenomena of Patagonia, 558.
Glyptodon, 286.
Gneiss, 1, 4, 9, 12, 18, 24, 25, 48, 50. 52, 53, 61, 67, 90, 93, 129, 133, 135, 148, 165, 166, 167, 168, 170, 251, 259, 266, 296, 312, 313, 322, 323, 324, 338, 340, 360, 372, 405, 445, 449, 458, 461, 462, 463, 474, 500, 501, 549, 550.
God, Botocudos, destitution of belief in, 599.
Goitre in Brazil, 330.
Gold, 297, 448, 471, 497, 503, 513; at Cantagallo, 50; on the Rio Mangarahy, 77; on Rio do Castello, 59; in drift at Sucuriú, 153; of vicinity of Minas Novas, ancient method of working, 160; abundance of, in Minas Nova region, not yet exhausted, 156, 161; probable existence of, in tertiary deposits in Brazil, 162; mode of occurrence of, in Parahyba do Norte, 448; at Villa das Lavras da Mangabeira, 463; mines of Jaraguá, 513; of Brazil, age of, 532; distribution of, in Brazil, 532; mode of occurrence at Morro Velho, 537; of Morro Velho invisible in the ore, 541; mode of occurrence of, at Rossa Grande, 542; red, 542, 545; of Gongo Soco, 541; color of, 542. (See Chap. XVIII.)
Gorgonia gracilis Verrill, 209.
Goyabeira, 47.
Granite, 295, 500.
Granja, 460.
Graphite, 448, 458, 472, 525.
Grazina (*Phaëton*), 180.
Greenstone, 328.
Green turtle, 112.
Grisolitas, 151.
Guarapary, harbor of, 63.
Guarapuava, gold at, 513.
Guayana, Tertiary rocks of, 493.
Guayamú, or Guáinumu (*Cardiosoma Guanhumi*), 239.
Gullies formed by rains, 159.
Gurirí (*Diplothemium maritimum* Mart.), 237; a social plant, 256.
Gypsum, 471.

H.

Halfeld, 414; on mastodon remains found near Paulo Affonso, 418.
Harbor of Ilhéos, 258; of Maceió, 425; of Pernambuco, 435.
Hares, 95.
Heliastræa aperta Verr., 207, 213, 269.
Heliconia, 94, 249.
Helmreichen, Dr. Virgilio, section across the Jequitinhonha valley, 138; on the Serra da Chapada, 306.
Heusser and Claraz, 534.
"Highlands of Brazil," 276.
Hills or Serras, isolated, on limestone plain, 310.
Holothurians, 62.
Hoplophorus, 286.
Hornstone, hills of, near Volta da Serra, 312.
Houses of Botocudos, 596.
Humboldt on name Brazil, x; on distribution of cocoa palm, 120; on social plants, 255.
Humpback whales, 181.
Hunt, Dr. T. Sterry, on Laurentian of Brazil, 550.
Hyatt, Alpheus, on cretaceous fossils from Maroïm, 385.
Hymenogorgia (*Gorgonia*) *quercifolia*, 62, 74, 196, 229.

I.

Icó, 460.
Iguape (*Yg* water, *gua* of varied color, *pé* way, Mart.), fertility of vicinity of, 271.
Ilex paraguayensis, 517.
Ilha do Paquetá, das Cobras, do Governador, Enxada, 6; do Lima, 45; Escalvada, Raza, 63; do Boi, decomposing gneiss, 70; Balêeiro, corals of 74; de Santa Barbara dos Abrolhos, 175 Redonda dos Abrolhos, 178; Itaparica, 267, coral banks of, 213; Grande, 258; Pequena, 258; Boyapeba, 265; dos Fradres, 267; Itamaracá, 437; de Fernando de Noronha, 437; do Bananal, 502; de Quieppe, 523.
Ilhas de Maricas, evidence of recent rise of, 35, 36; de Santa Anna, 41; dos Pacotes, 71.
Ilhéos, 258.
Imbuzeiro (*Spondias tuberosa* Arr.), 323.
Implements, human, in bone caverns, 286; of Botocudos, 592.
Indaiá palm, 144; (*Attalea compta* Mart.), 144, 166.
Ingá, 94, 116, 249; gneiss at, 446.
Inoceramus in cretaceous near Aracajú, 384.
Ipé (Tecoma), 94.
Ipecacuanha, 459.
Ipomœa littoralis, 237.
Iron ore, 58, 299, 301, 407, 447, 503, 515, 525.
Iron mines and works of Ypanema, 515.
Island of Victoria, 66; height of, 67.
Itabapuana (*Hy* water, *taba* village, *apuan* round?)
Itabirite and jacutinga auriferous, 533.
Itabirite, 534.
Itacolumi, height of, 3; (Maranhão), 484.
Itacolumite, 541, 545; the term loosely employed, 332; gold in, 533.
Itaipins reef, 257.
Itambé (*Itá* stone, and *çaimbé* rough),
Itanhaem (*Ita* stone, and *nheeng* speak, echoing rock).
Itapagipe, 334.
Itapitinga (reef), 258.
Itapuan (*ita* rock, *apuan* round), 346.
Itatiaiossú, Pico de, 2; height of, 3.
Itú (*Itú* or *Ytú* a fall), gold at, 510.

J.

Jacarandá (*Bignonia Braziliensis*), 93, 228, 246, 261, 459.
Jacaré, salt deposits of, 329.

Jacchus (*Hapale*) *leucocephalus*, 95.
Jaguaripe, 268.
Jack tree (*Artocarpus Braziliensis* Gom.), 245.
Jacobina, character of country in vicinity of, 312; climate of, 317.
Jacupemba (*Penelope marail*), 95.
Jacutinga, derivation of name, 533; described, 534, 535; gold in, free, 535; at Gongo Soco, 541.
James, Major O. C., 507, 514, 584.
Janahuba, 374.
Janchon, bad spirit of Botocudos, 599.
Jangadas, number of, in Province of Pernambuco, 432.
Januaria, salt licks of, 320.
Jaraguá, port of Maceió, 425; gold-mines of, Mawe's account of, 511.
Jatropha, 249.
Jequiá, tertiary bluffs of, 422.
Jiporock, tribe of, 593; Botocudo chief, 594.
Joazeiro, 327, 408.
Jundiahy, 513.
Jurassic rocks not yet known in Brazil, 554; of Andes, 554.
Jurupencem, head of navigation of Araguaya, 501.

K.

Kielmeyera, 291.
Kjoekkenmoeddings on the Illa do Governador, 6; at Santos, 6; at Santa Cruz, 6, 85.
Knobs, isolated, on lake plain, 315.
Koloshians of Alaska, custom of piercing the lip, 584.
Kuparack, skull of, described, 584.
Kyanite, 145.

L.

Lacerda, Dr. de, 337, 421.
Lagôa de Freitas, 11; de Maricá, 37; Saquarema, 37; Araruama, 38; Feia, 44; do Campello, 45; Marobá, 56; Jacuné, 83; Juparanãa, its depth, color of water, tertiary beds, 100; do Aviso, 103; de Monserras, 106; Tapada, 107; Maririeú, 115; Gavatá, 227; do Braco, near Belmonte, 236; Santa, 281; at Jacobina on the São Francisco, 409; da Pedra, mastodon remains in, 418; do Norte, Alagôas, 422; do Sul, Alagôas, 422; Paranaguá, 474.
Lagoons along coast, formation of, 44.
Lake Plain of Bahia, 314; lakes of, 314.
Lameirão, 76.
Lands of Rio das Velhas, 290.
Language of Botocudos, 602; simplicity of grammatical structure of, 603.
Lapa Vermelha, 282.
Laurentian rocks, 447, 449; resemblance between Brazilian gneisses and Laurentian of Canada, 549.
Laurus, 249.
Lead, sulphate of, chromate of, sulphide of, carbonate of, chloro-phosphate of, 448.
Leather-back turtle, 112.
Leaves, shedding of, by catinga woods in dry season, 322; rapid growth of, on approach of wet season, 322.
Lecythis, 94, 249.
Leite, Dr. França, his colony on the Rio Doce, 95.
Lengoas custom of piercing lip and ears, 584.
Lençôes, diamond washings at, 307
Linhares, fertility of lands near, 103.
Lizards, bones of, in caverns, 285.
Lepidotus, 347.
Lepidotus temnuras, 468.
Llianas, 250. L. Juparanãa (*Jui*, a frog, *paraná*, sea or lake), described, 99; compared with lakes of Alagôas, 422.
Liais on height of Piedade, 3; on São Francisco, 275; on the gold-mines of Brazil, 546.
Legs of Botocudo, thinness of, 579.
Licurí palm, 256, 423.
Lignite, 471.
Lime manufactured from corals, 213.
Limestone at Ypiranga, 15; at Cantagallo, 51; of coral reefs, 198; of São Francisco, 279; of Rio das Velhas, 281; at Malhada, 295; east of Chapada Diamantina; plain in Western Bahia, 310; aridity of, 317; at Estancia, 380; at Parahyba gold-mines, 447; near Crato, 464; almost entire absence of, from Laurentian of Brazil, 551.
Linhares, Rio Doce at, 97; massapé soils, 97; situation of, 99; tertiary beds at, 99.
Linckia ornithopus Lütken, 214.
Lip, perforation of, by Botocudos, how performed, 582; ornament, 583.
Littorina, 229.
Lixo, Recife do, 202.
Lizards, abundance of, on Abrolhos, 180.
Lode, Descubridora, Parahyba, 448; Lima, 447; Bôa Esperança, 448.
Loggerhead turtle (*Thalassochelys cauana*), abundance of, at Barra Secca, notes on, 108.
Lund, Dr. P. W., on bone caverns of Brazil, 281.

M.

Macahé and Campos Canal, 43.
Maçambamba, beach of, 39.
Macambira, 407, 441.
Maccapé, locality for fossil fishes in Ceará, 466.
Maceió Reef, 213; topography and geology of the vicinity, 422; city and harbor, 425.
Machacalis, village of, 167.
Macrauchenia, 286.
Macujé, discovery of diamonds at, 306.
Madrepora, absence of, from Brazilian polyp fauna, 214.
Madrijos, female whales with young calves, 184.
Mæandrina, absence of, from Brazilian polyp fauna, 214.
Mæstrichtien, 494.
Magalhães, Dr. Conto de, 502.
Magnesia, 471.
Maquiné, gold-mine of, 543.
Maize, 79, 142.
Malaphyr, 531.
Malhada, 265.
Mammals, number of fossil species discovered by Claussen and Lund in the bone caverns of Brazil, 285; remains of recent species of, found occurring with extinct species, 286.
Mammalian fauna richer in the post tertiary than now, 286; of true South American type, 287; not richer in genera formerly, 287; richness in giant forms, 288.
Mamão (*Papaya*), 217.
Man, remains of, in bone caverns, 286.

Manatí (*Manatus Americanus*), at Victoria, 75, at São Matheos, 122; in Rio Peruhype, 224.
Mandioca, 79, 99, 120, 123, 454, 458.
Mandubi, derivation of name, &c., 409.
Mangaba, or Mangabeira (*Hancornia speciosa* Gom.), 149, 374, 459, 465.
Manganese, 535, 536, 545.
Manguinhos, tertiary beds at, 55.
Mangrove swamps, mud of, 55; formation of, 220; unhealthiness of, 240; swamp at mouth of Rio Jequitinhonha, 239.
Mangroves, red (*Laguncularia racemosa*) and white (*Avicennia tomentosa*), 122; as agents in the silting up of swamps, 222; social plants, 256.
Manicina, absence of, from Brazilian polyp fauna, 214.
Maracassumé, gold-mines of, 485.
Maragogipe, 271.
Maranguape, 460.
Maranhão, population of province and city, 486; gold-mines, 546.
Maroïm, species of fly, noted for biting just at nightfall, 240; city of, 384; cretaceous of, 384.
Marsh, Prof. O. C., description of new crocodilians, 355; on occurrence of mosasaurians in Europe and America, 494.
Massapé soils, 99, 118, 395.
Mastodon, remains of, 261, 286, 298, 324, 325, 418, 419, 471, 573.
Mate, Paraguayan tea, 517.
Matta de São João, 368.
Mattas, 379, 429.
Mauritia vinifera, 290.
Mawe, Travels in Brazil, 51; on gold-mines of Jaraguá, 511.
McGrath, Dr., 437.
Megaptere, 181.
Megatherium, 286, 471.
Melania terebriformis Morris, 348, 350; *M. Nicolayana* Hartt, 350.
Melaphyr, 530.
Melastoma, 249, 291, 375; *M. Hookerianus*, 408, 423.
Mertensia dichotoma, 256.
Messalia Ortoni Gabb, 493.
Mestre Alvaro, 83.
Meteorolite of Bemdêgo, 325.
Meteorolites, shower of, in Ceará, 472.
Metamorphism no criterion of age, 280; signs of, in eozoic rock not disappearing toward the western part of plateau of Brazil, 500.
Miahype, 61.
Mica-slate, 135, 142, 148, 164, 252, 296, 298, 328, 445, 446, 458, 525, 551.
Mice, occurrence of bones of, in caverns of Lagôa Santa, 285.
Millepora alcicornis Linn, 40, 206; *var. M. digitata*, 206; *var. M. cellulosa*, 206; *var. M. fenestrata* Verr., 206; *M. Braziliensis* Verr., 195; *M. Nitida* Verr., 205.
Millepores, stinging properties of, 207.
Mimosa, 249, 250 408.
Minas da Caxoeria, 447.
Minas Geraes, a land-locked province, necessity of roads to the coast, 130.
Minas Novas, cause of failure of mines, 157; geology of vicinity of, 157.
Mining companies. Companhia Metallurgica do Assuruá, 308; Tasso Brazilian Gold-Mining Company (limited), 449; São João d'El-Rei Mining Company, 536; Rossa Grande Gold-Mining Company, 542; Dom Pedro North d'El-Rei Company, 542; East d'El-Rei Company, 544; Montes Aureos Gold-Mining Company (limited), 546.
Mispickel, the principal gold-bearer, 534, 540.
Moisture, existence of forests, dependence on, 321.
Molybdate of lead, 472.
Monkeys of Rio Doce, 95.
Monserrat, geology of, 334.
Monte Morêno, 65; Jutuquara, height, &c., 68; Parcoal, 226; Santo, 325.
Montes Pyreneos, height of, 501.
Moon, superstitions regarding, among Botocudos, 599.
Moraine, gigantic, of Amazonian glacier, 490.
Moraines of Pacatuba, 469.
Morainic deposits of Tijuca, 562.
Morro do Sapateiro, 48; de Mestre Alvaro, 81; de Agah, 59; de Nossa Senhora da Penha, 66; da Serra, 81, 83, 84; do Padre, 92; da Terra Alta, 97; do Kupan, 133; do Arião, 164; de São Paulo, 266; Redondo, 297; do Caldeirão-assú, pot-hole on, 316; do Conselho, sand-plains of vicinity of, 345; do Chaves, geology of, 404; dos Cabellos Brancos, 513; Velho gold-mine, 536; mode of extracting gold, 538; profits, 539, 540; depth of mines, 539; de Santa Anna, 542.
Mouchez, 226.
Moulding of drift-covered surfaces wholly different from that of surfaces exposed to the action of waves, 569.
Mountain ranges badly represented on maps, 278.
Mucury region, resemblance in topography, soils, &c., to the coffee regions near Rio, 133; topographical features due to moist climate and forests, 319.
Mud bottom bordering coral reef, 209.
Mundo Novo (Bahia), mastodon remains, 325; locality for fossil fishes in Ceará, 466.
Murás, custom of piercing lip, 584.
Murici (*Byrsonima*), 374.
Mussa Harttii Verr., 73, 196, 229.
Mussels, 73.
Mutum (*Crax*), 95.
Mygale, 180.
Mycetes Ursinus, 95.
Mylodon, 286.
Mystiça whale, 182.
Mythical reef, belief in accounted for, 189.

N.

Naegeli, Dr. H., 6.
Naknenuks, 579, 593.
Natal, 454.
Natica prælonga Leymerie, 385.
Natividade, 497.
Navigation by steam of São Francisco, 399, 421.
Navigation of the Araguaya, 501.
New Red Sandstone of Estancia, 379.
Nazareth, 268.
Néocomen, 555.
Neritina, 348; *N. pupa* Gabb, 493.
Nests of sea-urchins in rock, 36.
Nest of loggerhead turtle, mode of excavating, 109.

Neuwied, Prince Maximilian zu, 122; on journey from Ilheos to Bahia, 249; on country between Urubú and Faz. da Caxoeira, 257. See *Appendix*.
Nicolay, Rev. Mr., on turba deposits of Camamú, 264; on geology of Chapada Diamantina, 301; geology of interior of Bahia, 301.
Nœggerathia obtusa Carruthers, 525.
Noruega whale, 182.
Nova Almeida, 85.
Nuggets of gold, 297.
Numerals wanting in language of Botocudos, 604.

O.

Octopods, 63, 204.
Oculina, absence of, from Brazilian polyp fauna, 214.
Odontopteris Plantiana Carruthers, 524.
Oiteiros, 362.
"Olhos," 545.
Olivença, 248.
Oliveira, Sr., 437.
Onça, 95.
Oölitic rock at Aracaré, 396.
Opal, common, 327.
Ophiactis Krebsii Lütken, 197.
Ophiolepis paucispina Müll. and Troschel, 197.
Ophiomyxa flaccida Lütken, 197.
Ophionereis reticulata Lütken, 197, 203.
Ophiothrix violacea Müll. and Troschel, 197, 203.
Ophiura cinerea Lyman, 62, 197, 203.
Opossums, 94; great abundance of bones of, in bone caverns, 285.
Opuntia, 408.
Oreaster gigas, 197.
Organic remains, none in drift, 573.
Orgãos, Serra dos, height of, 3, 7.
Ornament of Botocudo, 590.
Orthoclase, 9, 446.
Orton, 493.
Os Busios, 40.
Os Possoes (Poçoes), 255.
Ostrich, American, 146.
Ottoni, Senator Theophilo B., his project to settle the Mucury and open a wagon-road from Santa Clara to Minas Novas, 131; mining right of in Minas Novas district, 162.
Ouro Preto, vicinity of auriferous, 536.
Ovulum gibbosum, 196.
Owls, instrumental in accumulating bones in caverns, 285.
Oysters, 73, 229.

P.

Paca (*Cœlogenys Paca*), 94.
Pachyderms, more abundant in post tertiary than at present, 287.
Pacific and Atlantic connected during the cretaceous, 391.
Palladium, 545; in gold of Gongo Soco, 542.
Palms of Bay of Rio, 7; of Rio Doce 94; cocoa (*Cocos nucifera*), 118; *Nayá*, 118; *Muri*, 118; *Timboré*, 118; Indaiá, 144; Carnahuba, 452.
Palm oil, 270.
Palmetto (*Euterpe*), 94; (*Euterpe edulis* Mart.), 216.
Palythoa, 40, 62, 192, 193.
Páo d'Arco (*Bignonia*), 94, 459.
Pão de Assucar (Rio), 8, 9; Victoria, 66; Rio de São Francisco, 410.
Páo Brazil (*Cesalpinia echinata*), 94, 228.
Parafuso, sand-beds at, 360.
Parahyba (city), population, commerce, &c. of, 442.
Parahyba do Norte, geology of gold-region of, 445.
Paraná, province of, 517.
Paranaguá, 518.
Parcel dos Abrolhos, 199; das Parêdes, 201.
Passiflora, 249.
Patagonia, drift of, 558.
Paulo Affonso, falls of, 414; compared with Niagara and the Salto Grande do Jequitinhonha, 418.
Peanut (*Arachis hypogœa*), 409.
Peat, 365, 509.
Pebble sheet of drift, 24.
Pectinia, 74; *P. Braziliensis* Edw. and Haime, 208.
Pedra Bonita, 32; Formosa, 53; Lisa, 54.
Penêdo, geology of vicinity of, 397; population, commerce, &c., 398.
Penn, Staff Commander, on the reefs near Cape São Roque, 188, on recife, 188.
Perigot, Dr., discoverer of the Brazilian coal-fields, 520.
Periperí, cretaceous rocks at, 359.
Pernambuco, rivers of, 430; convenience of, as a port, 432; situation of city, 433; harbor of, 435; population of city and province, 432; derivation of name, 432.
Philadelphia, 131; height above sea, 134; country in vicinity of, 134; richness of soil, fitness for colonization, 134; road, present state of, 135.
Phonolite of Fernando de Noronha, 438.
Physiognomy of Botocudos, 580.
Piassaba (*Attalea funifera* Mart.), 237, 247, 271, 336, 373.
Pico do Enchadão, 169; do Enchadinho, 169; de Itambé, 137; d'Itabira, 535; Piedade, Alto da Serra da, height of, 3.
Pine-apples, 431.
Pingoas d' agua, 164.
Piquí (*caryocar Brasiliensibus*), 149.
Piranha (*Pygocentrus*), derivation of name, 399; description of habits, 399; note on, by Gardner, 400; voracity of, 401, 402, 502.
Pirampeba (fish), 403.
Piso on Brazilian reef, 187.
Pisodus in cretaceous at Bahia, 355.
Pissis on gneiss of Corcovada and Copocabana, 10; on eozoic gneisses of Brazil, 548.
Pistacte, 297, 324, 327, 485.
Pitanga, section at, 368; diamond mines, 369.
Pitangueira (*Eugenia*), 19, 59.
Plain of tertiary at Carapina, 82.
Plains of sand at Victoria, 82; coast, formation of, 221; at mouth of Jequitinhonha, 221.
Planorbis Monserratensis Hartt, 351.
Plants, fossil, on Rio Pardo, 243; Abrolhos, 176; of carbonifera of Rio Grande do Sul, 524.
Plant, Nathaniel, Esq., report on coal-fields of Rio Jaguarão, 521.
Plataforma, cretaceous beds of, 347.
Plateau of Province of Bahia, 310; middle, of Bahia, 313; third or coast, of Bahia, 316; second, of Bahia, vegetation, climate of, 317; of Appodí, 451; of Brazil in São Paulo and

Paraná, 505; of Brazil, everywhere underlaid by gneiss, 550.
Platinum, 448; occurrence of in gold-mines of Brazil, 542.
Plexaurella dichotoma Köll, 62, 197.
Po-assú (Yg-apó-assú, *running water* or *big swamp*), 237; probable results of widening, 240.
Poçoes, 255.
Pohl, 496.
Pojichá, chief and tribe of Botocudos, 592, 594.
Pojuca tunnel, 371.
Polyps, distribution of, along Brazilian coast, 191.
Pompeo, 428.
Ponds on reefs, 210.
Ponta de Jecú, 64; do Tubarão, 65; Caixa de Pregos, 267; Garcia, 267.
Pontal at mouth of Rio Arassuahy, 163; da Barra do São Francisco, dunes at, 395.
Porcupines, bones of in caves at Lagôa Santa, 285.
Porites, in Bay of Rio, 192.
Porites solida Verrill, 203, 208.
Porphyry, 446, 449.
Porokum, skull of, 588.
Pororoca or Bore, 487.
Porto de Souza, geology of vicinity, 90; das Caixas, 6, 21; da Villa Nova, recent deposits at, 18; da Pedra, 76; de Souza, 89; richness of soil at, 91.
Porto Alegre (Rio Mucury), 128; (Rio Grande do Sul), 531.
Porto Seguro, 228; harbor and reef of, 228.
Porto das Piranhas, 412.
Post tertiary, 573.
Pot-holes, occurrence of diamonds in, 307; of Lake Plain of probable glacial origin, 315, 562, 571.
Prado, 225.
Praia Grande, geological section at, 13.
Precipices, character of, near Rio, 31.
Prionus cervicornis, larvæ of, eaten by Botocudos, 597.
Pristis, species of, in Rio Doce, 91.
Propriá, 404.
Protopithecus, 286.
Przewodowski, Mr., 272.
Psidium Guiava, 47.
Pteris caudata, 144, 252; social plant, 256.
Pterogorgia (*Gorgonia*) *gracilis* Verr., 81.
Pumice, 179, 248.
Pygocentrus (fish), 403.
Pyrites, magnetic iron, gold of, 534, 541; iron, associated with gold at Morro Velho mine, 537.

Q.

Quartz, auriferous, 448; veins, 448, 533.
Quartzite, 152, 296, 298; auriferous, 534.
Quadersandstein, 298.
Quixeramobim, 460.
Quina, 19, 459.
Quiricare, Rio, 117.

R.

Railroads of Brazil, 293; Cantagallo, 6, 19; Dom Pedro Segundo, geology of, 14; Pernambuco and São Francisco, 436; Santos and São Paulo, 506.
Rains on Lower São Francisco, 420; of Maranhão, 485.
Recife, meaning of word, 191, 432; city of, 432; do Lixo, 202; do Leste, 211; da Pedra Grande, 211; Itanhaem, 212; dos Itacolumis, 212, 227; das Timbebas, 212; do Porto Seguro, 212.
Reconcavo, 272.
Reefs, Coral, see Coral Reefs; stone, or consolidated beaches, of Guarapary, 62; at Barra Secca, 107, 113; Porto Seguro, 229; Santa Cruz, 232; Bahia, 342; Rio Vermelho, 344; Pernambuco, 434; Parahyba do Norte, 442, 454.
Reinhardt, Prof. G., 279; general conclusions with reference to the fossil fauna of bone caverns, 287; on fossil man of Brazil, 287.
Renilla Danæ, 192.
Rhacolepis buccalis, 468; *R. olfersii*, 468; *R. latus*, 468.
Rhea Darwini, 147.
Rhexia, species of, social plants, 256.
Riacho, meaning of word, 123; dos Porcos, 457; do Mundo Novo, 457; das Ostras, 124.
Ribeirão Diamantino, 149; d'Agôa Nova, 149; da Issara, 250; do Meio, 154.
Rice, 116, 261, 409, 410, 454.
Riedel, 281.
Rio, Bay of, 6; d'Agua Limpa, 154; d'Agua Suja, 137, 153; das Americanas, 126; Andarahy (*Andira* bat, *hy* river), diamond washings on, 306; Apiapitanga, 85; Apodí, 452; Araguaya, 501; Arassuahy (*Araassu-hy* great parrot water, or *Coaracy-cy* water of the sun, Mart.), 137, 163; gold in sands of, 163; d'Areia, that part of a river whose bed is composed of sand, so called in contrast with the part whose banks and bed are rocky, 126, 171, 242; da Barra Secca, 106; Benevente, 60; Bom Successo, discovery of gold in, 156; Bruscus, 447; Buranhaem, 227; Calhão, 137, 151; Camocim, 460; Candiota, coal-mines of, 521; Canindé, 474; Capanema (*cáa* leaf or wood, *panemo* sterile), 271; Capibaribe, 430; Capivary, 137, 145; Carahype, 83; Carapina, 76; Caravellas, 219; Caravellas, mouth of, 223; Carimataú, 452; Carunhanha (*Caryca* to run, *anhé* enough, "Fluvius sat rapidus," Martius), 295; do Castello, gold-washings on, 59; da Caxoeira, 227; Ceará-merim, 451; Chopotó, 87; Commandatuba (*Comanda* feijao, and *tyba* place), 242; de Contas, 260; Corrientes, 88; Cotinguiba (*Cotuc* to wash and *iba* tree, River of the *Sapindus* or soap tree?), 379; Cotinguiba or Cotindiba, 381; bar of, 382; Craminuan, 227; Crubixa, 75; Cuité, 89; Curipe, 76; Doce, description of Basin of, 86; width at Porto de Souza, 90; luxuriance of forest in, 93; freshets of, 98; productions of, 98; Bar of, 102, 104; below Linhares, 103; Company of, 104; American colony on, 105; salt trade of, difficulty of carrying on, 104, 105; da Fabrica, 219; Fanado, 137, 145; do Frade, 227; do Fundão, 248; Giboya, 250; Grande do Norte, 451; Itanhaem, 224; Imbuçahi, 364; Ipojuca, 430; Irapirang (*Yra* honey and *piranga* red), 381; Iriritiba, 60; Gravatá, 137, 148; Gualacho, 88; Guandú, 89; Guarapary, 63; Guaxindiba, 122; Gurgueia, 474; Iguassú, 44; Itabapuana, 52; Itacam-

birussú, 136; Itahúnas, 122; Itahype (*ita* stone, *hy* water, *pe* way, river among stones), 259; Itamarandiba, 137; Itapicurú, 323; Itapemerim, 58; Jacoruna or Jacuruna, (*Jacú* species of Penelope, *una* black), 268; Jacuahy, coal-basin on, 530; Jacuhy, 76; Jaguaribe, 458; Jaguaripe (river of the Onça), 267; Japaratuba, 379; Jecu, 64; Jequié, 266; Jequiriça (*juquira* salt, and *çabáa* bay, Mart.), 266; Jequitinhonha, 136; João de Tiba, 231; Joassema, 227; Johannes, 360; Jucurucú, 224; Juparanãa, 99; Jussiape, 259; Macabú, 44; Macucú, 18; da Mae d'Agua, 154; Mamanguape, 443; Mangarahyba, 76; Mangarahy, 77; Manhuassú, 90; Mariricú, 107, 223; Mearim or Meary, bore of, 486; do Meio (Medio), 77; Mucury, 125; Mugiquisaba, 235; Muriahe, 4, 47; Mutum, 91; Pampao, 126; Panca, 95; Pará, of Minas, 288; Paraguassú, 269, 303; Paraguay, valley of, worn in tertiary sandstones, 504; Parahyba do Sul, 3, 45; Parahyba do Norte, 441; Parahybuna, 4; Paranà, 518; Paranapanema, 518; Paraopeba, 289; Parapuca, 394; Pardo, 137, 237; Parnahyba, 473; Patipe, 241; Pero Cão, 63; Peruhype, 215; Piabanha, 4; Piauhy, 164, 379; Piracicaba, 88; Pirahy, 3; Pirahytinga, name given to upper part of the Parahyba do Sul, 3; Piranhas, 451; das Pirangas, 87; Pirapama, 430; Pirapitinga, 4; Piriqui-assú, 85; Piriqui-mirim, 85; do Peixe, 323; Piuma, 59; Pomba, 4; Potengi, 452; Poty, 474; Poxim (*poxim*, ugly), 241; Preto, 4, 125; Purús, 494; Real, 379, 380; dos Reis Magos, 85; de Salitre, 327; Salgado (Ceará), 457; da Salsa, 237; Santa Maria, 65; de São Francisco, 276, geology of, 276, explorers of, 275, rank of, among rivers, 288, courses described, 288; Upper, unhealthiness of, 292, navigability of, cost of removing obstructions from, 292, opening of, to steam navigation below Piranhas, 421, São Matheos, 117; de São Raffael, 101; São Sepe, coal-basin on, 529; Santa Antonia, 88; de Sant' Antonio, mastodon remains found in vicinity of, 298; de Santa Cruz (Esp. Sant.), 85; Santa Cruz (Bahia), 231; Santa Maria, 66, 77; Santa Rita, 143; Saçuhi-pequeno, 88; Saçuhi-grande, 88; Setubal, 137; country in vicinity of, 140; Serenhaem, 430; Serigi, 272; da Serra, 65; Setubinho, 140, 147; Sipó, 241; Soledade, 137; Taipe, 227; Tapajos, granite on, 500; sandstones of, 504; Tauhá, 76; Tibaje, 513; Tiete, 510; Tocantins, 501; Todos os Santos, 126; Trahiry, 451; Una (*black river*) of Espirito Santo, south of Victoria, 64; west of Victoria, 76; of Bahia, 246; of Pernambuco, 430; Upanema, 452; Urucú, 126; Uruguay, 518; agates of, 531; Vacaria, 137; Vasabarris, 379, 381; das Velhas, Liais survey of, 275, 286, course, velocity, obstruction, navigability, &c., 289, Liais's picture of the scenery on, 291, Verde, 328; Xingú, 503.

Rise, gradual, of coast within recent times, probably still in progress, 36; observation on, at Victoria, 72.

Roads, scarcity of wagon-roads in Minas Geraes, 130; wagon-road from Santa Clara to Philadelphia, 131.

Rock crystals, 531.

Rock salt said to occur on Rio Huallaga, 329.

Roccas, 214.

Rosewood, 93, 102.

Ruminants more abundant in post-tertiary than at present in Brazil, 287.

S.

Sabará, 290.

St. John, Mr. O. H., 275, 295, 475.

Saline streams, 304.

Saline deposits of Valley of São Francisco, 328; method of extracting the salt, 329.

Salines of Araruama, 38; of Maçambamba, 39.

Salt, cost of, in the interior, 130; commerce in, on Rio Doce, 104; of Fazenda Aldea, impurities of, 328; of São Francisco valley, note by Mr. Allen, 329; in Ceará, 471; trade in, of Mucury, 128; of Rio Grande do Norte, 455.

Salter on fossils of Silurian of Andes, 552.

Salto Grande do Jequitinhonha (village and falls), 170; do Paraná, 518.

Saltpetre in bone caves at Lagôa Santa, 284; of Rio de Salitre, 330.

Samambaia, 144; a pest in Brazil, growth of, aided by fires, 256.

Sambahiba, 374.

Sand-beaches, formation of, 219.

Sand, blowing, on São Francisco, 332.

Sands blown over drift at Bahia, 346; of Taboleiros, age of, 377.

Sandstone, tertiary, 47, 48, 55, 56, 57, 60, 100, 101, 113, 123, 124, 150, 243, 270, 277, 301, 305, 311, 332, 371, 379, 458, 464, 474, 475, 484, 489, 501; cretaceous, 175, 346, 347, 348, 397, 398; palæozoic, 243; triassic, 379.

Santa Catharina, corals of, 192; S. Clara, 128; S Cruz, 85; S. Euzebia, 310; S. Luzia, 452; S. Maria, 76.

Santo Amaro, 272.

Santos, 506.

São Christorão (city), 381; S. Fidelis, 49; S. João da Barra, 45; S. João d'El-Rei, 536; S. João d'Ipanêma, 515; S. Matheos, 80, 121; S. Miguel, 167; S. Paulo (city), 506, 510; S. Vicente, 534.

Sapucahy, cretaceous limestones at, 383.

Sapucaia (*Lecythis*), 94.

Saurians, remains of, at Monserrate, 347.

Sawfish in Rio Doce, 91.

Schieber, Mr. George, 126, 129.

Schinus terebinthifolia, 19, 423.

Schists, hornblendic, 446.

Schlobach, engineer, on height of Philadelphia, 134.

School, 151, 327.

Sciurus æstuans, 286.

Scorpions, 375.

Sea-anemones, 62.

Sea breeze at Penêdo, 420.

Sea-level, ancient, at Victoria, 71.

Sea-urchins, nests excavated in rocks by, 36.

Sebastião Leme do Prado, discoverer of gold-mines of Minas Novas, 156.

Sellow, 530.

Senonien, 556.

Sergipe d'El-Rei, 381.

Sergipian group, 556.

Seriema (*Dicholophus cristatus*), 254.

Serpentine, occurrence of, in eozoic limestone at Pirahy, 549.

Serra dos Aimorés, 84; das Almas, 254; das

Araras, 499; de Araripe (*Arara-ipe*, locality of), 278, 457, 464; Arassoiava or Guaraçoiava, 516; de Aratanha, glacial phenomena of, 469; de Areré, 461; do Assuruá, 308; da Borborema, 440, 451; do Boquerão, 304; de Buenos Ayres, 291; dos Cairiris Novos, 451; dos Cairiris Velhos, 428, 440; do Caldeirão da Onça, 303; de Caytelé, 296; do Condurú, 263; de Curumatahy (*Curimatá* fish, 499, and *hy* water or river), 291; dos Dous Irmãos, 278, 476; do Espinhaço, 136, 277; do Feijoal, 168; do Gado Brabo, 327; de Garanhuns, cotton of, 431; da Garça, 291; Geral, 498; Grande, 456; do Grão Mogor, 136, 139; das Ibiturunas, 87; d'Itabayana, 379, 381; d'Itacambira, 136; de Itaparica, 419; de Itaraca (*Ita* stone, and *aca* horned), 247; de Itaubira (*Ita* stone, *berab* flaming), 298; de Jaraguá, 511; das Lages, 298; da Lapa, 331; da Lua Cheia, 168, 169; near Macahé, height of, 42; do Macaco, 162; Mangabeira, 304; da Mantiqueira, general description of, geological structure, &c., 2; da Mantiqueira, 86; do Mar, 1, 14, 505, 547; da Mata da Corda, 291; do Matuca, 52; do Mocambo, 303; dos Montes Altos, 296; do Morro Queimado, 21; d'Olho d'Agua, sandstone at, 414; da Onça, 48; dos Órgãos, topographical and geological description of, 15; das Panellas, 168; do Paranan, 499; do Paraúna (*Pará* river, *una* black, Black-river), 291; das Pedras d'Agua, 312; das Pedras Brancas, 303; de Pereira, 462; da Piedade, 289, 535; do Pintor, 308; do Rio do Peixe, 323; de São Fidelis, 49; de São João, 41; de São Romão e Santa Paz, 52; da Saude, 303; do Sincorá, 299, 306; Sobrado, 137; da Sussuarana (*Suassú* deer, *rana* false, — Suassuarana puma, or Felis concolor), 250; da Tabatinga, 473; da Teixeira, 440, 446; da Terra Dura, 316; de Tiuba, 324; do Tombador, 316; da Topa, 475; dos Vertentes, 501; da Vigia, 168; da Villa Velha, 296; da Ybiapaba, derivation of name, 456; structure of, 457.
Serrado, 147.
Sertão, 429, 459.
Shales, cretaceous, of Monserrate, 349; with fish remains near Propriá, 404.
Siderastræa stellata Verr., 62, 74, 193, 203, 207, 214; *var. conferta* Verr., 194.
Sienite, 324.
Sincorá, diamond-washings at, 307.
Silurian rocks of Brazil, 551; limited range of animals during, 552; of Bolivar and Peru, 552; fossils at, 552.
Silver always found in Brazilian gold, 542.
Slate, conglomerate, 242.
Slates on Rio da Mae d'Agua, 154; (auriferos) of Minor Novas and Chapada, 157; decomposition of, 158.
Slides of rocks on Bahia Railroad, 373.
Sloths, 94.
Smilodon neogæus, 286.
Smith, Mr. S. I., on Brazilian Crustacean, 203.
Social plants, variety of, in Brazil, 255.
Soils of the Rio Doce region, fertility of, 96–103; thinness of on Lake Plain of Bahia, 314; of Rio São Matheos, 117, 118; of tertiary lands of Mucury, 133; of Urucú, 133; of vicinity of Calháo, 152.
Solanaceæ, 250.
Spar, double-refracting, 525
Sphargis coriacea, 112.
Sphenopteris, 525.
Specular iron, 472.
Spiders of Abrolhos, prey on lizards and birds, 180.
Spix and Martius on vegetation of sand plain near Rio, 19; on journey from Malhada to Cachoeira, 295; on dead corals in lake near Ilhéos, 259; on bemdêgo meteorolite, 325.
Squirrel, 95.
Staurotide, 252.
Steamboat lines from Bahia, 338.
Stone implements of Botocudos, 592; reefs, see *Reefs*.
Sturgeons, none in South America, 403.
Stratified deposits, absence of, in connection with drift, 566.
Striæ, glacial, reason for not finding in Brazil, 57.
Striated surfaces, absence of, in Brazil, 562; not reported in Patagonia and Chili.
Structure, want of, in drift, 564.
Stunden (e. g. in der 7th Stunde), 530.
Submerged border of coral reefs of Lixo, life of, 205.
Subterranean streams, 498
Sugar, 120, 261, 272, 380, 393, 431, 441, 442; method of preparing at Muriahé, 48; cane, 99, 452, 458, 459; plantations, 45, 48, 52, 271, 452.
Sugary quartz, 545.
Sulphate of magnesia, 330.
Sulphide of antimony, 472.
Sulphides occurring with gold, 534.
Sulphur said to occur in the Province of Rio Grand do Norte, 455.
Sumidouro cavern, remains of man in, 286.
Swamps of Rio Maricú, 116; between the Rios Peruhype and Caravellas, 219; between Rios Jequitinhonha and Pardo, 238; journey in, 239.
Symphyllia Harttii, 196.

T.

Tabatinga clay, 363.
Table land between São Francisco and Tocantins basins, 277; of São Paulo, 509.
Table-topped hills or Serras of the São Francisco valley, 331; of Ereré, Obydos, Cupatí, Almeyrim, 489, 490.
Taboleiros, 303, 304, 312, 362, 365, 373, 464
Talcose slates, auriferous, probably Silurian, 551.
Talhadão, 417.
Tamanduatahy, peat-bogs near, 509.
Tapanhoacanga, 536, 559.
Tapera da Cima, deposits of iron ore near, 332.
Tapir (*Tapirus Americanus*), 94.
Tapuyos, 578.
Taquara (Bamboo), 216.
Taquara-assú requires humidity and considerable elevation, 84; social plant, 256.
Taquara lisa, 141.
Tea, Paraguayan, 517; Chinese, culture of, in Brazil, 517.
Teiú lizard (*Teius monitor*), 111.
Tellina Amazoniensis Gabb., 493.
Termo do Jardim, 457
Terra roxa, 514.

Terrace form of the province of Bahia, 301.
Tertiary, Cantagallo Railroad, 20; between Os Buzios and Macahé, 41; Barreiras do Sirí, 56; Linhares, 99; Lagôa Juparanãa, 100; São Matheos, 117; Itahúna, 123; Santa Clara, 129; Jequitinhonha Valley, 139; between Peruhype and Mucury, 215; near Porto Seguro, 225; between Porto Seguro and Santa Cruz, 231; Rio Pardo, 243; fossils of, reported from Bay of Bahia, 269; Reconcavo, 272; vicinity of Camassarí, 361; lands of Bahia, 373; Alagoinhas, 375; Alagôas, 422; Maceió, 424; Pernambuco, 429; Parahyba do Norte, 443; Piauhy, 474; Maranhão, 484; Amazonas, 492; Pebas, 493; western part of Amazonas-Paraguay watershed, 503; of Brazil, *Résumé*, 557.
Thalassochelys cauana, abundance of, at Barra Secca, mode of depositing eggs, &c., 108.
Thoracosaurus Bahiensis Marsh, 357.
Tide pools on the reefs, life of, 197.
Tidsskrift Lütken's, 281.
Tijuca, peak of, height, 12; drift of valley of, 26.
Tillandsia, 21, 251.
Tobacco, 336, 441, 454.
Topaz, 151, 553.
Topography of gneiss region in the interior of Bahia, Sergipe, and Alagôas, 318; of tertiary, near Porto Seguro, 225; of tertiary near Pojuca tunnel, 371.
Torgjusen, Jacob, 202.
Tourmaline, 135, 145.
Tortoise, cretaceous, 358.
Toxodon, 286.
Traipú, geology of vicinity, 406.
Trancozo, 227.
Tramroad, steam, of the Paraguassú, 270, 338.
Trap of Abrolhos, 176; decomposition of, 177; hills, range of, in Rio Grande do Sul, 530.
Travelled boulders, 570.
Tres Irmãos, 10.
Triassic sandstones of Estancia, 379, 554.
Tripneustes, absence of, from Brazilian radiate fauna, 198.
Trumpet used by Botocudos, 592.
Turba deposits at Camamú, 262.
Turbonella minuscula Gabb., 493.
Tucum (*Astrocaryum tucuma*), 116.
Tupi names in Brazil, ix.
Tupinambas, lip ornaments, 584.
Turí, gold-mines of, 485
Turtles (sea), 108.
"Turtles," fossil or Septaria, at Maroïm, 393.
Turner, Engineer, 371.

U.

Ubá grass (*Gynerium parvifolium* Nees), 45, 94, 395; social plant, 256.
Uca una, 204.
Ulloa, on the Island of Fernando de Noronha, 439.
Unio, 101.
Unio (*Anodon?*) *Totium-Sanctorum*, Hartt, 348, 351.
Urubú, 257; change in geological structure, climate, and vegetation below, on São Francisco, 331.
Urucú (*Bixa Orellana* Linn.) used as paint by Botocudos; colony on Mucury, 133.

V.

Vaccinium, 118.
Vadelli, 301.
Valença, 266.
Valley of São Simão, 168; of the Calhão Arassuahy, 150.
Valleys of vicinity of Alto dos Bois, 156; without outlets, 303.
Vanilla, 517.
Vegetation of sand-plains near Victoria, 62, 65; of Swamp of Rio Maricicú, 116; on the Rio São Matheos below the city, 121; of chapada near Santa Rita, 144; of Alto dos Bois, 146; of Chapada of Minas Novas, 147; of vicinity of Calhão, 152; of vicinity of Sucuriú, 154; of Jequitinhonha less luxuriant than that of the Doce, 172; of the Abrolhos, 180; of country between Mucury and Peruhype, 216; of sand-plain at Belmonte, 237; on Rio Pardo, 244; of Pardo, 245; of Campos, 290; of Serra da Villa Velha, 296; of interior of Bahia, 317; of Campos of Alagoinhas, 375; near mouth of São Francisco, 395; of Rio Grande do Norte, 452.
Veins in decomposed rock never traceable into drift, 565; of granite, 164; (quartz), in Minas Novas region, 157; auriferous, at Chapada, 158; auriferous, of Parahyba, 448.
Venus flexuosa in sands on the Cantagallo Railroad Extension, 19.
Verrill, on resemblance between marine fauna of West Indies and Brazil, 198.
Victoria, harbor of, 66, 80; (Conquista), 255.
Villa da Barra do São Matheos, 122; da Barra do Itabapuana, 54; Viçosa, 218; da Barra do Jardim, 465; do Crato, 464; das Lavras da Mangabeira, 463; Nova da Rainha, 324; Nova near Penêdo, 397; do Rio de Contas, 298; São Bernardo, 461; da Serra, 84; do Sucuriú, 153; Nova, São Francisco, sandstones at, 397.
Vinhatico (*Acacia*), 94, 261.
Vivipara (*Paludina*) *Lacerdæ* Hartt, 350; *V.* (*Paludina*) *Williamsii* Hartt, 351.
Vocabularies, Botocudo, 605.
Voluta, 203, 214.
Von Eschwege, 501, 515, 535, 559
Von Tschudi on Botocudo tribes, 592; on distribution of cocoa-palm in Brazil, 119; on the Colony of Santa Leopoldina, 77.

W.

Ward, Mr. Thomas, 277, 279, 501.
Waste of land at mouth of the Amazonas, 491.
Water, scarcity of, in interior of Bahia, 256.
Wax of Carnahuba, 453.
Wave action, contrast between the rock surfaces produced by wave action and those formed by the glaciers, 342; during subsidence, drift not referable to, 563, 567; during rise of land, drift not referable to, 563, 565.
Weddel, 541.
Widmannstadtian figures shown by Bemdêgo meteorolite, 326.
Weiss, 530.
Wet season, few animals seen, 155.
Whale, flesh of, used for food, 185.

Whale fishery of Brazil and of the Abrolhos, 181; duration of fishery, 182; Armaçoes or trying hours, 183; launches and boats used in, number of men employed, wages, difficulties of fishing, imperfect methods, 183; at Bahia, 185.
Whalebone, 184.
Wheat, 142.
White water rivers, 223.
White ants, hills of, 257.
Winds of campos, 254; on São Francisco below falls, 421.
Wilson, Mr Hugh, 269, 338.
Williams, Mr. C. H., on Indian rock drawings, 326.
Williamson, Mr. E., on the geology and gold-mines of Parahyba dó Norte, 443
Wounds, facility with which they heal among Botocudos, 598.
Wyman, Prof. Jeffries, on Botocudo skull, 585.

Y.

Yellow fever, 337.
Ypiranga, limestone at, 15.

Z.

Zinc, sulphide of, 448, 472.
Zinebra, 172.
Zizyphus Joazeiro, 408.
Zoanthus, 62, 192.

THE END.

Cambridge: Electrotyped and Printed by Welch, Bigelow, & Co.

www.ingramcontent.com/pod-product-compliance
Lightning Source LLC
LaVergne TN
LVHW021051110826
845150LV00001B/48